General addition rule	$P(A \text{ or } B) = P(A) + P(B) - P(A \text{ and } B)$
Permutation rule	$P_{n,r} = \dfrac{n!}{(n-r)!}$
Combination rule	$C_{n,r} = \dfrac{n!}{r!\,(n-r)!}$
Mean of a discrete probability distribution	$\mu = \Sigma x P(x)$
Standard deviation of a discrete probability distribution	$\sigma = \sqrt{\Sigma(x-\mu)^2 P(x)}$

Binomial Distribution Formulas   where r = number of successes; p = probability of success; $q = 1 - p$

Formula for a binomial probability distribution	$P(r) = \dfrac{n!}{r!(n-r)!}\,p^r q^{n-r}$
Mean for a binomial distribution	$\mu = np$
Standard deviation for a binomial distribution	$\sigma = \sqrt{npq}$

Confidence Intervals

Confidence interval for a mean (large samples)	$\bar{x} - z_c \dfrac{\sigma}{\sqrt{n}} < \mu < \bar{x} + z_c \dfrac{\sigma}{\sqrt{n}}$
Confidence interval for a mean (small samples)	$\bar{x} - t_c \dfrac{s}{\sqrt{n}} < \mu < \bar{x} + t_c \dfrac{s}{\sqrt{n}}$
Confidence interval for a proportion (where $np > 5$ and $nq > 5$)	$\dfrac{r}{n} - \sqrt{\dfrac{\frac{r}{n}\left(1-\frac{r}{n}\right)}{n}} < p < \dfrac{r}{n} + \sqrt{\dfrac{\frac{r}{n}\left(1-\frac{r}{n}\right)}{n}}$

Sample Size

Sample size for estimating means	$n = \left(\dfrac{z_c \sigma}{E}\right)^2$
Sample size for estimating proportions	$n = p(1-p)\left(\dfrac{z_c}{E}\right)^2$ with preliminary estimate for p
	$n = \dfrac{1}{4}\left(\dfrac{z_c}{E}\right)^2$ with no preliminary estimate for p

Understandable Statistics

Understandable Statistics Third Edition

Concepts and Methods

Charles Henry Brase
Regis College

Corrinne Pellillo Brase
Arapahoe Community College

D. C. HEATH AND COMPANY
Lexington, Massachusetts   Toronto

Acquisitions Editor   Mary Lu Walsh

Development Editor   Ann Marie Jones

Production Editor   Kim Rieck Fisher

Designer   Henry Rachlin

Production Coordinator   Mike O'Dea

Cover photograph   Fenway Park in Boston by Richard Pasley/Stock, Boston

Published simultaneously in Canada.

Printed in the United States of America.

International Standard Book Number: 0-669-12181-9

Library of Congress Catalog Card Number: 86-81382

*This book is dedicated to the memory of
a great teacher, mathematician, and friend,*

Burton W. Jones
Professor Emeritus, University of Colorado

Preface

This text is designed for students with a minimal background in mathematics. The same pedagogical features that made the first two editions of the text so readable are preserved in the third edition.

We have retained the *graduated approach to problem solving*. Our experience in class testing this approach convinces us that the format we have adopted will work for a wide variety of students in a statistics course. The exercises and problems touch on applications appropriate to a broad range of interests. Complicated calculations are simplified by using step-by-step tabular procedures well suited to hand-held calculators. Key features include:

- *Detailed examples* which show students how to select and use appropriate procedures.

- *In-text exercises* that immediately follow an example. Each exercise gives the student a chance to *work* with a new concept before another is presented. The student must examine and analyze characteristic features of a problem similar to the preceding example. Completely worked out solutions occur beside each exercise to give immediate reinforcement to the learning process.

- *Section problems* that require the student to use all the new concepts mastered in the preceding section. They reinforce the material of the examples and exercises, providing additional exercise in practical application of the principles covered. Key steps to the solutions of odd-numbered problems are contained in Appendix II of the text.

- *Chapter problems*   Comprehensive problems at the end of each chapter cover each topic introduced in the chapter. Many problems require material and concepts from several sections. Most importantly, the student must decide what technique to apply to a problem. As in actual applications, the problem does not indicate which section the student should look to for the method of solution.

New in the third edition:

- *Many more section problems and many new problems*   Old problems have been extensively revised. There are more problems utilizing raw data. These new problems give the professor a greater choice in problem assignment and provide more practice for the student.

- *Expanded treatment of sampling distributions*   The material on sampling distributions is now self-contained in a separate chapter (Chapter 7) with an entire new section devoted to distributions of sample statistics. The material on the Central Limit Theorem has been revised.

- *New topics* We have added two sections introducing some popular Exploratory Data Analysis (EDA) techniques. These topics are Stem-and-Leaf displays and Box-and-Whisker plots. A new section explains the use of *P values* in hypothesis testing.

- *Revisions and additions* New formulas better designed for computational ease have been introduced in the material on linear regression and correlation. We have also included confidence bounds on *y*. Material on *random variables* and probability distributions as well as that on estimation and *confidence intervals* has also been revised for greater clarity.

- *Greater flexibility in choosing the level of probability* Section 4.3, Trees and Counting Techniques, is a new section placed *after* the introduction to probability. For courses in which a very light treatment of probability is desired, this new section may be omitted and Section 4.2, Rules of Probability, can be covered lightly. For courses needing a more extensive treatment of probability, the professor will find *more examples* and problems in Section 4.2. In addition the trees and counting techniques with accompanying probability applications will give the student greater skills and tools for solving problems in probability.

- *Using Computers sections* Computers are certainly a powerful tool for the statistician. In many institutions it is now possible to introduce the beginning statistics student to statistical software packages. At the end of most chapters, a section called *Using Computers* is included. These sections feature published raw data and situations from a variety of fields. The problems can be solved using almost any appropriate statistical software computer package. However, we have tailored the exercises so that they can easily be completed utilizing the software package *ComputerStat* that accompanies this text as a computer supplement.

- *ComputerStat* is a computer software package designed to accompany this text. Disks are available for the Apple® IIe or IIc (with 80-column video display) and IBM PC®. ComputerStat is a menu-driven package of 22 computer programs written in the language BASIC. ComputerStat is an interactive package designed to be very user friendly. The data input and output reflect the procedures the students use when they do a problem with pencil, paper, and calculator. The computer disks together with a user's manual are available from D. C. Heath and Company.

- *Instructor's Guide* Contains key steps and answers to even-numbered problems, *two* (revised) forms of chapter tests, and transparency masters for overhead projectors.

    Understandable Statistics: Concepts and Methods, third edition, is designed to be flexible. It offers the professor a choice of teaching possibilities. In most one-semester courses, it will not be practical to cover all the material in depth. However, depending on the emphasis of the course, the professor may choose to cover various topics.

    For a course with more *emphasis on descriptive statistics* all of Chapters 2 and 3 should be covered. For a course with more *emphasis on probability,* all of Chapter 4 should be covered.

Apple is a registered trademark of Apple Computer, Inc.
IBM PC is a registered trademark of International Business Machines Corp.

For a streamlined course designed to emphasize *inferential statistics*, Section 2.1, Random Samples; Section 2.3, Histograms and Frequency Distributions; Section 3.1, Mode, Median, and Mean; and Section 3.3, Measures of Variation, are all that are essential from Chapters 2 and 3. A light treatment of probability including Section 4.1, What Is Probability, part of Section 4.2, Some Probability Rules, and Section 4.4, Introduction to Random Variables and Probability Distributions, will enable the course to proceed quickly to Chapter 5, The Binomial Distribution, and on to the chapters on inferential statistics. To allow more time for Chi-square and ANOVA topics and some for nonparametric statistics, Chapter 5, The Binomial Distribution, and later sections involving proportions may be omitted.

Acknowledgments

It is our pleasure to acknowledge the assistance of colleagues and friends. We especially thank Frederick T. Daly, S.J., Diane Wagner, Gene Delay, Lee Shannon, and Harry Taylor, all of Regis College and Patricia Hauss and Jack Pepper of Arapahoe Community College. These colleagues have given us encouragement and useful suggestions for improving the text. We are also especially grateful to Louis Hoelzle, Bucks County Community College; Mel A. Mitchell, Clarion University of Pennsylvania; Albert Parish, College of Charleston; Lawrence Riddle, Emory University; Dorothy Ryan, Bunker Hill Community College; Mary McGowan Sullivan, Curry College; Steve Turner, Mountain Bell Corporate Finance Department (Denver, Colorado); and Clifford H. Wagner, Fitchburg State College, Massachusetts, who each gave valuable input in the creation of new material and revision of old.

In addition, we wish to thank the editorial and production staff of D. C. Heath for their timely advice and high quality work. Finally, we wish to thank our students. We sincerely believe that this is a text you can read and understand.

Charles Henry Brase
Corrinne Pellillo Brase

Table of Prerequisite Material

Chapter	Prerequisite Chapters	Prerequisite Sections
1 Getting Started	None	
2 Organizing Data	1	
3 Averages and Variation	1	2.1, 2.3
4 Elementary Probability Theory	1	2.1, 2.3, 3.1, 3.3
5 The Binomial Distribution	1	2.1, 2.3, 3.1, 3.3, 4.1, 4.2, 4.4 with 4.3 useful but not essential
6 Normal Distributions	1	2.1, 2.3, 3.1, 3.3, 4.1, 4.2, 4.4
7 Introduction to Sampling Distributions		
(omit 7.3)	1, 6	2.1, 2.3, 3.1, 3.3, 4.1, 4.2, 4.4
(include 7.3)	5 also	
8 Estimation		
(omit 8.3)	1, 6	2.1, 2.3, 3.1, 3.3, 4.2, 4.4, 7.1, 7.2
(include 8.3)	5 also	7.3 also
9 Hypothesis Testing		
(omit 9.6)	1, 6	2.1, 2.3, 3.1, 3.3, 4.1, 4.2, 4.4, 7.1, 7.2
(include 9.6)	5 also	7.3
10 Regression and Correlation	1	2.1
(omit 10.4)	1	2.1, 4.1, 7.1
(include 10.4)		4.2, 4.4, 9.1 also
11 Chi-Square and F Distributions		
Sections 11.1 and 11.2	1	2.1, 2.3, 3.1, 3.3, 4.1, 4.2, 4.4, 7.1
Section 11.3	1	2.1, 2.3, 3.1, 3.3, 4.1, 4.2, 4.4, 7.1, 8.1, 9.1
Section 11.4	1	2.1, 2.3, 3.1, 3.3, 4.1, 4.2, 4.4, 7.1, 9.1
12 Nonparametric Statistics		
Section 12.1	1	2.1, 3.1, 3.3, 7.1, 9.1, 9.6
Section 12.2	1	2.1, 3.1, 3.3, 6.1, 7.1, 9.1
Section 12.3	1	2.1, 3.1, 3.3, 7.1, 9.1

Contents

1 Getting Started   **1**

Section 1.1   What Is Statistics?   1
Section 1.2   Calculators and Computers in Statistics   6
Section 1.3   Where Will I Use Statistics?   7

Summary   8
Important Words and Symbols   8
Chapter Review Problems   8

2 Organizing Data   **9**

Section 2.1   Random Samples   10
Section 2.2   Graphs   16
Section 2.3   Histograms and Frequency Distributions   24
Section 2.4   Stem-and-Leaf Displays   36

Summary   44
Important Words and Symbols   44
Chapter Review Problems   44
Using Computers   48

3 Averages and Variations   **50**

Section 3.1   Mode, Median, and Mean   50
Section 3.2   Percentiles and Box-and-Whisker Plots   58
Section 3.3   Measures of Variation   67
Section 3.4   Mean and Standard Deviation of Grouped Data   78

Summary   88
Important Words and Symbols   88
Chapter Review Problems   89
Using Computers   92

4 Elementary Probability Theory — 94

Section 4.1   What Is Probability?   94
Section 4.2   Some Probability Rules   101
Section 4.3   Trees and Counting Techniques (Optional)   118
Section 4.4   Introduction to Random Variables and Probability Distributions   132

Summary   143
Important Words and Symbols   144
Chapter Review Problems   144
Using Computers   149

5 The Binomial Distribution — 151

Section 5.1   Binomial Experiments   151
Section 5.2   The Binomial Distribution   156
Section 5.3   Additional Properties of the Binomial Distribution   164

Summary   172
Important Words and Symbols   173
Chapter Review Problems   173
Using Computers   176

6 Normal Distributions — 178

Section 6.1   Graphs of Normal Probability Distributions   178
Section 6.2   Standard Units and the Standard Normal Distribution   188
Section 6.3   Areas Under the Standard Normal Curve   198
Section 6.4   Areas Under Any Normal Curve   208

Summary   218
Important Words and Symbols   219
Chapter Review Problems   219
Using Computers   222

7 Introduction to Sampling Distributions — 224

Section 7.1   Sampling Distributions   224
Section 7.2   The Central Limit Theorem   231

Section 7.3   Normal Approximation to the Binomial Distribution   242

Summary   248
Important Words and Symbols   249
Chapter Review Problems   249
Using Computers   252

8   Estimation   254

Section 8.1   Estimating μ with Large Samples   254
Section 8.2   Estimating μ with Small Samples   266
Section 8.3   Estimating p in the Binomial Distribution   275
Section 8.4   Choosing the Sample Size   283

Summary   291
Important Words and Symbols   291
Chapter Review Problems   291
Using Computers   294

9   Hypothesis Testing   297

Section 9.1   Introduction to Hypothesis Testing   297
Section 9.2   Tests Involving the Mean μ 309
Section 9.3   Tests Involving the Difference of Two Means
               (Independent Samples)   320
Section 9.4   Tests Involving Small Samples   329
Section 9.5   Tests Involving Paired Differences (Dependent Samples)   342
Section 9.6   Tests Involving Proportions and Differences of Proportions   354
Section 9.7   The P Value in Hypothesis Testing   365

Summary   373
Important Words and Symbols   374
Chapter Review Problems   374
Using Computers   379

10   Regression and Correlation   383

Section 10.1   Introduction to Paired Data and Scatter Diagrams   383
Section 10.2   Linear Regression and Confidence Bounds for Prediction   389

Section 10.3   The Linear Correlation Coefficient   408
Section 10.4   Testing the Correlation Coefficient   420

Summary   427
Important Words and Symbols   427
Chapter Review Problems   427
Using Computers   430

11 Chi-Square and *F* Distributions 432

Section 11.1   Chi Square: Tests of Independence   432
Section 11.2   Chi Square: Goodness of Fit   444
Section 11.3   Testing and Estimating Variances and Standard Deviations   451
Section 11.4   ANOVA: Comparing Several Sample Means   461

Summary   478
Important Words and Symbols   478
Chapter Review Problems   479
Using Computers   482

12 Nonparametric Statistics 485

Section 12.1   The Sign Test   485
Section 12.2   The Rank Sum Test   496
Section 12.3   Spearman Rank Correlation   503

Summary   516
Important Words and Symbols   517
Chapter Review Problems   517

Appendix I   Tables A1

1. Random Numbers   A1
2. Factorials   A2
3. Binomial Coefficients   A3
4. Binomial Probabilities   A4
5. Areas of a Standard Normal Distribution   A9
6. Student's *t* Distribution   A10
7. Critical Values of Pearson Product-Moment Correlation, *r*   A11

8. The χ^2 Distribution   A12

9. The *F* Distribution   A13

10. Critical Values for Spearman Rank Correlation, r_S A16

Appendix II   Answers to Odd-Numbered Problems   **A17**

Index   **A67**

1 Getting Started

To guess is cheap,
To guess wrongly is expensive

Old Chinese Proverb

Important decisions often involve so many complicated factors that a complete analysis is not practical or even possible. We are often forced into the position of making a guess, but as the proverb implies, a blind guess is not the best solution. Statistical methods, such as we will learn in this book, can help us make the best "educated guess."

What Is Statistics?

All of us have a built-in system of inference that helps us make decisions; without it we would be lost. Of course, we also have a built-in set of prejudices that affects our decisions. A definite advantage of statistical methods is that they can help us make decisions without prejudice. Moreover, statistics can be used for making decisions when faced with uncertainties. For instance, if we wish to estimate the proportion of people who will have a severe reaction to a new flu shot without giving the shot to everyone who wants it, the use of statistics is appropriate.

The general prerequisite for statistical decision making is the gathering of *numerical* facts. Procedures for evaluating numerical data together with rules of inference are central topics in the study of statistics. In short, we may say that *statistics is the study of how to collect, organize, analyze, and interpret numerical information.*

The statistical procedures you will learn in this book should supplement your built-in system of inference—that is, the results from statistical procedures and good sense should dovetail. Of course, statistical methods in themselves have no power to work miracles. These methods can help us make some decisions, but not all conceivable decisions. Remember, a properly applied statistical procedure is no more accurate than the data, or facts, upon which it is based. Finally, statistical results should be interpreted by one who understands not only the methods, but also the subject matter to which they have been applied.

In any investigation it is important to ask the right kind of questions. What are the key points of the problem and its solution? In this section we will examine a variety of problems. Then, thinking like statisticians, we will investigate some important concepts. As a start let's look at a few important terms.

The term *population* refers to all measurements or observations of interest. For example, if we want to know the average height of people who have climbed Mt. Everest, then the population consists of the heights of all those people. Now suppose we want to know the average height of twenty-year-old females in the United States. In this case the population consists of the heights of *all* twenty-year-old females in the United States.

Often it is not feasible to study the entire population. How would you measure the height of each twenty-year-old female in the United States? In such cases we look at samples. A *sample* is simply a part of the population. But not every sample is useful. The sample must represent the population. A *random sample* is such a representative sample; in the next chapter we will study these samples in much more detail. In the meantime we can think of a random sample as a sample determined completely by chance.

Generally speaking, statistical problems are those which require us to draw a random sample of observations from a larger population. Then statistical methods are used to form conclusions about the population based on the information in the sample. Let's look at some examples of populations and samples.

EXAMPLE   1

The department of tropical agriculture is doing a study of pineapples in an experimental field. In this case the data under consideration are the individual weights of all pineapples in the field.

a) The *population* is the weights of all the pineapples in the field.

b) A random collection of 100 pineapples is taken from the field. Each pineapple is weighed. The 100 weights form a random *sample* from the population of all weights.

• **Comment:**   When referring to a population or sample, be sure to give the *quantity* being measured or counted. For instance, in Example 1 it is not sufficient to say that the population consists of all pineapples in the field. We must also state the quantity to be measured. Unless we do, we won't know whether to consider weight, diameter, length, sugar content, acidity level, time to mature, or any other of the many possible measurements that could be made on pineapples.

Throughout this text you will encounter exercises embedded in the reading material. These exercises are included to give you an opportunity to work immediately with new ideas. Cover the answers on the right side (an index card will fit this purpose). After you have thought about or written down *your own response,* check the answers. If there are several parts to an exercise, check each part before you continue. You will be able to answer most of the exercise questions, but don't skip them—they are important.

—————————————— Exercise 1 ——————————————

Television station QUE wants to know the proportion of TV owners in Virginia who watch the station's new program at least once a week. The station asked a random group of 1,000 TV owners in Virginia if they watch the program at least once a week.

a) What is the population?

a) The population consists of the response (does, or does not, watch the new program) from each TV owner in Virginia.

b) What is the sample?

b) The sample consists of the response of each of the 1,000 TV owners in Virginia who were questioned.

Sometimes we do not have access to the entire population, and at other times the difficulties of working with an entire population are prohibitive. The benefit of statistical methods is that they allow us to draw conclusions about populations based only on information from samples. The handicap of these methods is that our conclusions are uncertain—probabilistic, if you will. Probabilism is one of the aspects of statistical thinking that may be unfamiliar to you since you dealt with certainties in your previous mathematical experience. For instance, if you are asked to solve the equation $x + 2 = 5$, you know the solution is $x = 3$. There is no uncertainty; you can check the solution in the equation. We do not demand absolute certainty from statistical methods. That is generally too much to ask. But we do have a measurable degree of probabilistic confidence that the conclusions of a method are valid.

In future work the probability attached to a conclusion will represent the amount of confidence we should have in the conclusion. Probability will be one of our basic tools in the study and application of statistical methods. In Chapter 4 we will study some elementary probability theory, but in the meantime let's take an intuitive point of view. Thus, if the probability that some statement is true is 0.97, then the statement is true about 97% of the time. The statement is false about 3% of the time.

Now let's look at some examples and see where these new terms fit into place. These examples do not begin to cover the complete range of statistical applications, but they are an indication of the types of problems that can be solved using the methods of this book.

EXAMPLE 2

If you travel a long distance by air, chances are that you will not have a direct flight. You usually need to change airplanes, and you will often change airlines in the process of making connections. Normally an agent for your first airline makes arrangements for the connecting flights, and you pay this agent a lump sum for your ticket. If more than one airline company is involved, the companies decide among themselves how much money each should receive.

In the past the division of the money resulted in a great deal of clerical work. Then three airlines decided to use statistical methods to determine how total revenue should be split. During a four-month trial period they took random samples from the overall population of all interairline tickets. From that sample they determined the proportion of total revenues to be distributed to each airline. They were able to estimate that the degree of error in their statistical process was not more than 0.07%—that is, $700 in $1,000,000. On the basis of this work more airlines have used statistical methods in settling interairline accounts. Some of the larger airlines estimate a clerical savings of $75,000 each year over the old methods.

EXAMPLE 3

In 1778 Captain James Cook discovered what we now call the Hawaiian Islands. He gave the islanders a present of several goats, and over the years these animals multiplied into wild herds totaling several thousand. They eat almost anything, including the famous silver sword plant, which was once unique to Hawaii.

At one time the silver sword grew abundantly on the island of Maui (in Haleakala, a national park on that island, the silver sword can still be found), but each year there seemed to be fewer and fewer plants. The disappearance of these plants could have been due to many things (e.g., tourists picking them illegally), but a biologist hypothesized that the goats were mainly to blame.

To determine the effect of goats on the vegetation, the rangers set up stations in remote areas of Haleakala. Very few tourists came to these areas, but they were home to many goats. At each station the rangers found two plots with about the same area, plant count, soil, and climatic conditions. One plot was carefully fenced; the other was not. At regular intervals a plant count was made in each plot.

In this example the population can be thought of as the plant count for the entire park. The samples are plant counts from the experiment stations. The claim is that goats are in fact detrimental to plant life in the park because they reduce the plant count by one fourth or more. Using statistical methods, the claim was confirmed with a high degree of confidence.

───────────────── Exercise 2 ─────────────────

Mountain Joy Company produces ice axes, which are used by mountain climbers to catch themselves in case of a fall on a steep glacier. Most ice axes have a steel head and a fiberglass shaft. A problem with this piece of equipment is that the shaft sometimes breaks after a hard fall.

The specifications for manufacturing ice axes require that 99% of all ice axes made should hold at least 400 lb on the shaft. The production manager wants to be as certain as possible that under present production methods the shafts can support at least 400 lb. To test an ice ax, a force of 400 lb is applied to the shaft. If the shaft does not break, it passes the test. It is possible to weaken the shaft when the force is applied, yet not break it. The manager does not wish to sell weakened ice axes, so she does not test them all.

a) What does the manager wish to test?	a) The manager wishes to test if 99% of ice axes made under present production methods hold at least 400 lb.
b) What is the population?	b) The population is the strength (in pounds) of each ice ax shaft produced under the present method.
c) Since the manager does not wish to test *each* ice ax, what should she do?	c) The manager should use a random sample of ice axes and test it.

We've seen several examples and exercises in which samples are to be used to predict some aspect of a population. A natural question is: How large must these samples be? We will give specific answers to this question in later chapters, but in general, the more certain you wish to be, the larger the sample must be.

In the next three chapters we will begin a more thorough study of random samples, populations, and the way to organize and summarize numerical information in samples and populations. The term *descriptive statistics* refers to this organization of data.

In Chapters 4, 5, and 6 we cover elementary probability theory and some basic probability distributions. By Chapter 7, we will use the previous groundwork to begin *inferential statistics*—that is, methods of using a sample to obtain information about a population.

Section Problems 1.1

1. The students at Eastmore College are concerned about the level of student fees. They took a random sample of 30 colleges and universities throughout the nation and obtained information about the student fees at these institutions. From this information they concluded that their student fees are higher than those of most colleges in the nation.
   a) What is the population?
   b) What is the sample?

2. The quality control department at Healthy Crunch, Inc. wants to estimate the shelf life of all Healthy Crunch granola bars. A random sample of ten bars was tested and the shelf

life determined. From the sample results, a shelf life for all Healthy Crunch granola bars was estimated.
a) What is the population?
b) What is the sample?

3. An insurance company wants to determine the time interval between the arrival of an insurance payment check and the time the check clears. A central payment office processes the payments for a five-state region. A random sample of 32 payment checks from this region was received and processed. The time interval between receipt and check clearance was determined for each check. From this information the company estimated the time interval necessary for all checks sent to this office to clear.
a) What is the population?
b) What is the sample?

4. If you were going to apply *statistical methods* to analyze teacher evaluations, which question form would be better?

Form A: In your own words, tell how this teacher compares with other teachers you have had.

Form B: Use the following scale to rank your teacher as compared with other teachers you have had.

1	2	3	4	5
worst	below average	average	above average	best

   Calculators and Computers in Statistics

Calculators are a tremendous aid to statistics students. A simple four-function calculator with memory and square root key will be adequate to do the calculations in this text. More sophisticated calculators have statistical function keys, and others are programmable.

The question of rounding always arises. Basically, you should try to avoid rounding in intermediate steps. However, this restriction is often difficult to abide by, so we urge you to carry as many digits as is convenient, and at least two more than you want in your final answer. You can expect slight variations in answers due to rounding.

Professional statisticians use computers, and there are many statistical software packages available. Some require large computers and others are designed for microcomputers. The authors of this text have developed an inexpensive and straightforward statistical package called ComputerStat as a supplement to this text. Disks and a user's manual are available from the publisher, D. C. Heath and Company, for the Apple IIe, IIc, or IBM PC. The programs are designed for beginning statistics students with no computer experience. They are written in the computer language BASIC and are interactive. Data are entered while a program is running, so there is no need to create separate data files.

Students should remember that although computers can process data efficiently, the user must determine which processes are appropriate, and then interpret the results in a meaningful way.

In this text we do not attempt to explain how to use statistical software. That is best done in manuals and supplements for the particular software you are using. However, at the end of many chapters is a section called Using Computers. Real world problems with data or situations from published sources are included. The problems can be solved using any appropriate statistical software package, or in most cases, with a programmable calculator and tables.

Section Problems 1.2

1. Review your calculator instructions manual so you can do calculations with confidence. Do some of the computation exercises in your manual that use the operations $+$, $-$, $\times$, $\div$, memory, and $\sqrt{}$.

2. If there are computer programs for statistics available to you, find out how you can use them. Plan to spend a little time reviewing the user's manual that goes with the software package.

SECTION 1.3   Where Will I Use Statistics?

Open any newspaper or magazine, or turn on the TV. Before long you will see a chart of numbers, the results of an opinion poll, a graph, or some comment about how this year compares with last year with regard to crime rate, employment, availability of goods, and so forth. Sports reports are full of statistics. Even the ads tell you to use a certain toothpaste because four out of five dentists recommend it.

We are bombarded by statistical information. Important decisions about the allocation of funds or services are made based on statistical information. You need to know some statistics simply to understand and evaluate the statistical information you are given.

If you look at college catalogues you will find that courses in statistics are required or recommended in areas such as psychology, sociology, computer science, biology, nursing, business, linguistics, economics, political science, education, pre-medicine, and pre-law.

In short, in almost any field you may find that you are required to present data in a meaningful way. You may be required to use some of the statistical methods you will learn about in this text. Even though you might not need to make statistical reports, you will more than likely read them or even be a subject in the report. For instance, in 1990 the U.S. Census Bureau will ask each of us to participate in creating a statistical profile of the United States by giving data about ourselves, our family, income, residence, and so on.

Section Problems 1.3

1. Read a newspaper and take note of articles or displays using statistics.

2. Next time you watch TV, listen to the ads and see how statistics are used to convince you to buy a product.

3. Go to the library and browse through a journal in a field that interests you. Note the use of statistics.

4. Next time someone asks you to respond to a survey, ask that person about the kinds of statistical reports that will be generated from the survey responses.

Summary

In this chapter we saw that statistics is the study of how to collect, organize, analyze, and interpret numerical information. We investigated some types of problems where statistics can be used. In these situations we saw examples of *populations* and *samples*. It is important to remember that the main role of inferential statistics is to draw conclusions about a population based on information obtained from a sample.

Important Words and Symbols

	Section*		Section
Statistics	1.1	Descriptive statistics	1.1
Population	1.1	Inferential statistics	1.1
Sample	1.1		

*Indicates section of first appearance.

Chapter Review Problems

1. Find a newspaper article that uses statistics. Is the data from the entire population, or just from a sample? What is the population? What is the sample?

2. A radio talk show asked listeners to respond either yes or no to the question: Is the candidate who spends the most on a campaign the most likely to win? Fifteen people called in and nine said yes. What is the implied population? What is the sample? Can you detect any bias in the selection of the sample?

2 Organizing Data

Get the facts first, and then you can distort them as much as you please.

Mark Twain

There are two kinds of *facts* or data we can collect: population data and sample data. Population data are complete. The registrar of Arapahoe Community College can tell you exactly how many students are taking a part-time load this term. The registrar has information about the course load of *every* student. In other words, data are available for the *entire population* of students registered this term. When you have population data, you don't need to make educated guesses about it.

In many instances, however, it is either impossible or too costly to gather data about every item or individual in the population of interest. For instance, if a canning company wants to test their final canned products for spoilage, they do not open *every* can. They would have no cans to market if they did. Instead, they take a sample from the total number of cans produced and test them. If the sample passes the various quality control tests, the entire day's production is assumed to be satisfactory.

The canning company is using inferential statistics—that is, they are using a sample to draw conclusions about the entire population. In order to do this, the sample must be representative of the entire population.

In the next section you'll see how to choose such a sample. The last three sections are devoted to organizing data, either from a population or a sample.

SECTION 2.1	Random Samples

Eat lamb—20,000 coyotes can't be wrong!

This slogan is sometimes found on bumper stickers in the western United States. The slogan indicates the trouble that ranchers have experienced in protecting their flocks from predators. Modern methods of predator control have changed considerably from those of earlier days, and to a certain extent the changes have come about through a closer examination of the sampling techniques used.

If we are to use information obtained from a sample to draw conclusions about a population, the sample must be representative of the *entire* population. We cannot simply choose our favorite items or the ones easiest to find. A type of sample that is representative of the entire population is a *random sample*. One of the properties of a random sample is that each member of the population has an equal chance of being included in the sample. But that is not enough. Another necessary condition is that *each* sample of the same size is equally likely to be selected.

> • **Definition:**   A *simple random sample* of n measurements from a population is one selected in such a manner that every sample of size n from the population has equal probability of being selected and every member of the population has equal probability of being included in the sample.

Consider the population of all coyotes in the western United States. The sample of the population which the ranchers observe is largely the coyotes that prefer to live near a ranch. It seems that many coyotes who choose to live near a ranch also like to eat lamb. In fact, most of the coyotes the ranchers observed appeared to be existing solely on sheep.

Based on their experience with this sample of the coyote population the ranchers concluded that *all* coyotes are dangerous to their flocks. Coyotes should be eliminated! The ranchers used a special poison bait to get rid of the coyotes. Not only was this poison distributed on ranch land, but with government cooperation it was also widely distributed on public lands.

The results were not all that great for the ranchers. The overall coyote population dropped where the poison was applied, but the ranchers had almost as much trouble as ever. They were losing almost as many sheep to coyotes as before. Why?

——————————— Exercise 1 ———————————

a) Do you think that the sample of coyotes the ranchers observed could be thought of as a random sample of the *entire* coyote population? Explain.

a) No; the ranchers only observed coyotes near their ranches.

b) If a sample is not chosen at random from the population, is it safe to use the results of that sample to describe the *entire* population?

b) No.

c) Do you think the idea of reducing the *entire* coyote population could be statistically justified on the basis of the ranchers' experience with the coyotes?

c) We don't think it could be statistically justified one way or another because the sample of coyotes was not a true random sample.

d) Biological field reports indicate that coyotes who eat sheep are fairly consistent in their preference for sheep, whereas the majority of coyotes who live in the wilderness stick to foods they find in the wild. When the poison was widespread, which group of coyotes would tend to get poisoned: those that ate sheep or those that ate what they could find in the wild?

d) When there is poison bait scattered throughout the wilderness, it is the wilderness coyote that eats it; the sheep-eating coyote doesn't bother with it.

Today there is an effort to selectively hunt specific coyotes that are known sheep killers. The important thing to learn from the above discussion is that if statistical methods are to be reliable, we must have a *true random sample*.

_____ Exercise 2 _____

Why don't the following procedures give a random sample for the entire population of New York City?

a) Select every third woman entering a beauty shop.

a) This sample is biased toward women who can afford to and like to go to beauty shops.

b) Select every third man entering a bar.

b) This sample is biased toward men who can afford to and prefer to visit bars.

c) Select every third person coming out of a boxing match at Madison Square Garden.

c) This sample is likely to be biased toward people who are sports-minded and can afford a ticket to Madison Square Garden.

In all cases there would probably be very few small children, elderly people, or poor people.

We've seen several sampling procedures that do not produce a random sample. How do we get random samples? Suppose you need to know if the emission system of the latest shipment of Toyotas satisfies pollution control standards. You want to pick a random sample of 30 cars from this shipment of 500 cars and test them. One way to pick a random sample is to number the cars 1 through 500. Write these numbers on cards, mix up the cards, and then draw 30 numbers. The sample will consist of the cars with the chosen numbers. If you mix the cards sufficiently, this procedure produces a random sample.

An easier way to select the numbers is to use a random number table. You can make one for yourself by writing the digits 0 through 9 on separate cards and mixing up these cards in a hat. Then draw a card, record the digit, return the card, and mix up the cards again. Draw another card, record the digit, and so on. However, Table 1 in Appendix I is a ready-made random number table. Let's see how to pick our random sample of 30 Toyotas by using this random number table.

EXAMPLE   1

Use a random number table to pick a random sample of 30 cars from a population of 500 cars.

Again we assign each car a different number between 1 and 500. Then we use the random number table to choose the sample. The table in the Appendix has 45 rows and eight columns of five digits each; it can be thought of as a solid mass of digits that has been broken up into rows and columns for user convenience.

You read the digits by beginning anywhere in the table. We dropped a pin on the table, and the head of the pin landed in row 16, column 2. We'll begin there and list all the digits in that row. If we need more digits, we'll move on to row 17, and so on. The digits we begin with are: 32880   37826   28232   58464   37726   51581   43138   30411   72660   04269   38533   73194   03771   83313   05359. Since the highest number assigned to a car is 500, and this number has three digits, we regroup our digits into blocks of 3: 328   803   782   628   232   584   643   772   651   581   431   383   041   172   660   042   693   853   373   194   037   718   331   305   359.

To construct our random sample we use the first 30 car numbers we encounter in the random number table when we start at row 16, column 2. The first car will be #328. There are no car numbers 803, 782, and 628, so we skip these numbers. The second car is #232. We skip the numbers 584, 643, 772, 651, and 581 since they are too large. The fourth and fifth cars in the sample are #431 and #383. The next group of three digits is 041, which is the same as 41, so we include car #41 in the sample. The other cars in the sample listed in row 16 are #172, #42, #373, #194, #37, #331, #305, #359. To get the rest of the cars in the sample we continue using the random number table in this fashion. If we encounter a number we've used before, we'll skip it.

Another important use of random number tables is *simulation*. We use the word simulation to refer to methods of providing arithmetical imitations of "real" phenomena. Simulation methods have been productive in studies involving nuclear reactors, cloud formation, cardiology (and medical science in general), highway design, production control, shipbuilding, airplane design, war games, economics, electronics, and countless other studies. A complete list would probably include something from every aspect of modern life. In Exercise 3 you'll perform a brief simulation.

_____ Exercise 3 _____

FIGURE 2-1   Direction of a Pollen Grain as It
Changes Position

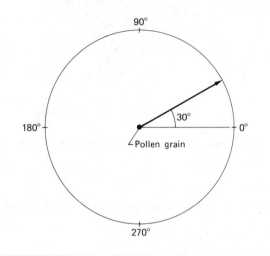

A single pollen grain is floating on a surface of water. Through a microscope we can observe the Brownian motion (random motion of the pollen grain caused by the impact of water molecules). As part of a project Carlos is to chart the course of a pollen grain as it moves on a drop of water. Rather than watch it through a microscope and measure the direction of each movement (an almost impossible task without a slow motion camera since the pollen grain moves so rapidly), he is instructed to simulate the motion for seven motions. Let's use a random number table to simulate the observed direction of the pollen grain for seven position changes.

For each position change, imagine the pollen grain to be in the center of a circle marked off in degree measure in such a way that 0° is east, 90° is north, 180° is west, and 270° is south (*see* Figure 2-1). An arrow pointing from the pollen grain indicates the direction of the pollen grain as it changes position.

a) Let's agree that measurement of the direction of the pollen grain to the nearest degree is accurate enough. Then our problem will be solved if we obtain seven random numbers, each of which is between 0 and 359. Since the highest possible number, 359, has three digits, we break up the digits in the random number table into blocks of _____ (one, two, three, four) digits.

b) Begin at row 5, column 7, and read the table row by row to find the seven random directions.

a) Three

b) The digits grouped in threes are 933   072   009   040   811   583   048   358   949   585   763   228   222, etc. The first number is 933, and since we are only interested in numbers between 0 and 359, we skip 933. The next number is 072, or 72, and the number after that is 009, or 9. In this way we find the random directions to be 72°, 9°, 40°, 48°, 358°, 228°, 222°.

We have been reading the random number table by rows, but there are other schemes we can follow when we use the random number table. We can read the digits by column instead of by row or we can read diagonally (Figure 2-2). But once you have picked the starting point, use a single scheme to read the random number table until you have completed the sample. In this way you can pick a random sample.

FIGURE 2-2   Schemes by Which to Read the Random Number Table

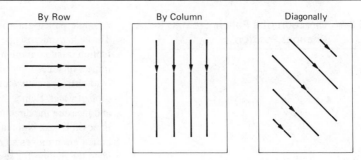

Section Problems 2-1

1. a) In your own words explain the meaning of the terms *random numbers* and *random samples*.
    b) Why are random samples so important in statistics?

2. Use a random number table to get a list of eight random numbers from 1 to 976. Explain your work.

3. Use a random number table to get a list of six random numbers from 1 to 17,431. Explain your work.

4. Use a random number table to get a list of five random numbers from 614 to 2,419. Explain your work.

5. Use a random number table to get a list of seven random numbers from 1 to 99. Explain your work.

6. Use a random number table to get a list of six random numbers from 324 to 1,077. Explain your work.

7. Use a random number table to get a list of five random numbers from 119 to 964. Explain your work.

8. Consider the population of all students at your school.
    a) Explain how you could get a random sample of ten students from this population.
    b) List some ways of getting samples from this population that are *not* random samples. Explain why each of these samples is *not* a random sample.

9. Suppose you are given the number 1 and each of the other students in your statistics class calls out consecutive numbers until each person in class has his or her own number. Explain how you could get a random sample of four students from your statistics class.
    a) Explain why the first four students walking into the classroom would not necessarily form a random sample.
    b) Explain why four students coming in late would not necessarily form a random sample.
    c) Explain why four students sitting in the back row would not necessarily form a random sample.
    d) Explain why the four tallest students would not necessarily form a random sample.

10. How would you use a random number table to get a random sample of 12 cars in the student parking area?

11. How would you use a random number table to get a random sample of 10 of the next 150 customers entering a restaurant?

12. How would you use a random number table to get a random sample of 16 men's dress shirts taken from the next 350 shirts coming off a production line?

13. How would you use a random number table to get a random sample of seven gasoline stations from the telephone directory yellow pages of your area?

14. How would you use a random number table to get a random sample of 12 stereo speakers from a sound equipment warehouse?

15. Professor More is designing a multiple-choice test. There are to be ten questions. Each question is to have five choices for answers. The choices are to be designated by the letters *a*, *b*, *c*, *d*, and *e*. Professor More wishes to use a random number table to determine which letter choice should contain the correct answer for a question. Using the number correspondence 1 for *a*, 2 for *b*, 3 for *c*, 4 for *d*, and 5 for *e*, use a random number table to determine the letter choice for the correct answer in each of the ten questions.

16. Professor More also uses true–false questions. She wishes to place 20 such questions on the next test. To decide whether to place a true statement or false statement in each of the 20 questions, she uses a random number table. She selects 20 digits from the table. An odd digit tells her to use a true statement. An even digit tells her to use a false statement. Use a random number table to pick a sequence of 20 digits, and describe the corresponding sequence of 20 true–false questions. What would the test key for your sequence look like?

17. Even in very calm air, a microscopic dust particle will exhibit random motion due to bombardment by air molecules. Use a random number table to simulate ten position changes of a microscopic dust particle. (Imagine the particle to be at the center of a small sphere and use latitude and longitude markings for a position pointer as in Figure 2-3.)

FIGURE 2-3

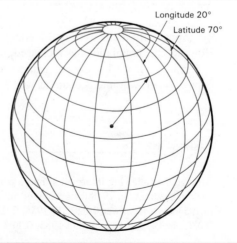

Longitude 20°
Latitude 70°

Graphs

Millionaires are booming in the United States! One national news magazine recently published these figures: in 1948 there were 13,000 millionaires; in 1953 there were 27,000; in 1958, 52,000; in 1963 the number jumped to 75,000; by 1968 the number increased to 117,000; in 1973 there were 180,000; by 1978, 250,000; and by 1983, 390,000.

The last sentence is a mass of numbers that is difficult to read. No matter what type of data we have, sample or population, it is important to be able to organize and present the data to other people. A table is a far more effective way to present information than is a sentence full of numbers. A properly labeled table frees the reader from the task of scanning the text to discover what the table represents. The table caption and the column headings identifying the table entries make a table a self-contained body of information. How are the data organized within the table? Organization depends on the data and on what you are trying to show. In the case of our data about millionaires it is natural to organize the table chronologically, or by year (Table 2-1).

TABLE 2-1 Americans Worth at Least One Million Dollars

Year	Number of Millionaires
1948	13,000
1953	27,000
1958	52,000
1963	75,000
1968	117,000
1973	180,000
1978	250,000
1983	390,000

Apparently there were considerably more millionaires in 1983 than in 1948. However, the population has increased, too, from about 147 million to an estimated 235 million. If we list the percentage of millionaires in these years, we will take population growth into account as well as the growth in the number of millionaires.

_____ Exercise 4 _____

Arrange the following information into a table with an appropriate caption and appropriate column headings. In 1948 the percent of millionaires in the United States was about 0.01%; in 1953 the percent was 0.02%; in 1958, 0.03%; in 1963, 0.04%; in 1968, 0.06%; in 1973, 0.08%; in 1978, 0.11%; and in 1983, 0.17%.

a) Make the table.

a) The information telling us that the table gives the percentage of millionaires in the population of the United States for various years should be noted either in the table caption or column headings.

TABLE 2-2 Growing Percentage of U.S. Millionaires

Year	Percentage of Population
1948	0.01
1953	0.02
1958	0.03
1963	0.04
1968	0.06
1973	0.08
1978	0.11
1983	0.17

b) Which five-year interval shows the greatest increase in percentage?

b) The years 1978–1983, because the percentage jumped from 0.11% to 0.17%, a change of 0.06%

c) Inflation has reduced the purchasing power of one million dollars ($1,000,000) in 1948 to less than 0.30 (30%) of that amount ($300,000) in 1983. The 1983 millionaire would need more than 3.3 million dollars to have the same purchasing power as the 1948 millionaire. Does your table take the changing buying power into account?

c) No. We would need more information to take purchasing power into account.

Newspapers and magazines seem to prefer graphs to tables when they present information. Graphs make any trend more obvious. For instance, which information display in Figure 2-4 emphasizes the fact that there are almost *five* times as many widows as widowers among the millionaire set: *a* or *b?* The graph enables the reader to compare facts and figures quickly. Bar graphs, pictographs, and pie charts are frequently used graphs.

FIGURE 2-4 Millionaires

a) 7,300 widowers and 36,100 widows

b) Widowers
7,300

Widows
36,100

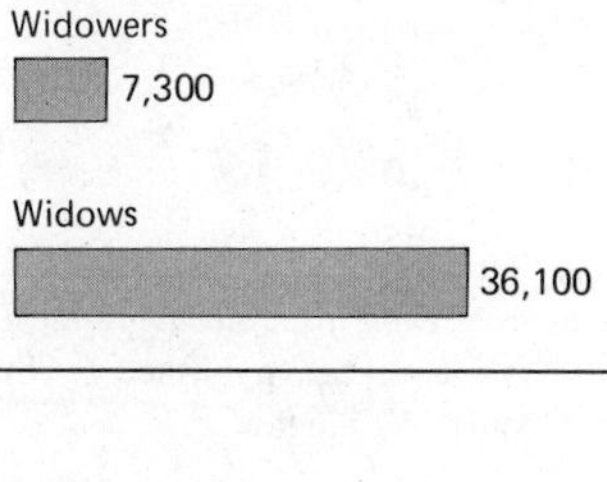

FIGURE 2-5   Days Before Death (Sitting Quietly in Shade at 110 °F)

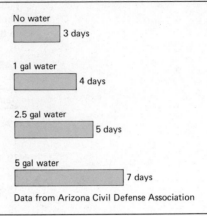

No water
3 days

1 gal water
4 days

2.5 gal water
5 days

5 gal water
7 days

Data from Arizona Civil Defense Association

Let's first consider *bar graphs*. The bars can be vertical or horizontal, but they should be of *uniform width* and be *uniformly spaced*. The length of a bar represents the quantity we wish to compare under various conditions. In Figure 2-5 we are comparing the number of days one can survive at 110 °F temperature with various amounts of water. The length of the bar represents days one can survive.

The following example shows you how to construct a bar graph.

EXAMPLE   2

Hikers are often cautioned to carry extra jackets so they will be prepared for the effect of wind chill. Suppose you are hiking in 50 °F temperature and a wind comes up. A breeze of 5 mph makes the effective temperature 48 °F. If the wind picks up to 10 mph, then the temperature is equivalent to 40 °F. A 15 mph wind drops the effective temperature to 36 °F, and a 20 mph wind drops the temperature to freezing. A 25 mph wind makes the effective temperature only 30 °F.

Before we make a bar graph of this information, we'll make a table of wind and effective temperature. (Even though you may not want to exhibit the table, make one anyway to organize the data. Then the graph will be easier to make.)

TABLE 2-3 Wind Chill at 50 °F

Wind Speed (mph)	Equivalent Temperature (°F)
Calm	50
5	48
10	40
15	36
20	32
25	30

Source: Data from *Surviving the Unexpected Wilderness Emergency,* by Gene Fear, published by Survival Education Association, Tacoma, Washington.

FIGURE 2-6   Wind Chill at 50 °F

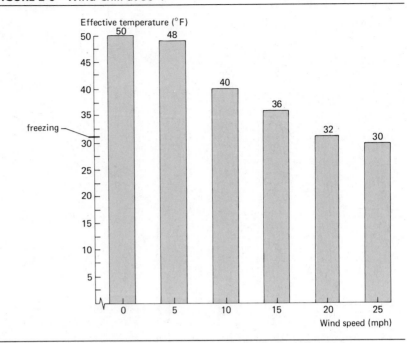

To make the bar graph (Figure 2-6) we'll list the wind speed on the horizontal axis and the effective temperature on the vertical one. Each bar will be centered over its wind speed, and the height of the bar will represent the effective temperature. Note again that the bars have the same width and the spacing between all the bars is the same. Both axes are labeled and have scale markings.

_____ Exercise 5 _____

Sunshine Travel Agency offers a rain insurance policy on their Hawaii tours. It costs an extra $100. If you buy the optional policy and it rains during more than 15% of the days of your trip, you will be reimbursed for meals and lodging during the extra rainy days (beyond the 15% and up to five days). You are planning a trip to Hawaii and you are debating about taking the insurance. You obtain the following rainfall information from the U.S. National Oceanic and Atmospheric Administration.

TABLE 2-4 Average Monthly Rainfall in Honolulu, Hawaii (1941–1980)

Month	Jan.	Feb.	Mar.	Apr.	May	June	July	Aug.	Sept.	Oct.	Nov.	Dec.
Rainfall (in.)	4.40	2.46	3.18	1.36	0.96	0.32	0.60	0.76	0.67	1.51	2.99	3.64

a) Make a bar graph of this information with month on the horizontal axis and rainfall on the vertical.

a) *See* Figure 2-7.

FIGURE 2-7 Average Monthly Rainfall, Honolulu, Hawaii (1941–1980)

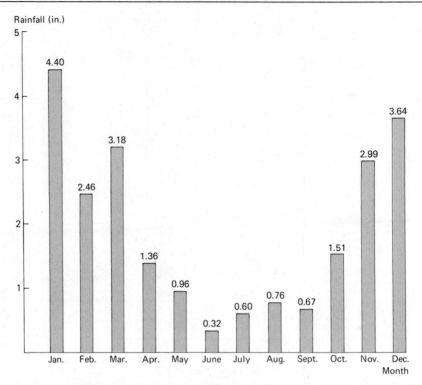

b) There is the rainy season and there is the dry season. From the graph, which six months would you say make up the rainy season?

c) Without the rain insurance, which winter month (November, December, January, February) would be best for your trip?

b) October, November, December, January, February, and March.

c) February, since it has the least average rainfall.

When you read bar graphs, be careful of changing scales. For instance, in Figure 2-8 you see two graphs showing the life expectancy of a man born in 1920 and one born in 1970. The change in life expectancy over the 63-year period illustrated in Figure 2-8 is large: from 54 years to 71, an increase of 17 years. But part (a) makes it seem that the life expectancy has more than tripled. Notice the squiggle ∿ at the bottom of the vertical axis. This is to inform you that the years 0 to 49 have been omitted. In the second bar graph no years were skipped, and the picture immediately gives an accurate impression. Many magazine articles use a changing scale, and you should watch for omitted values. If you omit values, be sure to give the reader fair warning with a squiggle ∿ at the beginning of the axis.

FIGURE 2-8   Male Life Expectancy from Birth

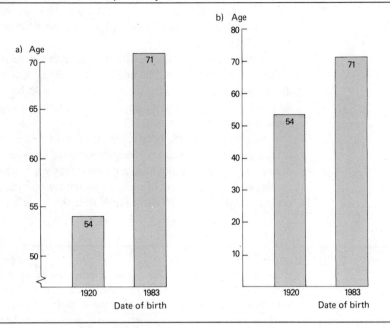

Sometimes pictures are used instead of solid bars. Such a graph is called a *pictogram*. A pictogram can give an accurate and sometimes more interesting display of data, but again there are aspects that can mislead a reader who is not alert. If the size of the picture changes, the results can be quite misleading. For instance, in Figure 2-9 we see two pictograms of the same information.

FIGURE 2-9   The Shrinking United States Dollar (1960–1985)

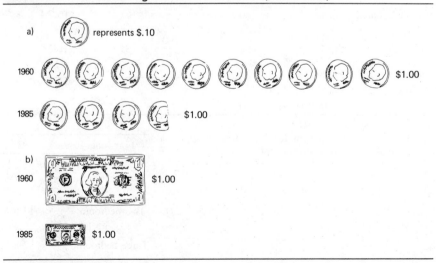

The 1985 dollar was worth only 37% of the 1960 dollar, but part (b) of Figure 2-9 makes the situation look worse than it is. Not only did the width shrink by 37%, but the height did as well. The total area of the 1985 bill is much less than 37% of the 1960 dollar bill.

In pictograms the pictures should all be the same size. The number of pictures rather than the size should be changed to indicate a changing quantity. In addition, the amounts represented by the pictograms should be labeled (e.g., $1.00) so that the reader does not have to count the number of pictures and to compensate for rounding off.

Another popular pictorial representation of data is the *circle graph,* or pie chart. It is relatively safe from misinterpretation and is especially useful for showing the division of a total quantity into its component parts. The components are represented by proportional segments of a circle, which are often labeled by corresponding percents of total. Exercise 6 shows how to make a circle graph.

_____ Exercise 6 _____

Homesweet College has no on-campus housing. However, there is a special housing service to help out-of-town students find approved living quarters. The current list includes 720 rental units as described in the first two columns of Table 2-5.

TABLE 2-5 Available Housing for Students

Unit	Number	Fractional Part	Percent	Number of Degrees Occupied in Circle
Room, no kitchen	360	$\frac{360}{720} = 0.5$	50%	$0.50 \times 360° = 180°$
Efficiency apartment	130	$\frac{130}{720} = 0.181$	18.1%	$0.181 \times 360° = 65°$
One-bedroom apartment	120	$\frac{120}{720} = 0.167$	16.7%	$0.167 \times \_\_\_ = \_\_\_$
Two-bedroom apartment	90	$\frac{90}{720} = 0.125$	12.5%	$\_\_\_ \times \_\_\_ = \_\_\_$
Three-bedroom apartment	20	_____	_____	$\_\_\_ \times \_\_\_ = \_\_\_$
Total Number = 720				

a) Fill in the missing parts of Table 2-5. Remember, the central angle of a circle is 360°.

a)

	Fractional Part	%	Number of Degrees
One-bedroom,	$\frac{120}{720}$	16.7%	$0.167 \times 360° = 60°$
Two-bedroom,	$\frac{90}{720}$	12.5%	$0.125 \times 360° = 45°$
Three-bedroom,	$\frac{20}{720}$	2.8%	$0.028 \times 360° = 10°$

b) Do the degrees in the last column add up to 360°?

c) Fill in the missing degrees and percents in the circle graph in Figure 2-10 (i.e., the degrees for one-bedroom apartments and for rooms).

b) They always should; however, due to rounding errors, the sum might be slightly more or less than 360°.

c) *See* Figure 2-11.

FIGURE 2-10   Student Housing at Homesweet College

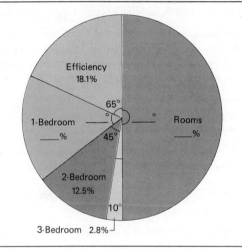

FIGURE 2-11   Completion of Figure 2-10

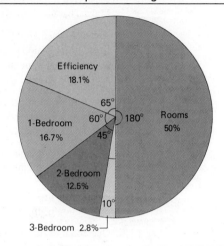

Section Problems 2.2

1. Over the last 20 years computer processing has become faster and cheaper. According to one magazine article (*U.S. News and World Report,* August 26, 1985) the time and cost of processing the same data on a large computer has gone from 29 seconds and 47 cents in 1965 to 4 seconds and 20 cents in 1975 to 0.4 seconds and 4 cents in 1985. Make a bar graph showing the cost of processing the data in 1965, 1975, and 1985. Then make another bar graph showing the time required to process the data in 1965, 1975, and 1985. As you compare the two bar graphs, can you tell which quantity has dropped the most: time or cost?

2. At Westgate Community College, a survey was done to determine when students are available for class. A questionnaire was given to a random sample of students. The instructions were to mark each of the time categories in which they would take classes. Many students marked more than one time category. Responses from the students in the sample indicated that 52 would take early morning classes, 85 would take mid-morning classes, 41 would take afternoon classes, and 37 would take evening classes. Make a bar graph to display student availability for each of the four time categories.

3. An allergy study involved a random sample of 1,000 people, each of whom had at least one parent with known allergies. Of the people in the sample, 230 had severe allergies, 360 had moderate allergies, 160 had mild allergies, and 250 had no apparent allergies. Make a bar graph showing the number of people in the sample and their indicated allergy levels.

4. Last year Jefferson Mortgage Company had 200 loan payments due January 1. Of these, 48 payments arrived before January 1, 36 arrived one day late, 33 arrived two days late, 37 arrived three days late, 26 arrived four days late, and 20 arrived 5 or more days late. Make a bar graph to display the number of payments arriving in each of the time periods.

5. Aid to Abused Children is a nonprofit organization. Funds are solicited from contributions. For each dollar contributed, 35 cents goes toward administrative personnel salaries, 20 cents toward publicity, and 45 cents toward centers taking care of abused children. Make a circle chart showing how the contributions are spent.

6. Opera Alive sponsors three major opera productions each year. Ticket sales provide 42% of the necessary funds, contributions provide 27%, and national grant money provides the rest. Draw a circle graph to show how production costs are met.

7. At Eastview College faculty members in the Physical Education Department are required to keep records of injuries suffered by students in classes involving participation in sports. Of the 57 injuries that occurred last year, 23 involved the knee; 14, the ankle; 4, the elbow; 10, the wrist; and 6, other parts of the body. Make a pie chart showing the distribution of the injuries.

8. One dollar of tuition at Plattsburg College was distributed in the following way: 27 cents went to faculty salaries, 8 cents to the library and media center, 9 cents to the physical plant (including utilities, cleaning, etc.), 30 cents to general college administration, 11 cents to counseling and career services, 8 cents to records and admissions, and 7 cents to miscellaneous expenses such as equipment, duplicating services, etc. Make a pie chart to show the distribution of the tuition dollar.

9. In the 1984 Summer Olympic Games the United States won 83 gold medals, West Germany won 17, Romania won 20, China won 15, Italy won 14, and Japan and Canada won 10 each. Make a pictogram showing the distribution of medals to these countries.

10. The Social Security tax is computed as a percent of a specified maximum wage. Over the years, that percent has increased. In 1940 the percent was 1%; by 1950 it increased to 1.5%; in 1960 it was 3%; in 1970 it was 4.8%; and by 1980 it was 6.13%. In 1988 it is scheduled to be 7.51%. Make a pictogram showing the percents of maximum wages contributed to Social Security for the years given.

SECTION 2.3   Histograms and Frequency Distributions

Healthy Crunch Cereal is about to take over sponsorship of the TV program "Space Voyage." The advertising manager has requested a report on the age distribution of the viewers so the spot ads can be tailored to appeal to the age groups with the most viewers. The viewer age report contains the graph in Figure 2-12 that was made from a random sample of viewers.

FIGURE 2-12   Viewer Age for "Space Voyage"

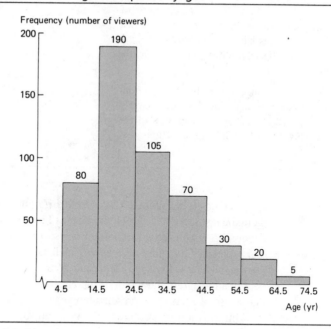

_____ Exercise 7 _____

Use the graph of the viewer age distribution for the program "Space Voyage" (Figure 2-12) to answer the following questions.

a) What does the height of each bar represent? How many viewers are represented in this graph?

a) The height of each bar represents the number of viewers of that age group. To get the total number of viewers, add the heights of the bars. There are 500 viewers represented.

b) What does the width of a bar represent?

b) The width represents the age group.

c) What ages are included in the group with the most viewers? Is this graph detailed enough to tell you exactly how many viewers are 21 years old?

c) The age group 14.5–24.5 has the most viewers. We cannot tell how many viewers are 21 years old. All we can say is that there are 190 viewers between the ages of 14.5 and 24.5 years.

d) From the information in this graph about the age of the viewers, which of the following ads do you think the manager might choose for "Space Voyage"?

Scene 1: A grandmother and first grader at the breakfast table. The grandmother says to the child, "Eat Healthy Crunch Cereal because it will make you grow."

d) Since the largest age group is between 14.5 and 24.5 the ad about the campers would probably be the best. This age group would not necessarily be interested in food that makes a first grader grow, or food that is eaten before one reads the stock market report.

(continued on page 26)

Scene 2: A middle-aged man reading the stock report. An empty bowl is on the table with an open box of Healthy Crunch Cereal beside it. The man puts down the paper and puts his hand on the box of Healthy Crunch as he says, "I eat Healthy Crunch even *before* I read the stock report."

Scene 3: Two young campers eating breakfast in front of their tent. A box of Healthy Crunch Cereal is clearly visible in the foreground. One camper says to the other, "Healthy Crunch will help us climb that mountain."

The graph of Figure 2-12 is a little different from the other bar graphs we looked at in the last section. A graph like that of Figure 2-12 is called a *histogram*. It differs from a bar graph in two important ways: the bars always touch, and the width of a bar represents a quantitative value, such as age. In a bar graph we could make the bars as wide as we wished, according to the visual impression we wanted to convey. But in a histogram the width of the bar has a meaning. For instance, in Figure 2-12 the width of each bar represents ten years.

Information is presented in condensed form in a histogram. The original data for the viewer age report included the *exact* number of 21-year-olds in the sample. In the histogram this number was grouped with the ages 14.5–24.5. We can tell how many viewers are in that age group, but we cannot tell exactly how many are 21 years old. However, the condensed information in the histogram can be assimilated more quickly than the same information in more detailed form.

If you are given many pieces of data, how do you condense the information to make a histogram? The task force to encourage car pools did a study of one-way commuting distances for workers in the downtown Dallas area. A random sample of 60 of these workers was taken. The commuting distances of the workers in the sample are given in Table 2-6.

The first thing to do is to decide how many bars or classes you want in the histogram. Five to 15 classes are usually used. If you use fewer than five classes, you risk losing too much information; but if you use more than 15 classes, the clarity

TABLE 2-6 One-Way Commuting Distances in Miles for 60 Workers in Downtown Dallas

13	47	10	3	16	20	17	40	4	2
7	25	8	21	19	15	3	17	14	6
12	45	1	8	4	16	11	18	23	12
6	2	14	13	7	15	46	12	9	18
34	13	41	28	36	17	24	27	29	9
14	26	10	24	37	31	8	16	12	16

of the diagram might be sacrificed for detail. Let the spread of the data and the purpose of the histogram be your guide when selecting the number of classes.

Next find a convenient class width. To do this, find the difference between the largest and smallest data values and divide by the number of classes.

$$\frac{\text{largest data value } - \text{ smallest data value}}{\text{desired number of classes}} \simeq \text{class width}$$

(The symbol $\simeq$ means approximately equal to.)

In this case let's use ten classes. The largest distance commuted is 47 mi and the smallest is 1 mi.

$$\frac{47 - 1}{10} = 4.6 \simeq 5$$

- **Comment:**   If you want the class width to be a whole number, always round *up* to the next whole number so that the classes cover the data. In the case of the commuters the class width will be 5 mi.

The lowest and highest values that can fit in a class are called the *lower class limit* and *upper class limit,* respectively. The *class width* is the difference between the lower class limit of one class and the lower class limit of the next class. Each class should have the same width, although it is not uncommon to see either the first or the last class width a little longer or shorter than the others. The center of the class is called the *midpoint*. This is found by adding the lower and upper class limits of one class and dividing by 2.

$$\text{midpoint} = \frac{\text{lower class limit } + \text{ upper class limit}}{2}$$

The midpoint is often used as a representative value of the entire class.

Now we can organize the commuting distance data into a *frequency table* (Table 2-7). Such a table shows the limits of each class, the frequency with which the data fall in a class, and the class midpoint. A tally will help us find the frequencies.

Now we're almost ready to make a histogram. But in a histogram we want the bars to touch. There is a space between the upper limit of one class and the lower limit of the next class. The halfway points of these intervals are called *class boundaries*. We use the class boundaries as the endpoints of the bars in the histogram. Then there is no space between the bars. Figure 2-13 is the histogram of commuter distances. The class boundaries are shown. (Sometimes only the class midpoints are labeled.)

TABLE 2-7 Frequency Table of One-Way Commuting Distances for 60 Downtown Dallas Workers (Data in Miles)

Class Lower Limit	Upper Limit	Tally	Frequency	Class Midpoint
1–5		丗 ||	7	3
6–10		丗 丗 |	11	8
11–15		丗 丗 |||	13	13
16–20		丗 丗 |	11	18
21–25		丗	5	23
26–30		||||	4	28
31–35		||	2	33
36–40		|||	3	38
41–45		||	2	43
46–50		||	2	48

FIGURE 2-13 One-Way Commuting Distances in Miles Driven by Downtown Dallas Workers

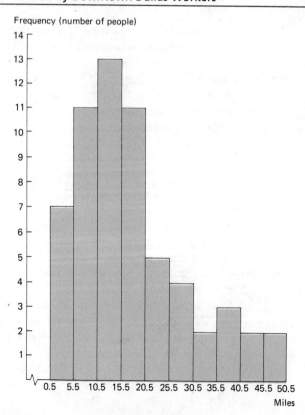

_____ Exercise 8 _____

One irate customer called Dollar Day Mail Order Company 40 times during the last two weeks to see if his order had arrived. Each time he called, he counted the number of rings before the phone was answered (Table 2-8).

TABLE 2-8 Number of Rings Before Phone Was Answered

20	10	8	7	3	5	15	6	9	5
6	18	13	18	1	19	10	19	2	6
4	17	16	9	3	20	15	8	14	19
20	7	14	6	3	17	2	14	4	11

a) What are the largest and smallest values in Table 2-8? If we want five classes, what should the class width be?

a) The largest value is 20; the smallest value is 1. The class width is

$$\frac{20 - 1}{5} = \frac{19}{5} \approx 4$$

b) Complete the following frequency table.

b)

TABLE 2-9 Frequency of Phone Rings Before Answer

Class Limits		Tally	Freq.	Midpoint
Lower	Upper			
1	–	4			
5	–	___			
___	–	12			
13	–	___			
___	–	___			

TABLE 2-10 Completion of Table 2-9

Class Limits		Tally	Freq.	Midpoint
Lower	Upper			
1	–	4	✝✝✝ |||	8	2.5
5	–	8	✝✝✝ ✝✝✝	10	6.5
9	–	12	✝✝✝	5	10.5
13	–	16	✝✝✝ ||	7	14.5
17	–	20	✝✝✝ ✝✝✝	10	18.5

c) Recall that the class boundary is halfway between the upper limit of one class and the lower limit of the next. Use this fact to find the class boundaries.

c)

TABLE 2-11 Class Boundaries

Class Limits		Class Boundaries	
Lower	Upper	Lower	Upper
1	4	0.5	4.5
5	8	4.5	8.5
9	12	8.5	12.5
13	16	___	___
17	20	___	___

Class Limits	Class Boundaries
13–16	12.5–16.5
17–20	16.5–20.5

d) Finish the histogram in Figure 2-14.   d) *See* Figure 2-15.

FIGURE 2-14   Rings Before Answer

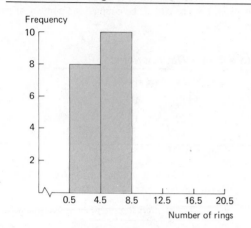

Number of rings

FIGURE 2-15   Completion of Figure 2-14

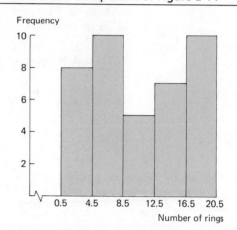

Number of rings

- **Comment:**   The use of class boundaries in histograms assures us that the bars of the histogram touch and that no data fall on the boundaries. Both of these features are important. However, a histogram displaying class boundaries looks somewhat awkward. For instance, the age range of 14.5 to 24.5 years shown in Figure 2-12 does not seem to be as convenient or natural a choice as an age range of 15 to 25 years. For this reason many magazines and newspapers do not use class boundaries as labels on a histogram. Instead, some use lower class limits as labels with the convention that a data value falling on the class limit is included in the next higher class (class to the right of the limit).

    When you use a computer program to create frequency tables and histograms, be sure to determine the convention that is being followed.

A histogram gives the impression that frequencies jump suddenly from one class to the next. If you want to emphasize the *continuous* rise or fall of the frequencies, you can use a *frequency polygon,* or *line graph.*

A frequency polygon is made by connecting in order the top midpoints of the bars in a histogram. For instance, in Figure 2-16 we have a histogram showing the age distribution of 1,000 diabetes patients ten years old or older chosen at random. The frequency of each class is indicated by a heavy dot over the midpoint. To make the corresponding frequency polygon we connect the heavy dots in order. The frequency polygon is shown in Figure 2-17.

A frequency polygon can be constructed without first drawing a histogram. Simply plot the class frequency over the class midpoint. Then connect the points in order.

FIGURE 2-16   Age Distribution of Male Diabetes
Patients 10 Years Old or Older

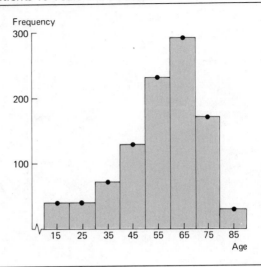

FIGURE 2-17   Frequency Polygon of Ages of Diabetes Patients

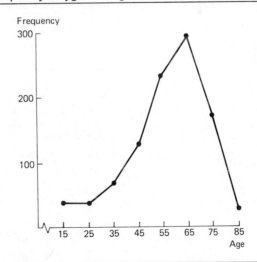

_____ Exercise 9 _____

The ski season in Aspen, Colorado, lasts from mid-November through mid-May. Figure 2-18 shows the high temperatures during the ski season. Make a frequency polygon from the histogram (*see* Figure 2-19).

FIGURE 2-18   High Temperatures During the
Ski Season, Aspen, Colorado (°F)

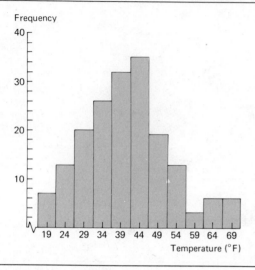

FIGURE 2-19   High Temperatures During the
Ski Season, Aspen, Colorado (°F)

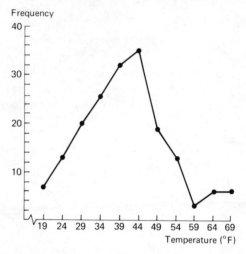

Temperatures from *Climatological Data* published by the U.S. Department of Commerce, 1973

FIGURE 2-20   Hourly Wages for Part-Time Student Workers at Rainy Vail College

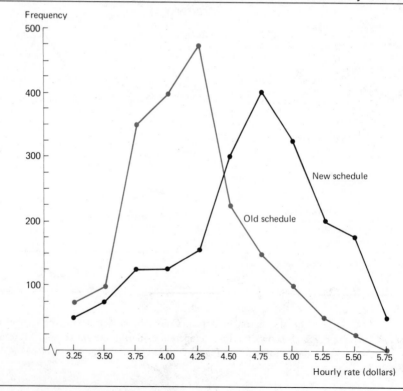

Frequency polygons are especially useful if you wish to compare two distributions. For instance, the President of Rainy Vail College is requesting a larger budget to pay 1,950 part-time student workers. The request is based on the fact that under the newly approved salary schedule student workers will be paid more. Under the old salary schedule a student was paid the same amount per hour no matter how many years he or she had been working. Under the new schedule a student receives a raise of 25¢ per hour for every year of job-related experience. The president's request contains the graph in Figure 2-20 to support the request for increased budget.

The number of student workers is the same under each schedule. However, under the new schedule, more students receive wages between $2.75 and $4.00 per hour than under the old plan. Consequently, if students are to work the same number of hours as they have in the past, more money will be needed to pay them.

Section Problems 2.3

1. Home Video Incorporated rents first-run movies for home video systems. A random sample of 32 days gave the following information about the number of movies rented each day at a local outlet store:

12	27	29	4	21	38	21	5
15	25	23	10	8	22	13	17
43	8	22	23	26	28	31	35
26	34	36	33	32	39	37	29

a) Make a frequency table using five classes.
b) Make a histogram from your table.
c) Make a frequency polygon.

2. The district manager of a suburban park and recreation area wants to estimate the amount of time people spend in their local park in July. A random sample of 50 local residents gave the following information on the estimated number of hours they spent in the regional park during the month of July:

8	5	9	12	15	18	2	3	5	4
2	8	6	10	13	14	8	10	9	13
21	2	4	3	7	7	6	7	6	16
19	11	3	3	8	4	11	10	9	8
6	6	6	7	8	12	6	7	20	7

a) Make a frequency table using seven classes.
b) Make a histogram.
c) Make a frequency polygon.

3. The American Association of Registered Dentists wants to estimate the number of fillings an adult (over 18 years of age) has accumulated. A random sample of 36 dental records gave the following information:

8	9	19	5	13	14	5	6	12	13
3	6	5	11	6	7	4	11	23	5
32	25	21	6	7	11	12	8	15	16
17	9	10	8	22	18				

a) Make a frequency table using six classes.
b) Make a histogram.
c) Make a frequency polygon.

4. A random sample of 42 auto insurance claims made on Any State Insurance Company gave the following information rounded to the nearest dollar:

705	800	1090	2408	2010	910	350
690	520	750	1000	860	170	1100
1150	1590	1841	1880	1942	2210	1886
1954	1710	1372	1200	1210	1351	1400
1790	1682	1310	1460	1550	1300	1463
1511	1190	1200	2350	510	1285	1700

a) Make a frequency table using eight classes.
b) Make a histogram.
c) Make a frequency polygon.

5. Nurses on the eighth floor of Community Hospital believe they need extra staffing at night. To estimate the night workload, a random sample of 35 nights was used. For each night the total number of room calls to the nurses' station on the eighth floor was recorded.

68	70	86	18	90	100	101
95	80	70	73	82	71	37
102	87	46	58	62	63	92
70	69	85	84	86	90	77
60	74	83	86	75	71	88

a) Make a frequency table using five classes.
b) Make a histogram.
c) Make a frequency polygon.

6. A random sample of 40 days gave the following information about the total number of people treated each day at the Community Hospital emergency room:

40	35	42	6	13	50	60	27
8	42	53	17	25	23	24	12
26	32	28	28	31	29	30	28
21	46	22	19	20	30	31	30
36	30	40	38	30	29	31	41

a) Make a frequency table using eight classes.
b) Make a histogram.
c) Make a frequency polygon.

7. A random sample of 30 customers at the grand opening of a new clothing store gave the following information about the age of each customer:

43	33	18	23	19	16
51	54	18	26	25	21
17	30	28	27	27	17
32	21	35	40	39	36
48	47	38	41	50	19

a) Make a frequency table using five classes.
b) Make a histogram.
c) Make a frequency polygon.

8. The manager of a large plant wishes to estimate the amount of gasoline that the workers use to commute to and from work each week. A questionnaire was given to a random sample of 40 employees on the payroll. This questionnaire asked the employee to record his or her gasoline usage to and from work for a designated week. Carpoolers split the total gasoline used by the number of people in the carpool. Here are the data in gallons of gasoline used per week:

10	4	25	15	17	6	8	14
5	20	21	16	3	30	17	14
8	12	24	15	23	18	12	7
9	11	10	17	21	13	18	13
18	14	22	6	12	16	19	26

a) Make a frequency table using six classes.
b) Make a histogram.
c) Make a frequency polygon.

9. The number of hamburgers sold by McDouglas hamburger stands in each of the 50 states during August last year was as follows (in units of 1,000 hamburgers):

86	70	38	100	115	135	51	72	53	65
110	136	56	25	42	60	93	23	58	96
108	131	148	150	73	156	82	71	173	200
147	68	92	15	119	110	172	183	66	65
52	97	212	63	88	93	88	95	31	44

a) Make a frequency table and a histogram for these data. Use 10 classes.
b) Make a frequency polygon for these data.

10. The number of houses sold each month by two realtors, Pete and Mildred, is shown below (for the last 36 months).

Pete

4	9	5	4	9	5	2	0	8	4	8	10
6	7	13	8	2	10	0	12	8	4	6	8
6	3	3	6	7	13	9	11	3	4	0	8

Mildred

6	9	0	1	2	3	6	3	9	8	7	5
5	3	8	1	0	5	0	4	8	9	5	7
4	1	1	6	2	3	1	8	5	3	13	5

a) Make a frequency table and a histogram for Pete and for Mildred. (Use five classes for each.)

b) By looking at the two histograms, can we say who sells more houses?

11. The agricultural experiment station at Centennial, Colorado, made the following chart of annual rainfall (to the nearest inch) from the year 1887 to the year 1955.

12	9	14	11	15	15	7	12	18
16	15	11	16	19	21	18	11	13
19	19	11	17	16	12	10	19	15
14	22	13	13	21	10	11	14	10
29	10	14	13	15	13	13	15	9
12	15	8	16	11	12	19	6	13
17	21	12	13	15	14	18	10	18
12	22	12	11	8	13			

a) Make a frequency table and a histogram using only five classes.

b) Make a frequency polygon from the histogram constructed in part (a).

SECTION 2.4   Stem-and-Leaf Displays

The stem-and-leaf display is just one of many useful ways of studying data in a field called Exploratory Data Analysis (often abbreviated as EDA techniques). John W. Tukey wrote one of the definitive books on the subject: *Exploratory Data Analysis* (Addison Wesley, 1977). Another very useful reference for these techniques is the book: *Applications, Basics, and Computing of Exploratory Data Analysis* by Paul F. Velleman and David C. Hoaglin (Duxbury Press, 1981). The EDA techniques are designed for easy implementation and can tell us a great deal about data groupings and extreme values. These techniques can help us ask the right kind of questions and give us new insights about data.

In this text we will introduce two of the EDA techniques: stem-and-leaf displays and, in Section 3.2, box-and-whisker plots. Let's first look at a stem-and-leaf display.

We know that frequency distributions and histograms provide a useful organization and summary of data. However, in a histogram, we lose most of the specific data values. A stem-and-leaf display is a device that organizes and groups data, but allows us to see as many of the digits in each data value as we wish. In the next example we will make a stem-and-leaf display.

EXAMPLE 3

Many airline passengers seem weighted down with their carry-on luggage. Just how much weight are they carrying? The carry-on luggage weights for a random sample of 40 passengers returning from a vacation to Hawaii were recorded in pounds (*see* Table 2-12).

TABLE 2-12 Weights of Carry-On Luggage in Pounds

30	27	12	42	35	47	38	36	27	35
22	17	29	3	21	0	38	32	41	33
26	45	18	43	18	32	31	32	19	21
33	31	28	29	51	12	32	18	21	26

To make a stem-and-leaf display we break the digits of each data value into *two* parts. The left group of digits is called a *stem,* and the remaining group of digits on the right is called a *leaf.* We are free to choose the number of digits to be included in the stem.

The weights in our example consist of two-digit numbers. For a two-digit number the stem selection is obviously the left digit. In our case the tens digits will form the stems and the units digits will form the leaves. For example, for the weight 12, the stem is 1 and the leaf is 2. For the weight 18, the stem is again 1 but the leaf is 8. In the stem-and-leaf display we list each possible stem once on the left and all its leaves in the same row on the right.

Figure 2-21 shows a stem-and-leaf display for the weights of carry-on luggage.

FIGURE 2-21   Weight of Carry-On Baggage

Stem	Leaves
0	3 0
1	2 7 8 8 9 2 8
2	7 7 2 9 1 6 1 8 9 1 6
3	0 5 8 6 5 8 2 3 2 1 2 3 1 2
4	2 7 1 5 3
5	1

FIGURE 2-22   Stem-and-Leaf Display of Carry-On Luggage Weights

leaf unit = 1 lb

3 | 2 represents 32

Stem	Leaves
0	3 0
1	2 7 8 8 9 2 8
2	7 7 2 9 1 6 1 8 9 1 6
3	0 5 8 6 5 8 2 3 2 1 2 3 1 2
4	2 7 1 5 3
5	1

FIGURE 2-23   Multiple Lines per Stem for Carry-On Luggage Weights

leaf unit = 1 lb

3 | 2 represents 32

0*	3 0
0.	
1*	2 2
1.	7 8 8 8 9
2*	2 1 1 1
2.	7 7 9 6 8 9 6
3*	0 2 3 2 1 2 3 1 2
3.	5 8 6 5 8
4*	2 1 3
4.	7 5
5*	1

From the stem-and-leaf display, we see that two bags weighed 27 lb, one weighed 3 lb, one weighed 51 lb, and so on. We see that most of the weights were in the 30-lb range, only two were less than 10 lb, and six were over 40 lb. Note that the length of the leaves gives the visual impression that a sideways histogram would present.

As a final step, we need to indicate the scale. This is usually done by indicating the unit value of a leaf. In this case the unit value is 1 lb. Figure 2-22 includes the scale used in the stem-and-leaf display of the carry-on luggage weights.

Figure 2-22 shows a basic stem-and-leaf display. Sometimes you will see the leaves ordered from smallest to largest, but this is not necessary in a basic display. There are no firm rules for selecting the group of digits for the stem. But whichever group you select, you must list all the possible stems from smallest to largest in the data collection.

Sometimes we may want to spread the data even more. In such cases we can use two or more lines for each stem. For instance, we could put leaves 0 through 4 on one line and 5 through 9 on the next line, as shown in Figure 2-23. The asterisk indicates possible digits 0–4 for the leaves, and the period indicates possible digits 5–9.

In this stem-and-leaf display, we see that the weight interval with the most data is between 30 and 34 lb. Depending on the data, each stem can be listed several times with various symbols designating the leaf range for the stem line.

Decimal points are usually omitted in the stems and leaves, but indicated in the unit designation as appropriate.

_____ Exercise 10 _____

Tel-a-Message is experimenting with computer-delivered telephone advertisements. Of primary concern is how much of the four-minute advertisement is heard. A study was done to see how long the advertisement ran before the listeners hung up. A random sample of 30 calls gave the information in Table 2-13.

TABLE 2-13 Time Spent Listening to Advertisement (in Minutes)

1.3	0.7	2.1	0.5	0.2	0.9	1.1	3.2
4.0	3.8	1.4	3.1	2.5	0.6	0.5	2.1
4.0	4.0	0.3	1.2	1.0	1.5	0.4	4.0
2.3	2.7	4.0	0.7	0.5	4.0		

a) We'll make a stem-and-leaf display using the first digit as the stem and the second as a leaf. What is the leaf unit?

b) List all the stem values.

a) The trailing digit is in the tenths position so

$$1 \text{ unit} = 0.1 \text{ min}$$

b) Stems: 0
       1
       2
       3
       4

c) Complete the stem-and-leaf display including the unit designation. (*Note:* Order of leaves is not important and will depend on whether you read across rows or down columns. We read across the rows for the answer.)

c) *See* Figure 2-24.

FIGURE 2-24 Time Before Hang-Up

1 unit = 0.1 min

1 | 3 represents 1.3 min

0	7 5 2 9 6 5 3 4 7 5
1	3 1 4 2 0 5
2	1 5 1 3 7
3	2 8 1
4	0 0 0 0 0 0

d) Looking at the stem-and-leaf display, what could you say about the time intervals before people hung up?

d) Most people hung up before the end of the advertisement. Of those people, most listened less than one minute. There were people who listened to the entire advertisement.

In the next example we show a stem-and-leaf display for data with three digits.

EXAMPLE 4

What does it take to win? If you're talking about basketball, one sports writer gave the answer. He listed the winning scores of the conference championship

FIGURE 2-25 Winning Scores of Conference Basketball Championship Games

leaf unit = 1 point

08 | 3 represents 083 or 83

08	3
09	2 8 7 9
10	6 1 3 6 9 5 4 2 5
11	8 2 7 9 2 1 0 2 6 7 2
12	8 5 0 5 4 6
13	2 5 1
14	3

games over the last 35 years. The scores follow below. To make a stem-and-leaf display, we'll use the first *two* digits as the stems (*see* Figure 2-25).

132	128	106	135	92	98	101
125	118	112	117	143	120	125
97	83	124	126	119	112	103
111	106	99	109	105	110	112
116	117	112	131	104	102	105

- **Comment:** Stem-and-leaf displays organize the data, let the data analyst spot extreme values, and are easy to create. In fact, they can be used to organize data so that frequency tables are easier to make. However, at this time histograms are used more frequently in formal data presentations while stem-and-leaf displays are used by data analysts to gain insights about the data.

Section Problems 2.4

1. The Reeling Angels, a popular recording group, gave a single concert in Littletown, Maine. The concert tickets sold out in the first few hours of availability. Scalper ticket agents purchased blocks of the best Section A tickets and resold them at much higher prices. A random sample of 30 people buying tickets from different scalpers showed the following prices paid per ticket (in dollars):

20	35	40	25	31
16	37	50	35	35
25	28	24	38	40
45	30	35	33	37
25	36	39	45	30
35	33	27	35	40

Make a stem-and-leaf display of the data.

2. Dream Travel Tours contracted with Slick Rent-a-Car to provide rental cars to travelers. Slick went out of business and did not provide the prepaid rental cars to 20 travelers. The travelers demanded refunds from Dream Travel. The times elapsed between refund requests and the actual refunds were as follows (in days):

26	49	22	58	43
37	48	32	21	54
42	63	27	33	47
45	67	21	48	57

Make a stem-and-leaf display of the data.

3. Fairfield College is considering a new student teacher evaluation form for all the classes on campus. One concern is the amount of time necessary to fill out the form. A random sample of 25 students were asked to fill out the form for the course they had at 10:00 A.M. The times (in minutes) to fill out the forms were as follows:

18	19	8	12	18
20	15	17	19	23
22	30	21	16	44
17	15	12	18	26
19	23	35	16	12

Make a stem-and-leaf display.

4. Some days you feel rich and other days you don't. Shawn Taloy looked at his checkbook balances for 30 days selected at random from the past year. The balances were as shown below (to the nearest ten dollars):

260	680	840	30	260
290	20	1170	50	140
60	70	210	480	570
200	150	330	460	220
10	490	120	90	210
170	1,200	130	290	260

Make a stem-and-leaf display of the data. You may omit the last digit of each number from the display if the scale on your display indicates the actual size of the numbers.

5. You want to buy a new car—the Felicita model. You look at the sticker prices of 15 Felicitas in stock and find that they vary (partly because of options, partly because of dealer). The prices of the cars (in dollars) are:

11,820	10,830	11,240	12,090	10,700	10,500
11,930	10,230	11,450	10,500	10,870	11,030
10,200	10,650	10,550			

Make a stem-and-leaf display. Consider using the first two digits as the stem and the next two as the leaves. It is permissible to simply omit the last digit if the scale on the diagram reflects the actual size of the numbers.

6. Professor Harris kept careful attendance records for his Science and Society course. For the 40 class meetings of the semester the percents of the class in attendance were:

92	89	97	99	98	95	87	84
91	81	75	82	85	84	87	90
98	100	73	85	89	92	73	60
75	87	89	91	83	97	100	86
87	84	82	96	73	75	92	100

Make a stem-and-leaf display using two lines per stem.

7. Management Efficiency Consultants are called upon to help companies become more efficient in their operations. One of the first studies they conduct is to determine the amount of time executives spend in meetings. For one company the data for 25 executives were as follows (in hours per week):

12	9	6	7	15
18	26	13	21	10
5	7	24	12	2
17	3	8	9	10
11	10	6	25	19

Make a stem-and-leaf display using two lines per stem.

8. Eastshore Community College requires all its students to take a basic math skills test before beginning any degree or certificate program. The scores range from 0 to 100. For 70 students interested in the Associate of Management Degree the scores were:

22	60	80	75	87	92	65	46	33	95
72	98	100	37	58	75	86	92	77	85
86	97	83	81	87	42	91	89	87	84
72	86	63	42	26	97	93	98	72	82
85	79	84	75	83	92	89	63	86	68
80	97	81	87	72	89	87	73	65	52
76	86	91	53	67	67	69	72	92	81

Make a stem-and-leaf display of the scores. If 70 is the minimal passing score, how many passed?

9. Making sure that patients get the prescribed medication in the correct dosage is an important task of the nurses at Memorial Community Hospital. At the hospital a survey involving 50 patients was done. The number of pills each patient was given per day was recorded as follows:

13	8	12	5	4	2	7	1	15	9
8	9	1	7	6	10	4	7	12	3
15	5	8	7	12	6	4	11	2	9
6	8	9	3	15	2	12	4	8	12
10	9	12	14	3	9	12	10	8	11

Make a stem-and-leaf display using two lines per stem.

10. Computer Services Incorporated offers customers a Computer Store option. Using a computer terminal and telephone modem, the customer contacts the service and can scan the catalogue and make orders from his or her own computer terminal. To use the service, the customer must pay a per-minute charge for the time connected. Last month a random

sample of 60 customers using the Computer Store option had connect times as follows (in minutes):

5	20	7	30	18	10	12	15	65	25
8	3	17	15	21	27	15	25	42	37
51	12	14	5	10	8	21	19	16	31
30	57	61	12	22	26	24	18	17	43
62	47	16	19	20	37	40	58	61	19
18	61	12	72	17	26	31	29	33	15

Make a stem-and-leaf display of the connect times.

Summary

When a sample is taken we must be careful to use a method of selection that will give us a sample that represents the entire population from which the sample is taken. A random sample is taken in such a way that each member of the population has equal chance of being included in the sample, and each sample of the same size as our sample has equal chance to be taken. Organizing and presenting data is the main purpose of that part of statistics called descriptive statistics. In this chapter we have studied tables, bar graphs, pictograms, circle graphs, histograms, frequency polygons, and stem-and-leaf displays. From the viewpoint of future applications, histograms are the most important because the area under a bar can represent the likelihood of data values falling in that class.

Important Words and Symbols

	Section		*Section*
Random sample	2.1	Class, lower limit, upper limit	2.3
Random number table	2.1	Class midpoint	2.3
Simulation	2.1	Histogram	2.3
Table	2.2	Frequency polygon	2.3
Bar graph	2.2	Class width	2.3
Pictograph	2.2	EDA	2.4
Pie chart or circle graph	2.2	Stem	2.4
Frequency distribution	2.3	Leaf	2.4
Frequency	2.3	Stem-and-leaf display	2.4

Chapter Review Problems

1. Existing oil reserves in the Western Hemisphere are estimated in Table 2-14. Make a circle chart for this data.

TABLE 2-14 Oil Reserves in
Western Hemisphere

Country	Billions of Barrels
United States	36.8
Canada	15.6
Mexico	59.3
South America	71.3

2. The troubleshooter at the college computer center has recorded the following time intervals between "downs" on the computer.

Time Between Computer Downs (in hours)

6	5	3	7	4	8	10	8	8	10
9	8	3	5	7	4	7	6	7	10
7	9	4	5	9	11	6	8	6	11
8	8	10	9	12	9	11	11	12	8
11	12	14	7	13	14	7	8	11	14

a) Make a stem-and-leaf display using two lines per stem.
b) Make a frequency table for the preceding data (use six classes).
c) Make a histogram.
d) Make a frequency polygon.

3. Arapahoe Community College holds evening classes on Colorado state history. The director of studies, wishing to know the age distribution of students in the class, obtained the following information (ages rounded to nearest year).

Ages of Evening Students Taking Colorado History

63	41	28	22	45	30	43	46	35	56	27	18
29	20	31	19	22	32	55	54	31	63	51	39
29	19	27	53	51	33	58	29	65	53	56	19

a) Make a stem-and-leaf display of the data.
b) Make a frequency table, histogram, and frequency polygon using five class intervals.
c) Make a frequency table, histogram, and frequency polygon using 12 class intervals.

4. The Director of the Campus Computing Center did a study to determine patterns of student computer usage. She found that 20% of the usage occurred during the first five weeks of the term, 30% during the middle five weeks, and 50% during the last five weeks. Display these data on a circle chart and on a bar graph.

5. As a project for her statistics class Laura was supposed to take a random sample of 30 cars in the student parking lot and determine the make and year of each car.
a) Describe how Laura could use a random number table to get a random sample of 30 cars in the parking lot.

b) After getting her random sample, Laura found that eight cars were Chevrolets, seven were Fords, five were Pontiacs, three were Toyotas, three were Chryslers, three were Oldsmobiles, and one was a Cadillac. Make a table for these data, and draw a bar graph. Draw a circle graph showing percentages of the total number of cars sampled.

c) The model years for each of the 30 cars are shown here:

83	74	71	73	69	77	85	76	82	70
75	78	81	73	76	84	75	77	86	83
73	80	76	78	70	85	75	81	77	78

Make a frequency table, histogram, and frequency polygon for these data (use five classes).

6. The administrative director of Hillman Memorial Hospital wants information about the length of stay of patients in the cardiac ward and the length of stay of patients in all other wards. The director decides to take a random sample of 40 patients from the cardiac ward and a random sample of 40 patients from all other wards. How could you use last year's file of hospital patients to get a random sample of 40 patients from all other wards? For the cardiac ward the sample gave the following information (in days):

4	8	5	5	6	8	2	4	6	4	3	5	3	8	10	9	6
4	7	9	5	8	5	10	5	3	7	6	5	13	6	5	7	11
10	5	10	11	12	13											

For all other wards, the sample gave (in days):

11	5	8	9	6	8	5	9	9	11	9	9	8	10	9
9	10	9	8	7	11	9	8	7	9	10	9	10	7	8
9	9	11	10	12	3	12	12	10	10					

a) For each set of data, construct a stem-and-leaf display using two lines per stem.

b) For each set of data, construct a frequency table (using a class width of three days).

c) Draw a histogram for the cardiac ward and a histogram for other wards.

d) Draw a frequency polygon for the cardiac ward and a frequency polygon for the other wards.

e) Looking at your graphs, do patients spend more time in the cardiac ward or in other wards?

7. Foreign investment in United States industry has been increasing. In 1955 foreign investments amounted to $13.4 billion; in 1960 investments totaled $18.4 billion; in 1965, $26.3 billion; in 1970, $44.8 billion; and in 1980, $160 billion.

a) Make a table representing the above information.

b) Make a bar graph.

c) Make a pictogram.

8. The president of Mammon Savings and Loan wants information about the size of savings accounts in the bank.

a) How could you use a random number table to draw a random sample of 50 customer accounts from bank files?

The results of a random sample of 50 accounts indicated the balances that follow (in hundreds of dollars):

25	31	21	24	22
45	21	35	22	15
48	39	5	28	26
1	9	42	5	11
15	32	53	44	20
34	2	28	32	57
38	28	23	7	25
21	10	2	31	9
50	8	11	1	22
22	20	29	22	10

b) Make a stem-and-leaf display.
c) Make a frequency table for the above data using lower class limits 0, 10, 20, 30, 40, and 50.
d) Make a histogram from your frequency table.
e) Make a frequency polygon.

9. The public relations and advertising office of the Allstart Battery Company wants information about the life of their car batteries. When a customer buys an Allstart battery he or she sends a guarantee certificate to the public relations office. A random sample of 42 customers were traced by means of return addresses on the guarantee certificates. Battery lives for the batteries purchased by these customers are given below (recorded to the nearest month).

29	13	19	34	24	29	45	41	23	22	38
32	50	46	56	26	14	25	39	33	49	35
28	25	26	21	27	34	30	36	18	24	34
31	41	33	23	34	26	21	40	48		

a) Use class midpoints 15, 20, 25, 30, 35, 40, 45, 50, and 55 to make a frequency table.
b) Make a histogram from your frequency table.
c) Make a frequency polygon.

10. Condense the data of problem 9 into the following time categories (in months): 12–24, 25–36, 37–48, 49 or over. Construct a circle graph for the percentage of batteries in each of these categories.

USING COMPUTERS

Professional statisticians in industry and research use computers to help them analyze and process statistical data. There are many computer programs available for statistics. Some commonly used statistical packages include Minitab®, SAS®, and SPSS.* There are many others as well. Any of these statistical packages may be used with this text.

The authors of this text have also written an inexpensive and straightforward computer statistics package called ComputerStat. The programs in ComputerStat are available on disks with an accompanying manual that explains how to use the programs and gives additional computer applications. The disks are intended for the Apple IIe, IIc, or the IBM PC. Programs are written in the computer language BASIC and are designed for the beginning statistics student with no previous computer experience. ComputerStat is available as a supplement from the publisher of this text.

Problems in this section may be done using ComputerStat or other statistical computer programs. Some programmable calculators have statistical function keys and software packages, and they can be used as well.

1. All computers have a random number generator. In the ComputerStat menu of topics, select Descriptive Statistics. Then use program S1, Random Samples, to solve the following problems.
   a) In a large condominium complex there are 473 condominiums numbered 1 to 473. You want to check the thermostat in a random sample of 50 different condominiums. Use the computer to find the numbers of your 50 condominiums.
   b) A theater is showing a new movie. After the movie you want to ask a random sample of 30 people for their opinions about the movie. As people buy their tickets, you write a number on the back. Since 278 tickets are purchased, you will use the numbers from 1 to 278. Just before the movie begins, you announce that a small prize will be given to the people whose numbers are called and who respond to a questionnaire after the movie. Use the computer to find the 30 different numbers you will call.
   c) The phone book indicates that there are 83 sporting goods stores in Kansas City. Using the alphabetical order of their appearance in the phone book you number the stores from 1 to 83. Use the computer to find a random sample of 10 different sporting goods stores you will call.
   d) You have a group of 18 people and you wish to split it into two groups for baseball teams. Use the computer to assign random teams.

2. Each business day the Dow Jones Information Retrieval Service gives closing price and volume of sales for all major stocks on the New York Stock Exchange. The phrase "volume leads price" is often heard in discussions about market activity. In fact, history has shown that an unusually high volume of sales of a stock generally indicates that an imminent change (either up or down) in stock price is about to take place. What is a high or low volume for a particular stock? What is an everyday or ordinary volume? How frequently do these volumes occur?

   Perhaps the best way to answer these questions is to track the market activity of your stock over a period of time. In particular, a frequency table of volumes over, say, a ten-week period

*Minitab is a registered trademark of Minitab, Inc. SAS is a registered trademark of SAS Institute, Inc. SPSS is a trademark of SPSS Inc.

would help answer some of these questions about volume. Table 2-15 lists volume of IBM (International Business Machines) stock (in hundreds of shares sold) for a ten-week period from July 8, 1985 through September 16, 1985.

In the ComputerStat menus of topics, select Descriptive Statistics. Then use program S2, Frequency Distribution, to do the following. Enter the given IBM volumes into the computer. Then have the computer create frequency tables using 3 classes; 5 classes; 8 classes; 10 classes. (*Note:* You enter the data only once!)

a) Compare the effect of lumping all the data together in only three classes with the opposite effect of thinning the data out into as many as 10 classes. Both these extremes have drawbacks. Explain them.

b) Looking at the frequency tables, what would you say is a high volume of sales? What is an everyday or ordinary volume? What is a low volume? What are the frequencies of days on which high, low, and ordinary volumes occur? In the given period of observation which volumes (or volume ranges) seem to occur more frequently? Which occur less frequently?

TABLE 2-15

Date	Volume (in hundreds)	Date	Volume (in hundreds)
07/08/85	7,926	08/12/85	9,801
07/09/85	11,995	08/13/85	7,458
07/10/85	16,873	08/14/85	5,437
07/11/85	10,439	08/15/85	7,235
07/12/85	9,868	08/16/85	8,049
07/15/85	15,577	08/19/85	5,115
07/16/85	16,013	08/20/85	9,251
07/17/85	18,856	08/21/85	7,090
07/18/85	17,296	08/22/85	7,848
07/19/85	13,891	08/23/85	6,385
07/22/85	10,756	08/26/85	6,119
07/23/85	18,788	08/27/85	7,868
07/24/85	14,653	08/28/85	5,960
07/25/85	15,388	08/29/85	5,051
07/26/85	14,205	08/30/85	6,251
07/29/85	10,173	09/03/85	10,198
07/30/85	11,398	09/04/85	7,373
07/31/85	13,830	09/05/85	9,892
08/01/85	13,609	09/06/85	10,619
08/02/85	6,668	09/09/85	11,088
08/05/85	6,704	09/10/85	9,448
08/06/85	10,754	09/11/85	8,360
08/07/85	9,543	09/12/85	13,885
08/08/85	11,173	09/13/85	12,230
08/09/85	6,910	09/16/85	6,873

Source: Data obtained electronically from the *Dow Jones News/Retrieval,* Dow Jones Historical Quotes, Princeton, 1985.

3 Averages and Variation

While the individual man is an insolvable puzzle,
in the aggregate he becomes
a mathematical certainty. You can,
for example, never foretell what any one man
will do, but you can say
with precision what an average number will be up to.

Arthur Conan Doyle, The Sign of Four

Sherlock Holmes spoke these words to his colleague Dr. Watson as the two were unraveling a mystery. The master detective was commenting upon the fact that if a single member is drawn at random from a population, we cannot predict *exactly* what that member will look like. However, there are some "average" features of the entire population that an individual is likely to possess. The degree of certainty with which we would expect to observe such average features in any individual depends upon our knowledge of the variation among individuals in the population. Sherlock Holmes has led us to two of the most important statistical concepts: average and variation.

SECTION 3.1   Mode, Median, and Mean

The average price of an ounce of gold is $423. The Zippy car averages 39 mpg on the highway. A survey showed the average shoe size for women is size 8.

In each of the preceding statements *one* number is used to describe the entire sample or population. Such a number is called an *average*. There are many ways to compute averages, but we will study only three of the major ones.

The easiest average to compute is the mode. The *mode* is the value or property which occurs most frequently in the data. For instance, if you count the number of letters in each word of the preceding paragraph, you will see that the mode is two letters. In other words, there are more words with exactly two letters than any other number of letters. (*See* Table 3-1.)

TABLE 3-1 Word Length of Paragraph Two

Number of Letters in Word	Number of Words
1	1
Mode → 2	10 ← *Greatest number of words*
3	6
4	8
5	4
6	5
7	2
8	2
9	1
10	2

Sometimes a distribution will not have a mode. If Professor Fair gives an *equal* number of A's, B's, C's, D's, and F's, then there is no modal grade. The data in Table 3-2 have no mode because no data value occurs more frequently than all the others.

TABLE 3-2 Number of Minutes Spent by Professor Adams's Students Using the School Computer Terminals Last Wednesday Afternoon

27	30	42	42	36	36	50

--- Exercise 1 ---

On the first day of finals 20 students at La Platta College were selected at random. They were asked how many hours they had slept the night before (rounded to the nearest hour). The results in hours were

8 6 5 6 4 3 5 8 7 7 5 6 2 0 5 7 6 6 7 8

a) Complete the following table.

TABLE 3-3 Hours Slept Before Finals

Hours Slept	Number of Students
0	1
1	0
2	1
3	1
4	_____
5	_____
6	_____
7	_____
8	_____

a)

TABLE 3-4 Completion of Table 3-3

Hours Slept	Number of Students
0	1
1	0
2	1
3	1
4	1
5	4
6	5
7	4
8	3

b) Is there a mode?

c) What is the modal number of hours slept?

b) Yes, there is one quantity that has the greatest frequency.

c) The modal number of hours slept is 6 hours.

The mode is an easy average to compute, but it is not too stable. For example, if one of the students in Exercise 1 had slept five hours instead of six on the night before the first finals, the mode would change to five. However, if you are interested in the *most common* value in a distribution, the mode is appropriate to use.

Another average is the *median,* or central value of an ordered distribution. When you are given the median, you know there are an equal number of values above and below it. To obtain the median, we order the data from the smallest value to the largest. Then we pick or construct the middle value.

The median of the following set of test scores for English literature is 75.

median
↓

50 51 60 64 65 70 | 75 | 80 81 85 90 95 97

6 below   6 above

There are as many test scores above as below the median.

For an even number of test scores the median must be constructed. It is not necessarily one of the given test scores. For instance, the following list has an even number of scores.

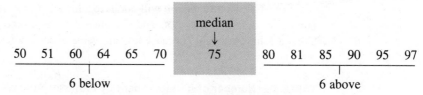

median
↓
?

51 60 64 69 70 75 ⌐?⌐ 78 80 85 90 91 95

middle values

To construct the median of a set of data with an even number of entries, add the two middle values and divide by 2.

$$\text{median} = \frac{\text{sum of two middle scores}}{2}$$

$$= \frac{75 + 78}{2} = 76.5$$

--- Exercise 2 ---

Belleview College must make a report to the budget committee about the average credit hour load a full-time student takes. (A 12 credit hour load is the minimum requirement for full-time status. For the same tuition students may take up to 20 credit hours.) A random sample of 40 students yielded the following information (in credit hours):

```
17   12   14   17   13   16   18   20   13   12
12   17   16   15   14   12   12   13   17   14
15   12   15   16   12   18   20   19   12   15
18   14   16   17   15   19   12   13   12   15
```

a) Organize the data from smallest to largest number of credit hours.

a) 12   12   12   12   12   12   12   12   12   12
    13   13   13   13   14   14   14   14   15   (15)
    (15) 15   15   15   16   16   16   16   17   17
    17   17   17   18   18   18   19   19   20   20

b) Since there are an _____ (odd, even) number of values, we add the two middle values and divide by 2 to get the median. What is the median credit hour load?

b) There are an even number of entries. The two middle values are circled in part (a).

$$\text{median} = \frac{15 + 15}{2} = 15$$

c) What is the mode of this distribution? Is it different from the median? If the budget committee is going to fund the school according to the average student credit hour load (more money for higher loads), which of these two averages do you think the college will use?

c) The mode is 12. It is different from the median. Since the median is higher, the school will probably use it and indicate that the average being used is the median.

The median is a more stable average than the mode, but it does not indicate the range of values above or below it. For instance, the median is 20 for both of the following groups of scores.

    a) 10   15   20   25   30
    b) 1   10   20   40   100

In the first group all scores are within ten points of the median; in the second group one score is 80 points above the median. The median uses the *position* rather than the specific value of each data entry.

An average that uses the exact value of each entry is the *mean* (sometimes called the arithmetic mean). To compute the mean, we add the values of all the entries and then divide by the number of entries.

$$\text{mean} = \frac{\text{sum of all the entries}}{\text{number of entries}}$$

The mean is the average usually used to compute a test average.

EXAMPLE 1

Linda needs at least a B in biology in order to graduate. She did not do too well on her first three tests; however, she did quite well on the last four. Here are her scores:

    58   67   60   84   93   98   100

Compute the mean and determine if Linda's grade will be a B (80 to 89 average) or a C (70 to 79 average).

$$\text{mean} = \frac{\text{sum of scores}}{\text{number of scores}} = \frac{58 + 67 + 60 + 84 + 93 + 98 + 100}{7} = \frac{560}{7} = 80$$

Since the average is 80, Linda will get the needed B.

• **Comment:** When we compute the mean, we sum the given data. There is a convenient notation to indicate the sum. Let x represent any value in the data set. Then the notation

$$\Sigma x \text{ (read, the sum of all given } x \text{ values)}$$

means that we are to sum all the data values. In other words, we are to sum all the entries in the distribution. The symbol Σ means *sum the following* and is capital sigma, the S of the Greek alphabet.

The symbol for the mean of a sample distribution of x values is denoted by $\bar{x}$ (read, x bar). This symbol is simply an x with a bar over it. If we let the letter n represent the number of entries in the data set, we have

$$\text{sample mean} = \bar{x} = \frac{\Sigma x}{n} \qquad (1)$$

───────── Exercise 3 ─────────

A fabric store manager is eager to see if the latest patterns for size 12 dresses show a longer hemline than last year's. If so, she can expect to sell more fabric since each pattern will call for more material. She took a random sample of ten dress patterns and found the finished lengths from back of neck to bottom of hem to be (in inches):

| 41.5 | 42 | 39 | 44 | 43.5 | 45 | 43 | 45 | 42 | 46 |

a) What is the value of n?

a) Since there are ten data entries, $n = 10$.

b) How do you find Σx? What is the value of Σx?

b) To find Σx, we add all the data entries together.

$$\Sigma x = 41.5 + 42 + 39 + 44 + 43.5 + 45$$
$$+ 43 + 45 + 42 + 46$$
$$= 431$$

c) Compute the mean, $\bar{x}$.

c) $\bar{x} = \dfrac{\Sigma x}{n} = \dfrac{431}{10} = 43.1$

d) Last year the mean length of size 12 dresses was 36 in. How much longer is the mean length now? Can the manager expect to sell more material per dress?

d) The difference is $43.1 - 36 = 7.1$ in. The manager can expect to sell more fabric for each dress.

We have seen three averages: the mode, the median, and the mean. For later work the mean is the most important of the averages. One disadvantage of the mean, however, is that it can be affected by exceptional values, as shown in the next exercise. In such cases the median would better represent the general level of the distribution.

_____ Exercise 4 _____

Rowdy Rho Fraternity is in danger of losing campus approval if they do not raise the mean grade point average of the entire group to at least 2.2 on a four-point scale. This term the averages of the members were:

1.8   2.0   2.0   2.0   2.0   1.9   1.8   2.3   2.5   2.3   1.9   2.2   2.0   2.3

a) What is the mean of the grade point averages?

a) mean $= \dfrac{\Sigma x}{n} = \dfrac{29.0}{14} = 2.07$

b) Rod made a 2.0 average this term because he was in the hospital six weeks. He believes he would have made a 3.9 average if he had been well. Recompute the mean with the first 2.0 replaced by 3.9. Would Rod have saved the fraternity if he had made a 3.9 grade point?

b) If we replace the first 2.0 by 3.9, the new mean is then

$$\text{mean} = \frac{\Sigma x}{n} = 2.21$$

If Rod had made a 3.9 instead of a 2.0, the fraternity would have been saved.

c) Suppose the college had required the fraternity to raise the *median* grade average to 2.2. Would Rod's potential 3.9 have saved the fraternity? What can you say about the effect of the exceptional value 3.9 on the median and mean?

c) The median of both distributions is 2.00. If Rod had made a 3.9, the medians would still be the same, indicating that half the members were still below a 2.00 average. Rowdy Rho would lose campus approval either way. The exceptional value 3.9 changed the mean, but did not change the median. In general, exceptional values will change the mean more than the median.

_____

The average you use depends on what you want to do with that average. If you want to know which value occurs most frequently in a distribution, use the mode. If a store wants to know which shirt size is most frequently requested, the mode is the proper average to use; and the store will know to carry more shirts of that size than any other. If you want to cut a distribution in half, use the median. A report showing the average salary of workers at Gator Tire Factory should show the median so that the top-level administrative salaries will not pull up the production-line salaries and make them look higher than they are. If you want each entry in the data to enter into the average, use the mean. As we shall see in later chapters, we use the mean if we want to estimate population average from a sample average. The mean uses all of the data entries and the mean can be analyzed more conveniently by statistical methods.

Section Problems 3.1

1. Response times for eight emergency police calls in Denver were measured to the nearest minute and found to be:

   7   10   8   5   8   6   8   9

   Find the mean, median, and mode.

2. The heights of ten varsity basketball players at Mountain View College were measured in inches and found to be:

   74   82   78   72   78   73   78   72   78   81

   Find the mean, median, and mode.

3. A random sample of 12 people gave their opinions about a new highway plan which would run the highway through a wildlife refuge in Florida. Opinions were given on a scale of 1 to 10 where 1 = strongly disagree with highway plan and 10 = strongly agree. The results of the survey were:

   3   1   3   2   8   3   10   5   3   7   9   1

   Find the mean, median, and mode of the responses.

4. A package delivery service provided the following information about the weights of 20 packages chosen at random (weights to the nearest ounce):

   50   55   18   21   64   32   21   52   33   41
   21   60   18   21   37   40   8   16   21   18

   Find the mean, median, and mode of the weights.

5. The number of hours of volunteer service given last month by nurses at a Red Cross emergency station was as follows:

   10   8   1   3   5   6   8   15   8   7   3   4
   6   3   8   9   8   2   4   16   11   12   7

   Find the mean, median, and mode of the hours of volunteer service.

6. Ten students took an aptitude exam in technical electronics. There were 1,000 possible points. The scores ran as follows:

   351   988   348   450   290   965   360   346   332   318

   a) Compute the mean of all ten scores.

b) Compute the median of all ten scores and compare your answer with that of part (a).

c) If all ten scores were to be used, which average (mean or median) would best describe *most* of the students?

d) There were two exceptionally high scores. Omit these two scores and compute the mean for the remaining eight scores. Compute the median for these eight scores. Compare your results and comment on the effect of a few extreme scores on the mean and the median.

7. Brookridge National Bank is a small bank in a rural Iowa town. The 12 people who work at the bank are: the president, the vice president (his son-in-law), eight tellers, and two secretaries. The annual salaries for these people in thousands of dollars are:

President: 93

Vice President: 80

Tellers: 15, 25, 14, 18, 21, 16, 19, 20

Secretaries: 12, 13

a) Compute the mean of all 12 salaries.

b) Compute the median of all 12 salaries and compare your answers with the mean of part (a). Which average best describes the salaries of the *majority* of employees?

c) Omit the salaries of the president and vice president. Calculate the mean and median for the remaining ten people.

d) Compare your answers from part (c) with those of parts (a) and (b). Comment on the effect of extreme values on the mean and median.

8. A reporter for the *Honolulu Star Bulletin* was doing a news article about car theft in Honolulu. For a given ten-day period the police reported the following numbers of car thefts:

9   6   10   8   10   8   4   8   3   8

Then for the next three days, for an unexplained reason, the number of car thefts jumped to 36, 51, and 30.

a) Compute the mean, median, and mode for the first ten-day period.

b) Compute the mean, median, and mode for the entire 13-day period.

c) Comment on the effect of extreme values on the mean, median, and mode in this problem.

9. The College Astronomy Club has been counting meteoroids each night for the past week. Between the hours of 10:00 P.M. and 1:00 A.M. the meteoroid count for each night was:

15   12   15   10   17   18   15

Then for the next two nights there was a meteoroid shower and the club counted 57 and 62 meteoroids.

a) Compute the mean, median, and mode for the meteoroid counts on the first seven nights.

b) Compute the mean, median, and mode for all nine nights.

c) Comment on the effect of extreme values on the mean, median, and mode in this problem.

10. At an athletic event, members of the college weightlifting team lifted the following weights (in pounds):

| 190 | 173 | 200 | 188 | 190 | 175 |
| 190 | 180 | 200 | 177 | 179 | 190 |

   a) Find the mean, median, and mode of these numbers.
   b) In your own words define the terms mean, median, and mode. Discuss how your definitions fit into the context of this problem. How do the mean, the median, and the mode each represent in a different way an estimate of the team's weightlifting capacity?

11. In an effort to estimate the size of elk herds spending the winter in Rocky Mountain Park, the rangers used a small airplane to spot-count herds of elk in the park. They found 15 groups of elk and recorded the group sizes as follows:

| 21 | 15 | 19 | 16 | 18 | 17 | 17 | 20 |
| 25 | 7 | 16 | 10 | 16 | 18 | 12 | |

   a) Compute the mean, median, and mode of the group sizes.
   b) In your own words define the terms mean, median, and mode. Discuss how your definitions fit into the context of this problem. How do the mean, the median, and the mode each represent in a different way an estimate of the group size of elk wintering in Rocky Mountain Park?

SECTION 3.2   Percentiles and Box-and-Whisker Plots

If we are told that 81 is the median score on a biology test, we know that no more than half the scores lie above 81 and no more than half lie below 81. Translating this information into the language of percents, we can say that, at most, 50% of the scores are above 81 and, at most, 50% are below. The median is an example of a percentile; in fact, it is the 50th percentile. The general definition of the Pth percentile follows.

> • **Definition:**   The *Pth percentile* is a measurement such that, at most, $P\%$ of the data fall below the Pth percentile and, at most, $(100 - P)\%$ fall above it.

In Figure 3-1 we see the 60th percentile marked on a histogram. We see that 60% of the data lie below the mark and 40% lie above it.

FIGURE 3-1   A Histogram with the 60th Percentile Shown

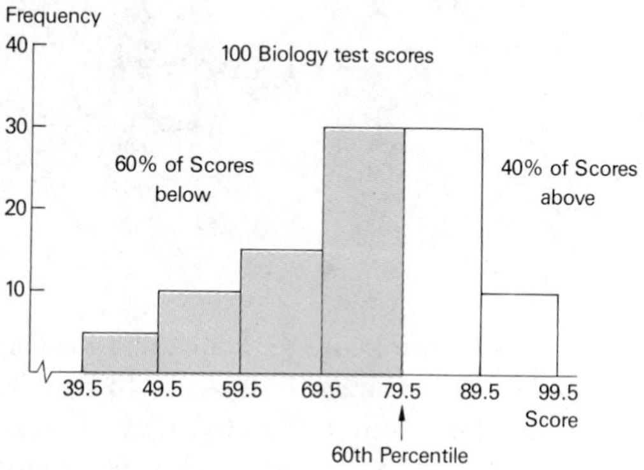

Exercise 5

You took the English Achievement Test to obtain college credit in Freshman English by examination. Your score was in the 89th percentile.

a) What percent of scores were below your score?

b) What percent were above your score?

a) At most 89% of the scores were below your score.

b) At most $(100 - 89)\% = 11\%$ were above your score.

Three percentiles that are of particular interest are the 25th percentile, known as the *lower quartile;* the 50th percentile, known as the *middle quartile* or *median;* and the 75th percentile, known as the *upper quartile*.

Rounding conventions used to find upper and lower quartiles vary. So instead of computing quartiles, we will show you how to compute hinges. *Hinges* are simply the medians of the lower and upper halves of a data bank. Lower and upper hinge values will be very close to lower and upper quartile values, respectively, but not necessarily the same.

The median value together with the upper and lower hinge values and the extreme low and high data values give us a very useful five-number summary of the data. We will use these five numbers to create a graphical sketch of the data called a *box-and-whisker plot*.

We already know how to find the median. Now let's study the process of finding the lower and upper hinge values.

To find hinges, we proceed as we did to find the median. First we order the data from smallest to largest. Then we rank the data two ways: from smallest to largest and, for convenience, from largest to smallest. Next we find the ranks of the median and hinge using the following formulas.

For n pieces of data,

$$\text{median rank} = \frac{n + 1}{2}$$

$$\text{hinge rank} = \frac{\text{number of values below median} + 1}{2}$$

Once we know the hinge and median ranks, we use the *corresponding data values* as the hinges or median. To find the lower hinge and median values we use the ranking from smallest to largest. To find the upper hinge value we use the ranking from largest to smallest. If a rank ends in .5, we are to average the data values with adjacent ranks. The next example demonstrates the process of finding hinges.

EXAMPLE   2

Renata College is a small college offering baccalaureate degrees in liberal arts and business. The Development Office (fund-raising) did a salary survey of alumni who graduated two years ago and have jobs. Sixteen alumni responded to the survey the first week. Table 3-5 shows their annual salaries (in thousands of dollars).

TABLE 3-5 Annual Salaries in Thousands of Dollars

28.5	29.5	22.0	20.5	26.8	19.2	13.7	24.1
18.3	17.9	23.6	27.0	33.5	24.6	23.8	26.1

To find the hinges we first order the data and rank it from smallest to largest as well as from largest to smallest.

TABLE 3-6 Salaries of Alumni Graduating Two Years Ago
(in thousands of dollars)

Salary	Rank Up	Rank Down	Salary	Rank Up	Rank Down
13.7	1	16	24.1	9	8
17.9	2	15	24.6	10	7
18.3	3	14	26.1	11	6
19.2	4	13	26.8	12	5
20.5	5	12	27.0	13	4
22.0	6	11	28.5	14	3
23.6	7	10	29.5	15	2
23.8	8	9	33.5	16	1

Now we find the median rank. Since we have $n = 16$ data values then

$$\text{median rank} = \frac{n + 1}{2} = \frac{17}{2} = 8.5$$

Next we find the median value. It is the mean of data values with rank 8 and rank 9. These values are shaded in Table 3-6.

$$\text{median value} = \frac{23.8 + 24.1}{2} = 23.95$$

To find the hinges, we first find their ranks. Since the median rank is 8.5, the number of values below the median is 8.

$$\text{hinge rank} = \frac{\text{number of values below median} + 1}{2} = \frac{8 + 1}{2} = 4.5$$

The rank value 4.5 tells us to average data values with ranks 4 and 5. These values are circled in Table 3-6. Therefore, we find the value of the lower hinge to be

$$\text{lower hinge value} = \frac{19.2 + 20.5}{2} = 19.85$$

We found the rank of the lower hinge to be 4.5. If we rank the data from largest to smallest, we can use this same rank to find the value of the upper hinge. The rank 4.5 tells us to average the data values with ranks 4 and 5 *from the highest data value*. These values are boxed in Table 3-6.

$$\text{upper hinge value} = \frac{26.8 + 27}{2} = 26.90$$

When we include the extreme low and high data values with the hinge and median values, we have a five-number summary of the data that gives us some idea of the data spread. A *box-and-whisker plot* gives us a graph of the data using these five values.

> To make a *box-and-whisker plot*, draw a box from hinge to hinge. Then include a solid line through the box to show the median. Finally, draw solid lines called whiskers from each hinge to the corresponding extreme values.

Figure 3-2 shows a box-and-whisker plot for the salaries of the alumni of Renata College. A quick glance at the box-and-whisker plot tells us that:

a) 50% of the salaries lie in the box part and range from about 20 to 27 thousand dollars. The box tells us where the middle half of the data lies.

FIGURE 3-2   Box-and-Whisker Plot of Salaries of Renata College Alumni Graduating Two Years Ago

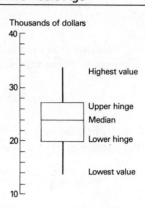

b) The median is about 24 thousand and is slightly closer to the top of the box.

c) The whiskers are about the same length, which says that the range from each hinge to the corresponding extreme value is about the same.

Any box plot quickly gives general information about overall level, amount of spread in both the middle 50% of the data and all the data, and location of median with respect to the hinges. Box-and-whisker plots can help us quickly compare two sets of data, as shown in the next exercise.

—————— Exercise 6 ——————

The Renata College Development Office also sent a salary survey to alumni who graduated five years ago. Again, questions about annual salary were asked. The responses received the first week are summarized in the box-and-whisker plot of Figure 3-3. The plot for the alumni graduating two years ago is repeated.

FIGURE 3-3   Box-and-Whisker Plots for Alumni Salaries (in thousands of dollars)

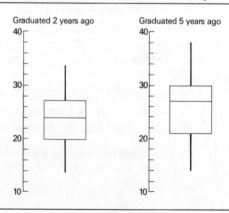

a) From Figure 3-3 estimate the median and the extreme values of the salaries of alumni graduating five years ago. What is the range of the middle half of the salaries?

b) Compare the two box plots of Figure 3-3 and make comments about the salaries of alumni graduating two and five years ago.

a) The median seems to be $27,000. The extremes are $14,000 and $38,000. The middle half of the data is enclosed by the box with low side at $21,000 and high side at $30,000.

b) The salaries of the alumni graduating five years ago have a larger spread, but begin slightly higher and extend to levels about $5,000 above those graduating two years ago. The middle half of the data is also more spread out, with higher boundaries and a higher median.

To make a box-and-whisker plot, we need five pieces of data: the extreme high, the extreme low, the median, and the two hinges. Remember that the hinges are just the medians of the upper and lower halves of the data. In the next exercise we ask you to make a box-and-whisker plot from raw data.

--- Exercise 7 ---

Packing and shipping costs for tomatoes could be reduced if more pounds of tomatoes would fit in a crate. This feat could be achieved if the tomatoes were squarer. Researchers at Valley Growers were given the task of developing a strain of tomatoes as square as possible. A squareness index ranging from 0 to 20 was developed with 0 representing perfectly round and 20 perfectly square. A random sample of 15 tomatoes from the fifth stage of a hybrid development process were measured for squareness. Their indexes are listed in Table 3-7.

TABLE 3-7 Squareness Index for Sample of 15 Tomatoes

12	10	13	7	14	11	15	15
9	8	8	5	13	12	10	

a) First, order and rank the data from smallest to largest and largest to smallest.

a) **TABLE 3-8** Squareness Index

Index	Rank Up	Rank Down	Index	Rank Up	Rank Down
5	1	15	11	8 *(median rank)*	8
7	2	14	12	9	7
8	3	13	12	10	6
8	4 *(lower hinge rank)*	12	13	11	5
9	5	11	13	12	4 *(upper hinge rank)*
10	6	10	14	13	3
10	7	9	15	14	2
			15	15	1

b) Find the median and hinge rank.

b) Since $n = 15$, then

$$\text{median rank} = \frac{n + 1}{2} = \frac{16}{2} = 8$$

$$\text{hinge rank} = \frac{\text{\# of data below median} + 1}{2}$$

$$= \frac{7 + 1}{2} = 4$$

c) Find the median and hinge values.

c) The median value is the value with rank 8. That value is 11. The hinge rank is 4 so

$$\text{lower hinge value} = 8$$

Counting four data values from the highest value gives us the upper hinge value.

$$\text{upper hinge value} = 13$$

d) Using the extremes, hinges, and median, make a box-and-whisker plot.

d) **FIGURE 3-4**   **Box-and-Whisker Plot for Squareness Index**

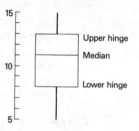

e) Comment on the meaning of Figure 3-4.

e) Looking at the box-and-whisker plot, we immediately conclude:

  i) The middle half of the data is represented by the box part of the figure, and ranges from 4 to 13.
  ii) The median is not centered in the box, but is closer to the upper hinge.
  iii) The lower whisker is longer than the upper whisker. This indicates a greater data spread on the lower end of the distribution.

• **Comment:** Sometimes, especially in a computer printout of a box-and-whisker plot, the box will be horizontal. Since boxes can be difficult for some printers to draw, endpoints of the box may be shown instead of the entire box.

We have developed the skeletal box-and-whisker plot. Variations will include fences, which are marks placed on either side of the box and representing various portions of the box. Values that lie outside the fences are called *outliers*. These values seem to stand by themselves, away from most of the data. They might be exceptional values and deserve closer study. For a more complete discussion of outliers and other variations of box plots see *Applications, Basics, and Computing of Exploratory Data Analysis* by Paul Velleman and David Hoaglin (Duxbury Press, 1981).

Section Problems 3.2

1. Angela took a general aptitude test and scored in the 82nd percentile for aptitude in accounting. What percent of the scores were below her score? What percent were above?

2. One standard for admission to Redfield College is that the student must rank in the upper quartile of his or her graduating high school class. What is the minimal percentile rank of a successful applicant?

3. The town of Butler, Nebraska, decided to give a teacher competency exam and defined the passing scores to be those in the 70th percentile or higher. The raw test scores ranged from 0 to 100. Was a raw score of 82 necessarily a passing score?

4. Clayton and Timothy took different sections of Introduction to Economics. Each section had a different final exam. Timothy scored 83 out of 100 and had a percentile rank in his class of 72. Clayton scored 85 out of 100 but his percentile rank in his class was 70. Who performed better with respect to the rest of the students in the class: Clayton, or Timothy?

5. At Center Hospital there is some concern about the high turnover of nurses. A survey was done to determine how long (in months) nurses had been in their current positions. The responses of 20 nurses were (in months):

23	2	5	14	25	36	27	42	12	8
7	23	29	26	28	11	20	31	8	36

a) Rank the data up and down.
b) Make a box-and-whisker plot of the data.

6. Another survey was done at Center Hospital to determine how long (in months) clerical staff had been in their current positions. The responses of 20 clerical staff members were (in months):

25	22	7	24	26	31	18	14	17	20
31	42	6	25	22	3	29	32	15	72

a) Rank the data up and down.
b) Make a box-and-whisker plot.
c) Compare this plot to the one in problem 5. Discuss the location of the medians, the location of the middle half of the data bank, and the distance from the hinges to the extreme values.

7. The Dow Jones Industrial Average is a measure used by investors to track the general stock market. From July 1, 1984 to July 1, 1985 the number of consecutive days the Dow Jones Industrial Average went up were:

1	1	3	3	6	1	1	4
2	1	1	1	3	1	4	1
4	2	1	1	1	2	1	1
3	1	1	2	3	1	2	4
1	1	1	3	1	1	2	1
2	2	1	1	2	1	1	2
1	7	2	1	3	3	5	1
3	3	1	1	1	2	5	

(*Note:* The high for the period was 1337.14 which occurred on July 1, 1985. The low was 1086.57 which occurred on July 24, 1984.)
a) Find the median and lower hinge and lowest extreme value. Are they all the same?
b) Find the upper hinge and the upper extreme value.
c) Make a box-and-whisker plot.

8. Counts of white cells in blood were done for 30 blood samples. The number of white cells (in thousands) per cubic millimeter of blood in each sample was:

3.8	4.9	5.0	7.3	6.1	7.7	5.8	8.1	8.2	7.6
5.7	5.2	9.5	10.1	10.0	9.6	8.7	7.5	4.9	8.9
6.3	7.2	8.1	9.2	5.7	8.4	6.9	4.2	7.1	4.9

a) Rank the data up and down.
b) Make a box-and-whisker plot.

9. Many travelers make airline reservations and then do not show up for their scheduled flights. Air Connect Airlines took a random sample of 40 flights and recorded the number of "no-shows" who were using discount fares. The results were:

3	5	2	5	1	4	0	7	8	10
12	7	5	0	2	7	5	8	6	12
6	10	18	16	21	9	10	3	9	7
9	10	15	8	4	6	5	7	9	9

a) Rank the data up and down.
b) Make a box-and-whisker plot.

10. Air Connect decided to require discount fares to be paid in advance and a fee to be charged for last-minute (less than 24 hours) cancellations or "no-shows." When this policy was in effect another random sample of 40 flights was selected and the number of "no-shows" who were using discount fares was recorded. The results were:

1	7	6	2	3	2	1	0	3	9
3	1	5	0	7	2	6	3	3	5
2	9	11	3	6	2	0	4	7	8
12	0	3	7	6	2	1	1	2	4

a) Rank the data up and down.
b) Make a box-and-whisker plot.
c) Compare this plot with the one in Problem 9. Discuss the location of the medians, the location of the middle half of the data banks, and the distance from the hinges to the extreme values.

11. Advantage VCR's carry a one-year warranty. However, the terms of the warranty require that a registration card be sent in. One of the questions on the registration card asks the owner's age. For a random sample of 50 registration cards, the ages given were:

38	27	42	31	31	35	21	36	37	32
45	47	32	26	23	19	27	38	47	51
52	55	46	33	27	22	25	27	31	34
33	37	40	41	32	47	28	24	43	45
55	27	35	39	17	32	37	43	42	51

a) Rank the data up and down.
b) Make a box-and-whisker plot of the ages.

12. McElroy Discount Fashions claims to have a fairly low mark-up percent on the items they sell. A random sample of 50 items showed the mark-up percent over cost to be (in percents):

28	35	47	42	51	15	72	33	27	22
37	36	51	29	36	72	51	55	68	53
49	37	45	48	52	71	75	31	25	27
46	58	49	61	37	67	27	72	85	21
59	55	42	75	78	72	77	72	42	65

a) Rank the data up and down.
b) Make a box-and-whisker plot of the percent mark-ups.

SECTION 3.3 Measures of Variation

An average is an attempt to summarize a set of data in just one number. We have studied several averages. As some of our examples have shown, an average taken by itself may not always be very meaningful. We need a statistical cross reference. This cross reference should be a measure of the *variance*, or spread, of the data.

The range is one such measure of variance. The *range* is the difference between the largest and smallest values of a distribution. For example, the distance between rows in the various sections of Flicker Auditorium are (in inches):

14   15   18   20   35

The range of these distances is

$$\text{range} = \text{largest value} - \text{smallest value}$$

$$= 35 - 14 = 21 \text{ in.}$$

The range indicates the variation between the smallest and largest entries, but it does not tell us how much other values vary from one another. We need a different measure of variation, as the next example shows.

EXAMPLE 3

You are trying to decide which record club to join: Discount Disks or Selecta Record. Both have the same bonuses for new members and both require members to buy one record a month from the monthly selections for at least one year. They both charge the same price for records, and they both advertise that the number of records in the monthly selections has mean 31 and range 79. The only advertised difference is that Discount Disks has no membership fee and Selecta Record costs $5.00 to join. Which club would you join? Before you make up your mind, look at the additional information in Table 3-9.

TABLE 3-9 Number of Record Selections Per Month

Month	Selecta Record	Discount Disks
January	100	80
February	30	1
March	30	70
April	30	2
May	21	70
June	23	1
July	21	1
August	21	70
September	24	3
October	21	70
November	30	2
December	21	2
	Mean = 31	Mean = 31
	Range = 79	Range = 79

The mean and range are not enough to tell you how much the number of monthly record selections varies from the advertised mean. Discount Disks could be a disappointment since some months they give you only one selection. Selecta Record is more consistent in their offerings since you always have at least 21 records from which to choose.

A measurement that will give you a better idea of how the data entries differ from the mean is the *standard deviation*. The formula for the standard deviation differs slightly depending on whether you are using an entire population or just a sample. At the moment we will compute the standard deviation for sample data only. When we have sample data, we use the letter *s* to denote the standard deviation. The formula for the sample standard deviation is

$$\text{sample standard deviation} = s = \sqrt{\frac{\Sigma(x - \bar{x})^2}{n - 1}} \qquad (2)$$

where *x* is any entry in the distribution, $\bar{x}$ is the mean, and *n* is the number of entries.

Notice that the standard deviation uses the difference between each entry *x* and the mean $\bar{x}$. This quantity $(x - \bar{x})$ will be negative if the mean $\bar{x}$ is greater than the entry *x*. If you take the sum

$$\Sigma(x - \bar{x})$$

then the negative values will cancel the positive values, leaving you with a variation measure of 0 even if some entries vary greatly from the mean.

In the formula for the standard deviation, the quantities $(x - \bar{x})$ are squared before they are summed. This device eliminates the possibility of having some negative values in the sum. So, in the formula, we have the quantity

$$\Sigma(x - \bar{x})^2$$

Then we divide this sum by $n - 1$ to get the quantity under the square root sign:

$$\frac{\Sigma(x - \bar{x})^2}{n - 1}$$

If we had the *entire* population, we would divide by *N*, the population size, and would thus have the mean of the values $(x - \bar{x})^2$. However, a random sample may not include the extreme values of a population, so to make the standard deviation computed from the sample larger, we divide by the smaller value $n - 1$. Then the sample standard deviation is the best estimate for the standard deviation of the entire population.

These three steps have given us a quantity called the *variance* of a sample, denoted by s^2:

$$\text{sample variance} = s^2 = \frac{\Sigma(x - \bar{x})^2}{n - 1} \qquad (3)$$

The record data in Example 2 was initially in months, but the variance s^2 of this data would be in *square months*. Square months—what's that? We obviously need to take the square root of the variance. This brings us to the standard deviation of a sample,

$$\text{sample standard deviation} = s = \sqrt{\frac{\Sigma(x - \bar{x})^2}{n - 1}}$$

The next example shows how to use this formula.

EXAMPLE 4

Big Blossom Greenhouse was commissioned to develop an extra large rose for the Rose Bowl Parade. A random sample of blossoms from Hybrid A bushes yielded these diameters (in inches) for mature peak blossoms:

2   3   4   5   6   8   10   10

There are several steps involved in computing the standard deviation, and a table will be helpful (*see* Table 3-10). Since $n = 8$, we take the total sum of the entries in the first column of Table 3-10 and divide by 8 to find the mean $\bar{x}$.

$$\bar{x} = \frac{\Sigma x}{n} = \frac{48}{8} = 6.0$$

TABLE 3-10 Diameter of Rose Blossoms (in inches)

Column I	Column II	Column III
x	$x - \bar{x}$	$(x - \bar{x})^2$
2	$2 - 6 = -4$	$(-4)^2 = 16$
3	$3 - 6 = -3$	$(-3)^2 = 9$
4	$4 - 6 = -2$	$(-2)^2 = 4$
5	$5 - 6 = -1$	$(-1)^2 = 1$
6	$6 - 6 = 0$	$(0)^2 = 0$
8	$8 - 6 = 2$	$(2)^2 = 4$
10	$10 - 6 = 4$	$(4)^2 = 16$
10	$10 - 6 = 4$	$(4)^2 = 16$
$\Sigma x = 48$		$\Sigma(x - \bar{x})^2 = 66$

Using this value for $\bar{x}$ we obtain the second column of the table. We square each value in the second column to obtain Column III, and then we add the values in Column III. To get the variance, we divide the sum of Column III by $n - 1$. Since $n = 8$, $n - 1 = 7$.

$$s^2 = \frac{\Sigma(x - \bar{x})^2}{n - 1} = \frac{66}{7} = 9.43$$

Finally, the standard deviation is obtained by taking the square root of the variance.

$$s = \sqrt{s^2} = \sqrt{9.43} = 3.07$$

(Generally you can use a table of square roots or a calculator to compute a square root.)

_____ Exercise 8 _____

Big Blossom Greenhouse gathered another random sample of mature peak blooms from Hybrid B. The eight blossoms had these widths (in inches):

5   5   5   6   6   6   7   8

a) Again we will construct a table so we can find the mean, variance, and standard deviation more easily. In this case what is the value of n? Find the sum of Column I in Table 3-11 and compute the mean. Complete Columns II and III of the table.

a) $n = 8$. The sum of Column 1 is $\Sigma x = 48$, so the mean is

$$\bar{x} = \frac{48}{8} = 6 \text{ in.}$$

TABLE 3-11

I	II	III
x	$x - \bar{x}$	$(x - \bar{x})^2$
5	-1	1
5	-1	1
5	-1	1
6	0	0
6	0	0
6	0	0
7	1	1
8	2	4
$\Sigma x =$ ____		$\Sigma(x - \bar{x})^2 =$ _8_

TABLE 3-12 Completion of Table 3-11

I	II	III
x	$x - \bar{x}$	$(x - \bar{x})^2$
5	-1	1
5	-1	1
5	-1	1
6	0	0
6	0	0
6	0	0
7	1	1
8	2	4
$\Sigma x = 48$		$\Sigma(x - \bar{x})^2 = 8$

b) What is the value of $n - 1$? Divide the total sum of Column III by $n - 1$ to find the variance.

b) $n - 1 = 7$

$$\text{variance} = s^2 = \frac{\Sigma(x - \bar{x})^2}{n - 1} = \frac{8}{7} = 1.14$$

c) Use a calculator to find the square root of the variance. Is this the standard deviation?

c) $\sqrt{\text{variance}} = \sqrt{s^2} = \sqrt{1.14} \approx 1.07$ in. The square root of the variance *is* the standard deviation. (*Note:* We say $\sqrt{1.14} \approx 1.07$. The symbol $\approx$ means approximately equal. We use $\approx$ since 1.07 is not exactly equal to $\sqrt{1.14}$.)

Let's summarize and compare the results of Exercise 8 and Example 4. The greenhouse found the following blossom diameters for Hybrid A and Hybrid B:

Hybrid A: mean, 6.0 in.; standard deviation, 3.07 in.
Hybrid B: mean, 6.0 in.; standard deviation, 1.07 in.

In both cases the means are the same: 6 in. But the first hybrid has a larger standard deviation. This means that the blossoms of Hybrid A are less consistent than those of Hybrid B. If you want a rosebush that occasionally has 10-in. blooms and 2-in. blooms, use the first hybrid. But if you want a bush that consistently produces roses close to 6 in. across, use Hybrid B.

There is another formula for the standard deviation which gives the same results as those of Formula (2). It is easier to use with a calculator since there are fewer subtractions involved.

The computation formula depends on the fact that

$$\Sigma(x - \bar{x})^2 = \Sigma x^2 - \frac{(\Sigma x)^2}{n}$$

which can be proved with the aid of some algebra. The expression $\Sigma(x - \bar{x})^2$ is a sum of squares. Using the notation SS_x to indicate this sum of squares, we get the relation

$$SS_x = \Sigma(x - \bar{x})^2 = \Sigma x^2 - \frac{(\Sigma x)^2}{n}$$

Then the computation formula for the standard deviation s is

• **Computation Formula for the Sample Standard Deviation s**

$$s = \sqrt{\frac{SS_x}{n - 1}} \tag{4}$$

where $SS_x = \Sigma x^2 - \frac{(\Sigma x)^2}{n}$

To compute Σx^2, we *first square* all the x values, and then take the sum. To compute $(\Sigma x)^2$, we *first sum* the x values, and then square the total.

The next exercise shows you how to use the computation formula. The expression SS_x will be used later, both in the chapter on regression and correlation and in the section on ANOVA.

_____ Exercise 9 _____

Rockwood Library was having difficulty because some books were being kept out long after the due date. The original late fine was 5¢ per day. The mean overdue time was found to be 10.8 days with a standard deviation of 5.02 days. The librarian decided to change the late-fine rate to 35¢ per day. Table 3-13 contains data from a random sample of overdue books under the new fine system.

a) Complete Table 3-13.

a)

TABLE 3-13 Number of Days Books Are Overdue

x	x^2
5	25
5	25
6	____
6	____
6	____
7	____
7	____
8	____
9	____
10	____
$\Sigma x =$ ____	$\Sigma x^2 =$ ____

TABLE 3-14 Completion of Table 3-13

x	x^2
5	25
5	25
6	36
6	36
6	36
7	49
7	49
8	64
9	81
10	100
$\Sigma x = 69$	$\Sigma x^2 = 501$

b) Evaluate SS_x.

b) Since $SS_x = \Sigma x^2 - \dfrac{(\Sigma x)^2}{n}$ we need to find the values of Σx^2 and $(\Sigma x)^2$. From the total of Column II we see that $\Sigma x^2 = 501$. By squaring the total of Column I we get

$$(\Sigma x)^2 = (69)^2 = 4761$$

$$SS_x = \Sigma x^2 - \frac{(\Sigma x)^2}{n}$$

$$= 501 - \frac{4761}{10}$$

$$= 501 - 476.1$$

$$= 24.9$$

c) Evaluate the sample standard deviation by using the formula

$$s = \sqrt{\frac{SS_x}{n-1}}$$

c) Since $n = 10$, $n - 1 = 9$ and

$$s = \sqrt{\frac{24.9}{9}}$$

$$\approx \sqrt{2.77}$$

$$\approx 1.66$$

d) Does the new fine system appear to have lowered the mean overdue time? Does it appear to have reduced the standard deviation? Under which fine system does it appear that the overdue times cluster more closely about the mean?

d) The new fine system appears to reduce both the mean and standard deviation of overdue times. The overdue time seems more closely clustered about the mean of the new fine system since that system appears to have a much smaller standard deviation.

In almost all applications of statistics we work with a random sample of data rather than the entire population of *all* possible data values. However, if we do in fact have data for the entire population, we can compute the *population mean* μ (Greek letter mu, pronounced *mew*) and *population standard deviation* σ (Greek letter sigma) using the following formulas:

$$\mu = \frac{\Sigma x}{N} \qquad \text{population mean}$$

$$\sigma = \sqrt{\frac{\Sigma(x - \mu)^2}{N}} \qquad \text{population standard deviation}$$

where N is the number of data values in the population and x represents the individual data values of the population. We note that the formula for μ is the same as the formula for $\bar{x}$ (the sample mean) and the formula for σ is the same as the formula for s (the sample standard deviation), except that N is used instead of $n - 1$ and μ is used instead of $\bar{x}$ in the formula for σ.

In the formulas for s and σ we use $n - 1$ to compute s and N to compute σ. Why? The reason is that N (capital letter) represents the population size while n (lower case) represents the sample size. Since a random sample usually will not contain extreme data values (large or small), we divide by $n - 1$ in the formula for s to make s a little larger than it would have been had we divided by n. Courses in advanced theoretical statistics show that this procedure will give us the best possible value for the sample standard deviation s. If we have the population of all data values, then the extreme data values are of course present, so we divide by N instead of $N - 1$.

EXAMPLE 5

In Hawaii there is a species of goose called the Nene Goose. Before Captain Cook discovered Hawaii in 1778, the Nene Goose was abundant. However, after guns were introduced, the Nene Goose population decreased severely. A few of these geese were sent to the London Zoo just before the Nene Goose became extinct in Hawaii. In effect, the London Zoo had the population of *all* Nene Geese. If eight Nene Geese were all the zoo had, and the weights (in pounds) of these birds were

12.7	15.2	19.4	8.2	16.4	10.8	14.6	23.5

find the population mean μ and the population standard deviation σ of weights of Nene Geese. (*See* Table 3-15.)

$$\mu = \frac{\Sigma x}{N} = \frac{120.8}{8} = 15.1 \, \text{lb}$$

$$\sigma = \sqrt{\frac{\Sigma(x - \mu)^2}{N}} = \sqrt{\frac{162.86}{8}} = \sqrt{20.36} \simeq 4.51 \, \text{lb}$$

TABLE 3-15 Weights of Nene Geese

x	$x - \mu$	$(x - \mu)^2$
12.7	-2.40	5.76
15.2	0.10	0.01
19.4	4.30	18.49
8.2	-6.90	47.61
16.4	1.30	1.69
10.8	-4.30	18.49
14.6	-0.50	0.25
23.5	8.40	70.56

$\Sigma x = 120.8$
$N = 8$

$\Sigma(x - \mu)^2 = 162.86$

In the late 1950s the London Zoo had a much larger population of Nene Geese, and some of these geese were sent back to national parks in Hawaii. Today the Nene Goose is protected, and visitors to the islands can see Nene Geese in Haleakala National Park on Maui.

Section Problems 3.3

1. At the University of Colorado a random sample of five faculty gave the following information about the number of hours spent on committee work each week:

   3   6   4   1   5

   a) Find the range.
   b) Find the sample mean.
   c) Find the sample standard deviation.

2. A random sample of six credit card accounts gave the following information about the amount due on each card:

   $53.18 $71.12 $115.10 $27.30 $36.19 $66.48

   a) Find the range.
   b) Find the sample mean.
   c) Find the sample standard deviation.

3. A random sample of seven New York plays gave the following information about how long each play ran on Broadway (in days):

   12   45   36   118   50   7   20

   a) Find the range.
   b) Find the sample mean.
   c) Find the sample standard deviation.

4. For the past ten years the daily high temperature on New Year's day in the mountain town of Tin Cup, Colorado, was (in °F):

    3°   25°   − 8°   17°   − 2°   10°   − 12°   21°   4°   6°

    a) Find the range.
    b) Find the sample mean.
    c) Find the sample standard deviation.

5. Petroleum pollution in oceans is known to increase the growth of a certain bacteria. Brian did a project for his ecology class for which he made a bacteria count (per 100 milliliters) in nine random samples of sea water. His counts gave the following readings:

    17   23   18   19   21   16   12   15   18

    a) Find the range.
    b) Find the sample mean.
    c) Find the sample standard deviation.

6. In the process of tuna fishing, porpoises are sometimes accidentally caught and killed. A U.S. oceanographic institute wants to study the number of porpoises killed in this way. Records from eight commercial tuna fishing fleets gave the following information about the number of porpoises killed in a three-month period:

    2   6   18   9   0   15   3   10

    a) Find the range.
    b) Find the sample mean.
    c) Find the sample standard deviation.

7. A museum curator examined the Crown of Charlemagne and found the seven rubies to have the following weights (in carats):

    19.8   43.8   36.1   52.4   63.1   20.7   46.3

    a) Find the range.
    b) Since these numbers represent the *population* of all rubies in the crown, find the *population mean*.
    c) Find the *population standard deviation*.

8. Sir Charles Bradley wrote six novels before he died. The page lengths of these novels are:

    280   318   279   356   410   305

    a) Find the range.
    b) Since these numbers represent the *population* of all novels written by Sir Charles, find the *population mean*.
    c) Find the *population standard deviation*.

9. The neighborhood association of Cherry Hills Village took a survey of opinions about rent control in their neighborhood. In this opinion poll 1 = strongly against rent control and 10 = strongly in favor of rent control. A random sample of 14 people gave the following opinions:

1	1	1	2	1	10	1	10	10	8	10	2	10	8

a) Compute the range, sample mean, and sample standard deviation of opinion ratings about rent control.

Another questionnaire asked for opinions about moving a mailbox from one side of the street to the other. Again a random sample of 14 people gave the following opinions where 1 = strongly disagree and 10 = strongly agree:

5	5	5	4	5	5	5	6	5	5	6	5	6	5

b) Compute the range, sample mean, and sample standard deviation of these numbers.
c) Compare your answers for parts (a) and (b). Were the means about the same? Were the opinions on the two issues distributed differently? How did the range and standard deviation reflect this when the mean did not? Explain your answer.

10. June purchased a new home computer and has been having trouble with voltage spikes on the power line. Such voltage jumps can be caused by the operation of appliances such as clothes dryers and electric irons, or just by a power surge on the outside power line. Her friend Jim is an electronics technician and has obtained the following data about voltages when certain electric appliances are turned on and off. Remember, the normal line voltage is 110 volts. All measurements are taken from the line and measured in volts.

73	140	78	142	80	140	90	133

a) Compute the sample mean, sample standard deviation, and range.

Jim advised June to buy a device called a power surge protector which protects the computer from strong voltage spikes. Using the power surge protector Jim again measured voltages to the computer when the same appliances were turned on and off. The results in volts were:

100	120	108	114	105	117	103	114

b) Compute the sample mean, sample standard deviation, and range of the voltages using the power surge protector.
c) Compare your answers for parts (a) and (b). Were the means about the same? Were the voltage distributions different with and without the power surge protector? How did the standard deviation and range reflect this when the mean did not? Explain your answer.

11. A certain brand of nylon monofilament fishing line is known to deteriorate in very cold temperatures. A spool of ten-pound test monofilament line was left out overnight at Fairbanks, Alaska, when temperatures dropped to −35 °F. A random sample of six pieces of line gave the following breaking strengths in pounds:

10.1	6.2	9.8	5.3	9.9	5.7

a) Compute the sample mean, sample standard deviation, and range.

A second spool of this line that had not been subjected to extreme cold temperatures gave the following breaking strengths (in pounds) for a random sample of six pieces of line:

10.2	9.7	9.8	10.3	9.6	10.1

b) Compute the sample mean, sample standard deviation, and range for these values.
c) Compare your answers for parts (a) and (b) and comment on the observed differences. Which line had the more consistent performance? How was this reflected in the sample standard deviations? In the ranges?

12. Ralph and Gloria did a 4-H project to demonstrate ways to get better gasoline mileage. They kept the car windows rolled up to prevent air drag, used only moderate acceleration from a standstill, and kept their speed down in general. Ralph recorded the mean miles per gallon for five days selected at random from the period in which he drove the car. Gloria did the same. The results are shown below.

Ralph	22.3	21.2	20.8	19.8	23.8
Gloria	25.2	19.1	18.0	24.4	20.3

a) Find the range for Ralph and for Gloria.
b) Find the mean and sample standard deviation for each.
c) Who consistently seems to have gotten better mileage: Ralph or Gloria? Who had the smaller sample standard deviation?

SECTION 3.4   **Mean and Standard Deviation of Grouped Data**

If you have a great many data, it can be quite tedious to compute the mean and standard deviation. Even if you have a calculator, you must punch in the data. In many cases a close approximation to the mean and standard deviation is all that is needed, and it is not difficult to approximate these two values from a frequency distribution.

The basic plan is as follows:

1. Make a frequency table.
2. Compute the midpoint for each class and call it x.
3. Count the number of entries in each class and denote the number by f.
4. Add the number of entries from each class together to find the total number of entries n in the sample distribution.

Treat each entry of a class as though it falls on the midpoint (x) of that class. Then the midpoint times the number of entries in a class (xf) represents the sum of the observations in the class. The formulas for the mean and standard deviation are as follows:

- **Mean for a frequency distribution:**

$$\bar{x} = \frac{\Sigma xf}{n} \tag{5}$$

where x is the midpoint of a class,
  f is the number of entries in that class,
  n is the total number of entries in the distribution,
  the summation Σ is over all classes in the distribution.

- **Sample standard deviation for a frequency distribution:**

$$s = \sqrt{\frac{\Sigma(x - \bar{x})^2 f}{n - 1}} \tag{6}$$

where x is the midpoint of a class,
  f is the number of entries in that class,
  n is the total number of entries in the distribution,
  the summation Σ is over all classes in the distribution.

EXAMPLE   6

The manager of Pantry Queen Supermarket wants to hire one more checkout clerk. To justify his request to the regional manager, the manager chose a random sample of 50 customers and timed how long each stood in line before a clerk could begin checking the customer out. The written request contained the histogram in Figure 3-5.

FIGURE 3-5   Time in Minutes Before Checkout Begins

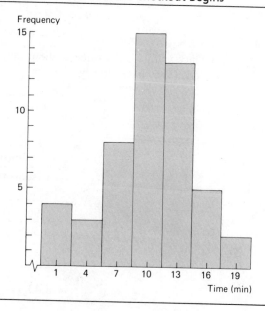

Approximate the mean and standard deviation of the distribution. First make a table with all the columns necessary to compute the mean and standard deviation (*see* Table 3-16). (Columns V, VI, and VII are filled in after the mean is computed.)

TABLE 3-16 Time in Minutes Before Checkout Begins

I	II	III	IV	V	VI	VII
	Freq.	Midpoint				
Class	f	x	xf	$x - \bar{x}$	$(x - \bar{x})^2$	$(x - \bar{x})^2 f$
0–2	4	1	4	−9.2	84.64	338.56
3–5	3	4	12	−6.2	38.44	115.32
6–8	8	7	56	−3.2	10.24	81.92
9–11	15	10	150	−0.2	0.04	0.60
12–14	13	13	169	2.8	7.84	101.92
15–17	5	16	80	5.8	33.64	168.20
18–20	2	19	38	8.8	77.44	154.88
	$\Sigma f = 50$		$\Sigma xf = 509$			$\Sigma(x - \bar{x})^2 f = 961.40$

Use the first four columns to find the mean. The value of n is found by summing Column II.

$$n = \Sigma f = 50$$

The mean is

$$\bar{x} = \frac{\Sigma xf}{n} = \frac{509}{50} \quad \text{(from the sum of Column IV)}$$

$$= 10.2 \, \text{min}$$

Once you have the value of the mean, you can complete Columns V, VI, and VII. (They have already been completed for our convenience.)

$$s = \sqrt{\frac{\Sigma(x - \bar{x})^2 f}{n - 1}}$$

$$= \sqrt{\frac{\text{sum of Column VII}}{(\text{sum of Column II}) - 1}}$$

$$= \sqrt{\frac{961.40}{50 - 1}}$$

$$= \sqrt{19.62}$$

$$= 4.43$$

_____ Exercise 10 _____

The Marathon Walk for World Peace has a 25-mile route. A random sample of participants walked the distances shown in Table 3-17.

TABLE 3-17 Distance Walked for World Peace

Distance (mi)	Number of Walkers
1–5	14
6–10	9
11–15	11
16–20	10
21–25	6

a) What is the value of n (the number of walkers in the sample)?

a) Add the column giving the number of walkers.

$$n = 50$$

b) Complete Table 3-18 and compute the mean.

b)

TABLE 3-18

Class	f (Freq.)	x (Midpoint)	xf
1–5	14	3	42
6–10	9	8	_____
11–15	11	_____	_____
16–20	10	_____	_____
21–25	6	_____	_____
	$\Sigma f = 50$		$\Sigma xf = $ _____

TABLE 3-19 Completion of Table 3-18

Class	f (Freq.)	x (Midpoint)	xf
1–5	14	3	42
6–10	9	8	72
11–15	11	13	143
16–20	10	18	180
21–25	6	23	138
			$\Sigma xf = 575$

$$\bar{x} = \frac{\Sigma xf}{n} = \frac{575}{50} = 11.5$$

c) Complete Table 3-20.

c) The last row of Table 3-20 should read

TABLE 3-20 Continuation of Table 3-18

$x - \bar{x}$	$(x - \bar{x})^2$	$(x - \bar{x})^2 f$
−8.50	72.25	1011.50
−3.50	12.25	110.25
1.50	2.25	24.75
6.50	42.25	422.50
11.50	_____	_____

11.50   132.25   793.50

We sum the $(x - \bar{x})^2 f$ column. That sum is 2362.50.

$$s = \sqrt{\frac{\Sigma(x - \bar{x})^2 f}{n - 1}} = \sqrt{\frac{2362.50}{49}} \approx \sqrt{48.21}$$

$$\approx 6.94$$

Which column do you sum to compute the standard deviation? Sum that column and find the standard deviation.

In the case where our data are not grouped, but there are several repeated data values, we can use the techniques of grouped data to find the mean and standard deviation fairly quickly. In such cases, we do not need to find a midpoint of a class interval since each class consists of a single data value.

EXAMPLE 7

A random sample of all college football players gave the following information about recovery time from shoulder injuries where

$$x = \text{number of weeks for recovery}$$
$$f = \text{number of injured players}$$

TABLE 3-21 Recovery Times from Shoulder Injuries

Recovery Time x	1	2	3	4	5	6	7	8
Frequency f	5	8	12	19	7	4	3	2

Find the sample mean and then find the sample standard deviation of recovery times. We again use a table to organize the data and make computations easier.

TABLE 3-22 Recovery Times

x	f	xf	$x - \bar{x}$	$(x - \bar{x})^2$	$(x - \bar{x})^2 f$
1	5	5	-2.82	7.95	39.76
2	8	16	-1.82	3.31	26.50
3	12	36	-0.82	0.67	8.07
4	19	76	0.18	0.03	0.62
5	7	35	1.18	1.39	9.75
6	4	24	2.18	4.75	19.01
7	3	21	3.18	10.11	30.34
8	2	16	4.18	17.47	34.94
	$\Sigma f = 60$	$\Sigma xf = 229$			$\Sigma(x - \bar{x})^2 f = 168.99$

We first find the sample mean $\bar{x}$.

$$\bar{x} = \frac{\Sigma xf}{\Sigma f} = \frac{229}{60} \approx 3.82$$

Then we use the value for $\bar{x}$ to compute the entries for the last three columns of the table. To compute the sample standard deviation we note that

$$n = \Sigma f = 60$$

$$s = \sqrt{\frac{\Sigma(x - \bar{x})^2 f}{n - 1}} \approx \sqrt{\frac{168.99}{59}} \approx \sqrt{2.86} \approx 1.69$$

The values we get for $\bar{x}$ and s are exactly (within rounding error) the same as those we would have obtained if we had listed each of the 60 data values separately and computed $\bar{x}$ and s.

Sometimes we wish to average numbers, but we want to assign more importance or weight to some of the numbers. For instance, suppose your professor tells you that your grade will be based on a midterm and a final exam, each of which have 100 possible points. However, the final exam will be worth 60% of the grade and the midterm only 40%. How could you determine your average score to reflect these different weights? The average you need is the *weighted average*.

If we view the weight of a measurement as a "frequency," then we discover that the formula for the mean of a frequency distribution gives us the weighted average.

$$\text{weighted average} = \frac{\Sigma xw}{\Sigma w}$$

where w is the weight of the data value x.

EXAMPLE 8

Suppose your midterm test score is 83 and your final exam score is 95. Using the weights of 40% for the midterm and 60% for the final exam, compute the weighted average of your scores. If the minimum average for an A is 90, will you earn an A?

By the formula, we multiply each score by its weight and add the results together. Then we divide by the sum of all the weights. Converting the percents to decimal notation, we get

$$\text{weighted average} = \frac{83(0.40) + 95(0.60)}{0.40 + 0.60}$$

$$= \frac{33.2 + 57}{1} = 90.2$$

Your average is high enough to earn an A.

In the following exercise we see that the sum of the weights is not always one.

___ Exercise 11 ___

In an investment portfolio, stocks are rated on a scale of 1 to 10 for dividend earning, security, and capital growth potential. On the scale 1 equals very poor and 10 equals excellent. In one investment strategy favoring security the dividend rating is given a weight of 2, the security a weight of 5, and the capital growth potential is given a weight of 3.

a) Stock A has the ratings shown in Table 3-23. Complete the table and find the weighted average rating of the stock.

TABLE 3-23 Stock A Rating

	Rating x	Weight w	xw
Dividend	7	2	——
Security	8	5	——
Growth	4	3	——
		$\Sigma w =$ ——	$\Sigma xw =$ ——

b) Stock B is also being considered for the portfolio. It has the ratings shown in Table 3-24. Complete the table and find the weighted average for Stock B. How does Stock B compare to Stock A for this particular investment portfolio?

TABLE 3-24 Stock B Rating

	Rating x	Weight w	xw
Dividend	2	2	——
Security	7	5	——
Growth	10	3	——
		$\Sigma w =$ ——	$\Sigma xw =$ ——

a) $\Sigma w = 10$

The last column has entries 14, 40, and 12, and the sum $\Sigma xw = 66$.

$$\text{weighted average} = \frac{\Sigma xw}{\Sigma w}$$

$$= \frac{66}{10}$$

$$= 6.6$$

b) $\Sigma w = 10$

The last column has entries 4, 35, and 30, and the sum $\Sigma xw = 69$.

$$\text{weighted average} = \frac{\Sigma xw}{\Sigma w}$$

$$= \frac{69}{10}$$

$$= 6.9$$

Stock B has a slightly higher weighted average than Stock A. For the weights assigned to this investment strategy, Stock B will be the better stock. However, if the weights were changed for a different investment portfolio, Stock A might have the higher average and be the better choice.

Section Problems 3.4

1. In the United States life expectancy of a male child born between 1979 and 1981 varies by state (including the District of Columbia) from 64.55 to 74.08 years. In the following table life expectancies are grouped by years.

Life Expectancy for Men, x	64–67	68–71	72–75
Frequency of States, f	1	38	12

   a) Estimate the mean life expectancy in years for all the states.
   b) Estimate the sample standard deviation of life expectancy in years.

2. The life expectancy of a female child born between 1979 and 1981 also varies by state (including the District of Columbia) from 73.70 to 80.33 years. In the following table life expectancies are grouped by years.

Life Expectancy for Women, x	73–75	76–78	79–81
Frequency of States, f	1	36	14

a) Estimate the mean life expectancy in years for all the states.

b) Estimate the sample standard deviation of life expectancy in years.

3. For both men and women, the lowest life expectancy occurs in the District of Columbia. That is the only entry in each of the lowest classes of problems 1 and 2. Remove that entry and recalculate an estimate for the mean and sample standard deviation of life expectancies:

a) for men.

b) for women.

4. For a random selection of 20 water wells near Custer, Wyoming, the distance from ground level to water level is shown (in feet) by the following frequency table:

Distance from Ground to Water (ft), x	12–14	15–17	18–20	21–23	24–26
Number of Wells, f	1	3	8	2	6

Using the midpoints x of the depth intervals, estimate (a) the mean depth and (b) the standard deviation.

5. The members of a random sample of 61 Pittsburgh policemen were asked how many years (rounded to the nearest year) they had served on the city's police force. The results were grouped into five-year intervals and are shown below.

Number of Years Service, x	1–5	6–10	11–15	16–20	21–25
Number of Policemen, f	20	12	14	10	5

Using the midpoints x of the time intervals, estimate (a) the mean time and (b) the standard deviation for the number of years of service.

6. Jan is a political science student who is studying the incidence of senators in her state legislature who are marked absent when a bill is voted on in the senate. Jan used state government documents to obtain a random sample of 60 bills which were voted on in the senate. For each bill Jan obtained the following information, where x is the number of senators absent when the vote on a bill was taken and f is the number of bills for which this level of absenteeism occurred.

Number Absent, x	0–4	5–9	10–14	15–19	20–24
Number of Bills, f	7	18	23	8	4

Use the midpoints of the x intervals to estimate (a) the mean and (b) the sample standard deviation for the number of senators absent.

7. A psychology test to measure memory skills was given to a random sample of 43 students. The results follow, where x is the student score and f is the frequency with which students obtained this score.

x	0–10	11–21	22–32	33–43	44–54
f	1	12	18	9	3

Use the midpoints to estimate (a) the mean and (b) the sample standard deviation of scores.

8. The president of the alumni association at Muddybanks College presented the following tabulation of alumni contributions which have been received so far this year:

$ Amount, x	5	10	15	20	50	100
Number of Alumni Making This Contribution, f	710	350	92	51	12	16

Find (a) the mean and (b) the sample standard deviation for the contributions.

9. Market surveys are sometimes used to set the market price of a new consumer item. A random sample of 335 adults in Chicago was shown a new type of home video-telephone not yet on the consumer market. After inspecting and testing the new video-telephone, each person was asked how much he or she would be willing to pay for this video-telephone. The results follow, where x is the cost and f is the frequency of people who said that the given cost was the highest amount they would pay for the item.

x	$50	$150	$250	$350	$450	$550
f	12	67	115	95	40	6

Find (a) the mean and (b) the sample standard deviation for the amount these people are willing to pay for a home video-telephone.

10. A magazine distributor has purchased a new computer software package to sort customer accounts and do billings. In a study of the effectiveness of the new software package, a random sample of different billings, each with 1,000 customers, was used. In the following, x is the number of seconds required to sort and compute billings for a batch of 1,000 customers and f is the number of batches of 1,000 customers requiring a given computing time.

x	6	7	8	9	10	11
f	2	7	11	16	23	14

a) Compute the sample mean $\bar{x}$ of computing times for a batch of 1,000 customers.
b) Compute the sample standard deviation s of computing times for a batch of 1,000 customers.

11. A random sample of readings at a radioactive spill site gave the following information where x is the radiation level in microcuries and f is the number of readings at a given radiation level.

x	5	10	15	20	25	30	35
f	6	18	11	5	3	1	1

a) Compute the sample mean $\bar{x}$ of radiation levels.
b) Compute the sample standard deviation s of radiation levels.

12. An illegal pain-killing drug is sometimes given to race horses in an effort to get them to run faster than normal. A veterinarian working for the racing commission is studying

this drug to determine how long the drug stays in the horse's blood stream. A random sample of horses were given a shot of 10 cc of the drug. In the following table x is the time in hours for the drug level in the blood stream to go back to zero and f is the number of horses.

x	2	4	6	8	10	12	14	16
f	1	3	8	11	9	6	4	2

a) Compute the sample mean $\bar{x}$ of recovery times for this drug.
b) Compute the sample standard deviation s of recovery times.

13. Allied Air Lines is doing a study of late arrival times for the New York to Atlanta run. A random sample of late flights gave the following information where x is the number of minutes late and f is the number of flights.

x	10	20	30	40	50	60	70	80	90
f	15	19	10	7	3	2	2	1	1

a) Compute the sample mean $\bar{x}$ of late arrival times.
b) Compute the sample standard deviation s of late arrival times.

14. A wildlife research team observed a random sample of 70 Alaskan wolf dens and counted the number of wolf pups per den. The following is a frequency tabulation of the data:

Number of Pups, x	0	1	2	3	4	5	6	7	8	9	10	11
Number of Dens with This Many Pups, f	5	3	6	7	9	12	10	8	0	5	4	1

Find (a) the mean number of pups per den, $\bar{x}$, and (b) the sample standard deviation, s.

15. In your biology class your final grade is based on several things: a lab score, scores on two major tests, and your score on the final exam. There are 100 points available for each score. However, the lab score is worth 25% of your total grade, each major test is worth 22.5%, and the final exam is worth 30%. Compute the weighted average for the following scores: 92 on the lab, 81 on the first major test, 93 on the second major test, and 85 on the final exam.

16. Suppose the weighting for the activities in your biology class changed so that the lab score was still worth 25%, but each of the two major tests were also worth 25%, and the final was worth only 25%. Compute the weighted average for the same set of scores as those in problem 15: 92 on lab, 81 on the first major test, 93 on the second major test, and 85 on the final exam. Is the weighted average different from that in problem 15? Since the weights are all the same, did you really need to use a weighted average, or could you simply have taken the mean of the four scores?

17. At General Hospital nurses are given performance evaluations to determine eligibility for merit pay raises. The supervisor rates them on a scale of 1 to 10 (10 being the highest rating) for several activities: promptness, record keeping, appearance, and bedside manner with patients. Then an average is determined by giving a weight of 2 for promptness, 3 for record keeping, 1 for appearance, and 4 for bedside manner with patients. What is the average rating for a nurse with ratings of 9 for promptness, 7 for record keeping, 6 for appearance, and 10 for bedside manner?

18. The alumni club of Jefferson College gives a $10,000 award to the most outstanding athlete each year. Since competitive sports include football, basketball, baseball, swimming, and tennis, the rating scale is a way of comparing athletes who participate in any of the sports. Each candidate is rated on a scale of 1 to 10 (with 10 being the best) for: individual performance, win–loss record of team, grade point average, and sportsmanship. The ratings are then averaged using a weight of 5 for individual performance, 2 for win–loss record of team, 1 for grade point average, and 3 for sportsmanship.

   a) One athlete had the following ratings: 9 for individual performance, 7 for team record, 6 for grade point average, 8 for sportsmanship. Compute the weighted average of the ratings.

   b) Another athlete had ratings of 8 for individual performance, 9 for team record, 5 for grade point average, and 9 for sportsmanship. Compute the weighted average of these ratings. Which athlete had the higher average rating?

Summary

In order to characterize numerical data we use both averages and variation. An average is an attempt to summarize the data into just one number: the average. We studied three important averages: the mode, the median, and the mean. However, an average alone can be misleading; we really need another statistical cross reference. The measure of data spread, or variation, satisfies the purpose. The two variations that we looked at most carefully were the range and the standard deviation. A box-and-whisker plot gives a good visual impression of the range of the data and the location of the middle half of the data.

In later work the average we will use most is the mean; the measure of variation we will use most is the standard deviation. Because the mean and standard deviation are so important, we learned how to estimate these values from data already organized in a frequency distribution.

Important Words and Symbols

	Section		Section
Average	3.1	Sample variance, s^2	3.3
Mode	3.1	Sample standard deviation, s^2	3.3
Median	3.1	Sum of the squares, SS_x	3.3
Mean, $\bar{x}$	3.1	Square of the sum $(\Sigma x)^2$	3.3
Summation symbol, Σ	3.1	Population size N	3.3
Percentile	3.2	Population mean, μ	3.3
Median rank	3.2	Population standard deviation,	
Hinge rank	3.2	σ	3.3
Hinge	3.2	Mean of grouped data	3.4
Whisker	3.2	Standard deviation of grouped	
Box-and-whisker plot	3.2	data	3.4
Variation	3.3	Weighted average	3.4
Range	3.3		

Chapter Review Problems

1. In Colorado a driver's license expires on the driver's birthday every four years. A driver is not permitted to renew the license except during the 90 days prior to and including the driver's birthday during the year in which the license expires. A random sample of 100 renewals showed the number of days before expiration that a license was renewed. The data is grouped such that x represents the days before expiration of license and f represents the frequency.

x	0–29	30–59	60–89
f	65	25	10

   a) Estimate the sample mean $\bar{x}$ of the days before expiration that a renewal is made.
   b) Estimate the sample standard deviation s.

2. A random sample of medical malpractice claims that had to be paid last year showed the following amounts (in millions of dollars):

3.2	1.4	0.7	0.2	0.5	2.1
0.8	1.7	1.6	0.5	0.6	0.2

   a) Make a box-and-whisker plot for this data.
   b) Compute the sample mean and median. Which is larger?
   c) Compute the range and sample standard deviation.

3. When people purchase letter quality printers for their computers, some consideration must be given to the speed at which the printer operates. The speed is usually given in characters per second (cps), with 7 cps roughly equivalent to typing 60 words per minute. Faster cps's are desirable so that new drafts or revisions can be quickly produced. A survey involving 30 letter quality printers showed speeds of printing to be (in cps):

10	30	12	45	20	15	18	12	14	55
32	21	14	12	25	15	30	25	24	20
12	16	30	15	24	36	48	45	52	40

   a) Make a box-and-whisker plot of the data.
   b) Make a frequency table using three classes. Then estimate the mean and sample standard deviation.
   c) If you have a statistical calculator or computer, use it to find the actual sample mean and sample standard deviation.

4. Professor Cramer determines a final grade based on attendance, two papers, three major tests, and a final exam. Each of these activities has a total of 100 possible points. However, the activities carry different weights. Attendance is worth 5%, each paper is worth 8%, each test is worth 15%, and the final is worth 34%.
   a) What is the average for a student with 92 on attendance, 73 on the first paper, 81 on the second paper, 85 on test 1, 87 on test 2, 83 on test 3, and 90 on the final exam?
   b) Compute the average for a student with the above scores on the papers, tests, and final exam, but with a score of only 20 on attendance.

5. An elevator is loaded with 16 people and is at its load limit of 2,500 lb. What is the mean weight of these people?

6. A new brand of cigarettes called Gasp has just come on the market. A random sample of eight Gasp cigarettes shows the following nicotine contents for each cigarette (in milligrams):

   1.2   1.6   1.2   1.5   1.8   1.2   1.6   1.9

   a) Find the mean, median, and mode.
   b) Find the sample standard deviation and the range.

7. The following table shows the periods of time a sample of 66 Pitkin County prisoners had to wait in jail prior to the beginning of their trials.

Number of Months, x	0	1	2	3	4	5	6	7	8
Number of Prisoners, f	20	17	9	5	0	6	3	4	2

   a) Find the mean number of months waited.
   b) Find the sample standard deviation of the waiting times.

8. The Softex Plastic Company tested a random sample of 12 pieces of new plastic for breaking strength. They obtained the following measurements in units of 1,000 pounds per square inch:

   2.0   2.3   4.1   3.3   1.9   3.5   3.4   1.8   4.0   2.7
   3.9   2.6

   a) Make a box-and-whisker plot.
   b) What is the mean breaking strength?
   c) What is the median?
   d) What is the range?
   e) What is the sample standard deviation?

9. Radar was used to check speeds on a random sample of 20 cars in rush-hour traffic on an Atlanta expressway at 7:30 A.M. with the following results (in miles per hour):

   50   60   48   60   56   55   60   58   50   45
   55   40   45   66   50   60   55   38   55   60

   a) Make a box-and-whisker plot of the data.
   b) Find the mean, median, and mode.
   c) Comment on the statement: "On the average, cars do not exceed the speed limit (55) during the early morning rush hour." Which average is being referred to in this statement? For which averages is this statement *not* true?

10. The circumferences (distance around) of 94 blue spruce trees selected at random in Roosevelt National Forest were measured. The results to the nearest inch were grouped in Table 3-25. Using the midpoints of the tree circumference classes, find the mean and standard deviation.

TABLE 3-25 Circumferences of Blue Spruce Trees

Circumference (in.)	Midpt.	Number of Trees
10–24	17	6
25–39	32	20
40–54	47	52
55–69	62	16

11. The following set of numbers consists of the ages (rounded to the nearest year) of ten people selected at random who were convicted of armed robbery last year in Brooks County, Mississippi:

    19   24   26   23   19   27   46   52   27   27

    a) Find the mean, median, and mode.
    b) Find the range and sample standard deviation.

12. The number of tourists visiting Silver City, Colorado, each day during the first week of July last year were:

    310   418   512   452   365   570   432

    a) Find the mean and median.
    b) Is there a mode?
    c) Find the range.

13. a) Is it possible that the range and standard deviation can be equal? If your answer is yes, give an example of a data set where they are equal.
    b) Is it possible that the mean, median, and mode can all be equal? Is it possible they can all be different? If your answer is yes to either question, give an example to prove your answer.

14. A racing car manufacturer makes two models, the Zapper and the Zonker, which are alike in every respect except body design and styling. To determine if the body design of the Zapper has less wind resistance, both cars are tested at the Indianapolis Speedway. The following table gives lap times in minutes on the 2.5-mi track:

Zapper	1.0	0.9	1.0	0.8	0.9	1.0	0.9	1.0
Zonker	1.3	1.2	1.0	0.9	1.1	0.9	1.4	1.3

    a) Find the mean lap time $\bar{x}$ for both the Zapper and the Zonker.
    b) Find the sample standard deviation s for both.
    c) Which model had better average lap times? Was its performance consistently better? How is this reflected in the standard deviation of lap times?

15. A performance evaluation for new sales representatives at Office Automation Incorporated involves several ratings done on a scale of 1 to 10, with 10 the highest rating. The activities rated include: new contacts, successful contacts, total contacts, dollar volume of sales, and reports. Then an overall rating is determined by using a weighted average. The weights are 2 for new contacts, 3 for successful contacts, 3 for total contacts, 5 for dollar value of sales, and 3 for reports. What would the overall rating be for a sales representative with ratings of 5 for new contacts, 8 for successful contacts, 7 for total contacts, 9 for dollar volume of sales, and 7 for reports?

USING COMPUTERS

Professional statisticians in industry and research use computers to help them analyze and process statistical data. There are many computer programs available for statistics. Some commonly used statistical packages include Minitab, SAS, and SPSS. There are many others as well. Any of these statistical packages may be used with this text.

The authors of this text have also written a computer statistics package called ComputerStat. The programs in ComputerStat are available on disks with an accompanying manual that explains how to use the programs and gives additional computer applications. The disks are intended for the Apple IIe, IIc, or the IBM PC. Programs are written in the computer language BASIC and are designed for the beginning statistics student with no previous computer experience. ComputerStat is available as a supplement from the publisher of this text.

Problems in this section may be done using ComputerStat or other statistical computer programs.

1. Trade winds are one of the beautiful features of island life in Hawaii. The following data represent total air movement in miles each day over a weather station in Hawaii as determined by a continuous anemometer recorder. The period of observation is January 1 to February 15, 1971.

26	14	18	14	113	50	13	22
27	57	28	50	72	52	105	138
16	33	18	16	32	26	11	16
17	14	57	100	35	20	21	34
18	13	18	28	21	13	25	19
11	19	22	19	15	20		

Source: United States Department of Commerce, National Oceanic and Atmospheric Administration, Environmental Data Service. *Climatological Data, Annual Summary, Hawaii,* vol. 67, no. 13. Asheville: National Climatic Center, 1971. 11, 24.

In the ComputerStat menu of topics, select Descriptive Statistics. Then use program S3, Averages and Variation for Ungrouped Data, to solve the following problem.

a) Enter the data into the computer as sample data.

b) Use the computer to find the sample mean, median, and (if it exists) mode.

c) Use the computer to find the range, sample variance, and sample standard deviation.

d) As a topic in Exploratory Data Analysis (EDA) we studied the box-and-whisker plot. Use the five-number summary provided by the computer to make a box-and-whisker plot of total air movement over the weather station.

e) There are four exceptionally high data values: 113, 105, 138, and 100. The strong winds of January 5 (113 reading) brought in a cold front that dropped snow on Haleakala National Park (at the 8,000 ft elevation). The residents were so excited that they drove up to see the snow and caused such a massive traffic jam that the Park Service had to close the road. The winds of January 15 and 16 (readings 105 and 138) brought in a storm that created more damaging funnel clouds than any other storm to that date in the recorded history of Hawaii. The strong winds of January 28 (reading 100) accompanied a storm with funnel clouds that did over $100,000 in damage. Eliminate these values (i.e. 100, 105, 113, 138) from the data bank and redo parts (a), (b), (c), and (d). Compare your results with those previously obtained. Which average is most affected? What happens to the standard deviation? How do the two box-and-whisker plots compare?

2. Sometimes natural beauty takes a form that is very austere. The summit of Longs Peak in Colorado (elevation 14,256 ft) is very beautiful, and in winter very austere. In the following data x = hourly peak gusts in miles per hour of winter wind on the summit of Longs Peak and f = frequency of occurrence for a sample period of 1,518 observation hours.

In the ComputerStat menu of topics, select Descriptive Statistics. Then use program S4, Mean, Variance, and Standard Deviation for Grouped Data, to solve the following problem.

a) Using the class limits, compute each class midpoint. *Hint:* Compute the first class midpoint. To get the others, just keep adding 10 to the previous class midpoint.

b) Enter the class midpoints and class frequencies into the computer as sample data.

c) Use the computer to find the approximate sample mean, sample variance, and sample standard deviation of hourly peak gusts of winter wind on the summit of Longs Peak.

Hourly Peak Gust mph	Total Hours of Occurrence Number
0 – 9	141
10 – 19	217
20 – 29	376
30 – 39	290
40 – 49	195
50 – 59	143
60 – 69	61
70 – 79	37
80 – 89	21
90 – 99	11
100 – 109	12
110 – 119	3
120 – 129	5
130 – 139	3
140 – 149	3

(In addition to the values shown in the table, there were three incidences where the wind gusts exceeded 150 mph.)

Source: Glidden, D. E. *Winter Wind Studies in Rocky Mountain National Park*. Estes Park, Colo.: Rocky Mountain Nature Association, 1982, 23.

4 Elementary Probability Theory

We see that the theory of probabilities is at bottom only common sense reduced to calculation; it makes us appreciate with exactitude what reasonable minds feel by a sort of instinct, often without being able to account for it.

Pierre Simon Laplace

This is how the great mathematician Laplace described the theory of mathematical probability. The discovery of the mathematical theory of probability was shared by two Frenchmen: Blaise Pascal and Pierre Fermat. These seventeenth-century scholars were attracted to the subject by the inquiries of the Chevalier de Méré, a gentleman gambler.

Although the first applications of probability were to games of chance and gambling, today the subject seems to pervade almost every aspect of modern life. Everything from the orbits of spacecraft to the social behavior of woodchucks is described in terms of probabilities.

SECTION 4.1   What Is Probability?

We encounter statements in terms of probability all the time. An excited sports announcer claims that Sheila has a 90% chance of breaking the world record in the upcoming 100-yd dash. Henry figures that if he guesses on a true–false question, the probability of getting it right is 1/2. The Right to Health Lobby claims the probability is 0.40 of getting an erroneous report from a medical lab in one low-cost health center. They are consequently lobbying for a federal agency to license and monitor all medical laboratories.

When we use probability in a statement, we're using a *number between 0 and 1* to indicate the likelihood of an event. We'll use the notation $P(A)$ (read, P of A) to denote the probability of event A. The closer to 1 the probability assignment is, the more likely the event is to occur. If the event A is certain to occur, then $P(A)$ should be 1.

It is important to know what probability statements mean and how to compute or assign probabilities to events because probability is the language of inferential statistics. For instance, suppose a college counselor claims that 70% of first-year students receive counseling to help plan their schedules. Because of the high percentage of students needing help, he is requesting that an additional counselor be hired. You want to test the counselor's claim. In doing so you pick a random sample of 30 first-year students and find that only three of them got help from a counselor. Can you challenge the counselor's claim on the basis of this random sample in which only 10% of the students got counselor help with their schedules? To answer this question we need to find the *probability* of picking a random sample of first-year students in which only 10% got counselor help. This is the kind of question we will consider in hypothesis testing (Chapter 9).

In the meantime we need to learn how to find probabilities or assign them to events. There are three major methods we will use. One is *intuition*. The sports announcer probably used Sheila's performances in past track events and his own confidence in her running ability as a basis for his prediction that she has a 90% chance of breaking the world record. In other words, the announcer feels that the probability is 0.90 that Sheila will break the world record.

The Right to Health Lobby used another method to arrive at their probability statement. They took the *relative frequency* with which erroneous lab reports occurred. From a random sample of $n = 100$ they found $f = 40$ erroneous lab reports. From this they computed the relative frequency of erroneous lab reports via formula (1).

- **Probability formula for relative frequency:**

$$\text{probability of an event} = \text{relative frequency} = \frac{f}{n} \qquad (1)$$

where f is the frequency of an event,
$\quad n$ is the sample size

In the case of the lab reports we have

$$\text{relative frequency} = \frac{f}{n} = \frac{40}{100} = 0.40$$

The relative frequency of erroneous lab reports was used as the *probability* of erroneous reports.

The technique of using the relative frequency of an event as the probability of that event is a common way of assigning probabilities and will be used a great deal in later chapters. The underlying assumption we make is that if events occurred a certain percentage of times in the past, they will occur about the same percentage of times in the future.

Henry used the third method of assigning probabilities when he determined the probability of correctly guessing the answer to a true–false question. Essentially he used the probability formula for *equally likely outcomes*.

- **Probability formula when outcomes are equally likely:**

$$\text{probability of an event} = \frac{\text{number of outcomes favorable to event}}{\text{total number of outcomes}} \qquad (2)$$

In Henry's case there are a total of two outcomes. His answer will be either correct or wrong. Since he is guessing, we assume the outcomes are equally likely, and only one is "favorable" to being correct. So, by formula (2),

$$P(\text{correct answer}) = \frac{\text{no. of favorable outcomes}}{\text{total no. of outcomes}} = \frac{1}{2}$$

We've seen three ways to assign probabilities: intuition, relative frequency, and, when outcomes are equally likely, a formula. Which do we use? Most of the time it depends on the information that is at hand or that can be feasibly obtained. Our choice of methods also depends on the particular problem. In Exercise 1 you will see three different situations, and you will decide which way to assign the probabilities. *Remember, probabilities are numbers between 0 and 1, so don't assign probabilities outside this range.*

_____ Exercise 1 _____

Assign a probability to the indicated event on the basis of the information provided. Indicate the technique you use: intuition, relative frequency, or the formula for equally likely outcomes.

a) The director of the Readlot College Health Center wishes to open an eye clinic. To justify the expense of such a clinic, the director reports the probability that a student selected at random from the college roster needs corrective lenses. He took a random sample of 500 students to compute this probability and found that 375 of them needed corrective lenses. What is the probability that a Readlot College student selected at random needs corrective lenses?

a) In this case we are given a sample size of 500 and we are told 375 of these students need glasses. It is appropriate to use a relative frequency for the desired probability.

$$P(\text{student needs glasses}) = \frac{f}{n} = \frac{375}{500} = 0.75$$

b) The Friends of the Library host a fund-raising barbecue. George is on the cleanup committee. There are four members on this committee, and they draw lots to see who will clean the grills. Assuming that each member is equally likely to be drawn, what is the probability George will be assigned the grill cleaning job?

b) There are four people on the committee and each is equally likely to be drawn. It is appropriate to use the formula for equally likely events. George can be drawn in only one way, so there is only one outcome favorable to that event.

$$P(\text{George}) = \frac{\text{no. of favorable outcomes}}{\text{total no. of outcomes}} = \frac{1}{4}$$

c) Joanna photographs whales for Sea Life Adventure films. On her next expedition she is to film blue whales feeding. Her boss asks her what she thinks the probability of success will be for this particular assignment. She gives an answer based on her knowledge of the habits of blue whales and the region she is to visit. She is almost certain she will be successful. What specific number do you suppose she gave for the probability of success, and how do you suppose she arrived at it?

c) Since Joanna is almost certain of success, she should make the probability close to 1. We would say P(success) is above 0.90 but less than 1. We think the probability assignment was based on intuition.

No matter how we compute probabilities, it is useful to know what outcomes are possible in a given setting. For instance, if you are going to decide the probability that Hardscrabble will win the Kentucky Derby, you need to know which other horses will be running.

The set of all possible outcomes of an experiment is the *sample space*. If you toss a coin, there are only two possible outcomes: heads or tails. The sample space for that experiment consists of the two outcomes.

It is especially convenient to know the sample space in the case where all outcomes are equally likely since we can compute probabilities of various events by using formula (2).

$$P(\text{event } A) = \frac{\text{no. of outcomes favorable to } A}{\text{total no. of outcomes}} \qquad (2)$$

To use this formula we need to know the sample space so that we can determine which outcomes are favorable to the event in question as well as the total number of outcomes.

EXAMPLE 1

Human eye color is controlled by a single pair of genes (one from the father and one from the mother) called a genotype. Brown eye color, B, is dominant over blue eye color, ℓ. Therefore, in the genotype $B\ell$, consisting of one brown gene B and one blue gene ℓ, the brown gene dominates. A person with a $B\ell$ genotype has brown eyes.

If both parents are brown-eyed and have genotype $B\ell$, what is the probability that their child will have blue eyes? What is the probability the child will have brown eyes?

To answer these questions we need to look at the sample space of all possible eye color genotypes for the child. They are given in Table 4-1.

TABLE 4-1 Eye Color Genotypes for Child

Father \ Mother	B	ℓ
B	BB	Bℓ
ℓ	ℓB	ℓℓ

The four possible genotypes for the child are equally likely, so we can use formula (2) to compute probabilities. Blue eyes can occur only with the ℓℓ genotype, so there is only one outcome favorable to blue eyes. By formula (2)

$$P(\text{blue eyes}) = \frac{\text{no. of favorable outcomes}}{\text{total no. of outcomes}} = \frac{1}{4}$$

Brown eyes occur with the three remaining genotypes: BB, Bℓ, and ℓB. By formula (2)

$$P(\text{brown eyes}) = \frac{\text{no. of favorable outcomes}}{\text{total no. of outcomes}} = \frac{3}{4}$$

_____ Exercise 2 _____

Professor Gutierrez is making up a final for a course in literature of the Southwest. He wants the last three questions to be of the true–false type. In order to guarantee that the answers do not follow his favorite pattern, he lists all possible true–false combinations for three questions on slips of paper and then picks one at random from a hat.

a) Finish listing the outcomes in the given sample space.

TTT	FTT
TTF	FTF
TFT	____
TFF	____

a) The missing outcomes are FFT and FFF.

b) What is the probability that all three items will be false? Use the formula

$$P(\text{all F}) = \frac{\text{no. of favorable outcomes}}{\text{total no. of outcomes}}$$

b) There is only one outcome, FFF, favorable to all false, so

$$P(\text{all F}) = \frac{1}{8}$$

c) What is the probability that exactly two items will be true?

c) There are three outcomes that have exactly two true items: TTF, TFT, and FTT. Thus,

$$P(\text{two T}) = \frac{\text{no. of favorable outcomes}}{\text{total no. of outcomes}} = \frac{3}{8}$$

There is one more important point about probability assignments. The sum of all the probabilities assigned to outcomes in a sample space must be *one*. This makes sense. If you think the probability is 0.65 that you will win a tennis match, you assume the probability is 0.35 that your opponent will win. This fact is particularly useful, for if the probability that an event occurs is denoted by p and the probability that it *does not* occur is denoted by q, we have

$$p + q = 1 \qquad \text{since the sum of the probabilities}$$

of the outcomes must be 1 or

$$q = 1 - p \tag{3}$$

For an event A, the event *not A* is called the *complement of A*. To compute the probability of the complement of A, we use formula (3) and find

$$P(\text{not } A) = 1 - P(A)$$

EXAMPLE 2

The probability that a college student without a flu shot will get the flu is 0.45. What is the probability that a college student will *not* get the flu if the student has not had the flu shot?

In this case we have

$$P(\text{will get flu}) = p = 0.45$$
$$P(\text{will not get flu}) = q = 1 - p = 1 - 0.45 = 0.55$$

———————————— Exercise 3 ————————————

A veterinarian tells you that if you breed two cream-colored guinea pigs, the probability that an offspring will be pure white is 0.25. What is the probability that it will not be pure white?

a) $P(\text{pure white}) + P(\text{not white}) =$ _____   a) 1

b) $P(\text{not pure white}) =$ _____   b) $1 - 0.25$, or 0.75

The important facts about probabilities we have seen in this section are:

1. The probability of an event A is denoted by $P(A)$.
2. The probability of any event is a number between 0 and 1. The closer to 1 the probability is, the more likely the event is.
3. The sum of the probabilities of outcomes in a sample space is 1.
4. Probabilities can be assigned by three methods: intuition, relative frequencies, or the formula for equally likely outcomes.
5. The probability that an event occurs plus the probability that the same event does not occur is 1.

Section Problems 4.1

√1. In your own words carefully answer the question: What is probability? List three methods of assigning probabilities.

2. List examples of where probability might be applied in business, medicine, social science, and natural science. Why do you think probability will be useful in the study of statistics?

3. Which of the following numbers cannot be the probability of some event?

   a) 0.71   b) 4.1   c) 1/8   d) −0.5
   e) 0.5   f) 0   g) 1   h) 150%

4. a) Explain why −0.41 cannot be the probability of some event.
   b) Explain why 1.21 cannot be the probability of some event.
   c) Explain why 120% cannot be the probability of some event.
   d) Can the number 0.56 be the probability of an event? Explain.

5. Kris told Joe the probability of her car starting tomorrow morning is just about 0.
   a) What method did Kris use for assigning this probability to the event that her car will start: intuition, relative frequency, or the formula for equally likely outcomes?
   b) If Kris is correct and the probability of her car starting is close to 0, would you say the car is likely to start or not? What would you say if the probability were close to 1?

6. The Quick Turn Taxi Company wants to estimate the probability of a taxi being in a wreck during a one-month period. Last month 22 out of their fleet of 200 taxis were involved in a wreck. What would you estimate the probability to be?

7. In an effort to place people in jobs best suited to their personalities, an employment service gives its applicants a psychology test to measure self-confidence for salesmanship. The results of the test are broken down into only three outcomes: superior, average, inferior. Of 1,000 people who took the test, it was found that 115 were superior, 519 were average, and 366 were inferior.
   a) What is the sample space of test results?
   b) If a person is chosen at random, what would you say the probability is for that person to be superior? average? inferior?
   c) Do the probabilities of part (b) add up to 1? Why should they?

8. The Aim n' Shoot Camera Company wants to estimate the probability that one of their new cameras is defective. A random sample of 400 new cameras shows 24 are defective.
   a) How would you estimate the probability that a new Aim n' Shoot camera is defective? What is your estimate?
   b) What is your estimate for the probability that an Aim n' Shoot camera is *not* defective?
   c) Either a camera is defective or it is not. What is the sample space in this problem? Do the probabilities assigned to the sample space add up to 1? (*Hint:* See parts (a) and (b).)

9. The city council has three liberal members (one of whom is the mayor) and two conservative members. One member is selected at random to testify in Washington, D.C.
   a) What is the probability this member is liberal?
   b) What is the probability this member is conservative?
   c) What is the probability the mayor is chosen?

10. A botanist has developed a new hybrid cotton plant that can withstand insects better than other cotton plants. However, there is some concern about the germination of seeds from the new plant. To estimate the probability that a seed from the new plant will germinate, a random sample of 3,000 seeds were planted in warm, moist soil. Of these seeds, 2,430 germinated.
    a) How would you estimate the probability that a seed will germinate? What is your estimate?
    b) How would you estimate the probability that a seed will *not* germinate? What is your estimate?
    c) Either a seed germinates, or it does not. What is the sample space in this problem? Do the probabilities assigned to the sample space add up to 1? Should they add up to 1? Explain.
    d) Are the outcomes in the sample space of part (c) equally likely?

11. a) If you roll a single die and count the number of dots on top, what is the sample space of all possible outcomes? Are the outcomes equally likely?
    b) Assign probabilities to the outcomes of the sample space of part (a). Do the probabilities add up to 1? Should they add up to 1? Explain.
    c) What is the probability of getting a number less than 5 on a single throw?
    d) What is the probability of getting 5 or 6 on a single throw?

12. John runs a computer software store. Yesterday he counted 127 people who walked by his store, 58 of whom came into the store. Of the 58, only 25 bought something in the store.
    a) Estimate the probability that a person who walks by the store will enter the store.
    b) Estimate the probability that a person who walks into the store will buy something.
    c) Estimate the probability that a person who walks by the store will come in *and* buy something.
    d) Estimate the probability that a person who comes into the store will buy nothing.
    e) Estimate the probability that a person who walks by the store will buy nothing.

SECTION 4.2   Some Probability Rules

You roll two dice. What is the probability you will get a five on each die? You draw two cards from a well-shuffled, standard deck without replacing the first card before drawing the second. What is the probability they will both be aces?

It seems that these two problems are nearly alike. They are alike in the sense that in each case you are to find the probability of two events occurring *together*. In the first problem you are to find

$$P(5 \text{ on 1st die } and \text{ 5 on 2nd die})$$

In the second you want

$$P(\text{ace on 1st card } and \text{ ace on 2nd card})$$

The two problems differ in one important aspect, however. In the dice problem the outcome of a five on the first die does not have any effect on the probability of getting a five on the second die. Because of this, the events are *independent*. In general, two events are independent if the occurrence or nonoccurrence of one does not change the probability that the other will occur.

In the card problem the probability of an ace on the first card is 4/52 since there are 52 cards in the deck and four of them are aces. If you get an ace on the first card, then the probability of an ace on the second is changed to 3/51 since one ace has already been drawn and only 51 cards remain in the deck. Therefore, the two events in the card-draw problem are *not* independent. They are, in fact, *dependent* since the outcome of the first draw changes the probability of getting an ace on the second card.

Why does the *independence or dependence* of two events matter? The type of events determines the way we compute the probability of the two events happening together. If two events A and B are *independent,* then we use formula (4) to compute the probability of the event A *and* B.

$$P(A \text{ and } B) = P(A) \cdot P(B) \quad \text{for independent events} \tag{4}$$

If the events are *dependent,* then we must take into account the change in the probability of one event caused by the occurrence of the other event. We use either formula (5) or (6) to compute the probability of A *and* B when A and B are dependent.

$$P(A \text{ and } B) = P(A) \cdot P(B, \textit{ given} \text{ that } A \text{ has occurred}) \left.\begin{array}{l} \\ \\ \end{array}\right\} \quad \begin{array}{l} \text{for dependents} \\ \text{events} \end{array} \quad \begin{array}{l} (5) \\ (6) \end{array}$$
$$P(A \text{ and } B) = P(B) \cdot P(A, \textit{ given} \text{ that } B \text{ has occurred})$$

We will use either formula (5) or (6) according to the information available.

Formulas (4), (5), and (6) constitute the *multiplication rules* of probability. They help us compute the probability of events happening together when the sample space is too large for convenient reference or when it is not completely known.

Let's use the multiplication rules to complete the dice and card problems. We'll compare the results with those obtained by using the sample space directly.

EXAMPLE 3

Suppose you are going to throw two fair dice. What is the probability of getting a five on each die?

Solution Using the Multiplication Rule: The two events are independent, so we should use formula (4). $P(5 \text{ on 1st die } and \text{ 5 on 2nd die}) = P(5 \text{ on 1st}) \cdot P(5 \text{ on 2nd})$. To finish the problem, we need only compute the probability of getting a five when we throw one die.

There are six faces on a die, and on a fair die each is equally likely to come up when you throw the die. Only one face has five dots, so by formula (2) for equally likely outcomes,

$$P(5 \text{ on die}) = \frac{1}{6}$$

Now we can complete the calculation.

$$P(5 \text{ on 1st die } and \text{ 5 on 2nd die}) = P(5 \text{ on 1st}) \cdot P(5 \text{ on 2nd})$$

$$= \frac{1}{6} \cdot \frac{1}{6}$$

$$= \frac{1}{36}$$

Solution Using Sample Space: The first task is to write down the sample space. Each die has six equally likely outcomes, and each outcome of the second die can be paired with each of the first. The sample space is shown in Figure 4-1. The total number of outcomes is 36, and only one is favorable to a five on the first die *and* a five on the second. The 36 outcomes are equally likely, so by formula (2) for equally likely outcomes,

$$P(5 \text{ on 1st } and \text{ 5 on 2nd}) = \frac{1}{36}$$

FIGURE 4-1 Sample Space for Two Dice

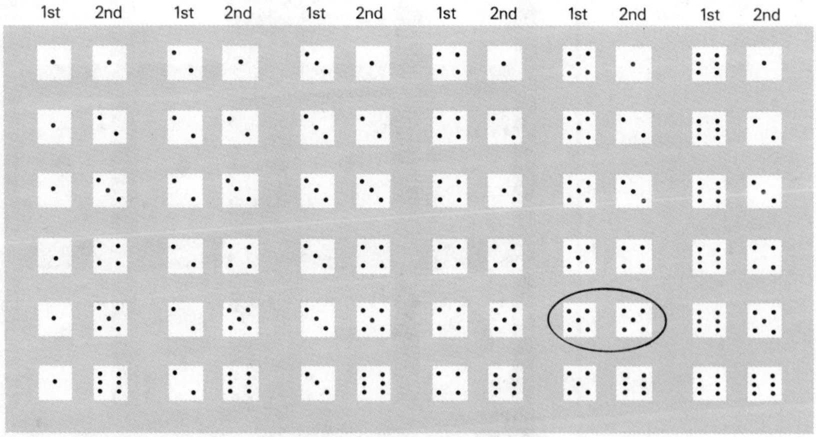

The two methods yield the same result. The multiplication rule was easier to use because we did not need to look at all 36 outcomes in the sample space for tossing two dice.

EXAMPLE 4

Compute the probability of drawing two aces from a well-shuffled deck of 52 cards if the first card is not replaced before the second card is drawn.

Multiplication Rule Method: These events are *dependent*. The probability of an ace on the first card is 4/52, but on the second card the probability of an ace is only 3/51 if an ace was drawn for the first card. An ace on the first draw changes the probability for an ace on the second draw. By the multiplication rule for dependent events we have

$$P(\text{ace on 1st } and \text{ ace on 2nd}) = P(\text{ace on 1st}) \cdot P(\text{ace on 2nd,}$$
$$given \text{ ace on first})$$

$$= \frac{4}{52} \cdot \frac{3}{51} = \frac{12}{2,652} = 0.0045$$

Sample Space Method: We won't actually look at the sample space, because each of the 51 possible outcomes for the second card must be paired with each of the 52 possible outcomes for the first card. This gives us a total of 2,652 outcomes in the sample space! We'll just think about the sample space and try to list all the outcomes favorable to the event of aces on both cards. The 12

FIGURE 4-2 Outcomes Favorable to Drawing Two Aces

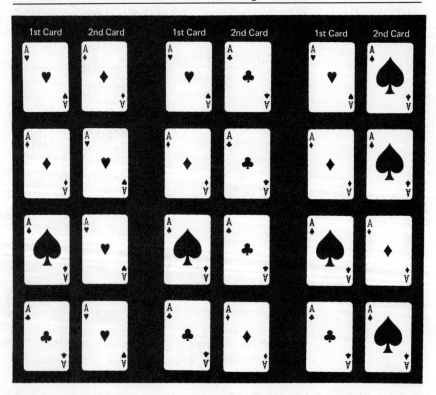

favorable outcomes are shown in Figure 4-2. By the formula for equally likely outcomes,

$$P(\text{ace on 1st card } and \text{ ace on 2nd card}) = \frac{12}{2,652} = 0.0045$$

Again, the two methods agree.

The multiplication rules apply whenever we wish to determine the probability of two events happening *together.* To indicate together, we use *and* between the events. But before you use a multiplication rule to compute the probability of *A and B*, you must determine if *A* and *B* are independent or dependent events. Let's practice using the multiplication rule.

────────────── Exercise 4 ──────────────

A quality control procedure for testing Ready-Flash flash bulbs consists of drawing two bulbs at random from each lot of 100 without replacing the first bulb before drawing the second. If both are defective, the entire lot is rejected. Find the probability that both bulbs are defective if the lot contains 10 defective flash bulbs. Since we are drawing the bulbs at random, assume each bulb in the lot has an equal chance of being drawn.

a) What is the probability of getting a defective bulb on the first draw?

a) The sample space consists of all 100 bulbs. Since each is equally likely to be drawn and there are 10 defective ones,

$$P(\text{defective bulb}) = \frac{10}{100} = \frac{1}{10}$$

b) The first bulb drawn is not replaced so there are only 99 bulbs for the second draw. What is the probability of getting a defective bulb on the second draw if the first bulb was defective?

b) If the first bulb is defective, then there are only nine defective bulbs left among the 99 remaining bulbs in the lot.

$P(\text{defective bulb on 2nd draw, } given \text{ defective}$

$$\text{bulb on 1st}) = \frac{9}{99} = \frac{1}{11}$$

c) Are the probabilities computed in parts (a) and (b) different? Does drawing a defective bulb on the first draw change the probability of getting a defective bulb on the second draw? Are the events dependent?

c) The answer to all these questions is yes.

d) Use the formula for dependent events,

$$P(A \text{ and } B) = P(A) \cdot P(B, \text{ given } A \text{ has occurred})$$
to compute *P*(1st bulb defective *and* 2nd bulb defective)

d) $P(\text{1st defective } and \text{ 2nd defective}) = \dfrac{1}{10} \cdot \dfrac{1}{11}$

$$= \frac{1}{110}$$

$$= 0.009$$

_____ Exercise 5 _____

Andrew is 55 and the probability that he will be alive in ten years is 0.72. Ellen is 35 and the probability that she will be alive in ten years is 0.92. Assuming that the life span of one will have no effect on the life span of the other, what is the probability they will both be alive in ten years?

a) Are these events dependent or independent?

a) Since the life span of one does not affect the life span of the other, the events are independent.

b) Use the appropriate multiplication rule to find

P(Andrew alive in 10 yr *and* Ellen alive in 10 yr)

b) We use the rule for independent events:

$$P(A \text{ and } B) = P(A) \cdot P(B)$$

P(Andrew alive *and* Ellen alive)

$$= P(\text{Andrew alive}) \cdot P(\text{Ellen alive})$$

$$= (0.72)(0.92) = 0.66$$

One of the multiplication rules can be used any time we are trying to find the probability of two events happening *together*. Pictorially, we are looking for the probability of the shaded region in Figure 4-3.

FIGURE 4-3   The Event *A* and *B*

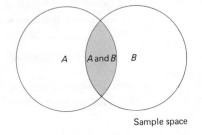

Sample space

FIGURE 4-4   The Event *A* or *B*

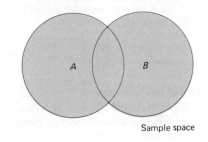

Sample space

Another way to combine events is to consider the possibility of one event *or* another occurring. For instance, if a sports car saleswoman gets an extra bonus if she sells a convertible or a car with leather upholstery, she is interested in the probability that you will buy a car that is a convertible *or* has leather upholstery. Of course, if you bought a convertible with leather upholstery, that would be fine, too. Pictorially, the shaded portion of Figure 4-4 represents the outcomes satisfying the *or* condition. Notice that the condition *A or B* is satisfied by any one of the following conditions:

1. Any outcome of *A* occurs.
2. Any outcome of *B* occurs.
3. Any outcome in *A* and *B* occurs.

It is important to distinguish between the *or* combinations and the *and* combinations because we apply different rules to compute the probabilities of the combinations.

_____ Exercise 6 _____

Indicate how each of the following pairs of events are combined. Use either the *and* combination or the *or* combination.

a) Satisfying the humanities requirement by taking a course in the history of Japan or by taking a course in classical literature

    a) Use the *or* combination.

b) Buying new tires and aligning the tires

    b) Use the *and* combination.

c) Getting an A not only in psychology, but also in biology

    c) Use the *and* combination.

d) Having at least one of these pets: cat, dog, bird, rabbit

    d) Use the *or* combination.

Once you decide that you are to find the probability of an *or* combination rather than an *and* combination, what formula do you use? Again it depends on the situation. If you want to compute the probability of drawing either a jack or a king on a single draw from a well-shuffled deck of cards, the formula is simple:

$$P(\text{jack } or \text{ king}) = P(\text{jack}) + P(\text{king}) = \frac{4}{52} + \frac{4}{52} = \frac{8}{52} = \frac{2}{13}$$

since there are four jacks and four kings in a deck of 52 cards.

If you want to compute the probability of drawing a king or a diamond on a single draw, the formula is a bit more complicated. We have to take the overlap of the two events into account so that we do not count the outcomes twice. We can see the overlap of the two events in Figure 4-5.

$$P(\text{king}) = \frac{4}{52} \qquad P(\text{diamond}) = \frac{13}{52} \qquad P(\text{king } and \text{ diamond}) = \frac{1}{52}$$

If we simply add $P(\text{king})$ and $P(\text{diamond})$, we're including $P(\text{king } and \text{ diamond})$ twice in the sum. To compensate for this double summing, we simply subtract $P(\text{king } and \text{ diamond})$ from the sum. So

$$P(\text{king } or \text{ diamond}) = P(\text{king}) + P(\text{diamond}) - P(\text{king } and \text{ diamond})$$

$$= \frac{4}{52} + \frac{13}{52} - \frac{1}{52}$$

$$= \frac{16}{52} = \frac{4}{13}$$

We say the events A and B are *mutually exclusive* if they cannot occur together. This means A and B have no outcomes in common or, put another way, $P(A \text{ and } B) = 0$. Formula (7) is the addition rule for *mutually exclusive* events A and B.

FIGURE 4-5 Drawing a King or a Diamond from a Standard Deck

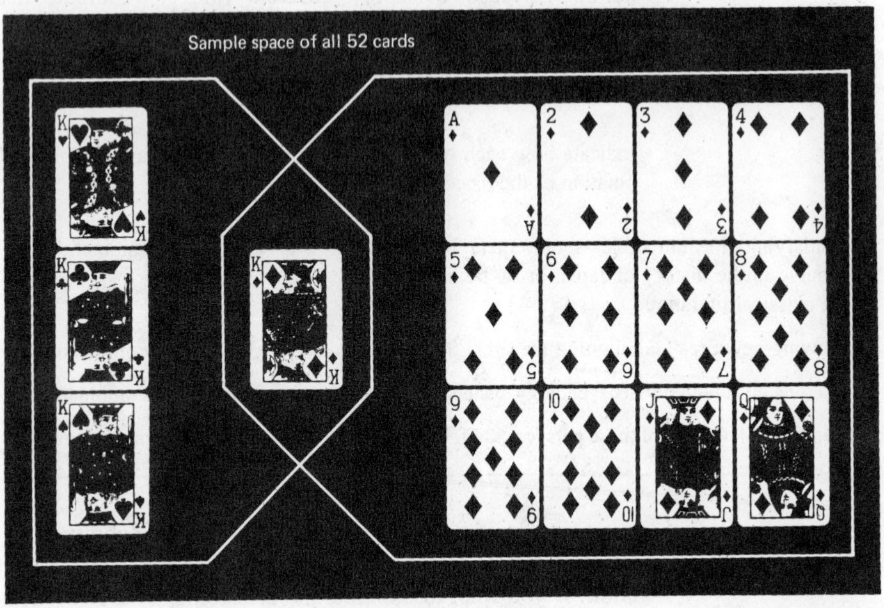

$$P(A \, or \, B) = P(A) + P(B) \text{ when } A \text{ and } B \text{ are mutually exclusive} \qquad (7)$$

If the events are not mutually exclusive, we must use the more general formula (8), which is the general addition rule for any events A and B.

$$P(A \, or \, B) = P(A) + P(B) - P(A \, and \, B) \qquad (8)$$

You may ask: Which formula should we use? The answer is: Use formula (7) only if you know that A and B are mutually exclusive (i.e., cannot occur together); if you do not know whether A and B are mutually exclusive or not, then use formula (8). Formula (8) is valid either way. Notice that when A and B are mutually exclusive, then $P(A \, and \, B) = 0$, so formula (8) reduces to formula (7).

_____ Exercise 7 _____

The Cost Less Clothing Store carries seconds in slacks. If you buy a pair of slacks in your regular waist size without trying them on, the probability the waist will be too tight is 0.30 and the probability it will be too loose is 0.10.

a) Are the events "too tight" and "too loose" mutually exclusive?

a) The waist cannot be both too tight and too loose at the same time, so the events are mutually exclusive.

b) If you choose a pair of slacks at random in your regular waist size, what is the probability that the waist will be too tight *or* too loose?

b) Since the events are mutually exclusive,

$$P(\text{too tight } or \text{ too loose})$$
$$= P(\text{too tight}) + P(\text{too loose})$$
$$= 0.30 + 0.10$$
$$= 0.40$$

_____ Exercise 8 _____

Professor Jackson is in charge of a program to prepare people for a high school equivalency exam. Records show that 80% of the students need work in math, 70% need work in English, and 55% need work in both areas.

a) Are the events "need math" and "need English" mutually exclusive?

a) These events are not mutually exclusive since some students need both. In fact,

$$P(\text{need math } and \text{ need English}) = 0.55$$

b) Use the appropriate formula to compute the probability that a student selected at random needs math *or* needs English.

b) Since the events are not mutually exclusive, we use formula (8).

$$P(\text{need math } or \text{ need English})$$
$$= P(\text{need math}) + P(\text{need English})$$
$$- P(\text{need math } and \text{ English})$$
$$= 0.80 + 0.70 - 0.55$$
$$= 0.95$$

The addition rule for mutually exclusive events can be extended so it applies to the situations in which we have more than two events that are each mutually exclusive to all the other events.

EXAMPLE 5

Laura is playing Monopoly. On her next move she needs to throw a number bigger than 8 on the two dice in order to land on her own property and pass Go. What is the probability Laura will roll a number bigger than 8?

Solution: When two dice are thrown the largest value that can come up is 12. Consequently, the only values larger than 8 are 9, 10, 11, and 12. These outcomes are mutually exclusive since only one of these values can possibly occur on one throw of the dice. The probability of throwing more than 8 is the same as

$$P(9 \text{ or } 10 \text{ or } 11 \text{ or } 12)$$

Since the events are mutually exclusive,

$$P(9 \text{ or } 10 \text{ or } 11 \text{ or } 12) = P(9) + P(10) + P(11) + P(12)$$

$$= \frac{4}{36} + \frac{3}{36} + \frac{2}{36} + \frac{1}{36}$$

$$= \frac{10}{36} = \frac{5}{18}$$

To get the specific values of $P(9)$, $P(10)$, $P(11)$, and $P(12)$, we used the sample space for throwing two dice (*see* Figure 4-1). There are 36 equally likely outcomes—for example, those favorable to 9 are 6,3; 3,6; 5,4; and 4,5. So $P(9) = 4/36$. The other values can be computed in a similar way.

The multiplication rule for independent events also extends to more than two independent events. If you toss a fair coin three times, the three events are independent. To compute the probability of getting three heads in the three tosses, we use the extended multiplication rule for independent events and the fact that $P(\text{head}) = 1/2$ for a fair coin.

$$P(\text{head } and \text{ head } and \text{ head}) = P(\text{head}) \cdot P(\text{head}) \cdot P(\text{head})$$

$$= \frac{1}{2} \cdot \frac{1}{2} \cdot \frac{1}{2}$$

$$= \frac{1}{8}$$

In the next example and exercise we look at a sample space consisting of several events and compute various probabilities by using the sample space and the rules of probability.

EXAMPLE 6

At Hopewell Electronics all 140 employees were asked about their political affiliation. The employees were grouped by type of work, as executives or production workers. The results with row and column totals are shown in Table 4-2.

TABLE 4-2 Employee Type and Political Affiliation

Type of Employee	Political Affiliation			
	Democrat	Republican	Independent	Row Total
Executive	5	34	9	48
Production Worker	63	21	8	92
Column Total	68	55	17	140 Grand Total

Suppose an employee is selected at random from the Hopewell employees. Let us use the following notation to represent different events: E = executive; PW = production worker; D = Democrat; R = Republican; I = Independent.

a) Compute $P(D)$ and $P(E)$.

To find these probabilities, we look at the entire sample space.

$$P(D) = \frac{\text{no. of Democrats}}{\text{no. of employees}} = \frac{68}{140} = 0.486$$

$$P(E) = \frac{\text{no. of executives}}{\text{no. of employees}} = \frac{48}{140} = 0.343$$

b) Compute $P(D, \text{ given } E)$.

For the conditional probability we restrict our attention to the portion of the sample space satisfying the condition of being an executive.

$$P(D, \text{given } E) = \frac{\text{no. of executives who are Democrats}}{\text{no. of executives}} = \frac{5}{48} = 0.104$$

Notice that $P(D)$ and $P(D, \text{ given } E)$ are not equal. The events Democrat and executive are *not* independent.

c) Compute $P(D \text{ and } E)$.

This probability is not conditional, so we must look at the entire sample space.

$$P(D \text{ and } E) = \frac{\text{no. of executives who are Democrats}}{\text{total no. of employees}} = \frac{5}{140} = 0.036$$

Let's recompute this probability using the rules of probability for dependent events.

$$P(D \text{ and } E) = P(E) \cdot P(D, \text{given } E) = \frac{48}{140} \cdot \frac{5}{48} = \frac{5}{140} = 0.036$$

The results using the rules are consistent with those using the sample space.

d) Compute $P(D \text{ or } E)$.

From part (c) we know that the events Democrat or executive are not mutually exclusive, since $P(D \text{ and } E) \neq 0$. Therefore

$$P(D \text{ or } E) = P(D) + P(E) - P(D \text{ and } E)$$

$$= \frac{68}{140} + \frac{48}{140} - \frac{5}{140} = \frac{111}{140} = 0.793$$

_____ Exercise 9 _____

Using Table 4-2 let's consider other probabilities regarding the type of employees at Hopewell and their political affiliation. This time let's consider the production worker and the affiliation of Independent. Suppose an employee is selected at random.

a) Compute $P(I)$ and $P(PW)$.

a) $P(I) = \dfrac{\text{no. of independents}}{\text{total no. of employees}}$

$\quad = \dfrac{17}{140} = 0.121$

$P(PW) = \dfrac{\text{no. of production workers}}{\text{total no. of employees}}$

$\quad = \dfrac{92}{140} = 0.657$

b) Compute $P(I, \text{ given } PW)$. This is a conditional probability. Be sure to restrict your attention to production workers since that is the condition given.

b) $P(I, \text{ given } PW) = \dfrac{\text{no. of independent production workers}}{\text{no. of production workers}}$

$\quad = \dfrac{8}{92} = 0.087$

c) Compute $P(I \text{ and } PW)$. In this case look at the entire sample space and at the number of employees who are both Independent and in production.

c) $P(I \text{ and } PW) = \dfrac{\text{no. of independent production workers}}{\text{total no. of employees}}$

$\quad = \dfrac{8}{140} = 0.057$

d) Use the multiplication rule for dependent events to calculate $P(I \text{ and } PW)$. Is the result the same as that of part (c)?

d) By the multiplication rule

$$P(I \text{ and } PW) = P(PW) \cdot P(I, \text{ given } PW)$$

$$= \dfrac{92}{140} \cdot \dfrac{8}{92} = \dfrac{8}{140} = 0.057$$

The results are the same.

e) Compute $P(I \text{ or } PW)$. Are the events mutually exclusive?

e) Since the events are not mutually exclusive,

$$P(I \text{ or } PW) = P(I) + P(PW) - P(I \text{ and } PW)$$

$$= \dfrac{17}{140} + \dfrac{92}{140} - \dfrac{8}{140}$$

$$= \dfrac{101}{140} = 0.721$$

Section Problems 4.2

In problems 1–4 use the appropriate addition and multiplication rules. Check to make sure that your answers are consistent with those obtained from using the sample space of Figure 4-1.

1. You roll two fair dice: a green one and a red one.
   a) Find $P(5 \text{ on green die } and \text{ 3 on red die})$.
   b) Find $P(3 \text{ on green die } and \text{ 5 on red die})$.
   c) Find $P((5 \text{ on green die } and \text{ 3 on red die}) \text{ or } (3 \text{ on green die } and \text{ 5 on red die}))$.

2. You roll two fair dice: a green one and a red one.
   a) Find P(1 on green die *and* 2 on red die).
   b) Find P(2 on green die *and* 1 on red die).
   c) Find P((1 on green die *and* 2 on red die) *or* (2 on green die *and* 1 on red die)).

3. Two fair dice are rolled: a green one and a red one.
   a) What is the probability of getting a sum of 6?
   b) What is the probability of getting a sum of 4?
   c) What is the probability of getting a sum of 6 *or* 4?

4. Two fair dice are rolled: a green one and a red one.
   a) What is the probability of getting a sum of 7?
   b) What is the probability of getting a sum of 11?
   c) What is the probability of getting a sum of 7 *or* 11?

5. You draw two cards from a standard deck of 52 without replacing the first one before drawing the second. (A standard deck contains four suits: hearts, diamonds, clubs, and spades. In each suit there are 13 cards: those numbered 2 through 10, one jack, one queen, one king, and one ace.)
   a) Find P(ace on 1st card *and* king on 2nd).
   b) Find P(king on 1st card *and* ace on 2nd card).
   c) Find the probability of drawing an ace and a king in either order.

6. You draw two cards from a standard deck of 52 without replacing the first one before drawing the second.
   a) Find P(3 on 1st card *and* 10 on 2nd).
   b) Find P(10 on 1st card *and* 3 on 2nd).
   c) Find the probability of drawing a 10 and a 3 in either order.

7. You draw two cards from a standard deck of 52, but before you draw the second card, you put the first one back and reshuffle the deck.
   a) Find P(ace on 1st card *and* king on 2nd).
   b) Find P(king on 1st card *and* ace on 2nd).
   c) Find the probability of drawing an ace and a king in either order.

8. You draw two cards from a standard deck of 52, but before you draw the second card, you put the first one back and reshuffle the deck.
   a) Find P(3 on 1st card *and* 10 on 2nd).
   b) Find P(10 on 1st card *and* 3 on 2nd).
   c) Find the probability of drawing a 10 and a 3 in either order.

9. You draw one card from a standard deck of 52.
   a) What is the probability that the card is the ace of spades?
   b) What is the probability that the card is a spade?
   c) What is the probability that the card is an ace or a spade?

10. You draw one card from a standard deck of 52.
    a) What is the probability that the card is a diamond?
    b) What is the probability that the card is the queen of diamonds?
    c) What is the probability that the card is a queen *or* a diamond?

11. June wants to become a policewoman. She must take the physical exam and then the written one. Records of past cadets indicate that the probability of passing the physical exam is 0.82. Then the probability that a cadet passes the written exam given he or she has passed the physical exam is 0.58. What is the probability that June passes both exams?

12. The dean of women at Brookfield College found that 20% of the female students are majoring in engineering. If 53% of the students at Brookfield are women, what is the probability that a student chosen at random will be a woman engineering major?

13. A new grading policy has been proposed by the dean of the College of Education for all education majors. All faculty and students in education were asked to give their opinion about the new grading policy. The results are shown below.

Opinion

	Favor	Neutral	Oppose	Row Total
Students	353	75	191	619
Faculty	11	5	18	34
Column Total	364	80	209	653

Suppose someone is selected at random from the School of Education (either student or faculty). Let us use the following notation for events: S = student, F = faculty, Fa = favor, N = neutral, O = oppose. In this notation, $P(O,\ given\ S)$ represents the probability that the person selected opposes the new grading policy, given that the person is a student.

a) Compute $P(Fa)$; $P(Fa,\ given\ F)$; $P(Fa,\ given\ S)$.
b) Compute $P(F\ and\ Fa)$.
c) Are the events F = faculty and Fa = favor policy independent? Explain.
d) Compute $P(S)$; $P(S,\ given\ O)$; $P(S,\ given\ Fa)$; $P(S,\ given\ N)$.
e) Compute $P(S\ and\ Fa)$; and $P(S\ and\ O)$.
f) Are the events S = student and Fa = favor policy independent? Explain.
g) Compute $P(Fa\ or\ O)$. Are these events mutually exclusive?

14. In a sales effectiveness seminar, a group of sales representatives tried two approaches to selling a customer a new automobile: the aggressive approach and the passive approach. From 1,160 customers the following record was kept:

Sales Result

Approach	Sale	No Sale	Row Total
Aggressive	270	310	580
Passive	416	164	580
Column Total	686	474	1,160

Suppose a customer is selected at random from the 1,160 participating customers. Let us use the following notation for events: A = aggressive approach, Pa = passive approach, S = sale, N = no sale. So $P(A)$ is the probability that an aggressive approach was used, and so on.

a) Compute $P(S)$; $P(S,\ given\ A)$; $P(S,\ given\ Pa)$.
b) Are the events N = no sale and A = aggressive approach independent? Explain.
c) Compute $P(N)$; $P(N,\ given\ A)$; $P(S,\ given\ A)$.

d) Are the events S = sale and Pa = passive approach independent? Explain.

e) Compute $P(A \text{ and } S)$; $P(Pa \text{ and } S)$.

f) Compute $P(A \text{ or } S)$.

15. The table below shows how 558 people applying for a credit card were classified according to home ownership and length of time in present job.

Length of Time in Present Job

Home Status	Less than 2 Years	2 or More Years	Row Total
Owner	73	194	267
Renter	210	81	291
Column Total	283	275	558

Suppose a person is chosen at random from the 558 applicants. Let O = event this person owns home; R = event this person rents; L = event this person has had present job less than two years; and M = event this person has had present job two years or more.

a) Compute $P(O)$; $P(O, \text{ given } L)$; $P(O, \text{ given } M)$.

b) Compute $P(O \text{ and } M)$; $P(O \text{ or } M)$.

c) Compute $P(R)$; $P(R, \text{ given } L)$; $P(R, \text{ given } M)$.

d) Compute $P(R \text{ and } L)$; $P(R \text{ or } L)$.

e) Are the events O = own home and M = present job two or more years independent? Explain. Are the events mutually exclusive? Explain.

16. In a small rural community in West Virginia those adults seeking work or having work were classified as follows:

Employment Status

	Unemployed	Employed	Row Total
Men	206	412	618
Women	386	305	691
Column Total	592	717	1,309

Suppose a person is chosen at random from those seeking work or having work. Let M = event this person is a man; W = event this person is a woman; U = event this person is unemployed; and E = event this person is employed.

a) Compute $P(U)$; $P(U, \text{ given } M)$; $P(U, \text{ given } W)$.

b) Compute $P(E)$; $P(E, \text{ given } M)$; $P(E, \text{ given } W)$.

c) Compute $P(M)$; $P(M, \text{ given } E)$; $P(M, \text{ given } U)$.

d) Compute $P(W)$; $P(W, \text{ given } E)$; $P(W, \text{ given } U)$.

e) Compute $P(M \text{ and } E)$; $P(W \text{ or } U)$.

f) Are the events W = woman and U = unemployed mutually exclusive? Explain.

g) Are the events M = man and E = employed independent? Explain.

17. Brookridge College did a survey of former students who were incoming freshmen eight years ago. Of these former students, some graduated from college and some dropped out and did not graduate. Using national averages, the college established a salary cutoff point for upper middle income. From the surveys they were able to determine present salaries of the former students participating in the survey. The classification for those holding jobs today is shown below.

Salary Level

Degree Status	Below Upper Middle Income	At or Above Upper Middle Income	Row Total
College Graduate	105	406	511
College Dropout	351	291	642
Column Total	456	697	1,153

Suppose a former student is drawn at random from the people classified above. Let G = event this person graduated; D = event this person dropped out of college; B = event salary is below upper middle income; and A = event salary is at or above upper middle income.

a) Compute $P(A)$; $P(A, \text{ given } G)$; $P(A, \text{ given } D)$.
b) Compute $P(B)$; $P(B, \text{ given } G)$; $P(B, \text{ given } D)$.
c) Compute $P(G)$; $P(G, \text{ given } A)$; $P(G, \text{ given } B)$.
d) Compute $P(D)$; $P(D, \text{ given } A)$; $P(D, \text{ given } B)$.
e) Compute $P(G \text{ and } A)$; and $P(D \text{ or } B)$.
f) Are the events A = above upper middle income salary and G = graduated from college independent? Explain.
g) Are the events B = below upper middle income salary and D = dropped out of college mutually exclusive? Explain.

18. Airlines have many different fares and service categories. Records of the last several years show that 12% of the tickets Columbia Airways sells are first class, 28% are excursion fares, 35% are executive commuter fares, and 25% are coach. If a person boarding a Columbia flight is selected at random, what is the probability that the person holds a first class *or* executive commuter ticket? What is the probability that the person has a coach *or* excursion fare ticket?

19. Wing Foot is a shoe franchise commonly found in shopping centers across the United States. Wing Foot knows that their stores will not show a profit unless they gross over $93,000 per year. Let A be the event that a new Wing Foot store grosses over $93,000 its first year. Let B be the event that a store grosses over $93,000 its second year. Wing Foot has an administrative policy of closing a new store if it does not show a profit in *either* of the first two years. The accounting office at Wing Foot provided the following information: 65% of *all* Wing Foot stores show a profit the first year; 71% of *all* Wing Foot stores show a profit the second year (this includes stores that did not show a profit in the first year); however, 87% of Wing Foot stores that showed a profit the first year also showed a profit the second year. Compute the following:

a) $P(A)$
b) $P(B)$

c) *P*(*B, given A*)

d) *P*(*A and B*)

e) *P*(*A or B*)

f) What is the probability that a new Wing Foot store will not be closed after two years? What is the probability that a new Wing Foot store will be closed after two years?

20. Cut-Rate Sam's Appliance Store has received a shipment of 12 toasters. Unbeknownst to Sam, eight are defective.

a) Pete is the first person to buy one of these toasters. What is the probability it is defective?

b) Iris comes in after Pete and buys a toaster. What is the probability that her toaster is defective if Pete's was defective? If Pete's was not defective?

c) What is the probability that both Pete's and Iris's toasters are defective?

21. AnyState Insurance Company issues new contracts to their auto insurance customers every 36 months. The AnyState accounting office found that 8% of all their new contract customers made an auto accident claim during the first contract. During the second contract, about 8% of the customers again made auto accident claims. However, 23% of the customers who made accident claims during the first contract also made accident claims during the second contract.

Let *A* be the event that a customer makes an auto accident claim during the first contract, and let *B* be the event that a customer makes a claim during the second contract. Compute the following:

a) *P*(*A*)

b) *P*(*B*)

c) *P*(*B, given A*)

d) *P*(*A and B*)

e) *P*(*A or B*)

f) What is the probability that a customer makes it through both contract periods without making an auto accident claim?

22. The Eastmore Program is a special program to help alcoholics. In the Eastmore Program an alcoholic lives at home, but undergoes a two-phase treatment plan. Phase I is an intensive group-therapy program lasting ten weeks. Phase II is a long-term counseling program lasting one year. Eastmore Programs are located in most major cities, and past data gave the following information, based on percentages of success and failure collected over a long period of time: the probability that a client will have a relapse in phase I is 0.27; the probability that a client will have a relapse in phase II is 0.23. However, if a client did not have a relapse in phase I, then the probability that this client will not have a relapse in phase II is 0.95. If a client did have a relapse in phase I, then the probability that this client will have a relapse in phase II is 0.70.

Let *A* be the event that a client has a relapse in phase I and *B* be the event that a client has a relapse in phase II. Let *C* be the event that a client has no relapse in phase I and *D* be the event that a client has no relapse in phase II. Compute the following:

a) *P*(*A*), *P*(*B*), *P*(*C*), *P*(*D*)

b) *P*(*B, given A*), *P*(*D, given C*)

c) *P*(*A and B*), *P*(*C and D*)

d) *P*(*A or B*)

e) What is the probability that a client will go through both phase I and phase II without a relapse?

f) What is the probability that a client will have a relapse in both phase I and phase II?

g) What is the probability that a client will have a relapse in either phase I or phase II?

23. The state medical school has discovered a new test for tuberculosis. (If the test indicates a person has tuberculosis, the test is positive.) Experimentation has shown that the probability of a positive test is 0.82, given a person has tuberculosis. The probability is 0.09 that the test registers positive, given the person does not have tuberculosis. Assume that in the general population the probability that a person has tuberculosis is 0.04. What is the probability that a person chosen at random will
   a) have tuberculosis and a positive test?
   b) not have tuberculosis?
   c) not have tuberculosis and have a positive test?

SECTION 4.3   Trees and Counting Techniques

When outcomes are equally likely, we compute the probability of an event by using the formula

$$P(A) = \frac{\text{number of outcomes favorable to the event } A}{\text{number of outcomes in the sample space}}$$

The probability formula requires that we be able to determine the number of outcomes in the sample space. In the problems we have done in previous sections this task has not been difficult because the number of outcomes was small or the sample space consisted of fairly straightforward events. The tools we present in this section will help you count the number of possible outcomes in larger sample spaces or those formed by more complicated events.

A *tree diagram* helps us display the outcomes of an experiment consisting of a series of activities. The total number of outcomes corresponds to the total number of final branches in the tree. Perhaps the best way to learn to make a tree diagram is to see one. In the next example we will see a tree diagram and analyze its parts.

EXAMPLE 7

Jacqueline is in the nursing program and is required to take a course in psychology and one in anatomy and physiology (*A* and *P*) next semester. She also wants to take Spanish II. If there are four sections of psychology, two of *A* and *P*, and three of Spanish, how many different class schedules can Jacqueline choose from? (Assume that the times of the sections do not conflict with each other.) Figure 4-6 shows a tree diagram for Jacqueline's possible schedules.

Let's study the tree diagram and see how it shows Jacqueline's schedule choices. There are four branches from Start. These branches indicate the four possible choices for psychology sections. No matter which section of psychology Jacqueline chooses, she can choose from the two available *A* and *P* sections. Therefore, we have two branches leading from *each* psychology branch. Finally, after the psychology and *A* and *P* sections are selected, there are three choices for Spanish II. That is why there are three branches from *each A* and *P* section.

FIGURE 4-6 Tree Diagram for Selecting Class Schedules

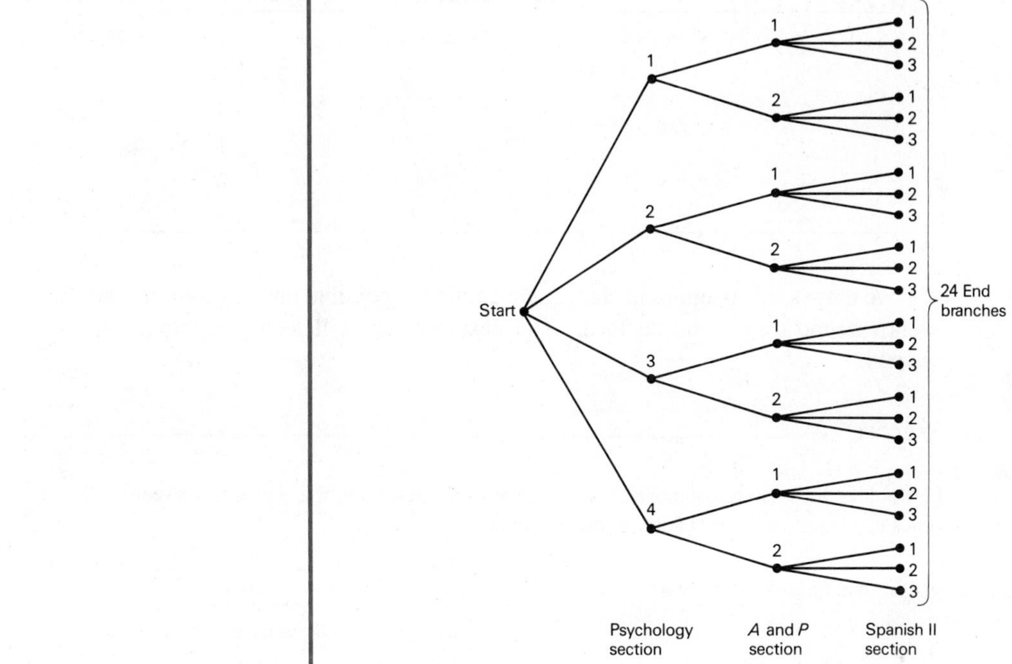

The tree ends with a total of 24 branches. This number of end branches tells us the number of possible schedules. The outcomes themselves can be listed from the tree by following each series of branches from Start to the end. For instance, the top branch from Start generates the schedules shown in Table 4-3.

TABLE 4-3 Schedules Utilizing Section 1 of Psychology

Psychology Section	*A* and *P* Section	Spanish II Section
1	1	1
1	1	2
1	1	3
1	2	1
1	2	2
1	2	3

Following the second branch from Start, we see all of the possible schedules utilizing Section 2 of psychology (*see* Table 4-4).

The other 12 schedules can be listed in a similar manner.

TABLE 4-4 Schedules Utilizing Section 2 of Psychology

Psychology Section	A and P Section	Spanish II Section
2	1	1
2	1	2
2	1	3
2	2	1
2	2	2
2	2	3

We draw a tree diagram in stages, indicating the possible outcomes for the first event, second event, and so forth. The next exercise will lead you through the process.

—————————— Exercise 10 ——————————

Louis plays three tennis matches. Use a tree diagram to list the possible win and loss sequences Louis can experience for the set of three matches.

a) On the first match Louis can win or lose. From Start indicate these two branches.

a) **FIGURE 4-7** *W* = Win, *L* = Lose

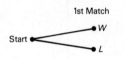

b) Regardless of whether Louis wins or loses the first match, he plays the second, and can again win or lose. Attach branches representing these two outcomes to *each* of the first match results.

b) **FIGURE 4-8** *W* = Win, *L* = Lose

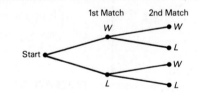

c) Louis may win or lose the third match. Attach branches representing these two outcomes to *each* of the second match results.

c) **FIGURE 4-9** *W* = Win, *L* = Lose

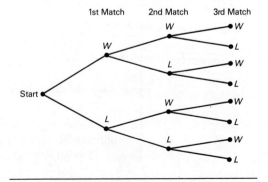

d) How many possible win-lose sequences are there for the three matches?

e) Complete this list of win–lose sequences.

1st	2nd	3rd
W	W	W
W	W	L
W	L	W
W	L	L
—	—	—
—	—	—
—	—	—
—	—	—

d) Since there are eight branches at the end, there are eight sequences.

e) The last four sequences all involve a loss on Match 1.

1st	2nd	3rd
L	W	W
L	W	L
L	L	W
L	L	L

Tree diagrams help us display the outcomes of an experiment involving several stages. If we label each branch of the tree with an appropriate probability, we can use the tree diagram to help us compute probabilities of an outcome displayed on the tree. One of the easiest ways to illustrate this feature of tree diagrams is to use an experiment of drawing balls out of an urn. We do this in the next example.

EXAMPLE 8

Suppose there are five balls in an urn. They are identical except in color. Three of the balls are red and two are blue. You are instructed to draw out one ball, note its color, and set it aside. Then you are to draw out another ball and note its color. What are the outcomes of the experiment? What is the probability of each outcome?

The tree diagram shown in Figure 4-10 will help us answer these questions.

FIGURE 4-10 Tree Diagram for Urn Experiment

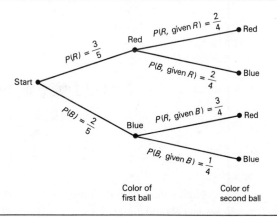

Notice that since you did not replace the first ball before drawing the second one, the two stages of the experiment are dependent. The probability associated with the color of the second ball depends on the color of the first ball. For instance, on the top branches, the color of the first ball drawn is red, so we compute the probabilities of the colors on the second ball accordingly. The tree diagram helps us organize the probabilities.

From the diagram, we see that there are four possible outcomes to the experiment. They are:

$$RR = \text{red on 1st } and \text{ red on 2nd}$$
$$RB = \text{red on 1st } and \text{ blue on 2nd}$$
$$BR = \text{blue on 1st } and \text{ red on 2nd}$$
$$BB = \text{blue on 1st } and \text{ blue on 2nd}$$

To compute the probability of each outcome, we will use the multiplication rule for dependent events. As we follow the branches for each outcome, we will find the necessary probabilities.

$$P(R \text{ on 1st } and \text{ } R \text{ on 2nd}) = P(R) \cdot P(R, \text{ given } R)$$
$$= \frac{3}{5} \cdot \frac{2}{4}$$
$$= \frac{3}{10}$$

$$P(R \text{ on 1st } and \text{ } B \text{ on 2nd}) = P(R) \cdot P(B, \text{ given } R)$$
$$= \frac{3}{5} \cdot \frac{2}{4}$$
$$= \frac{3}{10}$$

$$P(B \text{ on 1st } and \text{ } R \text{ on 2nd}) = P(B) \cdot P(R, \text{ given } B)$$
$$= \frac{2}{5} \cdot \frac{3}{4}$$
$$= \frac{3}{10}$$

$$P(B \text{ on 1st } and \text{ } B \text{ on 2nd}) = P(B) \cdot P(B, \text{ given } B)$$
$$= \frac{2}{5} \cdot \frac{1}{4}$$
$$= \frac{1}{10}$$

Notice that the probabilities of the outcomes in the sample space add to 1, as they should.

_____ Exercise 11 _____

Repeat the urn experiment with the five balls, three of which are red and two of which are blue. This time *replace* the first ball before drawing the second.

a) Draw a tree diagram for the outcomes of this experiment. Show the probabilities of each stage on the appropriate branch. *Hint:* Are the stages dependent or independent?

a) **FIGURE 4-11**   Tree Diagram for Urn Experiment (With Replacement)

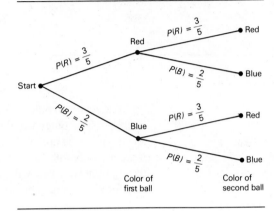

b) List the four possible outcomes of the experiment.

b) red on 1st *and* red on 2nd
red on 1st *and* blue on 2nd
blue on 1st *and* red on 2nd
blue on 1st *and* blue on 2nd

c) Use the multiplication rule for independent events and the probabilities shown on your tree to compute the probability of each outcome.

c) $P(\text{1st } R \text{ and 2nd } R) = P(R){\cdot}P(R)$

$$= \frac{3}{5} \cdot \frac{3}{5}$$

$$= \frac{9}{25}$$

$P(\text{1st } R \text{ and 2nd } B) = P(R){\cdot}P(B)$

$$= \frac{3}{5} \cdot \frac{2}{5}$$

$$= \frac{6}{25}$$

$P(\text{1st } B \text{ and 2nd } R) = P(B){\cdot}P(R)$

$$= \frac{2}{5} \cdot \frac{3}{5}$$

$$= \frac{6}{25}$$

$P(\text{1st } B \text{ and 2nd } B) = P(B){\cdot}P(B)$

$$= \frac{2}{5} \cdot \frac{2}{5}$$

$$= \frac{4}{25}$$

d) Do the probabilities of the outcomes in the sample space add up to 1?

e) Compare the tree diagram of this exercise with that of the previous example in which the first ball was not replaced before the second was drawn. Are the outcomes the same? Are the probabilities of the corresponding outcomes the same?

d) Yes, as they should.

e) The outcomes are the same: *RR*, *RB*, *BR*, and *BB*. However, because one experiment did not permit replacement of the first ball before the second was drawn, and the other experiment required replacement, the corresponding probabilities of the second stages of the experiment are different. The corresponding probabilities of the final outcomes are also different.

When an outcome is composed of a series of events, tree diagrams tell us how many possible outcomes there are. They also help us list the individual outcomes and organize the probabilities associated with each stage of the outcomes. However, if we are only interested in the number of outcomes created by a series of events, the multiplication rule will give us the total more directly. We state the multiplication rule for an outcome composed of a series of two events.

- **Multiplication Rule of Counting**   If there are n possible outcomes for event E_1 and m possible outcomes for event E_2, then there are a total of $n \times m$ or nm possible outcomes for the series of events E_1 followed by E_2.

The rule extends to outcomes created by a series of 3, 4, or more events. We simply multiply the number of outcomes possible for each step in the series of events to get the total number of outcomes for the series.

EXAMPLE 9

The Night Hawk is the new car model produced by Limited Motors, Inc. It comes with a choice of two body styles, three interior package options, and four different colors, and the choice of automatic or standard transmission. Select-an-Auto Car Dealership wants to carry one of each of the different types of Night Hawks. How many cars are required?

There are four items to select. We take the product of the number of choices for each item.

$$\begin{pmatrix} \text{no. of body} \\ \text{styles} \end{pmatrix} \begin{pmatrix} \text{no. of} \\ \text{interiors} \end{pmatrix} \begin{pmatrix} \text{no. of} \\ \text{colors} \end{pmatrix} \begin{pmatrix} \text{no. of transmission} \\ \text{types} \end{pmatrix}$$

$$(2)(3)(4)(2) = 48$$

Select-an-Auto must stock 48 cars in order to have one of each possible type.

_____ Exercise 12 _____

The Old Sage Inn offers a special dinner menu each night. There are two appetizers to choose from, three main courses, and four desserts. A customer can select one item from each category. How many different meals can be ordered from the special dinner menu?

a) Each special dinner consists of three items. List the item and the number of choices per item.

    a) appetizer – 2   main course – 3   dessert – 4

b) To find the number of different dinners composed of the three items, multiply the number of choices per item together.

    b) (2)(3)(4) = 24
    There are 24 different dinners that can be ordered from the special dinner menu.

Sometimes when we consider n items, we need to know the number of different ordered *arrangements* of the n items that are possible. The multiplication rule can help us find the number of possible ordered arrangements. Let's consider the classic example of determining the number of different ways in which eight people can be seated at a dinner table. For the first chair there are eight choices. For the second chair we have seven choices, since one person is already seated. For the third chair there are six choices, since two people are already seated. By the time we get to the last chair, there is only one person left for that seat. We can view each arrangement as an outcome of a series of eight events. Event 1 is *fill the first chair,* Event 2 is *fill the second chair,* and so forth. The multiplication rule will tell us the number of different outcomes.

$$\text{Choices for} \quad (8)(7)(6)(5)(4)(3)(2)(1) = 40{,}320$$

$$\underbrace{\uparrow \quad \uparrow \quad \uparrow \quad \uparrow \quad \uparrow \quad \uparrow \quad \uparrow \quad \uparrow}_{\text{chair position}}$$
$$\text{1st\ \ 2nd\ 3rd\ 4th\ 5th\ 6th\ 7th\ 8th}$$

In all, there are 40,320 different seating arrangements for eight people. It is no wonder that it takes a little time to seat guests at a dinner table!

The multiplication pattern shown above is not unusual. In fact, it is an example of the multiplication indicated by the factorial notation 8!.

$$! \text{ is read } factorial$$
$$8! \text{ is read } 8\ factorial$$
$$8! = 8 \cdot 7 \cdot 6 \cdot 5 \cdot 4 \cdot 3 \cdot 2 \cdot 1$$

In general, $n!$ indicates the product of n with each of the positive counting numbers less than n. By special definition $0! = 1$.

- **Factorial Notation** For a counting number n
$$n! = n(n-1)(n-2) \cdot \cdots \cdot 1$$
$$0! = 1$$
$$1! = 1$$

_____ Exercise 13 _____

a) Evaluate 3!.

b) How many different ways can three objects be arranged in order? How many choices do you have for the first position? for the second position? for the third position?

c) Verify step (b) with a three-stage tree diagram.

a) $3! = 3 \cdot 2 \cdot 1 = 6$

b) We have three choices for the first item, two for the second item, and one for the third item. By the multiplication rule, we have

$$(3)(2)(1) = 3! = 6 \text{ arrangements}$$

c) **FIGURE 4-12**

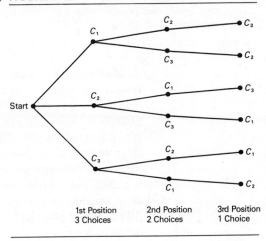

1st Position	2nd Position	3rd Position
3 Choices	2 Choices	1 Choice

We have considered the number of ordered arrangements of n objects taken as an entire group. Specifically, we considered a dinner party for eight and found the number of ordered seating arrangements for all eight people. However, suppose you have an open house and have only five chairs. How many ways can five out of the eight people seat themselves in the chairs? The formula we use to compute this number is called the *permutation formula*. We will simply state the formula and show you how to use it.

> • **Counting Rule for Permutations** The number of ways to *arrange in order* n distinct objects, taking them r at a time, is
>
> $$P_{n,r} = \frac{n!}{(n-r)!} \qquad (9)$$
>
> where n and r are whole numbers and $n \geqslant r$.

EXAMPLE 10

Let's compute the number of ordered seating arrangements we have for eight people in five chairs.

In this case we are considering a total of $n = 8$ different people and we wish to arrange $r = 5$ of these people. Substituting into formula (9) we have

$$P_{n,r} = \frac{n!}{(n-r)!} \qquad (9)$$

$$P_{8,5} = \frac{8!}{(8-5)!}$$

$$= \frac{8!}{3!}$$

$$= \frac{8 \cdot 7 \cdot 6 \cdot 5 \cdot 4 \cdot 3 \cdot 2 \cdot 1}{3 \cdot 2 \cdot 1}$$

$$= \frac{8 \cdot 7 \cdot 6 \cdot 5 \cdot 4 \cdot \cancel{3} \cdot \cancel{2} \cdot \cancel{1}}{\cancel{3} \cdot \cancel{2} \cdot \cancel{1}} \qquad \text{cancel like factors}$$

$$= 8 \cdot 7 \cdot 6 \cdot 5 \cdot 4$$

$$= 6{,}720$$

Note: On some calculators you can find the factorial key $x!$. In such cases you could calculate 8! and 3! directly on the calculator and divide the quantities, instead of canceling like factors. Also note that Table 2 of the appendix displays the values from 1 to 20 factorial. Some calculators also have a permutations key, often labeled $_nP_r$.

Exercise 14

The board of directors of Belford Community Hospital has twelve members. Three officers— president, vice president, and treasurer—must be elected from the members. How many different possible slates of officers are there? We will view a slate of officers as a list of three people with one person for president listed first, one person for vice president listed second, and one person for treasurer listed third. For instance, if Mr. Acosta, Ms. Hill, and Mr. Smith wish to be on a slate together, there are several different slates possible, depending on which one will run for president, which for vice president, and which for treasurer. Not only are we asking for the number of different groups of three names for a slate; we are also concerned about order, since it makes a difference which name is listed in which position.

a) What is the size of the group from which the slates of officers will be selected? This is the value of n.

a) $n = 12$

b) How many people will be selected for each slate of officers? This is the value of r.

b) $r = 3$

c) Each slate of officers is composed of three candidates. Different slates occur as we arrange the three candidates in the positions of president, vice president, and treasurer. For this reason, we need to consider the number of *permutations* of twelve items arranged in groups of three. Compute $P_{n,r}$.

c) $$P_{n,r} = \frac{n!}{(n-r)!}$$

$$P_{12,3} = \frac{12!}{(12-3)!} = \frac{12!}{9!}$$

$$= \frac{12 \cdot 11 \cdot 10 \cdot \cancel{9} \cdot \cancel{8} \cdot \cancel{7} \cdot \cancel{6} \cdot \cancel{5} \cdot \cancel{4} \cdot \cancel{3} \cdot \cancel{2} \cdot \cancel{1}}{\cancel{9} \cdot \cancel{8} \cdot \cancel{7} \cdot \cancel{6} \cdot \cancel{5} \cdot \cancel{4} \cdot \cancel{3} \cdot \cancel{2} \cdot \cancel{1}}$$

$$= 12 \cdot 11 \cdot 10 = 1{,}320$$

There are 1,320 different possible slates of officers.

In each of our previous counting formulas, we have taken the *order* of the objects or people into account. But what if order is not important? For instance, suppose we need to choose three members from the twelve-member board of directors of Belford Community Hospital to go to a convention. We are interested in *different groupings* of twelve people so that each group contains three people. The order is of no concern since all three will go to the convention. In other words, we need to consider the number of different *combinations* of twelve people taken three at a time. Our next formula will help us compute this number of different combinations.

> • **Counting Rule for Combinations**   The number of *combinations* of *n* objects taken *r* at a time is
>
> $$C_{n,r} = \frac{n!}{r!(n-r)!} \qquad (10)$$
>
> where *n* and *r* are whole numbers and $n \geq r$. Other commonly used notations for combinations include $_nC_r$ and $\binom{n}{r}$.

Notice the difference between the concepts of permutations and combinations. When we consider permutations we are considering groupings *and order.* When we consider combinations, we are considering only the number of different groupings. For combinations, order within the groupings is not considered. As a result, the number of combinations of *n* objects taken *r* at a time is generally smaller than the number of permutations of the same *n* objects taken *r* at a time. In fact, the combinations formula is simply the permutations formula with the number of permutations of each distinct group divided out. In the formula for combinations, notice the factor of *r*! in the denominator.

Now let's look at an example in which we compute the number of *combinations* of twelve people taken three at a time.

EXAMPLE   11

Three members from the group of twelve on the board of directors at Belford Community Hospital will be selected to go to a convention with all expenses paid. How many different groups of three are there?

In this case we are interested in *combinations* rather than permutations of twelve people taken three at a time. Using formula (10) we get

$$C_{n,r} = \frac{n!}{r!(n-r)!} \quad \text{or} \quad C_{12,3} = \frac{12!}{3!(12-3)!} = \frac{12!}{3!9!}$$

$$= \frac{12 \cdot 11 \cdot 10 \cdot 9 \cdot 8 \cdot 7 \cdot 6 \cdot 5 \cdot 4 \cdot 3 \cdot 2 \cdot 1}{(3 \cdot 2 \cdot 1)(9 \cdot 8 \cdot 7 \cdot 6 \cdot 5 \cdot 4 \cdot 3 \cdot 2 \cdot 1)}$$

$$= \frac{\overset{4}{\cancel{12}} \cdot 11 \cdot \overset{5}{\cancel{10}} \, (\cancel{9 \cdot 8 \cdot 7 \cdot 6 \cdot 5 \cdot 4 \cdot 3 \cdot 2 \cdot 1})}{(\cancel{3} \cdot \cancel{2} \cdot \cancel{1})(\cancel{9 \cdot 8 \cdot 7 \cdot 6 \cdot 5 \cdot 4 \cdot 3 \cdot 2 \cdot 1})} \qquad \text{cancel like factors}$$

$$= 4 \cdot 11 \cdot 5 = 220$$

There are 220 different possible groups of three to go to the convention. Since order is not considered, this number is much smaller than the number of different slates of three officers we computed in Exercise 14.

We have different formulas for permutations and combinations of n objects taken r at a time. How do you decide which one to use? Always ask yourself if order in the groups of r objects is relevant. If it is, use $P_{n,r}$. If order is not relevant, use $C_{n,r}$.

_____ Exercise 15 _____

In your political science class you are given a list of ten books. You are to select four to read during the semester. How many different *combinations* of four books are available from the list of ten?

a) Is the order in which you read the books relevant to the task of selecting the books?

a) No.

b) Do we use the number of permutations or combinations of ten books taken four at a time?

b) Since consideration of order in which the books are selected is not relevant, we compute the number of *combinations* of ten books taken four at a time.

c) How many books are available from which to select? How many must you read? What is the value of n? of r?

c) There are ten books among which you must select four to read. $n = 10$ and $r = 4$.

d) Compute $C_{10,4}$ to determine the number of different groups of four books from the list of ten.

d) $C_{n,r} = \dfrac{n!}{r!(n-r)!}$

$$C_{10,4} = \frac{10!}{4!(10-4)!}$$

$$= \frac{10!}{4!6!}$$

$$= \frac{10 \cdot 9 \cdot 8 \cdot 7 \cdot 6 \cdot 5 \cdot 4 \cdot 3 \cdot 2 \cdot 1}{(4 \cdot 3 \cdot 2 \cdot 1)(6 \cdot 5 \cdot 4 \cdot 3 \cdot 2 \cdot 1)}$$

$$= \frac{10 \cdot \overset{3}{\cancel{9}} \cdot \cancel{8} \cdot 7 (\cancel{6 \cdot 5 \cdot 4 \cdot 3 \cdot 2 \cdot 1})}{(\cancel{4} \cdot \cancel{3} \cdot \overset{1}{\cancel{2}} \cdot 1)(\cancel{6 \cdot 5 \cdot 4 \cdot 3 \cdot 2 \cdot 1})}$$

$$= 10 \cdot 3 \cdot 7 = 210$$

There are 210 different groups of four books to select from the list of ten.

We have introduced you to three counting formulas: the multiplication rule, the permutations rule, and the combinations rule. There are other rules that apply when the objects are not distinct. Many counting problems are easy to state and fairly difficult to solve. Some have you combine several counting rules. However, the problems for this section are all straightforward. Some ask you to use your counting abilities to compute probabilities.

Section Problems 4.3

1. a) Draw a tree diagram to display all the possible head–tail sequences that can occur when you flip a coin three times.
   b) How many sequences contain exactly two heads?
   c) *Probability extension:* Assuming the sequences are all equally likely, what is the probability that you will get exactly two heads when you toss a coin three times?

2. a) Draw a tree diagram to display all the possible outcomes that can occur when you flip a coin and then toss a die.
   b) How many outcomes contain a head and a number greater than four?
   c) *Probability extension:* Assuming the outcomes displayed in the tree diagram are all equally likely, what is the probability that you will get a head *and* a number greater than four when you flip a coin and toss a die?

3. There are six balls in an urn. They are identical except for color. Two are red, three are blue, and one is yellow. You are to draw a ball from the urn, note its color, and set it aside. Then you are to draw another ball from the urn and note its color.
   a) Make a tree diagram to show all possible outcomes of the experiment. Label the probability associated with each stage of the experiment on the appropriate branch.
   b) *Probability extension:* Compute the probability for each outcome of the experiment.

4. Repeat the experiment described in problem 3. However, replace the first ball before you draw the second one.
   a) Make a tree diagram to show all possible outcomes of the experiment. Label the probability associated with each stage of the experiment on the appropriate branch.
   b) *Probability extension:* Compute the probability for each outcome of the experiment.

5. Consider three true–false questions. There are two possible outcomes for each question: true or false.
   a) Draw a tree diagram showing all possible sequences of responses for the three questions. Does your tree diagram look similar to the one in problem 1? Why would you expect this result?
   b) *Probability extension:* Only one sequence will contain all three correct answers. Assuming you are guessing and all of the sequences are equally likely to occur when you guess, what is the probability of getting all three questions correct?

6. a) Make a tree diagram to show all the possible sequences of answers for three multiple-choice questions, each with four possible responses.
   b) *Probability extension:* Assuming that you are guessing the answers so that all outcomes listed in the tree are equally likely, what is the probability that you will guess the one sequence that contains all three correct answers?

7. Four wires (red, green, blue, and yellow) need to be attached to a circuit board. A robotic device is to attach the wires. The wires can be attached in any order, and the production manager wishes to determine which order would be fastest for the robot to use. Use the multiplication rule of counting to determine the number of all the possible sequences of assembly that must be tested. (*Hint:* There are four choices for the first wire, three choices for the second, two for the third, and only one for the fourth.)

8. A sales representative must visit four cities: Omaha, Dallas, Wichita, and Oklahoma City. There are direct air connections between each of the cities. Use the multiplication rule of counting to determine the number of different choices the sales representative has for the order in which to visit the cities. How is this problem similar to problem 7?

9. You have two decks of cards (52 cards per deck) and you draw one card from each deck.
   a) Use the multiplication rule of counting to determine the number of pairs of cards possible.
   b) There are four kings in each deck. How many pairs of kings are possible?
   c) *Probability extension:* Assuming all pairs are equally likely to be drawn, what is the probability of drawing two kings?

10. You toss a pair of dice.
    a) Use the multiplication rule of counting to determine the number of possible pairs of outcomes. (Recall that there are six possible outcomes for each die.)
    b) There are three even numbers on each die. How many outcomes are possible with even numbers appearing on each die?
    c) *Probability extension:* What is the probability that both dice will show an even number?

11. Barbara is a research biologist for Green Carpet Lawns. She is studying the effects of fertilizer type, temperature at time of application, and water treatment after application. She has four fertilizer types, three temperature zones, and three water treatments to test. Use the multiplication rule of counting to determine the number of different lawn plots she needs to have in order to test each fertilizer type, temperature range, and water treatment configuration.

12. The Deli Special lunch offers a choice of three different sandwiches, four kinds of salads, and five different desserts. Use the multiplication rule of counting to determine the number of different lunches that can be ordered using the Deli Special lunch option, if each lunch consists of one sandwich, one salad, and one dessert.

13. Compute $P_{5,2}$.   14. Compute $P_{8,3}$.

15. Compute $P_{7,7}$.   16. Compute $P_{9,9}$.

17. Compute $C_{5,2}$.   18. Compute $C_{8,3}$.

19. Compute $C_{7,7}$.   20. Compute $C_{8,8}$.

21. There are three nursing positions to be filled at Lilly Hospital. Position one is the day nursing supervisor; position two is the night nursing supervisor; and position three is the nursing coordinator position. There are 15 candidates qualified for all three of the positions. Use the permutation rule to determine the number of different ways the positions can be filled by these applicants.

22. In the Cash Now lottery game there are ten finalists who submitted entry tickets on time. From these ten tickets, three grand prize winners will be drawn. The first prize is one million dollars; the second prize is one hundred thousand dollars; and the third prize is ten thousand dollars. Use the permutation rule to determine the total number of different ways the winners can be drawn. (Assume tickets are not replaced after they are drawn.)

23. Matching questions are sometimes used on objective tests.
    a) If there are eight words to be matched with eight definitions (numbered 1 through 8), use the permutation rule to determine the number of possible word-to-definition matches. (Assume each word corresponds to exactly one definition.)

    b) If there are eight words but only five definitions, use the permutation rule to determine the number of possible word-to-definition matches.

24. To emphasize the importance of correct shelving of books, a librarian tells a group of students the number of possible orders in which just six books may be placed on a shelf. What is that number?

25. The University of Montana ski team has five entrants in a men's downhill ski event. The coach would like the first, second, and third places to go to the team members. In how many ways can the five team entrants achieve first, second, and third places?

26. During the Computer Daze special promotion, a customer purchasing a computer and printer is given a choice of three free software packages. There are ten different software packages from which to select. How many different combinations of software packages can be selected?

27. There are 15 qualified applicants for five trainee positions in a fast food management program. How many different groups of trainees can be selected? (*Hint:* Is order important? If not, use the formula for combinations.)

28. One professor grades homework by randomly choosing five out of twelve homework problems to grade.
    a) How many different groups of five problems are there from the twelve problems?
    b) *Probability extension:* Jerry did only five problems of one assignment. What is the probability that the problems he did comprised the group that was selected to be graded?
    c) Silvia did seven problems. How many different groups of five did she complete? What is the probability that one of the groups of five she completed comprised the group selected to be graded?

29. The qualified applicant pool for six management trainee positions consists of seven women and five men.
    a) How many different groups of applicants can be selected for the positions?
    b) How many different groups of trainees would consist entirely of women?
    c) *Probability extension:* If the applicants are equally qualified, and the trainee positions are selected by drawing the names at random so that all groups of six are equally likely, what is the probability that the trainee class will consist entirely of women?

SECTION 4.4   Introduction to Random Variables and Probability Distributions

For our purposes we will say a *statistical experiment* is any process by which an observation (or measurement) is obtained. Examples of statistical experiments are:

1) Counting the number of eggs in a robin's nest
2) Measuring daily rainfall in inches
3) Counting the number of defective light bulbs in a case of bulbs
4) Measuring the weight in kilograms of a polar bear cub

Let x represent a quantitative variable which is measured in an experiment. We are interested in the numerical values that x can take on. So x = number of eggs in a robin's nest or x = weight in kilograms of a polar bear cub would be examples of such quantitative variables. Furthermore, we say that the quantitative variable x is a *random variable* because the value that x takes on in a given experiment is a chance or random outcome. We will study two types of random variables: *discrete random variables* and *continuous random variables*.

> • **Definition:** When the observations of a quantitative random variable can take on only a finite number of values, or a countable number of values, we say that the variable is a *discrete random variable.**

In most of the cases we will consider, a discrete random variable will be the result of a count. For instance, the following are each examples of discrete random variables:

1) The number of students in a certain section of a statistics course this term. This value must be a counting number such as 25, or 57, or 139, and so forth. The values 25.34 or 25 1/2 are not possible.
2) The number of osprey chicks living in a nest. Again, fractional parts are not possible.
3) The number of students who vote in a given student body election.

> • **Definition:** When the observations of a quantitative random variable can take on any of the countless number of values in a line interval, we say that the variable is a *continuous random variable.*

For our purposes, we will see most continuous random variables occurring as the result of a measurement. For example, each of the following produces a continuous random variable:

1) The air pressure in an automobile tire. The air pressure could in theory take on any value from 0 lbs/in^2 (psi) to the bursting pressure of the tire. Values such as 20.126 psi, 20.12678 psi, and so forth, are possible.
2) The height of students in your statistics class. The height could in theory take on any value from a low of, say, three feet to a high of, say, seven feet.

---

*It is important to realize that the terms *countable* and *discrete* have different mathematical meanings. However, it is beyond the scope of this text to go into an extensive formal discussion of these terms. We think an intuitive (and perhaps naive) approach will be most useful for our purposes. This approach is in keeping with most introductory courses in statistics. The reader interested in a more rigorous discussion of the matter is referred to the advanced text, *Introduction to Mathematical Statistics,* by Robert V. Hogg and Allen T. Craig, which is published by Macmillan Publishing Co., Inc.

3) The number of miles per gallon fuel consumption of a car taken at random from the highway. In theory this could be any value from, say, one mile per gallon to, say, 75 miles per gallon.

The distinction between discrete and continuous random variables is important because of the different mathematical techniques associated with the two kinds of random variables. Although we will not discuss these techniques at great length in this book, the distinction is very important in the study of advanced mathematical statistics.

In general, measurements of quantities such as length, weight, volume, temperature, or time yield continuous random variables. If the temperature changes from 12 °C to 13 °C, for example, it must take on all the temperature values between 12 and 13. Temperatures cannot just jump from one reading to the next. Discrete random variables often come from counts, such as the number of passing scores on an exam or the number of weeds in a garden.

——————————— Exercise 16 ———————————

Which of the following random variables are discrete and which are continuous?

a) *Measure* the time it takes a student selected at random to register for the fall term.

a) Time can take on any value, so this is a continuous random variable.

b) *Count* the number of bad checks drawn on Upright Bank on a day selected at random.

b) The number of bad checks can be only whole numbers such as 1, 2, 3, etc. This is a discrete variable.

c) *Measure* the amount of gasoline needed to drive your car 200 miles.

c) We are measuring volume, which can assume any value, so this is a continuous random variable.

d) Pick a random sample of 50 registered voters in a district and find the number who voted in the last county election.

d) This is a count, so the variable is discrete.

A random variable has a probability distribution whether it is discrete or continuous. The *probability distribution* is simply an assignment of probabilities to the specific values of the random variable or to a range of values of the random variable.

The probability distribution of a *discrete* random variable has a probability assigned to *each* value of the random variable. The sum of these probabilities must be one. Let's look at a discrete probability distribution and its graph.

EXAMPLE 12

Dr. Fidgit developed a test to measure boredom tolerance. He administered it to a group of 20,000 adults between the ages of 25 and 35. The possible scores were 0, 1, 2, 3, 4, 5, and 6, with 6 indicating the highest tolerance for boredom. The test results for this group are shown in Table 4-5.

a) If a subject is chosen at random from this group, the probability that he or she will have a score of 3 is 6,000/20,000, or 0.30. In a similar way we can

TABLE 4-5	Boredom Tolerance Test Scores for 20,000 Subjects
Score	Number of Subjects
0	1,400
1	2,600
2	3,600
3	6,000
4	4,400
5	1,600
6	400

TABLE 4-6	Probability Distribution of Scores on Boredom Tolerance Test
Score	Probability
0	0.07
1	0.13
2	0.18
3	0.30
4	0.22
5	0.08
6	0.02

use the relative frequency to compute the probabilities for the other scores (Table 4-6). These probability assignments make up the probability distribution. Notice that the scores are mutually exclusive: no one subject has two scores. The sum of the probabilities of all the scores is 1.

b) The graph of this distribution is simply a histogram (*see* Figure 4-13) in which the height of the bar over a score represents the probability of that score. Notice that each bar is one unit wide, so the area of the bar over a score represents the probability of that score. Since the sum of the probabilities is 1, the area under the graph is also 1.

c) The Topnotch Clothing Company needs to hire someone with a score on the boredom tolerance test of 5 or 6 to operate the fabric press machine. Since the scores 5 and 6 are mutually exclusive, the probability that someone in

FIGURE 4-13   Graph of the Probability Distribution of Test Scores

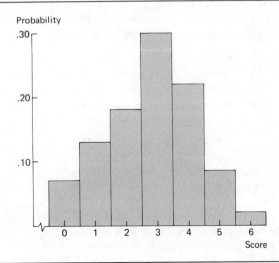

the group who took the boredom tolerance test made either a 5 or a 6 is the sum

$$P(5 \text{ or } 6) = P(5) + P(6)$$
$$= 0.08 + 0.02$$
$$= 0.10$$

One out of ten of the group who took the boredom tolerance test would quality for the position at Topnotch Clothing.

_____ Exercise 17 _____

One of the elementary tools of cryptanalysis (the science of code breaking) is to use relative frequencies of occurrence of different letters in the alphabet to break standard English alphabet codes. Large samples of plain text such as newspaper stories generally yield about the same relative frequencies for letters. A sample 1,000 letters long yielded the information in Table 4-7.

a) Use the relative frequencies to compute the omitted probabilities in Table 4-7.

a)

TABLE 4-7 Frequency of Letters in a 1,000-Letter Sample

Letter	Freq.	Prob.	Letter	Freq.	Prob.
A	73	____	N	78	0.078
B	9	0.009	O	74	____
C	30	0.030	P	27	0.027
D	44	0.044	Q	3	0.003
E	130	____	R	77	0.077
F	28	0.028	S	63	0.063
G	16	0.016	T	93	0.093
H	35	0.035	U	27	____
I	74	____	V	13	0.013
J	2	0.002	W	16	0.016
K	3	0.003	X	5	0.005
L	35	0.035	Y	19	0.019
M	25	0.025	Z	1	0.001

Source: From *Elementary Cryptanalysis: A Mathematical Approach,* by Abraham Sinkov. Copyright © 1968 by Yale University. Reprinted by permission of Random House, Inc.

TABLE 4-8 Completion of Table 4-7

Letter	Rel. Freq.	Prob.
A	$\dfrac{73}{1,000}$	0.073
E	$\dfrac{130}{1,000}$	0.130
I	$\dfrac{74}{1,000}$	0.074
O	$\dfrac{74}{1,000}$	0.074
U	$\dfrac{27}{1,000}$	0.027

b) Do the probabilities of all the individual letters add up to 1?

b) Yes

c) If a letter is selected at random from a newspaper story, what is the probability that the letter will be a vowel?

c) If a letter is selected at random,

$$P(a,e,i,o, \text{ or } u) = P(a) + P(e) + P(i)$$
$$+ P(o) + P(u)$$
$$= 0.073 + 0.130 + 0.074$$
$$+ 0.074 + 0.027$$
$$= 0.378$$

A probability distribution can be thought of as a relative frequency distribution. It has a mean and standard deviation. If we are referring to the probability distribution of a *population*, then we use the Greek letters μ for the mean and σ for the standard deviation. When we see the Greek letters used we know the information given is from the *entire population* rather than just a sample. If we have a sample probability distribution, we use $\bar{x}$ (*x* bar) and *s*, respectively, for the mean and standard deviation. For a given population, μ and σ are fixed numbers and are sometimes called the *parameters* of the population.

> ● **Definition:** The *mean* and the *standard deviation* of a *discrete population* probability distribution are found by using these formulas:
>
> $$\text{mean} \quad \mu = \Sigma x P(x)$$
>
> $$\text{standard deviation} \quad \sigma = \sqrt{\Sigma (x - \mu)^2 P(x)}$$
>
> where *x* is the value of a random variable,
> *P(x)* is the probability of that variable,
> the sum Σ is taken for all the values of the random variable.

The mean of a probability distribution is often called the *expected value* of the distribution. This terminology reflects the idea that the mean represents a "central point" or "cluster point" for the entire distribution. Of course, the mean or expected value is an average value and, as such, it *need not be a point of the sample space.*

EXAMPLE 13

Train for Tomorrow did a study to discover how many times adults changed jobs during their working careers between the ages of 25 and 65 years. After an extensive study, the results in Table 4-9 were published.

a) Find the expected number of times an adult changes jobs. The expected value is the mean.

$$\mu = \Sigma x P(x) = 5.6 \quad \text{(from Column III)}$$

An adult can expect to change jobs six times (to the nearest whole number) in a working career.

TABLE 4-9 Job Changes During a Working Career (Age 25–65)

x (No. of Job Changes)	P(x)	xP(x)	x − μ	(x − μ)²	(x − μ)²P(x)
0	0.01	0	−5.6	31.36	0.314
1	0.02	0.02	−4.6	21.16	0.423
2	0.04	0.08	−3.6	12.96	0.518
3	0.08	0.24	−2.6	6.76	0.541
4	0.11	0.44	−1.6	2.56	0.282
5	0.15	0.75	−0.6	0.36	0.054
6	0.25	1.50	0.4	0.16	0.040
7	0.20	1.40	1.4	1.96	0.392
8	0.09	0.72	2.4	5.76	0.518
9	0.05	0.45	3.4	11.56	0.578

$$\mu = \Sigma xP(x) = 5.6 \qquad \Sigma(x - \mu)^2 P(x) = 3.660$$

b) Find the standard deviation.

$$\sigma = \sqrt{\Sigma(x - \mu)^2 P(x)} = \sqrt{3.660} \quad \text{(from Column VI)}$$
$$\approx 1.9$$

_____ Exercise 18 _____

At a carnival you pay $2.00 to play a coin-flipping game with three fair coins. On each coin one side has the number 0 and the other side has the number 1. You flip the three coins at one time and you win $1.00 for every 1 that appears on top. Are your expected earnings equal to the cost to play? We'll answer this question in several steps.

a) In this game the random variable of interest counts the number of ones that show. What is the sample space for the values of this random variable?

a) The sample space is {0, 1, 2, 3} since 0 ones can come up, 1 one can appear, 2 ones can appear, or 3 ones can appear.

b) There are eight equally likely outcomes for throwing three coins. They are 000, 001, 010, 011, 100, 101, _____ , and _____ .

b) 110 and 111

c) Complete Table 4-10.

c)

TABLE 4-10

No. of Ones	Freq.	P(x)	xP(x)
0	1	0.125	0
1	3	0.375	_____
2	3	_____	_____
3	_____	_____	_____

TABLE 4-11 Completion of Table 4-10

x	Freq.	P(x)	xP(x)
0	1	0.125	0
1	3	0.375	0.375
2	3	0.375	0.750
3	1	0.125	0.375

d) The expected value is the sum

$$\mu = \Sigma x P(x)$$

Sum the appropriate column of Table 4-10 to find this value. Are your expected earnings less than, equal to, or more than the cost of the game?

d) The expected value can be found by summing the last column of Table 4-11. The expected value is $1.50. Since it cost $2.00 to play the game, the expected value is less than the cost. The carnival is making money. In the long run the carnival can expect to make an average of about 50 cents per player.

We have seen probability distributions of discrete variables and the formulas to compute the mean and standard deviation of discrete population probability distributions. Probability distributions of continuous random variables are similar except that the probability assignments are made to intervals of values rather than to specific values of the random variable. We will see an important example of a discrete probability distribution, the binomial distribution, in the next chapter and one of a continuous probability distribution in Chapter 6 when we study the normal distribution.

Section Problems 4.4

1. Which of the following are continuous variables, and which are discrete?
   a) number of traffic fatalities per year in the state of Florida
   b) distance a golf ball travels after being hit with a driver
   c) time required to drive from home to college on any given day
   d) number of ships in Pearl Harbor on any given day
   e) your weight before breakfast each morning
   f) cost of an adult ticket at each of the movie theaters in your town

2. Which of the following are continuous variables, and which are discrete?
   a) number of cows in a pasture
   b) number of electrons in a molecule
   c) voltage on a power line
   d) volume of milk given by a cow
   e) distance from Cape Canaveral to a point chosen at random in the Sea of Tranquility on the moon
   f) number of words in a book chosen at random

3. Which of the following are continuous variables, and which are discrete?
   a) amount of sleep you got last night
   b) home team score in a basketball game
   c) number of ducks sitting on a pond
   d) Btus absorbed by a solar panel
   e) volume of water in Lake Powell
   f) number of prisoners in the county jail

4. Which of the following are continuous variables, and which are discrete?
   a) closing price of a stock on the New York Stock Exchange
   b) speed of an airplane
   c) age of a college professor chosen at random

d) number of books in the college bookstore
e) weight of a football player chosen at random
f) number of lightning strikes in Rocky Mountain National Park on a given day

5. Osprey, Incorporated makes fishing rods. A new model fishing rod is ready to be put on the market. In an effort to establish a suggested retail price, Osprey, Inc. gave a demonstration of the new rod to a random sample of 392 fishermen. Suggested retail prices for the rod were listed at $15, $20, $25, $30, $35, $40, and $45. After examining the rod, each person was asked to circle the maximal price he or she would pay for such a rod. The results are shown below, where x = price and f = frequency with which this price would be paid.

x	15	20	25	30	35	40	45
f	18	48	72	94	86	51	23

a) If a fisherman is chosen at random from those examining the new rod, use relative frequencies to calculate the probability $P(x)$ that the maximal price he or she will pay for the rod is x = 15, 20, 25, 30, 35, 40, 45 dollars.
b) Use a histogram to graph the probabilities of part (a).
c) Assuming these fishermen represent the population of all fishermen, what do you estimate the probability is that a fisherman chosen at random will pay $40 or $45 for the new rod?
d) If a fisherman is chosen at random, what do you estimate the probability is that this fisherman will pay $15, $20, or $25 for the new rod?
e) Compute the expected value of the x distribution. Explain how the number you computed might be related to a suggested retail price for the new rod.
f) Compute the standard deviation of the x distribution.

6. A local taxi company is interested in the number of pieces of luggage a cab carries on a taxi run. A random sample of 260 taxi runs gave the following information where x = number of pieces of luggage and f = frequency which which taxi runs carried this many pieces of luggage.

x	0	1	2	3	4	5	6	7	8	9	10
f	42	51	63	38	19	16	12	10	6	2	1

a) If a taxi run is chosen at random from these 260 runs, use relative frequencies to find $P(x)$ for x = 0, 1, 2, 3, 4, 5, 6, 7, 8, 9, 10.
b) Use a histogram to graph the probability distribution of part (a).
c) Assuming these 260 taxi runs represent the population of all taxi runs in this area, what do you estimate the probability is that a randomly selected taxi run will have from 0 to 4 pieces of luggage (including 0 and 4)?
d) What do you estimate the probability is that a randomly selected taxi run will have from 6 to 10 pieces of luggage (including 6 and 10)?
e) Compute the expected value of the x distribution.
f) Compute the standard deviation of the x distribution.

7. The head nurse on the third floor of a community hospital is interested in the number of nighttime room calls requiring a nurse. For a random sample of 208 nights (9:00 P.M. to 6:00 A.M.) the following information was obtained where x = number of room calls

requiring a nurse and f = frequency with which this many calls occurred (i.e., number of nights).

x	36	37	38	39	40	41	42	43	44	45
f	6	10	11	20	26	32	34	28	25	16

a) If a night is chosen at random from these 208 nights, use relative frequencies to find $P(x)$ when x = 36, 37, 38, 39, 40, 41, 42, 43, 44, 45.
b) Use a histogram to graph the probability distribution of part (a).
c) Assuming these 208 nights represent the population of all nights at community hospital, what do you estimate the probability is that on a randomly selected night there will be from 39 to 43 (including 39 and 43) room calls requiring a nurse?
d) What do you estimate the probability is that there will be from 36 to 40 (including 36 and 40) room calls requiring a nurse?
e) Find the expected number of room calls requiring a nurse.
f) Find the standard deviation of the x distribution.

8. A psychologist has devised a test that measures the congeniality factor (desire to work with people). A random sample of 1,000 adults took the test. Their scores are given in Table 4-12 (higher scores mean more congeniality).

TABLE 4-12 Congeniality Test Scores for 1,000 Adults

x(Score)	Frequency of Adults with This Score
1	90
2	480
3	220
4	150
5	60

a) If a subject is chosen at random from those taking the test, the probability that he or she has a score of 1 is 90/1,000, or 0.09. In a similar way use the relative frequency to find the other probabilities in Table 4-13.
b) Use a histogram to graph the probability distribution of Table 4-13.

TABLE 4-13 Probabilities for Congeniality Test Scores

x	Probability of x
1	$P(1) = 0.09$
2	_____
3	_____
4	_____
5	_____

Assuming the sample probabilities are good estimates for the population probabilities, answer parts (c) through (f).

c) Columbian Airlines needs to hire people with a high congeniality factor to work as stewards, stewardesses, and ticket agents. If Columbian Airlines requires a congeniality factor of 4 or 5, what is the probability that a person chosen at random will meet the requirement?

d) The National Forest Service wants to hire people with low congeniality factors so they will be happy to staff isolated fire-watch towers. What is the probability that a person chosen at random will meet the requirement of a congeniality factor of 3 or less?

e) What is the expected value μ of the congeniality factor?

f) What is the standard deviation of the congeniality factor?

9. The Army gives a battery of exams to all new recruits. One exam measures a person's ability to work with technical machinery. This exam was given to a random sample of 360 new recruits. The results are shown below, where x is the score on a ten-point scale and f is the frequency of new recruits with this score.

x	1	2	3	4	5	6	7	8	9	10
f	28	42	79	83	51	36	18	12	7	4

a) If a person is chosen at random from those taking the test, use the relative frequency of test scores to estimate $P(x)$ as x goes from 1 to 10.

b) Use a histogram to graph the probability distribution of part (a).

c) The ground-to-air missile battalion needs people with a score of seven or higher on the exam. What is the probability that a new recruit will meet this requirement?

d) The kitchen battalion can use people with a score of three or less. What is the probability that a new recruit is not overqualified for this work?

e) Compute the expected value μ of exam scores.

f) Compute the standard deviation of exam scores.

10. Reggie Richman has a poor driving record and must take out a special auto insurance policy. He wants to insure his $28,000 sports car. Based on previous driving records, AnyState has made the following estimates: in a given year a total loss may occur with a probability of 0.11; a 50% loss may occur with a probability of 0.18; a 25% loss may occur with a probability of 0.28; a 10% loss may occur with a probability of 0.42. If AnyState will pay no benefits for any other partial losses, and AnyState wants to make a profit of $725, how much should they charge Reggie for insurance? (*Hint:* Let x represent the dollars lost in an accident.)

TABLE 4-14

% of Loss	x	$P(x)$	$xP(x)$
100	28,000	0.11	
50	14,000	0.18	
25	7,000	0.28	
10	2,800	0.42	
0	0	0.01	

Complete the $xP(x)$ column and compute μ, the expected loss AnyState will have to pay. Then add on the desired profit.

11. Mr. Dithers wants to insure his yacht for $80,000. The Big Rock Insurance Company estimates a total loss may occur with a probability of 0.005; a 50% loss with probability of 0.01; and a 25% loss with probability 0.05. If Big Rock will pay no benefits for any other partial loss, what premium should Mr. Dithers pay each year if Big Rock wants to make a profit of $250? (*Hint:* See problem 10.)

12. Alpha Alpha Moo sorority is having a car raffle to buy playground equipment for disadvantaged children. The sorority buys a used car for $3,000 and sells 3,750 raffle tickets at $1.50 per ticket.
    a) If you buy 30 tickets, what is the probability that you will win the car? What is the probability that you will not win the car?
    b) Your expected earnings can be found by multiplying the value of the car by the probability that you will win it. What are your expected earnings? Is it more or less than the amount you paid for 30 tickets? How much did you effectively contribute to Alpha Alpha Moo's charity?

13. The Old West Historical Museum in Black Hawk, Colorado, is trying to raise money for building restoration. They are selling raffle tickets for an all-expense-paid trip to Hawaii for two. A travel agent sold them the trip for $2,479.16, and friends of the museum sold 1,792 tickets at $4.00 per ticket.
    a) Karen is a friend of the museum who sold ten tickets to her father. What is the probability that her father will win the trip? What is the probability that he will not win?
    b) The expected earnings for Karen's father can be found by multiplying the cost of the trip by the probability that he will win the trip. What are his expected earnings? How much did he effectively contribute to the museum?

14. The college hiking club is having a fund raiser to buy a new toboggan for winter outings. They are selling Chinese fortune cookies for 35 cents each. Each cookie contains a piece of paper with a different number written on it. A random drawing will determine which number is the winner of a dinner for two at a local Chinese restaurant. The dinner is valued at $40.00. Since the fortune cookies were donated to the club, we can ignore the cost of the cookies. The club sold 816 cookies before the drawing.
    a) John bought 12 cookies. What is the probability that he will win the dinner for two? What is the probability that he will not win?
    b) John's expected earnings can be found by multiplying the cost of the dinner by the probability that he will win. What are John's expected earnings? How much did John effectively contribute to the club?

Summary

In this chapter we first examined the question: What is probability? We found that probabilities can be assigned to events by intuition, by the method of relative frequency, or by the method of equally likely outcomes. Next we studied some probability rules. The most important rules are the multiplication rules for independent and dependent events and the addition rules for mutually exclusive and general events. We also looked at some counting techniques useful in computing probabilities. These techniques included tree diagrams, the multiplication rule for counting, combinations, and permutations. The last section dealt with random variables and probability distributions. We learned that there are two types of random variables: discrete and continuous. We studied the mean μ, which is also the expected value of a distribution, and the standard deviation σ of a discrete probability population distribution.

Important Words and Symbols

	Section		Section
Relative frequency	4.1	Trees	4.3
Equally likely outcomes	4.1	Multiplication rule of counting	4.3
Sample space	4.1	Permutation rule	4.3
Probability of an event A,		Combination rule	4.3
$P(A)$	4.1	Random variable	4.4
Complement of A	4.1	Continuous	4.4
Mutually exclusive events	4.2	Discrete	4.4
Addition rules (for mutually		Population	4.4
exclusive and general events)	4.2	Parameters	4.4
Dependent events	4.2	Mean of a population	
A *and* B	4.2	probability distribution, μ	4.4
A *or* B	4.2	Standard deviation of a	
Independent events	4.2	population probability	
A, *given* B	4.2	distribution, σ	4.4
Multiplication rules (for		Expected value, μ	4.4
independent and dependent			
events)	4.2		

Chapter Review Problems

1. The physical education department at Muddybanks College wants to know the distribution of weights of freshman women taking physical education. The weights of 1,000 women are grouped by intervals of 15 pounds in Table 4-15.

TABLE 4-15 Weights of 1,000 Freshman Women

Weight Grouping	Number of Women
70–84	16
85–99	91
100–114	210
115–129	420
130–144	150
145–159	83
160–174	21
175–199	9

If a woman is picked at random from these 1,000, what is the probability she will weigh
a) between 70 and 84 pounds?
b) between 85 and 99 pounds?
c) between 100 and 114 pounds?
d) 115 pounds or more?
e) 144 pounds or less?

2. Suppose you throw a fair die and flip a fair coin. Let's represent the outcomes of 3 on the die and heads on the coin by 3H.
a) One outcome is 3H. What are the other outcomes? What is the sample space?
b) Are all outcomes in the sample space equally likely? Explain.
c) What is the probability of getting heads and a number less than 3?

3. Two cards are drawn at random from a standard deck. (A standard deck has 52 cards: 13 hearts, 13 diamonds, 13 clubs, and 13 spades.)
a) If the first card is replaced before the second is drawn, what is the probability that both cards will be hearts?
b) If the first card is not replaced before the second is drawn, what is the probability that both cards will be hearts?

4. The Polar Bear Ice Cream Company did a customer survey to determine which carton size customers preferred. A random sample of 800 Polar Bear customers gave the information in Table 4-16.

TABLE 4-16 Carton Size Preferences of 800 Polar Bear Customers

x (No. of Cups)	Number of Customers Preferring This Size
1 (a cup)	100
4 (a quart)	380
8 (a half gallon)	250
16 (a gallon)	70

If someone is chosen at random from this group of 800 Polar Bear customers, then answer the following.
a) What is the probability that he or she prefers the cup size? the quart size? the half gallon size? the gallon size?
b) What is the probability that this customer prefers either the gallon or half gallon size?
c) What is the expected value μ of the number of cups a customer will prefer?
d) What is the standard deviation for the number of cups a customer prefers?

5. a) Describe how you could use a relative frequency to estimate the probability that a thumbtack will land with its flat side down.
b) What is the sample space of outcomes for the thumbtack?
c) How would you make a probability assignment to this sample space if, when you drop 500 tacks, 340 land flat side down?

6. Allergic reactions to poison ivy can be miserable. Plant oils cause the reaction. Researchers at Allergy Institute did a study to see the effects of washing the oil off within five

minutes of exposure. A random sample of 1,000 people with known allergies to poison ivy participated in the study. Oil from the poison ivy plant was rubbed on a patch of skin. For 500 of the subjects, it was washed off *within* five minutes. For the other 500 subjects, the oil was washed off *after* five minutes. The results are summarized in Table 4-17.

TABLE 4-17 Time in Which Oil Was Washed off

Reaction	Within 5 Minutes	After 5 Minutes	Row Total
None	420	50	470
Mild	60	330	390
Strong	20	120	140
Column Total	500	500	1,000

Let's use the following notation for the various events: W = washing oil off within 5 minutes; A = washing oil off after 5 minutes; N = no reaction; M = mild reaction; S = strong reaction. Find the following probabilities for a person selected at random from this sample of 1,000 subjects.

a) $P(N)$; $P(M)$; $P(S)$

b) $P(N, \text{ given } W)$; $P(S, \text{ given } W)$

c) $P(N, \text{ given } A)$; $P(S, \text{ given } A)$

d) $P(N \text{ and } W)$; $P(M \text{ and } W)$

e) $P(N \text{ or } M)$ Are the events N = no reaction and M = mild reaction mutually exclusive? Explain.

f) Are the events N = no reaction and W = washing oil off within 5 minutes independent? Explain.

7. In one variation of chess the players each use a board and neither player can see the opponent's board. The only pieces on a player's board are his or her own. The opponent's pieces are "invisible." The players take turns and move as usual except that a third person tells each player if a piece has been taken or a move is illegal. Not including the king as an offensive piece, there are 15 other pieces with point values as shown in Table 4-18.

TABLE 4-18 Number and Point Value of Chessmen

Name of Piece	f (No. of Pieces)	x (Point Value of One Piece)
Queen	1	9
Bishop	2	3
Knight	2	3
Rook	2	5
Pawn	8	1

a) Let's consider the first piece you lose (not the king). If it is lost at random, what is the probability that it is: the queen? a bishop? a knight? a rook? a pawn?

b) What is the expected point value of your first loss?

c) Is the answer for part (b) also the expected point value of your second loss? Explain your answer.

8. In a game of craps you roll two fair dice. Whether you win or lose depends on the sum of the numbers occurring on the tops of the dice. Let x be the random variable which is the sum of the numbers on the tops of the dice.
   a) What values can x take on?
   b) What is the probability distribution of these x values (e.g., what is the probability that $x = 2$, 3, etc.)?
   c) Graph the probability distribution of x.
   d) What is the expected value of the x distribution?
   e) Find the standard deviation of the x distribution.
   f) What is the probability that on a single throw x will not be 2 or 7 or 11?

9. To estimate the number of books required per course, the student council at Bookworth College took a random sample of 100 courses and obtained the information in Table 4-19.

TABLE 4-19 Book Requirements for 100 Courses

x (No. of Required Books)	f (No. of Courses)
0	5
1	48
2	22
3	11
4	9
5	3
6	2

   a) Make a table and graph the x probability distribution.
   b) What is the expected value of the number of books required per course?
   c) If the average book costs $25 and you take five new courses, how much can you expect to pay for books?

10. Class records at Rockwood College indicate a student selected at random has probability 0.77 of passing French 101. For the student who passes French 101 the probability is 0.90 that he or she will pass French 102. What is the probability that a student selected at random will pass both French 101 and 102?

*11. There is money to send two out of eight city council members to a conference in Honolulu. All want to go, so they decide to choose the members to go to the conference by a random process. How many different combinations of two council members can be selected from the eight who want to go to the conference?

*12. Compute
   a) $P_{7,2}$ b) $C_{7,2}$ c) $P_{3,3}$ d) $C_{4,4}$

*13. Freeze Dry Food, Inc. packages all its foods in clear plastic that is sealed. The quality control for the packaging process checks for three items: 1) that the weight shown is correct; 2) that the label is correct; and 3) that the package is properly sealed. These three processes can be done in any order. A computer-operated device directs the

---

*These problems are from Section 4.3.

packages to the three inspection stations according to the backlog in that area. If there is a larger backlog in one area, products are sent to one of the other two areas first. In how many different ways can a package be cycled through the three inspection stations?

*14. A student must satisfy the literature, social science, and philosophy requirements this semester. There are four literature courses to select from, three social science courses, and two philosophy courses. Make a tree diagram showing all the possible sequences of literature, social science, and philosophy courses.

*15. There are five multiple-choice questions on an exam, each with four possible answers. Use the multiplication rule of counting to determine the number of possible answer sequences for the five questions. Only one of the sets can contain all five correct answers. If you are guessing, so that you are as likely to choose one sequence of answers as another, what is the probability of getting all five answers correct?

*16. A coin is tossed six times. Use the multiplication rule of counting to determine the number of possible head–tail sequences that can occur.

*17. To open a combination lock, you turn the dial to the right and stop at a number, then you turn it to the left and stop at a second number. Finally, you turn it back to the right and stop at a third number. If you used the correct sequence of numbers, the lock opens. If the dial of the lock contains ten numbers, 0 through 9, use the multiplication rule to determine the number of different combinations possible for the lock. (*Note:* The same number can be reused.)

*18. You have a combination lock. Again, to open it you turn the knob to the right and stop at a first number, then you turn it to the left and stop at a second number. Finally, you turn it to the right and stop at a third number. Suppose you remember that the three numbers for your lock are 2, 9, and 5, but you don't remember the order in which the numbers occur. How many permutations of these three numbers are possible?

USING COMPUTERS

Professional statisticians in industry and research use computers to help them analyze and process statistical data. There are many computer programs available for statistics. Some commonly used statistical packages include Minitab, SAS, and SPSS. There are many others as well. Any of these statistical packages may be used with this text.

The authors of this text have also written a computer statistics package called ComputerStat. The programs in ComputerStat are available on disks with an accompanying manual that explains how to use the programs and gives additional computer applications. The disks are intended for the Apple IIe, IIc, or the IBM PC. Programs are written in the computer language BASIC and are designed for the beginning statistics student with no previous computer experience. ComputerStat is available as a supplement from the publisher of this text.

Problems in this section may be done using ComputerStat or other statistical computer programs.

1. In addition to large grand prize drawings, the Colorado State Lottery also has instant cash prizes. For one lottery game about 18,000,000 tickets were printed. Of these, 80,640 qualified a player for an entry in the grand prize drawing. The remaining 17,919,360 tickets had no cash prize, a cash prize, or a prize of another lottery ticket (value $1). For these tickets the following information was given:

Value of Prize (in dollars)	Number of Winners Possible
0	13,846,016
1 (a free lottery ticket)	2,016,000
2	1,733,760
5	282,240
10	40,320
500	1,008
50,000	16

Source: Adapted from the information given for the game "The Sky's the Limit, 1985." Colorado Lottery, Pueblo.

a) If we consider only the instant prize winners, use the above information to calculate the probability of winning 0, 2, 5, 10, 500, or 50,000 dollars, or a one dollar lottery ticket.

b) In the ComputerStat menu of topics, select Probability Distributions and the Central Limit Theorem. Then use program S5, Expected Value and Standard Deviation of a Probability Distribution, to compute the expected value and standard deviation of this probability distribution.

c) Repeat parts (a) and (b) if we include the possibility of winning the one million dollar grand prize.

d) Proceeds from the Colorado State Lottery benefit state parks and recreation areas. Look at your answer to part (c). When you buy a Colorado State Lottery ticket, what are your expected winnings? How much have you effectively contributed to state parks and recreation? (Remember, each ticket costs one dollar.)

2. For those marriages that ended in divorce, the following information was given about the length of the marriage.

Length of Marriage to Nearest Year or Span of Years	Percentage
0	4.4
1	8.1
2	9.0
3	9.2
4	8.4
5	7.2
6	6.2
7	5.7
8	4.7
9	4.1
10–14	13.5
15–19	8.0
20–24	5.4
25–29	3.3
30–34	2.8

(*Note:* The last class is really open-ended: 30 or more years. However, in order to process the data, we need a right endpoint, so we will treat all years over 34 as 34.)

Source: Hacker, Andrew, ed. *A Statistical Portrait of the American People*. New York: Viking Press, 1983, 109. Copyright © 1983 by Andrew Hacker. Reprinted by permission of Viking Penguin, Inc.

The information opposite with midpoints of class intervals (where necessary) gives a probability distribution that a marriage ending in divorce will last 0, 1, 2, 3, 4, 5, 6, 7, 8, 9, 12, 17, 22, 27, or 32 years. In the menu of topics of ComputerStat, select Probability Distributions and the Central Limit Theorem. Then use program S5, Expected Value and Standard Deviation of a Probability Distribution, to find the expected value and standard deviation of this probability distribution.

5 The Binomial Distribution

He who has heard the same thing told by 12,000 eye-witnesses has only 12,000 probabilities, which are equal to one strong probability, which is far from certainty.

Voltaire

When it is not in our power to determine what is true, we ought to follow what is most probable.

Descartes

It is important to realize that statistics and probability do not deal in the realm of certainty. If there is any realm of human knowledge where genuine certainty exists, you may be sure that our statistical methods are not needed there. In most human endeavors and in almost all the natural world around us the element of chance happenings cannot be avoided. When we cannot expect something with true certainty we must rely on probability to be our guide. In this and the next chapter we will study two of the most important probability distributions of mathematical statistics. Their influence will be felt throughout our entire study of statistics.

Binomial Experiments

On a TV quiz show each contestant has a try at the Wheel of Fortune. The Wheel of Fortune is a roulette wheel with 36 slots, one of which is gold. If the ball lands in the gold slot, the contestant wins $50,000. No other slot pays. What is the probability that the quiz show will have to pay the fortune to three contestants out of 100?

In this problem the contestant and the quiz show sponsors are concerned about only two outcomes from the Wheel of Fortune: the ball lands on the gold, or the ball does not land on the gold. This problem is typical of an entire class of problems that are characterized by the feature that there are exactly two possible outcomes (for each trial) of interest. These problems are called *binomial experiments*, or *Bernoulli experiments*, after Jacob Bernoulli who studied them extensively in the late 1600s.

A *binomial experiment* must have these features:

1. There are a fixed number of trials. We denote this number by the letter n.
2. The n trials are independent and repeated under identical conditions.
3. Each trial has only two outcomes: success, denoted by S, and failure, denoted by F.
4. For each individual trial the probability of success is the same. We denote the probability of success by p and that of failure by q. Since each trial results in either success or failure, $p + q = 1$ so $q = 1 - p$.
5. The central problem of a binomial experiment is to find the probability of r successes out of n trials.

EXAMPLE   1

Let's see how the Wheel of Fortune problem meets the criteria of a binomial experiment. We'll take the criteria one at a time.

1. Each of the 100 contestants has a trial at the wheel so there are $n = 100$ trials in this problem.

2. Assuming the wheel is fair, the trials are independent since the result of one spin of the wheel has no effect on the results of other spins.

3. We are interested in only two outcomes on each spin of the wheel: the ball either lands on the gold or it does not. Let's call landing on the gold *success* (S) and not landing on the gold *failure* (F). In general, the assignment of the terms success and failure to outcomes does not imply good or bad results. These terms are assigned simply for the user's convenience.

4. On each trial the probability p of success (landing on the gold) is 1/36 since there are 36 slots and only one of them is gold. Consequently, the probability of failure is

$$q = 1 - p = 1 - \frac{1}{36} = \frac{35}{36}$$

on each trial.

5. We want to know the probability of three successes out of 100 trials so $r = 3$ in this example. It turns out that the probability the quiz show will have to pay the fortune to three contestants out of 100 is about 0.23. In the next section we'll see how this probability was computed.

_____ Exercise 1 _____

The registrar of a college noted that for many years the withdrawal rate from an introductory chemistry course has been 35% each term. If 80 students register for the course, what is the probability that 55 will complete it?

This is a binomial experiment. Let's see how it fits each of the specifications.

a) Each of the 80 students enrolled in the course can make the decision to withdraw or complete the course. The decision of each student can be thought of as a trial. How many trials are there? $n = $ _____?

a) There are 80 trials so $n = 80$.

b) In this problem we will assume that the decision one student makes about withdrawing from the course does not affect the decision of any other student. Under this assumption are the trials independent?

b) Yes.

c) A trial consists of a student deciding to withdraw or complete the chemistry course. How many possible outcomes are there to each trial?

c) Two: the decision to withdraw or the decision to complete the course.

d) Let's call completing the course success. Then withdrawing is failure. The probability of failure for each trial is $q = 0.35$. Use the fact that $q = 1 - p$ to find p, the probability of success for each trial.

d) Since $q = 0.35$ and $q = 1 - p$, p must equal 0.65.

e) In this problem we want to know the probability of $r = $ _____ successes out of $n = $ _____ trials.

e) $r = 55; n = 80$

A binomial experiment must satisfy all the criteria. One criterion that is fairly easy to check is whether each trial has exactly two possible outcomes (*see* Exercise 2).

_____ Exercise 2 _____

We want to determine if the following experiment is binomial:

A random sample of thirty men between the ages of 20 and 35 is taken from the population of Teliville. Each man is asked to name his favorite TV program.

a) The response of each of the 30 men is a trial, so there are $n = $ _____ trials.

a) $n = 30$

b) How many outcomes are possible on each trial? Can this be a binomial experiment?

b) There are many possible outcomes—as many as there are programs on TV in Teliville. This *cannot* be a binomial experiment since each trial has more than two possible outcomes.

Independence is another criterion the binomial trials must satisfy. In other words, the outcome of one trial cannot affect the outcome of any other trial. Example 2 illustrates a case in which the condition of independence is violated.

EXAMPLE 2

The intramural committee of a college consists of four women and six men. To select a chairperson and recorder they put all the members' names in a hat and draw two names without replacing the first name drawn. The first name drawn will be chairperson, and the second will be recorder. What is the probability that both offices will be held by women?

Analyze this problem to see if it is a binomial experiment. Each draw is a trial so there are two trials. The outcome under consideration is whether the name drawn is a woman or not. We'll define success to be that a woman's name is drawn.

These trials are not independent because the outcome of the first trial affects the probability of success on the second trial. The probability of a woman on the first draw is 4/10. Since the name of the first draw is not replaced, the probability of a woman on the second draw is 3/9 if a woman was selected on the first draw and 4/9 if a man was chosen on the first draw. In either case the probability of success on the second trial has been affected by the outcome of the first trial.

Since the trials are not independent, the problem is not a binomial experiment.

Anytime we make selections from a population *without replacement, we do not have independent trials*. However, replacement is often not practical. If the number of trials is quite small with respect to the population, we almost have independent trials and we can say the situation is closely approximated by a binomial experiment. For instance, suppose we select 20 tuition bills at random from a collection of 10,000 bills issued at one college and observe if the bill is in error or not. If 600 of the 10,000 bills are in error, then the probability that the first one selected is in error is 600/10,000, or 0.0600. If the first is in error, then the probability that the second is in error is 599/9,999, or 0.0599. Even if the first 19 bills selected are in error, the probability that the 20th is also in error is 581/9,981, or 0.0582. All of these probabilities round to 0.06, and we can say that the independence condition is approximately satisfied.

———————— Exercise 3 ————————————

Let's analyze the following binomial experiment to determine *p, q, n,* and *r:*

A medical research team claims that a certain blood type (call it *T*) occurs in 15% of the population. Suppose we choose 18 people at random from the population and test the blood type of each. What is the probability that three of these people have blood type *T*? (*Note:* Independence is approximated since 18 people is an extremely small sample with respect to the entire population.)

a) In this experiment we are observing whether or not a person has type *T* blood. We will say we have a success if the person has type *T* blood. What is failure?

a) Failure occurs if a person does not have type *T* blood.

b) The probability of success is 0.15 since 15% of the population has type *T* blood. What is the probability of failure, *q*?

b) The probability of failure is

$$q = 1 - p$$
$$= 1 - 0.15 = 0.85$$

c) In this experiment there are $n =$ _____ trials.

d) We wish to compute the probability of three successes out of 18 trials. In this case $r =$ _____ .

c) In this experiment $n = 18$.

d) In this case $r = 3$.

In the next section we will see how to compute the probability of r successes out of n trials when we have a binomial experiment.

Section Problems 5.1

Which of the following are binomial experiments? For those that are binomial experiments,
  a) what makes up a *trial*?
  b) what is a *success*? a *failure*?
  c) what are the values of n, p, q, and r?
For those that are not binomial experiments, what is it that keeps them from being so?

1. At Community Hospital, the nursing staff is large enough so that 80% of the time a nurse can respond to a room call within three minutes. Last night there were 73 room calls. Nurses responded to 62 of them within three minutes.

2. A travel agent has four different packages which he claims are very popular. He claims that 99% of all customers enjoy the Hawaii package, 95% enjoy the Europe package, 96% of the participants enjoy the Alaska package, and 97% of participants enjoy the New York package. Last month he sold 51 packages and got back reports indicating that 43 people enjoyed them.

3. The local police department claims to solve 90% of all crimes committed in their jurisdiction. The local newspaper reported that there have been 54 crimes in the past two months, of which 46 have been solved.

4. A new over-the-counter medication is advertised as reducing blood pressure for 99% of those who use it. A random sample of 100 people with high blood pressure took the medication and eight were observed to have reduced blood pressure.

5. A random sample of 58 people watching TV showed that 18 preferred sports, 21 preferred talk shows, 10 preferred late night movies, and 9 preferred murder mystery shows. It is claimed that of the people who watch TV, 30% prefer sports, 36% prefer talk shows, 17% prefer late night movies, and 17% prefer murder mysteries.

6. It is claimed that 70% of all cars on the Valley Highway are going faster than 55 miles per hour. Radar was used to observe the speeds of 22 cars selected at random on the highway. However, when 11 of the cars were observed, a police car was in plain view of the drivers. When the other 11 cars were observed, no police car was in sight. Of the 22 cars observed, 12 were going over 55 miles per hour.

7. Six firefighters volunteer for weekend duty so that they can have Thanksgiving Day off. However, the Chief says that only two are needed that weekend. To decide who works the weekend duty instead of Thanksgiving Day, they take turns drawing straws. Four straws are of one length and two are shorter. The straws are not replaced after being drawn. The firefighters who draw the short straws get to work on the weekend instead of on Thanksgiving Day.

8. An investor claims that there is a 73% chance that a bull market will begin if the volume of stocks traded on the New York Stock Exchange remains over one hundred million shares per day for three weeks. Otherwise, the probability is only 20% that a bull market will occur. In a five-year period, nine bull markets began.

9. At the request of a pharmaceutical company, a doctor is testing a new drug on a random sample of 50 patients with Rocky Mountain spotted fever. The company claims the drug will cure 90% of the patients who use it. The doctor finds that 43 of the patients are in fact cured.

10. The Redi Battery Company claims 99.97% of its new batteries have a potential of no less than 1.5 volts. A random sample of 10,000 Redi batteries shows that 9,200 have a potential of 1.5 volts; the others had a lower potential.

11. An agricultural school claims to have invented a hormone that will cause 85% of the laying hens receiving the hormone to produce at least two eggs per hen per day. The hormone was given to a random sample of 500 hens one day at Zent's farm, and 360 of the hens laid at least two eggs that day.

12. Steve and Kathy guide climbing parties up Longs Peak. They say that when the weather is good the probability of reaching the summit (without having to turn around for some reason) is 0.80, but when the weather is not so good the probability is 0.50. Last summer Steve and Kathy guided 20 climbs. Sometimes the weather was good and sometimes it was bad, but they made it to the summit 14 times in all.

SECTION 5.2   The Binomial Distribution

The central problem of a binomial experiment is to find the probability of r successes out of n trials. In this section we'll see how to find these probabilities.

Suppose you are taking a timed final exam. You have three multiple-choice questions left to do. Each question has four suggested answers and only one of the answers is correct. You have only five seconds left to do these three questions, so you decide to mark answers on the answer sheet without even reading the questions. Assuming that your answers are randomly selected, what is the probability that you get zero, one, two, or all three questions correct?

This is a binomial experiment. Each of the questions can be thought of as a trial so there are $n = 3$ trials. The possible outcomes on each trial are success S, indicating a correct response, or failure F, meaning a wrong answer. The trials are independent since the outcome of any one trial does not affect the outcome of the others.

What is the probability of success on any question? Since you are guessing and there are four answers from which to select, the probability of a correct answer is 0.25. The probability q of a wrong answer is then 0.75. In short, we have a binomial experiment with $n = 3$, $p = 0.25$, and $q = 0.75$.

Now what are the possible outcomes in terms of success or failure for these three trials? Let's use the notation SSF to mean success on the first question, success on the second, and failure on the third. There are eight possible combinations of S's and F's. They are

$$SSS \quad SSF \quad SFS \quad SFF \quad FSS \quad FSF \quad FFS \quad FFF$$

To compute the probability of each outcome, we can use the multiplication law since the trials are independent. For instance, the probability of success on the first two questions and failure on the last is

$$P(SSF) = P(S) \cdot P(S) \cdot P(F) = p \cdot p \cdot q = p^2 q = (0.25)^2(0.75) = 0.047$$

In a similar fashion we can compute the probability of each of the eight outcomes. These are shown in Table 5-1, along with the number of successes r associated with each trial.

TABLE 5-1 Outcomes for a Binomial Experiment with $n = 3$ Trials

Outcome	Probability of Outcome	r (No. of Successes)
SSS	$P(SSS) = P(S)P(S)P(S) = p^3 = (0.25)^3 = 0.016$	3
SSF	$P(SSF) = P(S)P(S)P(F) = p^2 q = (0.25)^2(0.75) = 0.047$	2
SFS	$P(SFS) = P(S)P(F)P(S) = p^2 q = (0.25)^2(0.75) = 0.047$	2
FSS	$P(FSS) = P(F)P(S)P(S) = p^2 q = (0.25)^2(0.75) = 0.047$	2
SFF	$P(SFF) = P(S)P(F)P(F) = pq^2 = (0.25)(0.75)^2 = 0.141$	1
FSF	$P(FSF) = P(F)P(S)P(F) = pq^2 = (0.25)(0.75)^2 = 0.141$	1
FFS	$P(FFS) = P(F)P(F)P(S) = pq^2 = (0.25)(0.75)^2 = 0.141$	1
FFF	$P(FFF) = P(F)P(F)P(F) = q^3 = (0.75)^3 = 0.422$	0

Now we can compute the probability of r successes out of three trials for $r = 0, 1, 2,$ or 3. Let's compute $P(1)$. The notation $P(1)$ stands for the probability of one success. For three trials there are three different outcomes that show exactly one success. They are the outcomes SFF, FSF, and FFS. So

$$\begin{aligned}
P(1) &= P(SFF \text{ or } FSF \text{ or } FFS) = P(SFF) + P(FSF) + P(FFS) \\
&= pq^2 + pq^2 + pq^2 \\
&= 3pq^2 \\
&= 3(0.25)(0.75)^2 \\
&= 0.423
\end{aligned}$$

In the same way we can find $P(0)$, $P(2)$, and $P(3)$. These values are shown in Table 5-2.

We have done quite a bit of work to determine your chances of $r = 0, 1, 2,$ or 3 successes on three multiple-choice questions if you are just guessing. And now we can see that there is only a small chance (about 0.016) that you will get them all correct.

The model we constructed in Table 5-2 to compute the probability of r successes out of three trials can be used for any binomial experiment with $n = 3$ trials. Simply

TABLE 5-2 $P(r)$ for $n = 3$ Trials, $p = 0.25$

r (No. of Successes)	$P(r)$ (Probability of r Successes in 3 Trials)		$P(r)$ for $p = 0.25$
0	$P(0) = P(FFF)$	$= q^3$	0.422
1	$P(1) = P(SFF) + P(FSF) + P(FFS) = 3pq^2$		0.423
2	$P(2) = P(SSF) + P(SFS) + P(FSS) = 3p^2q$		0.141
3	$P(3) = P(SSS)$	$= p^3$	0.16

change the values of p and q to fit the experiment. In Exercise 4 we will use this model again.

Exercise 4

Maria is doing a study on the issue of the quarter system versus the semester system. To obtain faculty input she mails out questionnaires to the faculty. The probability that a faculty member returns the completed questionnaire is 0.65. Three faculty members chosen at random from the foreign language department are sent questionnaires. Compute the probability that *exactly two* completed questionnaires are returned and the probability that *all three* are returned. We'll do these computations in steps.

a) In this problem what are the values of n, p, q?

a) $n = 3, p = 0.65, q = 1 - 0.65 = 0.35$

b) The probability that exactly two questionnaires will be returned is $P(\underline{\hspace{1cm}})$. In this case $r = \underline{\hspace{1cm}}$. By Table 5-2

$$P(2) = 3p^2q \quad \text{for } n = 3 \text{ trials}$$

Use this formula to compute $P(2)$.

b) We want $P(2)$ so $r = 2$.

$$P(2) = 3(0.65)^2(0.35) = 0.444$$

c) Use the appropriate formula from Table 5-2 to compute the probability that all three questionnaires will be returned.

c) $P(3) = p^3 = (0.65)^3 = 0.275$

Table 5-2 can only be used as a model for computing the probability of r successes out of *three* trials. How can we compute the probability of seven successes out of ten trials? We can develop a table for $n = 10$, but this would be a tremendous task because there are 1,024 possible combinations of successes and failures on ten trials. Fortunately, mathematicians have given us a direct formula to compute the probability of r successes for any number of trials.

- **Formula for the binomial probability distribution:**

$$P(r) = C_{n,r}\,p^r q^{n-r}$$

where $C_{n,r}$ is the *binomial coefficient*. Values of $C_{n,r}$ for various values of n and r can be found in Table 3 of Appendix I.

Those of you who studied the section on counting techniques (Section 4.3) will recognize the symbol $C_{n,r}$ as the symbol used for the number of combinations of n objects taken r at a time. For those who did not study Section 4.3, a brief description of the meaning of the symbol $C_{n,r}$ follows. Examples and exercises will show you how to do the calculations necessary to compute $C_{n,r}$.

In the meantime, let's look more carefully at the formula itself. There are two main parts. The expression $p^r q^{n-r}$ is the probability of getting one outcome with r successes and $n - r$ failures. The binomial coefficient $C_{n,r}$ counts the number of outcomes that have r successes and $n - r$ failures. For instance, in the case of $n = 3$ trials we saw in Table 5-1 that the probability of getting an outcome with one success and two failures was pq^2. This is the value of $p^r q^{n-r}$ when $r = 1$ and $n = 3$. We also observed that there were three outcomes with one success and two failures so $C_{3,1}$ is 3.

Table 3 of Appendix I gives the values of the binomial coefficient $C_{n,r}$ for selected values of n and r. However, you can compute $C_{n,r}$ directly from a formula. This formula is boxed to indicate it is optional. You may skip it and go on to examples where we use the binomial probability distribution to compute $P(r)$.

OPTIONAL

The formula for the computation of the binomial coefficient $C_{n,r}$ is

$$C_{n,r} = \frac{n!}{r!(n-r)!}$$

where $n!$ (read, n factorial) is the product of n with all the counting numbers less than n. 0! is defined to be 1.

Example:

$$C_{5,3} = \frac{5!}{3!(5-3)!} = \frac{5!}{3!(2!)} = \frac{5 \cdot \overset{2}{\cancel{4}} \cdot \cancel{3} \cdot \cancel{2} \cdot \cancel{1}}{\cancel{3} \cdot \cancel{2} \cdot \cancel{1}(\cancel{2} \cdot \cancel{1})} = 10$$

This means that there are ten ways to list three successes and two failures when we have five trials.

Note: Some calculators have a factorial key, usually indicated by !. But if yours does not, you can compute a factorial by doing the necessary multiplications. It is a good idea to cancel as much as possible (as shown in the example) before doing the multiplications. Otherwise, you can easily generate numbers too large for the calculator display. Table 2 of Appendix I has values for 1 factorial through 20 factorial.

Now let's look at some applications of the binomial distribution formula.

EXAMPLE 3

A biologist is studying a new hybrid tomato. It is known that the seeds of this hybrid tomato have probability 0.70 of germinating. The biologist plants ten seeds.

a) What is the probability that *exactly* eight seeds will germinate?

This is a binomial experiment with $n = 10$ trials. Each seed planted represents an independent trial. We'll say germination is success, so the probability for success on each trial is 0.70.

$$n = 10 \qquad p = 0.70 \qquad q = 0.30 \qquad r = 8$$

We wish to find $P(8)$, the probability of exactly eight successes.

$$P(r) = C_{n,r}p^r q^{n-r}$$

$P(8) = C_{10,8}(0.70)^8(0.30)^2 \qquad C_{10,8} = 45 \text{ from Table 3, Appendix I}$

$\qquad = 45(0.0576)(0.090)$

$\qquad = 0.233$

b) What is the probability that *at least* eight seeds will germinate?

In this case we are interested in the probability of eight or more seeds germinating. This means we are to compute $P(r \geq 8)$. Since the events are mutually exclusive, we can use the addition rule

$$P(r \geq 8) = P(r = 8 \quad or \quad r = 9 \quad or \quad r = 10) = P(8) + P(9) + P(10)$$

We already know the value of $P(8)$. We need to compute $P(9)$ and $P(10)$.

$P(9) = C_{10,9}(0.70)^9(0.30)^1$

$\qquad = 10(0.0404)(0.30) \qquad C_{10,9} = 10 \text{ from Table 3, Appendix I}$

$\qquad = 0.121$

For $r = 10$ we have

$P(10) = C_{10,10}(0.70)^{10}(0.3)^0$

$\qquad = 1(0.028)(1) \qquad$ since any number to the 0 power equals 1,

$\qquad = 0.028 \qquad$ and $C_{10,10}$ is also 1 from Table 3, Appendix I.

Now we have all the parts necessary to compute $P(r \geq 8)$.

$$P(r \geq 8) = P(8) + P(9) + P(10)$$
$$= 0.233 + 0.121 + 0.028$$
$$= 0.382$$

Even with a table for the values of the binomial coefficient $C_{n,r}$, the computation of $P(r)$ is long because we must raise p to the rth power and q to the $(n - r)$th power. If n is large and your calculator does not have an exponent button, these computations can become quite time-consuming. Table 4 of Appendix I gives values for $P(r)$ for selected p and values of n through 20. To use the table, you find the section labeled with your value of n. Then you find the entry in the column headed by your p and the row labeled by the r value of interest. In Exercise 5 you'll practice using the formula for $P(r)$ in one part and then use Table 4 (Appendix I) for $P(r)$ values in the second part.

_____ Exercise 5 _____

A rarely performed and somewhat risky eye operation is known to be successful in restoring the eyesight of 30% of the patients who undergo the operation. A team of surgeons has developed a new technique for this operation that has been successful for four out of six operations. Does it seem likely that the new technique is much better than the old? We'll use the binomial probability distribution to answer this question.

a) Each operation is a binomial trial. In this case
$$n = \underline{\quad\quad}, p = \underline{\quad\quad}, q = \underline{\quad\quad}.$$

a) $n = 6, p = 0.30, q = 1 - 0.30 = 0.70$

b) Use your values of n, p, and q, as well as Table 3 of Appendix I to compute $P(4)$ from the formula
$$P(r) = C_{n,r}p^r q^{n-r}$$

b) $P(4) = C_{6,4}(0.30)^4(0.70)^2$
$$= 15(0.0081)(0.490)$$
$$= 0.060$$

c) Compute the probability of *at least* four successes out of the six trials.
$$P(r \geqslant 4) = P(r = 4 \quad \text{or} \quad r = 5 \quad \text{or} \quad r = 6)$$
$$= P(4) + P(5) + P(6)$$

Use Table 4 of Appendix I to find values of $P(4)$, $P(5)$, and $P(6)$. Then use these values to compute $P(r \geqslant 4)$.

c) To find $P(4)$, $P(5)$, and $P(6)$ in Table 4, we look in the section labeled $n = 6$. Then we find the column headed by $p = 0.30$. To find $P(4)$ we use the row labeled $r = 4$. For the values of $P(5)$ and $P(6)$ use the same column but change the row headers to $r = 5$ and $r = 6$, respectively.
$$P(r \geqslant 4) = P(4) + P(5) + P(6)$$
$$= 0.060 + 0.010 + 0.001$$
$$= 0.071$$

d) Under the older operation technique the probability that at least four patients out of six regain their eyesight is _____ . Does it seem that the new technique is better than the old? Would you encourage the surgeon team to do more work on the new technique?

d) It seems the new technique is better than the old since, by pure chance, the probability of four or more successes out of six trials is only 0.071 for the old technique. We would encourage the team to do more work on the new technique.

Section Problems 5.2

In each of the following problems the binomial distribution will be used. Please answer the following questions and then complete the problem.

    i) What makes up a trial? What is a success? What is a failure?
    ii) What are the values of n, p, and q?

1. A quarter is flipped three times.
    a) Find the probability of getting exactly three heads.
    b) Find the probability of getting exactly two heads.
    c) Find the probability of getting two or more heads.
    d) Find the probability of getting exactly three tails. (*Hint:* Getting exactly three tails is equivalent to getting exactly zero heads.)

2. A quarter is flipped 11 times.
   a) Find the probability of getting exactly six heads.
   b) Find the probability of getting more than six heads.
   c) Find the probability of getting fewer than six heads.
   d) Find the probability of getting from four to nine heads (including four and nine).

3. Richard has just been given a ten-question multiple-choice quiz in his history class. Each question has five answers, of which only one is correct. Since Richard has not attended class recently, he doesn't know any of the answers. Assuming Richard guesses on all ten questions, find the indicated probabilities.
   a) What is the probability that he will answer all questions correctly?
   b) What is the probability that he will answer five or more questions correctly?
   c) What is the probability that he will answer none of the questions correctly?
   d) What is the probability that he will answer at least three questions correctly?

4. A new parking lot on campus would require the destruction of several dozen very large spruce trees that are too large to dig up and replant. The student newspaper stated that 70% of the student body opposes the new parking lot. A random sample of 20 students were asked about the proposed new parking lot. Assuming the student newspaper is correct, find the probability that
   a) all 20 students oppose the new lot.
   b) at least 15 oppose the new lot.
   c) no more than ten oppose the new lot.
   d) from ten to fifteen oppose the new lot (including ten and fifteen).

5. The college health center did a campus-wide survey of students and found that 15% smoke. A group of nine students randomly come together in the student cafeteria and sit at the same table. Assume the cafeteria does not have a no smoking section and that a student who ordinarily smokes will smoke at this time. Find the probability that
   a) no student at this table smokes.
   b) at least one student smokes.
   c) more than two smoke.
   d) from one to five smoke (including one and five).

6. The mayor said that local fire stations are so well equipped and staffed that they are able to get to a fire in five minutes or less about 80% of the time. If this is so, what is the probability that out of the next ten fire calls
   a) the fire department arrives in five or less minutes for each call?
   b) for at least three calls it takes more than five minutes for the fire department to arrive?
   c) it takes more than five minutes for the fire department to arrive for all ten calls?
   d) it takes five or less minutes for half of the calls?

7. A TV sports commentator claims that 45% of all football injuries are knee injuries. Assuming that this claim is true, what is the probability that in a game with five reported injuries
   a) all are knee injuries?
   b) none are knee injuries?
   c) at least three are knee injuries?
   d) no more than two are knee injuries?

8. After examining daily receipts over the past year, it was found that the Blue Parrot Italian Restaurant has been grossing over $1,200 a day for about 85% of its business days. Using this as a reasonably accurate measure, find the probability that the Blue Parrot will gross over $1,200

a) at least five days in the next seven business days.
b) at least five days in the next ten business days.
c) less than three days in the next five business days.
d) less than seven days in the next ten business days.

9. A certain type of penicillin will cause a skin rash in 10% of the patients receiving it.
   a) If this penicillin is given to a random sample of 15 patients, what is the probability that no more than two will have a skin reaction?
   b) If it is given to a random sample of nine patients, what is the probability that at least one will have a reaction?
   c) If it is given to a random sample of 12 patients, what is the probability that more than two will have a reaction?
   d) If it is given to a random sample of six patients, what is the probability that more than two will have a reaction?

10. A safety engineer claims that 20% of all automobile accidents are due to mechanical failure. If this is correct, what is the probability that out of eight automobile accidents, exactly three are due to mechanical failure?

11. The Tasty Bean Coffee Company claims their coffee is so good that you can distinguish it from any other coffee. Five different brands of coffee (one of them Tasty Bean) are set before tasters who are to pick the one that tastes the best. Suppose there is really no difference in the way any of these coffees taste; however, each of four tasters picks one coffee anyway (not knowing which is which, because the coffee is in identical cups). What is the probability that
    a) all four tasters choose Tasty Bean?
    b) none of them chooses Tasty Bean?
    c) at least three choose Tasty Bean?

12. In Summit County 65% of the voter population are Republicans. What is the probability that a random sample of ten Summit County voters will contain
    a) exactly eight Republicans?
    b) exactly two Republicans?

13. Only three out of four patients who have an artery bypass heart operation are known to survive five years without needing another bypass operation. Of seven patients who recently had such an operation, what is the probability that
    a) all will survive five years without needing another operation?
    b) at least four will survive five years without needing another operation?
    c) exactly four will survive five years without needing another operation?

14. It is estimated that one out of five students on campus who want work-study aid qualify for the work-study program. Twelve students who want work-study randomly enter the college work-study office and ask to be admitted to the program. What is the probability that
    a) all of them are qualified?
    b) more than half of them are qualified?
    c) none of them is qualified?

15. Sam is a computer software salesman who has a history of making successful sales calls one fourth of the time. What is the probability that he will
    a) be successful on exactly three of the next five calls?
    b) be successful on at least three of the next five calls?
    c) be successful on no more than two of the next five calls?

Additional Properties of the Binomial Distribution

Any probability distribution may be represented in graphical form. How should we graph the binomial distribution? Remember, the binomial distribution tells us the probability of *r* successes out of *n* trials. So, we'll place values of *r* along the horizontal axis and values of $P(r)$ on the vertical axis. The binomial distribution is a discrete probability distribution since *r* can assume only whole number values such as 0, 1, 2, 3, and so on. Therefore, a histogram is an appropriate graph of a binomial distribution. Let's look at an example to see exactly how we'll make these histograms.

EXAMPLE 4

A waiter at the Green Spot Restaurant has learned from long experience that the probability a lone diner will leave a tip is only 0.7. During one lunch hour he serves six people who are dining by themselves. Make a graph of the binomial probability distribution which shows the probabilities that 0, 1, 2, 3, 4, 5, or all 6 lone diners leave tips.

This is a binomial experiment of $n = 6$ trials. *Success* is achieved when the lone diner leaves a tip. So the probability of success is 0.7 and that of failure is 0.3.

$$n = 6 \qquad p = 0.7 \qquad q = 0.3$$

We want to make a histogram showing the probability of *r* successes when $r = 0, 1, 2, 3, 4, 5$, or 6. It is easier to make the histogram if we first make a table of *r* values and the corresponding $P(r)$ values (Table 5-3). We'll use Table 4 of Appendix I to find the $P(r)$ values for $n = 6$ and $p = 0.70$.

To construct the histogram we'll put *r* values on the horizontal axis and $P(r)$ values on the vertical axis. Our bars will be one unit wide and will be centered over the appropriate *r* value. The height of the bar over a particular *r* value tells the probability of that *r*. (*See* Figure 5-1.)

TABLE 5-3 Binomial Distribution for $n = 6$ and $p = 0.70$

r	$P(r)$
0	0.001
1	0.010
2	0.060
3	0.185
4	0.324
5	0.303
6	0.118

FIGURE 5-1   Graph of the Binomial Distribution for $n = 6$ and $p = 0.7$

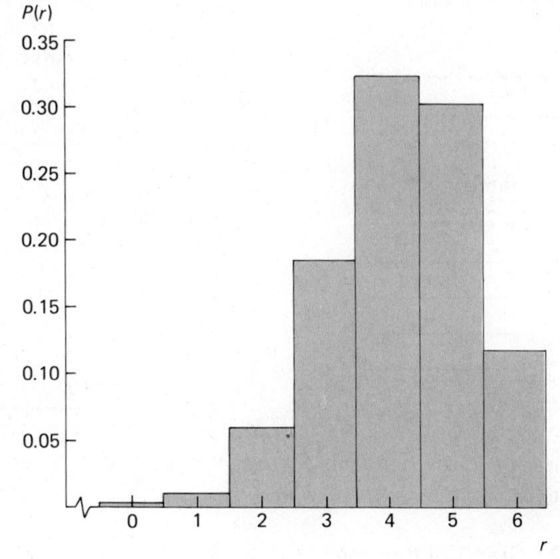

The probability of a particular value of r is given not only by the height of the bar over that r value, but also by the *area* of the bar. Each bar is only one unit wide so the area (area = height times width) equals its height. Since the area of each bar represents the probability of the r value under it, the sum of the areas of the bars must be 1. In Example 4 the sum turns out to be 1.001. It is not exactly equal to 1 because of round-off error.

Exercise 6 illustrates another binomial distribution with $n = 6$ trials. The graph will be different from that of Figure 5-1 because the probability of success p is different.

───────────── Exercise 6 ─────────────────────────────────

Jim enjoys playing basketball. He figures that he makes about 50% of the field goals he attempts during a game. Make a histogram showing the probability that Jim will make 0, 1, 2, 3, 4, 5, or 6 shots out of six attempted field goals.

a) This is a binomial experiment with $n =$ _____ trials. In this situation we'll say success occurs when Jim makes an attempted field goal. What is the value of p?

a) In this example $n = 6$ and $p = 0.5$.

b) Use Table 4 of Appendix I to complete Table 5-4 of $P(r)$ values for $n = 6$ and $p = 0.5$.

TABLE 5-4

r	$P(r)$
0	0.016
1	0.094
2	0.234
3	_____
4	_____
5	_____
6	_____

b)

TABLE 5-5 Completion of Table 5-4

r	$P(r)$
.	.
.	.
.	.
3	0.312
4	0.234
5	0.094
6	0.016

c) Use the values of $P(r)$ given in Table 5-4 to complete the histogram in Figure 5-2.

FIGURE 5-2   Graph of Binomial Distribution for $n = 6$ and $p = 0.5$

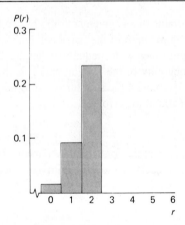

c)

FIGURE 5-3   Completion of Figure 5-2

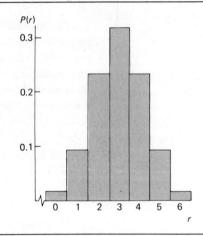

d) The area of the bar over $r = 2$ is 0.234. What is the area of the bar over $r = 4$? How does the probability that Jim makes two field goals out of six compare with the probability that he makes four field goals out of six?

d) The area of the bar over $r = 4$ is also 0.234. Jim is as likely to make two out of six field goals as he is to make four out of six.

In Example 4 and Exercise 6 we see the graphs of two binomial distributions associated with $n = 6$ trials. The two graphs are different because the probability of success p is different in the two cases. In Example 4, $p = 0.7$ and the graph is skewed to the left—i.e., the left tail is longer. In Exercise 6, p is equal to 0.5 and the graph is symmetrical—that is, if we fold it in half, the two halves coincide exactly. Whenever *p equals 0.5 the graph of the binomial distribution will be symmetrical no matter how many trials we have.* In Chapter 7 we will see that if the number of trials n is quite large, the binomial distribution is almost symmetrical even when p is not close to 0.5.

Two other features that help describe the graph of any distribution are the balance point of the distribution and the spread of the distribution about that balance point. The *balance point* is the mean μ of the distribution, and the *measure of spread* that is most commonly used is the standard deviation σ. The mean μ is the *expected value* of the number of successes.

For the binomial distribution there are two special formulas we can use to compute the mean μ and the standard deviation σ. These are much easier to use than the general formulas given in Section 4.4 for μ and σ of any probability distribution.

- **For the binomial distribution:**

$$\mu = np$$
$$\sigma = \sqrt{npq}$$

where n is the number of trials,
     p is the probability of success,
     q is the probability of failure ($q = 1 - p$).

EXAMPLE 5

Let's compute the mean and standard deviation for the distribution of Example 4 that describes the probabilities of lone diners leaving tips at the Green Spot Restaurant. In that example

$$n = 6 \qquad p = 0.7 \qquad q = 0.3$$

For the binomial distribution

$$\mu = np$$
$$= 6(0.7) = 4.2$$

The balance point of the distribution is at $\mu = 4.2$. The standard deviation is given by

$$\sigma = \sqrt{npq}$$
$$= \sqrt{6(0.7)(0.3)}$$
$$= \sqrt{1.26}$$
$$\approx 1.12$$

The mean μ is not only the balance point of the distribution; it is also the expected value of r. Specifically, in Example 4 the waiter can expect 4.2 lone diners out of six to leave a tip. (The waiter would probably round the expected value to four tippers out of six.)

──────────── Exercise 7 ────────────────────────

When Jim (of Exercise 6) shoots field goals in basketball games, the probability that he makes a shot is only 0.5.

a) The mean of the binomial distribution is the expected value of r successes out of n trials. Out of six throws, what is the expected number of goals Jim will make?

a) The expected value is the mean μ

$$\mu = np = 6(0.5) = 3$$

Jim can expect to make three goals out of six tries.

b) For six trials, what is the standard deviation of the binomial distribution of the number of successful field goals Jim makes?

b) $\sigma = \sqrt{npq} = \sqrt{6(0.5)(0.5)} = \sqrt{1.5} = 1.22$

In applications you do not want to confuse the expected value of r with certain probabilities associated with r. Exercise 8 illustrates this point.

──────────── Exercise 8 ────────────────────────

A satellite requires three solar cells for its power. The probability that any one of these cells will fail is 0.15, and the cells operate or fail independently.

Part I: In this part we want to find the least number of cells the satellite should have so that the *expected value* of the number of working cells is no smaller than three. In this situation n represents the number of cells, r is the number of successful or working cells, p is the probability that a cell will work, q is the probability that a cell will fail, and μ is the expected value that should be no smaller than three.

a) What is the value of q? of p?

a) $q = 0.15$ as given in the problem. p must be 0.85 since $p = 1 - q$.

b) The expected value μ for the number of working cells is given by $\mu = np$. The expected value of the number of working cells should be no smaller than three so

$$3 \le \mu = np$$

From part (a) we know the value of p. Solve the inequality $3 \le np$ for n.

b)
$$3 \le np$$
$$3 \le n(0.85)$$
$$\frac{3}{0.85} \le n \quad \text{divide both sides by 0.85}$$
$$3.53 \le n$$

c) Since n is between 3 and 4, would you round it to 3 or 4 to make sure that μ is at least 3?

c) n should be at least 3.53. Since we can't have a fraction of a cell, we had best make $n = 4$. For $n = 4$, $\mu = 4(0.85) = 3.4$. This value satisfies the condition that μ be at least 3.

Part II: In this part we want to find the smallest number of cells the satellite should have to be 97% sure that there will be adequate power—i.e., that at least three cells work.

a) The letter r has been used to denote the number of successes. In this case r represents the number of working cells. We are trying to find the number n of cells necessary to assure that (choose the correct statement)

(i) $P(r \geqslant 3) = 0.97$ or
(ii) $P(r \leqslant 3) = 0.97$

a) $P(r \geqslant 3) = 0.97$

b) We need to find a value for n so that

$$P(r \geqslant 3) = 0.97$$

Let's try $n = 4$. Then $r \geqslant 3$ means $r = 3$ or 4 so

$$P(r \geqslant 3) = P(3) + P(4)$$

Use Table 4 with $n = 4$ and $p = 0.85$ to find values of $P(3)$ and $P(4)$. Then compute $P(r \geqslant 3)$ for $n = 4$. Will $n = 4$ guarantee that $P(r \geqslant 3)$ is at least 0.97? (Table 4 is in Appendix I.)

b) $P(3) = 0.368$
$P(4) = 0.522$
$P(r \geqslant 3) = 0.368 + 0.522 = 0.890$

Thus $n = 4$ is *not* sufficient to be 97% sure that at least three cells will work. For $n = 4$ the probability that at least three will work is only 0.890.

c) Now try $n = 5$ cells. For $n = 5$

$$P(r \geqslant 3) = P(3) + P(4) + P(5)$$

since r can be 3, 4, or 5. Are $n = 5$ cells adequate? [Be sure to find new values of $P(3)$ and $P(4)$ since we now have $n = 5$.]

c) $P(r \geqslant 3) = P(3) + P(4) + P(5)$
$= 0.138 + 0.392 + 0.444$
$= 0.974$

Thus $n = 5$ cells are required if we want to be 97% sure that there will be at least three working cells.

In Part I and Part II we got different values for n. Why? In Part I we had $n = 4$ and $\mu = 3.4$. This means that if we put up lots of satellites with four cells, we can expect that an *average* of 3.4 cells will be working per satellite. But for $n = 4$ cells there is only a probability of 0.89 that at least three cells will work in any one satellite. In Part II we are trying to find the number of cells necessary so that the probability is 0.97 that at least three will work in any *one* satellite. If we use $n = 5$ cells, then we can satisfy this requirement.

Section Problems 5.3

1. A TV preference survey showed that 35% of all households watching evening cable TV are watching a sports-related program. Suppose a random sample of six households watching evening cable TV are contacted.
   a) Make a histogram showing the probability of $r = 0, 1, 2, 3, 4, 5, 6$ households watching a sports-related program.
   b) Find the mean μ of this probability distribution. How many of the six households do we expect to be watching a sports-related program?
   c) Find the standard deviation σ of the probability distribution.

2. An insurance company says 15% of all fires are caused by arson. A random sample of five fire insurance claims are under study.
   a) Make a histogram showing the probability of $r = 0, 1, 2, 3, 4, 5$ arson fires out of five fires.
   b) Find the mean μ of this probability distribution. What is the expected number of arson fires out of five fires?
   c) Find the standard deviation σ of the probability distribution.

3. Safety engineers say an automobile tire that has less than 1/32 inch of rubber tread is not safe. It has been estimated that 10% of all cars in Colorado have unsafe tires. The highway patrol recently stopped seven cars and did a routine safety check including a check of tire tread.
   a) Make a histogram showing the probability of r autos with unsafe tires for $r = 0, 1, 2, 3, 4, 5, 6, 7$.
   b) Find the mean μ of this probability distribution. How many cars do we expect to have unsafe tires?
   c) Find the standard deviation σ of the probability distribution.

4. National studies indicate that 45% of all senior citizens (those over 65 years of age) have high blood pressure. At a nursing station in a medical clinic 12 senior citizens have just completed a routine physical exam.
   a) Let r represent the number of senior citizens who have high blood pressure. Make a histogram showing the probability of r for $r = 0$ through 12.
   b) Find the mean μ of this probability distribution. What is the expected number of these senior citizens who have high blood pressure?
   c) Find the standard deviation σ of the probability distribution.

5. Over a long period of time it has been observed that Vicky has probability 0.7 of hitting a target with a single rifle shot. Suppose Vicky fires five shots at the target.
   a) Make a histogram showing the probability of r successes (hits) when $r = 0, 1, 2, 3, 4, 5$.
   b) Find the mean μ of this probability distribution. How many hits is Vicky expected to make?
   c) Find the standard deviation σ of this probability distribution.

6. The quality control inspector of a production plant will reject a batch of syringes if two or more defectives are found in a random sample of eight syringes taken from the batch. Suppose the batch contains 1% defective syringes.
   a) Make a histogram showing the probability of $r = 0, 1, 2, 3, 4, 5, 6, 7, 8$ defective syringes in a random sample of eight syringes.
   b) Find μ. What is the expected number of defective syringes the inspector will find?
   c) What is the probability that the batch will be accepted?
   d) Find σ.

7. Garfield College has an art appreciation course in the humanities program that is taken on a pass/fail basis. Over a long period of time, the art department has observed that about 80% of the students pass the course. Sixteen students are enrolled in a typical section of the course this term. Let's say that from a student's view the course was a success if he or she passed.
   a) Find the probability $P(r)$ of r successes for r ranging from 0 to 16.
   b) Make a histogram for the probability distribution of part (a).
   c) What is μ, the expected number of students who will pass?
   d) What is the standard deviation σ?

8. A new serum is claimed to be 70% effective in preventing the common cold. A random sample of 11 people are injected with the serum. The serum is said to be successful if a person survives the winter without a cold.
    a) Make a histogram showing the probability of r successes for all values of r from 0 to 11.
    b) Find the mean of the distribution of part (a). What is the expected number of people who will survive the winter without a cold?
    c) Find the standard deviation of the distribution of part (a).

9. A biologist has found that 40% of all brown bears are infected with trichinosis.
    a) What is the expected value of r, the number of infected brown bears, in a sample of 27?
    b) What is the standard deviation of the probability distribution of infected brown bears in the random sample of part (a)?

10. The National Forest Service uses satellites with infrared sensors to detect forest fires in remote wilderness areas. The satellites are very sensitive and do find all the hot spots. However, the satellites register geothermal hot spots, cloud cover, manmade bonfires, etc., as well as hot spots caused by forest fires. Over a period of years, the Forest Service has decided that only 75% of the hot spots reported by the satellite are forest fires. The usual procedure is to send out a reconnaissance plane to check on what the satellite has detected. Let us say that the satellite was successful if it reports a hot spot that is a forest fire.
    Recently the satellite reported nine hot spots in the remote Seward Peninsula of Alaska.
    a) Find the probability $P(r)$ of r successes for r ranging from 0 to 9.
    b) Make a histogram for the probability distribution of part (a).
    c) Based on the satellite information, what is the expected number μ of real forest fires on the Seward Peninsula?
    d) What is the standard deviation σ?

11. Alice is a social worker whose job is to counsel juvenile delinquents on parole from a detention center. Each of the juveniles Alice counsels is a first-time offender, but Alice knows from past work with them that about 35% will become second-time offenders. Let us say success means that a juvenile does not become a second-time offender. Alice has been given a group of 12 delinquents.
    a) Find the probability $P(r)$ of r successes for r ranging from 0 to 12.
    b) Make a histogram for the probability distribution of part (a).
    c) What is the expected number μ of juveniles in Alice's group who will not be second offenders?
    d) What is the standard deviation σ?

12. At the local community hospital, newborn babies are routinely given antibiotics to prevent postnatal infection. However, the antibiotics are known to give a minor skin rash to 25% of the babies. Six newborn babies have been given the antibiotics.
    a) Let r represent the number of babies that get a skin rash. Find $P(r)$ as r ranges from 0 to 6.
    b) Make a histogram of the probability distribution of part (a).
    c) What is the expected number μ of babies to get the rash?
    d) What is the standard deviation σ?

13. The probability that a single radar station will detect an enemy plane is 0.65. How many such stations are required to be 98% certain that an enemy plane will be detected by at

least one station? If four stations are in use, what is the expected number of stations that will detect an enemy plane?

14. Salvage Products is a mail-order firm that sells merchandise at a substantial discount because some of the items are damaged or have missing parts. Customers may not return merchandise. The catalogue lists rubber-coated raincoats at 80% off regular retail price. However, many of the raincoats leak. The catalogue states that 40% of the raincoats in stock leak. How many of these raincoats should you order to be 97% sure that at least one does not leak? If you order ten raincoats, what is the expected number that do not leak?

15. A large bank vault has several automatic burglar alarms. The probability is 0.55 that a single alarm will detect a burglar. How many such alarms should be used to be 99% certain that a burglar is detected by at least one alarm? Suppose the bank installs nine alarms. What is the expected number of alarms that would detect a burglar?

16. The owners of a motel in Florida have noticed that in the long run about 40% of the people who stop and inquire about a room for the night actually rent a room. How many inquiries must the owner answer to be 99% sure of renting at least one room? If 25 separate inquiries are made about rooms, what is the expected number that will result in room rentals?

Summary

A binomial experiment must meet the following criteria:

1. There are a fixed number of trials denoted by n.

2. The trials are independent and repeated under identical conditions.

3. Each trial has only two outcomes: success S and failure F.

4. For each trial the probability p of success remains the same. The probability of failure is $q = 1 - p$.

5. The central problem is to find the probability of r successes out of n trials.

The formula for the binomial probability distribution is

$$P(r) = C_{n,r} p^r q^{n-r}$$

where $C_{n,r}$ is the binomial coefficient as found in Table 3 of Appendix I. For certain values of p and n, $P(r)$ can be found directly in Table 4 of Appendix I.

The mean or expected value and standard deviation of the binomial distribution are given by the formulas

$$\mu = np$$

$$\sigma = \sqrt{npq} \quad \text{where } q = 1 - p$$

Important Words and Symbols

	Section		Section
Binomial experiment	5.1	Binomial probability	
Independent trials	5.1	distribution	
Success and failure in a		$P(r) = C_{n,r}p^r q^{n-r}$	5.2
binomial experiment	5.1	n factorial, $n!$	5.2
Probability of success $P(S)$		Mean for the binomial	
$= p$	5.1	distribution $\mu = np$	5.3
Probability of failure $P(F)$		Standard deviation for the	
$= q$	5.1	binomial distribution	
Binomial coefficient $C_{n,r}$	5.2	$\sigma = \sqrt{npq}$	5.3

Chapter Review Problems

1. It is known that 80% of all guinea pigs injected with a certain culture will contract red blood cell anemia. In a laboratory experiment ten guinea pigs are injected with the culture.
   a) Let r be the number of guinea pigs that contract the red blood cell anemia. Make a histogram for the r-distribution probabilities.
   b) What is the probability that six or more guinea pigs will get the red blood cell anemia?
   c) What is the expected number of guinea pigs that will get the red blood cell anemia?
   d) What is the standard deviation of the r probability distribution?

2. A stationery store has decided to accept a large shipment of ball-point pens if an inspection of 20 randomly selected pens yields no more than two defective pens.
   a) Find the probability that this shipment is accepted if 5% of the total shipment is defective.
   b) Find the probability that this shipment is not accepted if 15% of the total shipment is defective.

3. The student government claims that 85% of all students favor an increase in student fees to buy indoor potted plants for the classrooms. A random sample of 12 students produced two in favor of the project. What is the probability that two or fewer in the sample will favor the project, assuming the student government is correct? Do the data support the student government claim or does it seem the percentage favoring the increase in fees is less than 85%?

4. An employee of the Dry Gulch Distillery knows that 1,000 of 10,000 Dry Gulch whisky bottles have been filled with water. A random sample of six bottles is selected. What is the probability that
   a) at least one bottle is filled with water?
   b) none of the bottles is filled with water?

5. Seventy-five percent of the trees planted by Quick Grow Landscaping survive. Quick Grow has just planted 16 trees on campus.

    a) What is the probability that 12 or more survive?

    b) What is the probability that less than half survive?

    c) What is the expected number of trees that will survive?

6. The three engines of a jet airliner are arranged to operate independently. The probability of an in-flight engine failure is 0.05 for each single engine. Let r represent the number of engines that fail during a flight.

    a) Make a histogram of the probability distribution of r.

    b) What is the probability that no failures occur? What is the probability that more than one failure occurs?

    c) What is the expected value and standard deviation of the r probability distribution?

7. It is estimated that 75% of a grapefruit crop is good; the other 25% have rotten centers which cannot be detected unless the grapefruit are cut open. The grapefruit are sold in sacks of ten. Let r be the number of good grapefruit in a sack.

    a) Make a histogram of the probability distribution of r.

    b) What is the probability of getting no more than one bad grapefruit in a sack? What is the probability of getting at least one good grapefruit in a sack?

    c) What is the expected number of good grapefruit in a sack?

    d) What is the standard deviation of the r probability distribution?

8. Camp Wee-O-Wee has found that about 8% of young campers get poison ivy each season. If 273 children are registered for the summer season, about how many can be expected to get poison ivy?

9. A survey has found that about 17% of all M.D.s in the United States have changed their specialty at least once. In a city with 600 M.D.s what is the expected number who have changed specialties?

10. The Orchard Cafe has found that about 5% of the parties who make reservations don't show up. If 82 party reservations have been made, how many can be expected to show up? Find the standard deviation of this distribution.

11. The We Care Lawn Service has found that about one out of five people will respond favorably to a certain telephone sales pitch. Suppose 15 people are called. Let r be the number who respond favorably.

    a) Make a histogram for the r probability distribution.

    b) What is the expected number of people who will respond favorably? Find the standard deviation of the r distribution.

    c) What is the probability that at least three respond favorably? What is the probability that exactly three respond favorably?

12. The dropout rate at Rock High is 17%. If 1,500 students have enrolled, how many can be expected to drop out?

13. Jack Rabbit Car Wax has developed two new car waxes: Flopsie and Mopsie. To determine if consumers really prefer one over the other, both waxes were applied to the same car (Flopsie on the right side and Mopsie on the left). A random sample of 20 car owners were asked which side of the car had the best polish job. If there is actually no difference in consumer preference (i.e., 50% prefer Flopsie), what is the probability that at most five car owners state a preference for Flopsie?

14. When David drives from Columbus to Cincinnati, the probability is 0.75 that at any given time he is going faster than the speed limit. If there are four radar traps between

Columbus and Cincinnati, what is the probability David will be caught at least once (assume independence)?

15. Ten thousand thumbtacks were dropped on a table and 6,500 landed with the point up. If 500 tacks are dropped, how many would you expect to land point up?

16. There are three true–false questions on a psychology test. Rita is out of time and just guesses, so the probability of a correct answer is 50%.

   a) Draw a histogram showing the probability of $r = 0, 1, 2, 3$ correct answers for the three questions.

   b) What is the expected number of correct answers?

   c) Find the standard deviation of the r distribution.

USING COMPUTERS

Professional statisticians in industry and research use computers to help them analyze and process statistical data. There are many computer programs available for statistics. Some commonly used statistical packages include Minitab, SAS, and SPSS. There are many others as well. Any of these statistical packages may be used with this text.

The authors of this text have also written a computer statistics package called ComputerStat. The programs in ComputerStat are available on disks with an accompanying manual that explains how to use the programs and gives additional computer applications. The disks are intended for the Apple IIe, IIc, or the IBM PC. Programs are written in the computer language BASIC and are designed for the beginning statistics student with no previous computer experience. ComputerStat is available as a supplement from the publisher of this text.

Problems in this section may be done using ComputerStat or other statistical computer programs.

Although tables of binomial probabilities can be found in most libraries, such tables are often inadequate. Either the value of p (the probability of success on a trial) you are looking for is not in the table, or the value of n (the number of trials) you are looking for is too large for the table. In Chapter 7 we will study the normal approximation to the binomial. This approximation is a great help in many practical applications. Even so, we sometimes use the formula for the binomial probability distribution on a computer or programmable calculator to compute the probability we want.

The following percentages were obtained over many years of observation by the U.S. Weather Bureau. All data listed are for the month of December.

Location	Long Term Mean % of Clear Days in Dec.
Juneau, Alaska	18%
Seattle, Washington	24%
Hilo, Hawaii	36%
Honolulu, Hawaii	60%
Las Vegas, Nevada	75%
Phoenix, Arizona	77%

Adapted from *Local Climatological Data*, U.S. Weather Bureau publication, "Normals, Means, and Extremes" Table.

In the locations listed, the month of December is a relatively stable month with respect to weather. Since weather patterns from one day to the next are more or less the same, it is reasonable to use a binomial probability model.

1. Let r be the number of clear days in December. Since December has 31 days, $0 \leq r \leq 31$. In the ComputerStat main menu of topics, select Probability Distributions and Central Limit Theorem. Then use program S6, Binomial Coefficients and Probability Distributions, to find the probability $P(r)$ for each of the listed locations when $r = 0, 1, 2, \ldots, 31$.

2. For each location what is the expected value of the probability distribution? What is the standard deviation?

You may find that using cumulative probabilities and appropriate subtraction of probabilities will make the solution of problems 3 to 7 easier than adding probabilities.

3. Estimate the probability that Juneau will have at most seven clear days in December.

4. Estimate the probability that Seattle will have from five to ten (including five and ten) clear days in December.

5. Estimate the probability that Hilo will have at least 12 clear days in December.

6. Estimate the probability that Phoenix will have 20 or more clear days in December.

7. Estimate the probability that Las Vegas will have from 20 to 25 (including 20 and 25) clear days in December.

6 Normal Distributions

*One cannot escape the feeling that these mathematical formulas
have an independent existence and an intelligence of their own,
that they are wiser than we are, wiser even than their discoverers,
that we get more out of them than was originally put into them.*

Heinrich Hertz

*How can it be that mathematics, a product of human thought
independent of experience, is so admirably adopted to the objects
of reality?*

Albert Einstein

Heinrich Hertz was a pioneer in the study of radio waves. His work and the later
work of Maxwell and Marconi led the way to modern radio, television, and radar.
Albert Einstein is world renowned for his great discoveries in relativity and quantum
mechanics. Everyone who has worked in both mathematics and real-world applications cannot help but marvel how the "pure thought" of the mathematical sciences
can predict and explain events in other realms. In this chapter we will study the
single most important type of probability distribution in all of mathematical statistics:
the normal distribution. Reasons for the importance of the normal distribution are
that it applies to a wide variety of situations and other distributions tend to become
normal under certain conditions.

SECTION 6.1   Graphs of Normal Probability Distributions

One of the most important examples of a continuous probability distribution is the
normal distribution. This distribution was studied by the French mathematician
Abraham de Moivre (1667–1754) and later by the German mathematician Carl

Friedrich Gauss (1777–1855), whose work is so important that the normal distribution is sometimes called Gaussian. The work of these mathematicians provided a foundation upon which much of the theory of statistical inference is based.

Applications of a normal probability distribution are so numerous that some mathematicians refer to it as "a veritable Boy Scout knife of statistics." However, before we can do applications, we must examine some of the properties of a normal distribution.

A rather complicated formula, presented in advanced statistics books, defines a normal distribution in terms of μ and σ, the mean and standard deviation of the population distribution. It is only through this formula that we can verify if a distribution is normal. However, we can look at the graph of a normal distribution and get a good pictorial idea of some of the essential features of any normal distribution.

The graph of a normal distribution is called a *normal curve*. It possesses a shape very much like the cross section of a pile of dry sand. Because of its shape blacksmiths would sometimes use a pile of dry sand in the construction of a mold for a bell. Thus, the normal curve is also called a bell-shaped curve (*see* Figure 6-1).

FIGURE 6-1 A Normal Curve

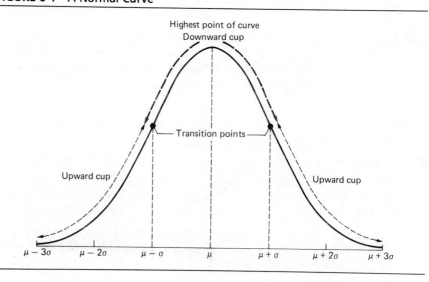

We see that a general normal curve is smooth and symmetrical about the vertical line over the mean μ. Notice that the highest point of the curve occurs over μ. If the distribution were graphed on a piece of sheet metal, cut out, and placed on a knife edge, the balance point would be at μ. We also see that the curve tends to level out and approach the horizontal (x-axis) like a glider making a landing. However, in mathematical theory such a glider would never quite finish its landing because a normal curve never touches the horizontal axis.

The parameter σ controls the spread of the curve. The curve is quite close to the horizontal axis at $\mu + 3\sigma$ and $\mu - 3\sigma$. Thus, if the standard deviation σ is large, the curve will be more spread out; if it is small, the curve will be more peaked. Figure 6-1 shows the normal curve cupped downward for an interval on either side of the mean μ. Then it begins to cup upward as we go to the lower part of the bell. The exact places where the transition between upward and downward cupping occur are above the points $\mu + \sigma$ and $\mu - \sigma$.

Let's summarize the important properties of a normal curve.

1. The curve is "bell-shaped" with the highest point over the mean μ.
2. It is symmetrical about a vertical line through μ.
3. The curve approaches the horizontal axis but never touches or crosses it.
4. The transition points between cupping upward and downward occur at $\mu + \sigma$ and $\mu - \sigma$.

_____ Exercise 1 _____

Each of the curves in Figure 6-2 fails to be a normal curve. Give reasons why these curves are not normal curves.

FIGURE 6-2

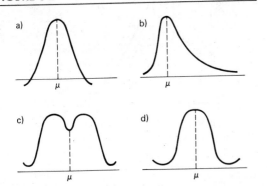

a)

b)

c)

d)

a) A normal curve gets closer and closer to the horizontal axis, but it never touches it or crosses it.

b) A normal curve must be symmetric. This curve is not.

c) A normal curve is bell-shaped with one peak. Since this curve has two peaks, it is not normal.

d) The tails of a normal curve must get closer and closer to the x-axis. In this curve the tails are going away from the x-axis.

_____ Exercise 2 _____

The points A, B, and C are indicated on the normal curve in Figure 6-3. One of these points is μ, one is $\mu + \sigma$, and one is $\mu - 2\sigma$.

FIGURE 6-3 A Normal Curve

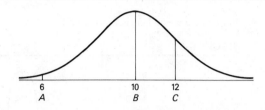

a) Which point corresponds to the mean? What is the value of μ?

b) Which point corresponds to $\mu + \sigma$? Use the values of $\mu + \sigma$ and μ to compute σ.

c) Which point corresponds to $\mu - 2\sigma$?

a) The mean μ is under the peak of the normal curve. The point B corresponds to the mean, so $\mu = 10$.

b) The point C where the curve changes from cupped down to cupped up is one standard deviation σ from the mean. The point C is $\mu + \sigma$. Since $\mu + \sigma = 12$ and $\mu = 10$, $\sigma = 2$.

c) Since $\mu = 10$ and $\sigma = 2$ we see that

$$\mu - 2\sigma = 10 - 2(2) = 6$$

Point A corresponds to $\mu - 2\sigma$.

The parameters which control the shape of a normal curve are the mean μ and the standard deviation σ. When both μ and σ are specified, a specific normal curve is determined. In brief, μ locates the balance point and σ determines the extent of the spread. Figure 6-4 shows some normal curves for different μ's and σ's.

FIGURE 6-4 Examples of Normal Curves

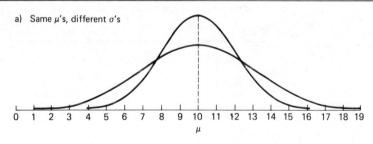

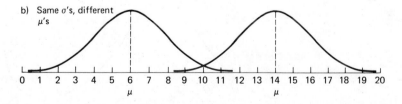

_____ Exercise 3 _____

Look at the normal curves in Figure 6-5.

FIGURE 6-5

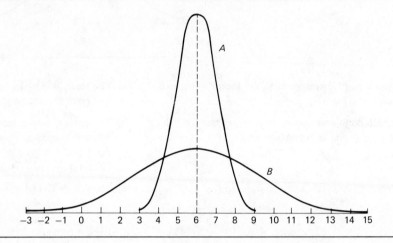

a) Do these distributions have the same mean? If so, what is it?

b) One of the curves corresponds to a normal distribution with $\sigma = 3$, and the other to one with $\sigma = 1$. Which curve has which σ?

a) The means are the same since both graphs have the high point over 6. $\mu = 6$.

b) Curve A has $\sigma = 1$, and curve B has $\sigma = 3$. (Since curve B is more spread out, it has the larger σ value.)

_____ Exercise 4 _____

Look at the normal curves in Figure 6-6.

FIGURE 6-6

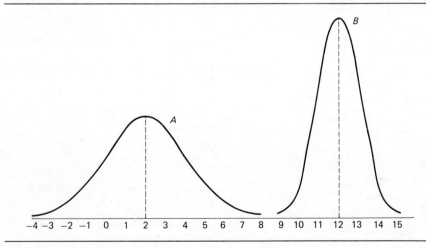

a) Do these distributions have the same mean? What is the mean of each distribution?

a) No. Curve A has $\mu = 2$; curve B has $\mu = 12$.

b) Is it true that of the two curves, the curve with the larger mean must also have the larger standard deviation? Explain your answer.

b) It is not true. In general, the mean and standard deviations have no influence on each other. So a curve with a large mean need not have a large standard deviation.

c) The standard deviations of these curves are not given, but the shape of the curves indicates that one curve has a larger standard deviation than the other. Which curve has the larger standard deviation?

c) Curve A is more spread out than curve B, so curve A has the larger standard deviation.

The total area under any normal curve studied in this book will *always* be *one*. The graph of the normal distribution is important because the portion of the *area* under the curve above a given interval represents the *probability* that a measurement will lie in that interval. The portions of the total area within one, two, and three standard deviations from the mean are shown in Figure 6-7. From the figure you can see that almost all the measurements from a normal distribution lie within three standard deviations of the mean.

FIGURE 6-7   Area Under a Normal Curve

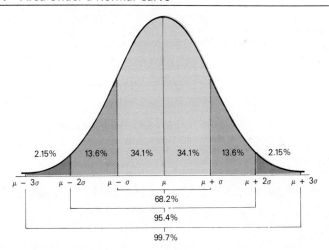

EXAMPLE 1

The playing life of a Sunshine radio is normally distributed with mean $\mu = 600$ hr and standard deviation $\sigma = 100$ hr. What is the probability that a radio selected at random will last from 600 to 700 hours?

The probability that the playing time will be between 600 and 700 hours is equal to the percentage of the total area under the curve that is shaded in Figure 6-8. Since $\mu = 600$ and $\mu + \sigma = 600 + 100 = 700$, we see that the shaded area is simply the area between μ and $\mu + \sigma$. The area from μ to $\mu + \sigma$ is 34.1% of the total area. This tells us that the probability a Sunshine radio will last between 600 and 700 playing hours is 0.341.

FIGURE 6-8 Distribution of Playing Times

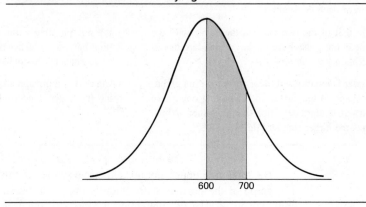

———— Exercise 5 ————

The yearly wheat yield per acre on a particular farm is normally distributed with mean $\mu = 35$ bushels and standard deviation $\sigma = 8$ bushels.

a) Shade the area under the curve in Figure 6-9 that represents the probability that an acre will yield between 19 and 35 bushels.

a)

FIGURE 6-9

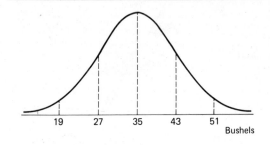

FIGURE 6-10 Completion of Figure 6-9

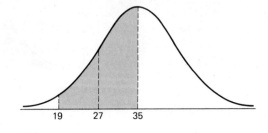

b) Is the area the same as the area between $\mu - 2\sigma$ and μ?

b) Yes, since $\mu = 35$ and $\mu - 2\sigma = 35 - 2(8) = 19$.

c) Use Figure 6-7 to find the per cent of area over the interval between 19 and 35.

c) The area between the values $\mu - 2\sigma$ and μ is 47.7% of the total area.

d) What is the probability that the yield will be between 19 and 35 bushels per acre?

d) It is 47.7% of the total area, which is 1. Therefore, the probability is 0.477 that the yield will be between 19 and 35 bushels.

Example 1 and Exercise 5 both involve intervals beginning with one number and ending with another from among the following:

$$\mu - 3\sigma \quad \mu - 2\sigma \quad \mu - \sigma \quad \mu \quad \mu + \sigma \quad \mu + 2\sigma \quad \mu + 3\sigma$$

The following facts are used for those problems:

1. About 68.2% of the area under a normal curve falls in the interval between $\mu + \sigma$ and $\mu - \sigma$.
2. About 95.4% of the area falls between $\mu + 2\sigma$ and $\mu - 2\sigma$.
3. About 99.7% of the area falls between $\mu + 3\sigma$ and $\mu - 3\sigma$.

These intervals and percents are important to remember because they serve as rough guides to normal distributions.

However, to find the probability that a measurement lies in an interval such as that between $\mu + 0.4\sigma$ and $\mu + 0.7\sigma$, we need more machinery. This machinery is developed in the next section. In Sections 6.3 and 6.4 we see how to find the probability that a measurement from a normal distribution lies in *any* specified interval.

Section Problems 6.1

1. Which, if any, of the curves in Figure 6-11 look(s) like a normal curve? If a curve is not a normal curve, tell why.

   FIGURE 6-11

   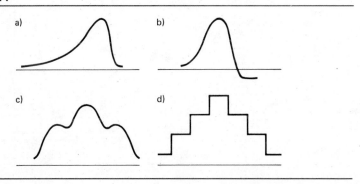

   a)   b)

   c)   d)

2. Look at the normal curve in Figure 6-12 and find μ, $\mu + \sigma$, and σ.

FIGURE 6-12

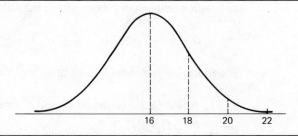

16   18   20   22

3. Look at the two normal curves in Figures 6-13 and 6-14. Which has the larger standard deviation? What is the mean of the curve in Figure 6-13? What is the mean of the curve in Figure 6-14?

FIGURE 6-13

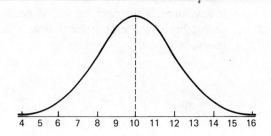

4   5   6   7   8   9   10   11   12   13   14   15   16

FIGURE 6-14

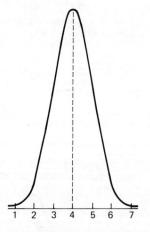

1   2   3   4   5   6   7

4. Sketch a normal curve
   a) with mean 15 and standard deviation 2.

    b) with mean 15 and standard deviation 3.
    c) with mean 12 and standard deviation 2.
    d) with mean 12 and standard deviation 3.
    e) Consider two normal curves. If the first one has a larger mean than the second one, must it have a larger standard deviation than the second one as well?

5. What percentage of the area under the normal curve lies
    a) to the left of μ?
    b) between $\mu - \sigma$ and $\mu + \sigma$?
    c) between $\mu - 3\sigma$ and $\mu + 3\sigma$?

6. What percentage of the area under a normal curve lies
    a) to the right of μ?
    b) between $\mu - 2\sigma$ and $\mu + 2\sigma$?
    c) to the right of 3σ?

7. Assuming the heights of college women are normally distributed, with mean 62 in. and standard deviation 3 in., answer the following questions. (*Hint:* Use problem 6 and Figure 6-7.)
    a) What percent of women are taller than 62 in.?
    b) What percent of women are shorter than 62 in.?
    c) What percent of women are between 59 in. and 65 in.?
    d) What percent of women are between 56 in. and 68 in.?

8. The incubation time for Rhode Island Red chicks is normally distributed with mean 21 days and standard deviation approximately one day. Look at Figure 6-7 and answer the following questions. If 1,000 eggs are being incubated, how many chicks do we expect will hatch
    a) in 19 to 23 days?
    b) in 20 to 22 days?
    c) in 21 days or fewer?
    d) in 18 to 24 days? (Assume all eggs eventually hatch.)
(*Note:* In this problem let us agree to think of a single day or a succession of days as a continuous interval of time.)

9. At a large government office center in Chicago, it has been found that the time employees spend on the phone each day for business purposes has a normal distribution with mean 47 minutes and standard deviation 10 minutes. Examine Figure 6-7 and answer the following questions. If an employee is selected at random, what is the probability that he or she spends
    a) from 27 to 67 minutes on the phone each day?
    b) from 37 to 57 minutes on the phone each day?
    c) 47 or fewer minutes on the phone each day?

10. A vending machine automatically pours soft drinks into cups. The amount of soft drink dispensed into a cup is normally distributed with mean 7.6 oz and standard deviation 0.4 oz. Examine Figure 6-7 and answer the following questions.
    a) Estimate the probability that the machine will overflow an 8 oz cup.
    b) Estimate the probability that the machine will not overflow an 8 oz cup.
    c) The machine has just been loaded with 850 cups. How many of these do you expect will overflow when served?

SECTION 6.2   Standard Units and the Standard Normal Distribution

Normal distributions vary from one another in two main ways: the mean μ may be located anywhere on the x-axis, and the bell shape may be more or less spread according to the size of the standard deviation σ. The differences among the normal distributions cause difficulties when we try to compute the area under the curve in a specified interval and, hence, the probability that a measurement will fall in that interval.

It would be a futile task to try to set up a table of areas under the normal curve for each different μ and σ combination. We need a way to standardize the distributions so that we can use *one* table of areas for *all* normal distributions. We achieve this standardization by considering how many standard deviations a measurement lies from the mean. In this way we can compare a value in one normal distribution with a value in another different normal distribution. The next situation shows how this is done.

Suppose Tina and Jack are in two different sections of the same course. Each section is quite large, and the scores on the midterm exams of each section follow a normal distribution. In Tina's section, the average (mean) was 64 and her score was 74. In Jack's section the mean was 72 and his score, 82. Both Tina and Jack were pleased that their scores were each ten points above the average of each respective section. But the fact that each was ten points above average does not really tell us how each did *with respect to the other students in the section.* We need to know more about the shape of the distribution. In Figure 6-15 we see the normal distribution of grades for each section.

FIGURE 6-15   Distributions of Midterm Scores

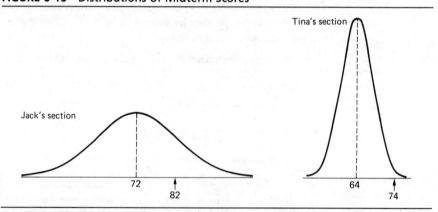

Tina's 74 was higher than most of the other scores in her section, while Jack's 82 is only an upper middle score in his section. Tina's score is far better with respect to her class than Jack's score with respect to his class.

The previous situation demonstrates that it is not sufficient to know the difference between a measurement (x value) and the mean of a distribution. We need also to consider the spread of the curve, or the standard deviation. What we really want to know is the number of standard deviations between a measurement and the mean. This "distance" takes both μ and σ into account.

There is a simple formula that we can use to compute the number z of standard deviations between a measurement x and the mean μ of a normal distribution with standard deviation σ.

$$\begin{pmatrix} \text{Number of standard deviations} \\ \text{between the measurement and} \\ \text{the mean} \end{pmatrix} = \begin{pmatrix} \dfrac{\text{Difference between the}}{\text{measurement and the mean}} \\ \text{Standard deviation} \end{pmatrix}$$

Written in symbols, this formula is

$$z = \frac{x - \mu}{\sigma}$$

- **Definition:** The *z value* or *z score* tells us the number of standard deviations the original measurement is from the mean. The *z* value is in *standard units*.

The mean is a special value of a distribution. Let's see what happens when we convert $x = \mu$ to a z value:

$$z = \frac{x - \mu}{\sigma}$$

$$= \frac{\mu - \mu}{\sigma} \quad \text{for } x = \mu$$

$$= 0$$

The mean of the original distribution is always zero in standard units. This makes sense because the mean is zero standard deviations from itself.

An x value in the original distribution that is *above* the mean μ has a corresponding z value that is *positive*. Again this makes sense since a measurement above the mean would be a positive number of standard deviations from the mean. Likewise, an x value *below* the mean has a *negative z* value. (*See* Table 6-1).

TABLE 6-1 x Values and Corresponding z Values

x Value in Original Distribution	Corresponding z Value or Standard Unit
$x = \mu$	$z = 0$
$x > \mu$	$z > 0$
$x < \mu$	$z < 0$

- **Note:** Unless otherwise stated, in the remainder of this book we will take the word *average* to be the arithmetic mean $\bar{x}$.

EXAMPLE 2

A pizza parlor franchise specifies that the average (mean) amount of cheese on a large pizza should be 8 oz and the standard deviation only 0.5 oz. An inspector picks out a large pizza at random in one of the pizza parlors and finds that it is made with 6.9 oz of cheese. Assume that the amount of cheese on a pizza follows a normal distribution. If the amount of cheese is more than *three* standard deviations below the mean, the parlor will be in danger of losing its franchise. (Remember, in a normal distribution we are unlikely to find measurements more than three standard deviations from the mean since 99.7% of all measurements fall within three standard deviations of the mean.)

How many standard deviations from the mean is 6.9? Is the pizza parlor in danger of losing its franchise?

Since we want to know the number of standard deviations from the mean, we want to convert 6.9 to standard z units.

$$z = \frac{x - \mu}{\sigma}$$

$$= \frac{6.9 - 8}{0.5}$$

$$= -2.2$$

So the amount of cheese on the selected pizza is only 2.2 standard deviations from the mean. Note that the fact that z is negative indicates that the amount of cheese was 2.2 standard deviations *below* the mean. The parlor will not lose its franchise based on this sample.

_____ Exercise 6 _____

A student has computed that it takes an average (mean) of 17 min with a standard deviation of 3 min to drive from home, park the car, and walk to an early morning class.

a) One day it took the student 21 min to get to class. How many standard deviations from the average is that? Is the z value positive or negative? Explain why it should be either positive or negative.

a) The number of standard deviations from the mean is given by the z value

$$z = \frac{x - \mu}{\sigma} = \frac{21 - 17}{3} = 1.33$$

The z value is positive. We would expect a positive z value since 21 min is *more* than the mean of 17.

b) Another day it took only 12 min for the student to get to class. What is this measurement in standard units? Is the z value positive or negative? Why should it be positive or negative?

b) The measurement in standard units is

$$z = \frac{x - \mu}{\sigma} = \frac{12 - 17}{3} = -1.67$$

Here the z value is negative, as we should expect, because 12 min is less than the mean of 17 min.

c) Another day it took 17 min for the student to go from home to class. What is the z value? Why should you expect this answer?

c) In this case the z value is

$$z = \frac{x - \mu}{\sigma} = \frac{17 - 17}{3} = 0$$

We expect this result because 17 min is the mean, and the z value of the mean is always zero.

We have seen how to convert from x measurements to standard units z. We can easily reverse the process if we know μ and σ for the original distribution. For when we solve

$$z = \frac{x - \mu}{\sigma}$$

for x, we get

$$x = \sigma z + \mu$$

EXAMPLE 3

In Example 2 we talked about the amount of cheese required by a franchise for a large pizza. Again, the mean amount of cheese required is 8 oz with a standard deviation of 0.5 oz. The parlor can lose its franchise if the amount of cheese on their large pizza is more than three standard deviations below the mean. What is the minimum amount of cheese that can be placed on a large pizza according to the franchise?

Here we need to convert $z = -3$ to information about x oz of cheese. We use the formula

$$
\begin{aligned}
x &= \sigma z + \mu \\
&= 0.5(-3) + 8 \\
&= 6.5 \text{ oz.}
\end{aligned}
$$

The franchise will not approve a large pizza with less than 6.5 oz of cheese.

In many testing situations we hear the terms *raw score* and z *score*. The raw score is just the score in the original measuring units and the z score is the score in standard units. Exercise 7 illustrates these different units.

_____ Exercise 7 _____

Marulla's z score on a college entrance exam is 1.3. If the raw scores have a mean of 480 and standard deviation of 70 points, what is her raw score?

Here we are given z, σ, and μ. We need to find the raw score x corresponding to the z score 1.3.

$$x = \sigma z + \mu$$
$$= 70(1.3) + 480$$
$$= 571$$

EXAMPLE 4

A tire manufacturer claims that the average life of its tires is 18,000 mi. The tire life is normally distributed with a standard deviation of 1,400 mi. The company will refund two thirds of the original purchase price if a tire wears out at a mileage two or more standard deviations below the mean. (The company should not have to give too many refunds since 95.4% of the tires should have a life within 2σ of μ.) Translate the condition for a two-thirds cost refund into actual miles on the tire.

Here we are interested in all z values less than or equal to -2 so $z \leq -2$. We need to translate this condition into a statement about x. The upper bound for z is -2 (*see* Figure 6-16).

FIGURE 6-16

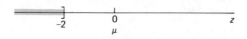

If $z = -2$, then

$$x = \sigma z + \mu = (1,400)(-2) + 18,000$$
$$= -2,800 + 18,000$$
$$= 15,200 \text{ mi}$$

So $z \leq -2$ means $x \leq 15,200$ mi (Figure 6-17).

FIGURE 6-17

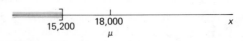

_____ Exercise 8 _____

A company called Camp Comfort makes down-filled sleeping bags. The amount of down in an adult sleeping bag has a mean of 32 oz and a standard deviation of 0.9 oz. Quality control specifies that if the down fill in a product is less than or equal to -2.7 standard deviations from the mean, the item must be sold as a second.

a) Does -2.7 standard deviations less than the mean translate to $z \leq -2.7$?

a) Yes.

b) The upper limit is $z = -2.7$. Translate this value into a weight in ounces for down in a sleeping bag.

b) $x = \sigma z + \mu = 0.9(-2.7) + 32$
$= -2.43 + 32 = 29.57$

c) $z \leq -2.7$ is the same as $x \leq$ _____

c) $z \leq -2.7$ is the same as $x \leq 29.57$

d) Indicate these values on the number lines in Figure 6-18.

d)

FIGURE 6-18

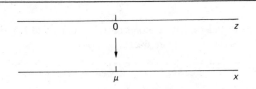

FIGURE 6-19 Completion of Figure 6-18

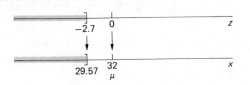

EXAMPLE 5

In a large class the final exam scores are normally distributed with a mean score of 75 points and a standard deviation of 12 points. The C exams have scores ranging from 63 to 87. What are these scores in standard units?

We have x from 63 to 87 points (Figure 6-20). This is the same as $63 \leq x \leq 87$.

FIGURE 6-20

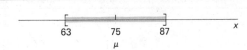

The upper x score is 87. This corresponds to an upper z score of

$$z = \frac{x - \mu}{\sigma} = \frac{87 - 75}{12} = \frac{12}{12} = 1$$

The lower x score is 63. This corresponds to a z score of

$$z = \frac{x - \mu}{\sigma} = \frac{63 - 75}{12} = \frac{-12}{12} = -1$$

The C students are those with scores within one standard deviation of the mean. That means that about 68.2% of the exams are C exams since the scores are normally distributed.

_____ Exercise 9 _____

The professor gives B's to students in the class described in Example 5 who have scores ranging from 88 to 93.

What are the z scores for the B students? Indicate the possible z scores on a number line.

The B students have x scores ranging from 88 to 93.

The upper bound is 93. That corresponds to a z score of 1.5 since

$$z = \frac{x - \mu}{\sigma} = \frac{93 - 75}{12} = \frac{18}{12} = 1.5$$

FIGURE 6-21

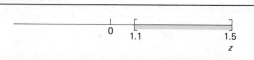

The lower bound x = 88 corresponds to a z score of

$$z = \frac{x - \mu}{\sigma} = \frac{88 - 75}{12} = \frac{13}{12} \approx 1.1$$

The B students have z scores ranging from 1.1 to 1.5, or $1.1 \le z \le 1.5$ (Figure 6-21).

If the original distribution of *x values is normal,* then the corresponding *z values have a normal distribution as well.* The z distribution has a mean of *zero* and a standard deviation of *one.* The normal curve with these properties has a special name.

• **Definition:** The *standard normal distribution* is a normal distribution with mean $\mu = 0$ and standard deviation $\sigma = 1$ (Figure 6-22).

FIGURE 6.22   The Standard Normal Distribution ($\mu = 0, \sigma = 1$)

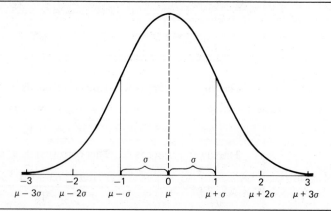

Any normal distribution of *x* values can be converted to the standard normal distribution by converting all *x* values to their corresponding *z* values. Let's look at the graphical interpretation of this transformation in Figure 6-23.

FIGURE 6-23   The Transformation of a Normal Distribution to the
Standard Normal Distribution

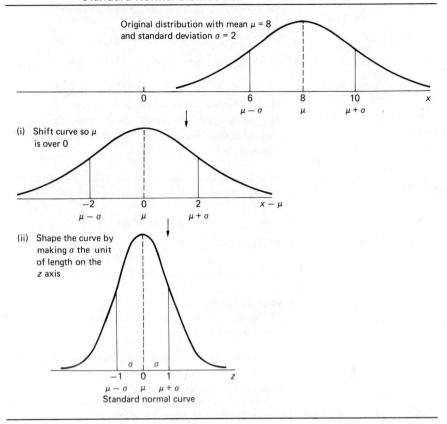

In this section we have studied standard units and the standard normal distribution. It is important to remember that every normal distribution can be transformed into a standard normal distribution by using the formula

$$z = \frac{x - \mu}{\sigma}$$

The resulting standard distribution will always have mean $\mu = 0$ and standard deviation $\sigma = 1$.

Section Problems 6.2

In these problems assume that all the distributions are *normal*. In all problems in Chapter 6, *average* is always taken to be the arithmetic mean, $\bar{x}$.

1. The college Physical Education Department offered an Advanced First Aid course last semester. The scores on the comprehensive final exam were normally distributed and the z scores for some of the students are shown below:

   Robert, 1.1   Jan, 1.7   Susan, -2.0

   Joel, 0.0   John, -0.8 Linda, 1.6

   a) Which of these students scored above the mean?
   b) Which of these students scored on the mean?
   c) Which of these students scored below the mean?
   d) If the mean score was $\mu = 150$ with standard deviation $\sigma = 20$, what was the final exam score for each student?

2. Scores on teacher evaluations at Hoople State College follow a normal distribution. The standardized scores for members of the Mathematics Department are:

   Dr. Lee, 0.3   Dr. Willis, 0.0

   Dr. Smith, -1.2 Mr. Lang, 2.3

   a) Which professors scored above the mean?
   b) Which scored below the mean?
   c) Which scored on the mean?
   d) If the mean score was $\mu = 73$ and the standard deviation was $\sigma = 10$, what was the raw score for each professor?

3. You Drive Car Rental Company keeps careful records of mileage and maintenance for each of its cars. For the fleet of one-year-old cars, the mileages on the various rental cars followed a normal distribution with mean 12,750 miles and standard deviation 1,540 miles. Find standardized z values for the following mileages of one-year-old rental cars.
   a) 13,900   c) 11,475   e) 14,000
   b) 12,750   d) 9,780

4. Air Connect is a commuter airline that serves the East Coast. The time interval from the time the planes begin to load passengers to the time of takeoff has been found to be normally distributed with mean 25 minutes and standard deviation 6 minutes. Find the standardized z values for the following passenger loading times:
   a) 21   c) 25
   b) 12   d) 38

5. Data collected over a period of years shows that the average daily temperature in Honolulu is 72 °F and the standard deviation is 10 °F. During one particular year, the average daily temperature ranged from 53° to 93°. Convert the interval $53° \leqslant x \leqslant 93°$ to an interval of z values.

6. Let x represent the life of a 60-watt light bulb. The x distribution has mean $\mu = 1,000$ hours, with standard deviation $\sigma = 75$ hours. Convert each of the following x intervals to standard z intervals.
   a) $450 \leqslant x \leqslant 1,350$ d) $500 \leqslant x$
   b) $900 \leqslant x \leqslant 1,100$ e) $x \leqslant 300$
   c) $990 \leqslant x \leqslant 1,010$ f) $x \leqslant 1,200$

7. Let x represent the average miles per gallon of gasoline that owners get from their new Nippon model automobile. For this model car the mean of the x distribution is advertised to be $\mu = 44$ mpg, with standard deviation $\sigma = 6$ mpg. Convert each of the following x intervals to standard z intervals.

   a) $x \geqslant 44$ d) $x \leqslant 49$
   b) $40 \leqslant x \leqslant 50$ e) $43 \leqslant x \leqslant 45$
   c) $32 \leqslant x \leqslant 39$ f) $x \leqslant 38$

8. Intelligence Quotient test scores (IQ scores) have a population mean $\mu = 100$ and population standard deviation $\sigma = 15$. Convert each of the following given standard z intervals to $x =$ IQ score intervals.

   a) $0 \leqslant z \leqslant 1$ d) $-1.96 \leqslant z$
   b) $0 \leqslant z \leqslant 0.42$ e) $z \leqslant 2.58$
   c) $-2.04 \leqslant z \leqslant 2.07$ f) $z \geqslant 0.94$

9. A high school counselor was given the following z intervals concerning a vocational training aptitude test. The test scores had a mean $\mu = 450$ points and standard deviation of $\sigma = 35$ points. Convert each z interval to an $x =$ test score interval.

   a) $-1.14 \leqslant z \leqslant 2.27$ d) $-1.96 \leqslant z \leqslant 1.96$
   b) $z \leqslant -2.58$ e) $z \leqslant 0$
   c) $1.645 \leqslant z$ f) $1.28 \leqslant z \leqslant 1.44$

10. David and Laura are both applying for a position on a ski patrol team. David took the Advanced First Aid course at his college. His score on the comprehensive final exam was 173 points. The final exam scores followed a normal distribution with mean 150 and standard deviation 25. Laura took an Advanced First Aid course at the manufacturing plant where she works. For the method of testing used there, her cumulative score on all exams was 88. The cumulative scores followed a normal distribution with mean 65 and standard deviation 10. Both courses are comparable in content and difficulty. There is only one position available on the ski patrol and David and Laura are equally qualified as skiers.

   a) Both David and Laura scored 23 points above their respective means. Does this mean they both gave the same performance in the First Aid course? Explain your answer.

   b) Would you choose David or Laura for the ski patrol position based on knowledge of first aid and skiing ability? Explain your answer by locating both z scores for performance in the First Aid course under a standard normal curve.

11. The general manager of a computer software sales company is reviewing the sales record of two sales representatives, Niko and Walter. In Niko's district the long-term mean sales have been \$4,173 each month with standard deviation \$671. In Walter's district the long-term mean sales have been \$5,520 each month with standard deviation \$450. In both districts the distribution of monthly sales is approximately normal.

   Recently Niko's sales have been \$5,884 each month in his district and Walter's sales have been \$6,000 each month in his district.

   a) Niko sold less than Walter in recent months. Does this mean Niko is not as good a sales representative? Explain.

   b) Convert the recent sales records for both Niko and Walter to standard z values. Locate these z values under a standard normal curve and compare them.

   c) If both Niko and Walter were up for promotion, and only one could be promoted, which would you choose? Explain.

12. Professors Adams and Riley are both teaching different sections of the same course, Political Science 201. Harold is in Professor Adams's class, and Mary is in Professor

Riley's class. On the midterm exam Harold got 193 points while Mary got 182 points. The scores of both exams were normally distributed, with Professor Adams's exam having a mean $\mu_1 = 180$ and standard deviation $\sigma_1 = 20$ points. For Professor Riley's exam, the mean was $\mu_2 = 165$ with standard deviation $\sigma_2 = 10$ points. If both professors grade on a normal curve, who did better, Harold or Mary? Explain your answer.

13. Two of your friends have written you letters explaining how they are training for the famous Boston Marathon. George is training on a hilly jogging loop along the Oregon coast. The jogging club in this area knows that for the general population of runners, the mean time to complete this loop is $\mu_1 = 167.4$ minutes, with standard deviation $\sigma_1 = 25.9$ minutes. Fred is training on a flat and longer jogging loop in the Miami, Florida, area. For the general population of runners it is known that the mean time to complete this loop is $\mu_2 = 143.1$ minutes, with standard deviation $\sigma_2 = 20.7$ minutes. George says his time on the Oregon loop is 91.5 minutes, and Fred says his time on the Miami loop is 86.2 minutes. Who do you think is in better condition, George or Fred? Explain your answer.

14. Jim scored 630 in a national bankers examination in which the mean is 600 and the standard deviation is 70. June scored 530 on the Hoople College bankers examination for which the mean is 500 and the standard deviation is 25. If Jim and June both apply for a job at Hoople State Bank and each examination has equal weight toward getting the job, who has the better chance? Explain your answer.

SECTION 6.3   Areas Under the Standard Normal Curve

In Section 6.2 we saw how we can convert *any* normal distribution to the *standard* normal distribution. We can change any *x* value to a *z* value and back again. But what is the advantage of all this work? The advantage is that there are extensive tables that show the area under the standard normal curve for almost any interval along the *z*-axis. The areas are important because they are equal to the *probability* that the measurement of an item selected at random falls in this interval. Thus, the *standard* normal distribution can be a tremendously helpful tool.

For instance, Sunshine Stereo guarantees their turntables for two years. The company statistician has computed that the turntable life is normally distributed with mean 2.3 yr and standard deviation 0.4 yr. What is the probability that a turntable will stop working during the guarantee period?

To answer questions of this type, we convert the given normal distribution to the standard normal distribution. Then we use a table to find the area over the interval in question and, hence, the probability an item selected at random will fall in that interval. Before we can carry out this plan, though, we must practice using Table 5 of Appendix 1 to find areas under the standard normal curve. Figure 6-24 illustrates areas which correspond to various interval descriptions.

In Section 6.1 we observed that for a normal distribution about 68.2% of the data falls within one standard deviation of the mean, about 95.4% falls within two standard deviations, and about 99.7% falls within three standard deviations of the mean. For the standard normal curve, the standard deviation equals one, and the mean is zero. The interval within one standard deviation of the mean is just the interval from −1 to 1. Figure 6-25 shows the standard normal curve and the areas within one, two, and three standard deviations of the mean.

Figure 6-25 shows us how to obtain certain areas under a standard normal curve. How do we find other areas under the standard normal curve? The most convenient way is to use a table. Because of the symmetry of the normal curve it is possible to obtain all the areas we will need if we have only a table of areas from $z = 0$ to $z = $ some positive number. The following sequence of examples will show how this can be done using Table 5 of Appendix I.

FIGURE 6-24 Areas Under a Standard Normal Probability Curve

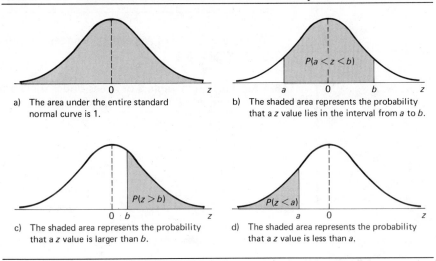

a) The area under the entire standard normal curve is 1.

b) The shaded area represents the probability that a z value lies in the interval from a to b.

c) The shaded area represents the probability that a z value is larger than b.

d) The shaded area represents the probability that a z value is less than a.

FIGURE 6-25 The Standard Normal Curve

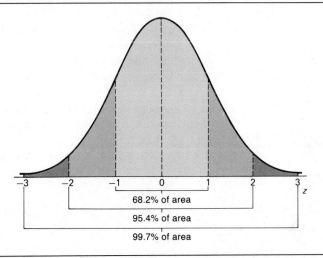

68.2% of area

95.4% of area

99.7% of area

EXAMPLE 6

Find the area under the standard normal curve between $z = 0$ and $z = 1$. This area is shown in Figure 6-26.

FIGURE 6-26 Area Between $z = 0$ and $z = 1$

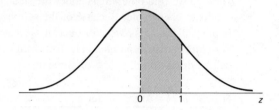

We will use Table 5 of Appendix I. For convenience we have included part of Table 5 in Table 6-2.

TABLE 6-2 Portion of Table 5 of Areas Under the Standard Normal Curve from $z = 0$ to the Indicated Value of z

z	0.00	0.01	0.02	0.03	0.04	0.05 . . .	0.09
.							
.							
.							
0.9	0.3159	0.3186	0.3212	0.3238	0.3264	0.3289 . . .	0.3389
1.0	0.3413	0.3438	0.3461	0.3485	0.3508	0.3531 . . .	0.3621
1.1	0.3643	0.3665	0.3686	0.3708	0.3729	0.3949 . . .	0.3830
.							
.							
.							
2.5	0.4938	0.4940	0.4941	0.4943	0.4945	0.4946 . . .	0.4952

In the upper left corner of the table we see the letter z. The column under z gives us the units value and tenths for z. The other column headings indicate the hundredths value of z. The entries in the table give the areas under the normal curve from the mean $z = 0$ to a specified value of z. To find the area from $z = 0$ to $z = 1$, we observe that if $z = 1$, then the units value of z is 1 and the tenths value is 0. So we look in the column labeled z for 1.0. The area from $z = 0$ to $z = 1$ is given in the corresponding row of the column with heading 0.00 since $z = 1$ is the same as $z = 1.00$. The area we read from the table for $z = 1.00$ is 0.3413. It is the shaded value in Table 6-2.

_____ Exercise 10 _____

In this exercise we will find the area under the standard normal curve from $z = 0$ to $z = 2.53$. We will use the abbreviated Table 6-2.

a) Shade the area we are to find in Figure 6-27.

FIGURE 6-27

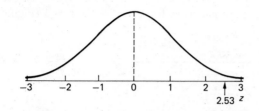

a)

FIGURE 6-28 Completion of Figure 6-27

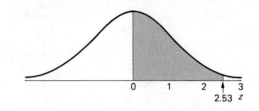

b) For $z = 2.53$ the units value is _____ and the tenths value is _____ , so we look in the column under z for the number _____ .

c) For $z = 2.53$ the hundredths value is _____ , so we look in the column headings for the number 0.03.

d) The area between $z = 0$ and $z = 2.53$ is given by the entry in the row beginning with 2.5 and in the column headed by 0.03. What is the area?

b) The unit value is 2 and the tenths value is 0.5, so we look in the z column for the value 2.5.

c) The hundredths value is 0.03.

d) The area is 0.4943.

Since the normal curve is symmetric about its mean, we can use Table 5 of Appendix I to find an area under the curve between a *negative z* value and 0 as in Figure 6-29.

FIGURE 6-29 Area Between $z = -2.34$ and 0

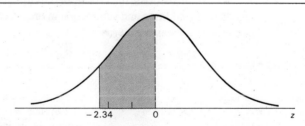

EXAMPLE 7

Find the area under the standard normal curve from $z = -2.34$ to 0. The area from $z = -2.34$ to 0 is the same as the area from $z = 0$ to 2.34. (*See* Figure 6-30.)

FIGURE 6-30 Area from $z = -2.34$ to 0 Equals Area from $z = 0$ to 2.34

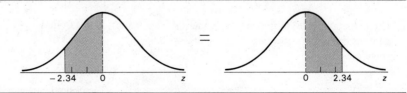

By Table 5, Appendix I, the area from 0 to 2.34 is 0.4904. Therefore, the area from $z = -2.34$ to 0 is also 0.4904.

EXAMPLE 8

Find the area under the standard normal curve from $z = -2$ to $z = 1$ (Figure 6-31).

FIGURE 6-31   Area from $z = -2$ to $z = 1$

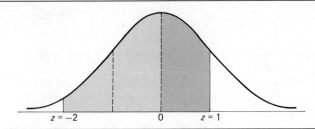

We do this in two parts. First we find the area from -2 to 0. Because the curve is symmetrical the areas to the left of 0 are the same as the corresponding areas to the right of 0. Therefore, the area from -2 to 0 is the same as the area from $z = 0$ to $z = 2$. By Table 5, we see that the area is 0.4772. From Example 6 we know the area from $z = 0$ to $z = 1$ is 0.3413. The area from $z = -2$ to $z = 1$ is the sum

$$(\text{area from } -2 \text{ to } 1) = (\text{area from } -2 \text{ to } 0) + (\text{area from } 0 \text{ to } 1)$$
$$\downarrow \qquad\qquad\qquad \downarrow \qquad\qquad\qquad \downarrow$$
$$0.8185 \qquad = \qquad 0.4772 \qquad + \qquad 0.3413$$

To find areas other than those between a given z value and 0, we use Table 5 together with addition and subtraction of areas between a given z value and 0. Figure 6-32 gives five basic patterns.

_____ Exercise 11 _____

a) Find the area under the standard normal curve between $z = 0$ and $z = 2.65$.

a) We look under the z column of Table 5 until we find 2.6, then we stay in this row and move to the right until we are in the column headed by 0.05. The area from $z = 0$ to $z = 2.65$ is given by the entry 0.4960.

b) Find the area under the standard normal curve between $z = -3$ and $z = 0$.

b) Since the area from $z = -3$ to $z = 0$ is the same as the area from $z = 0$ to $z = 3$, we look down the z column until we find 3.0. Then we move to the right in this row until we are in the column headed by 0.00. This entry is 0.4987, which is the area from $z = -3$ to $z = 0$.

c) Use parts (a) and (b) to find the area under the standard normal curve between $z = -3$ and $z = 2.65$ (Figure 6-33).

c) $\left(\begin{array}{c}\text{area from}\\-3 \text{ to } 2.65\end{array}\right) = \left(\begin{array}{c}\text{area from}\\-3 \text{ to } 0\end{array}\right) + \left(\begin{array}{c}\text{area from}\\0 \text{ to } 2.65\end{array}\right)$
$$\qquad\quad \downarrow \qquad\qquad\qquad \downarrow \qquad\qquad\qquad \downarrow$$
$$\qquad 0.9947 \qquad = \qquad 0.4987 \qquad + \qquad 0.4960$$

The desired area is 0.9947.

FIGURE 6-32   Patterns for Finding Areas Under the Standard Normal Curve

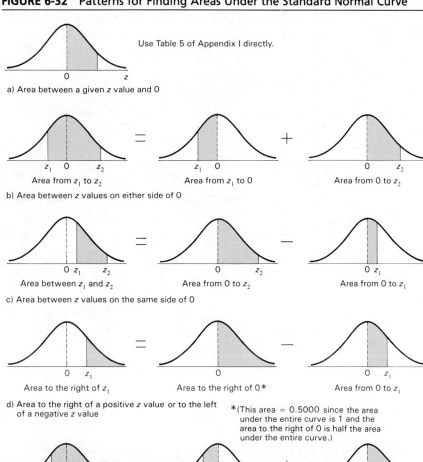

a) Area between a given *z* value and 0   Use Table 5 of Appendix I directly.

Area from z_1 to z_2 =   Area from z_1 to 0   +   Area from 0 to z_2

b) Area between *z* values on either side of 0

Area between z_1 and z_2 =   Area from 0 to z_2 −   Area from 0 to z_1

c) Area between *z* values on the same side of 0

Area to the right of z_1 =   Area to the right of 0*   −   Area from 0 to z_1

d) Area to the right of a positive *z* value or to the left of a negative *z* value

*(This area = 0.5000 since the area under the entire curve is 1 and the area to the right of 0 is half the area under the entire curve.)

Area to the right of z_1 =   Area from z_1 to 0   +   Area to the right of 0 which is 0.5000

e) Area to the right of a negative *z* value or to the left of a positive *z* value

FIGURE 6-33   Area Between *z* = −3 and *z* = 2.65

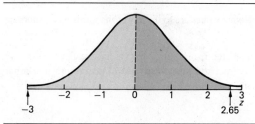

EXAMPLE   9

Find the area under the standard normal curve in Figure 6-34 from $z = 1$ to $z = 2.7$. Again we do this computation in two parts. But this time the z values are both on the same side of the mean. In such a case we do a subtraction instead of an addition.

FIGURE 6-34   Area from $z = 1$ to $z = 2.7$

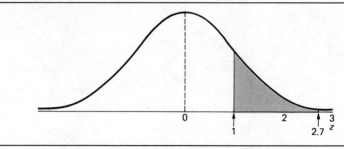

$$\begin{pmatrix} \text{area from} \\ \text{1 to 2.7} \end{pmatrix} = \begin{pmatrix} \text{area from} \\ \text{0 to 2.7} \end{pmatrix} - \begin{pmatrix} \text{area from} \\ \text{0 to 1} \end{pmatrix}$$

$$\downarrow \qquad\qquad \downarrow \qquad\qquad \downarrow$$

$$0.1552 \quad = \quad 0.4965 \quad - \quad 0.3413$$

The desired area is 0.1552.

_____ Exercise 12 _____

This time we are to find the area under the standard normal curve in Figure 6-35 between $z = -2.15$ and $z = -0.32$.

FIGURE 6-35   Area from $z = -2.15$ to $z = -0.32$

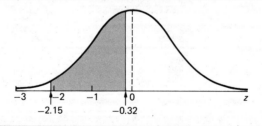

a) Are both z values on the same side of the mean? Will we add or subtract areas?

a) Both z values are to the left of the mean so we subtract.

b) Figure 6-35 shows the area we are to find. Which is smaller: the area from -2.15 to 0 or the area from -0.32 to 0? Which area do we subtract?

b) The area from -0.32 to 0 is the smaller area, so we subtract the area from -0.32 to 0 from the larger area.

c) The area from -2.15 to 0 is the same as the area from $z = 0$ to $z =$ _____ and the area from -0.32 to 0 is the same as the area from 0 to $z =$ _____ .

c) 2.15; 0.32

d) The area from 0 to $z = 2.15$ is _____ and the area from $z = 0$ to $z = 0.32$ is _____ . Use these areas and subtraction to find the area from $z = 0.32$ to $z = 2.15$, which is the same as the area from $z = -0.32$ to $z = -2.15$.

d)
$$\begin{pmatrix} \text{area from} \\ 0\,\text{to}\,2.15 \end{pmatrix} - \begin{pmatrix} \text{area from} \\ 0\,\text{to}\,0.32 \end{pmatrix} = \begin{pmatrix} \text{area from} \\ 0.32\,\text{to}\,2.15 \end{pmatrix}$$

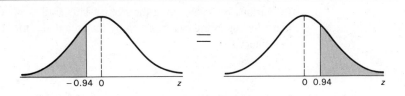

$$0.4842 \quad - \quad 0.1255 \quad = \quad 0.3587$$

EXAMPLE 10

Find the area under the standard normal curve to the left of $z = -0.94$.

We sketch the area and notice that the area to the left of -0.94 is the same as the area to the right of 0.94 (*see* Figure 6-36).

FIGURE 6-36 Shaded Areas are Equal

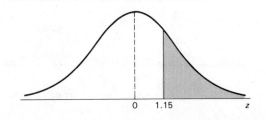

To find the area to the right of 0.94, we observe

$$\begin{pmatrix} \text{area to the} \\ \text{right of}\,0.94 \end{pmatrix} = \begin{pmatrix} \text{area to the} \\ \text{right of}\,0 \end{pmatrix} - \begin{pmatrix} \text{area from} \\ 0\,\text{to}\,0.94 \end{pmatrix}$$

$$= \quad 0.5000 \quad - \quad 0.3264$$

$$= 0.1736$$

Notice that the area to the right of zero is one half the area under the entire curve. Since the area under the entire curve is 1, the area to the right of zero is 1/2 or 0.5000. The area to the left of -0.94 equals that to the right of 0.94, so the desired area is 0.1736.

_____ Exercise 13 _____

Find the area under the standard normal curve to the right of $z = 1.15$.

a) Sketch the area to be found.

a) **FIGURE 6-37** Area to Be Found

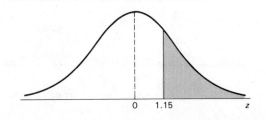

b) The area to the right of $z = 0$ equals _____ .

c) Find the area between $z = 0$ and $z = 1.15$.

d) Use the area of parts (b) and (c) to find the area to the right of $z = 1.15$.

b) The area to the right of $z = 0$ equals 0.5000.

c) By Table 5 of Appendix I, the area is 0.3749.

d) area to the right of 1.15

$$= \begin{pmatrix} \text{area to the} \\ \text{right of } 0 \end{pmatrix} - \begin{pmatrix} \text{area between} \\ 0 \text{ and } 1.15 \end{pmatrix}$$

$$\begin{matrix} \downarrow & & \downarrow \\ = \quad 0.5000 & - & 0.3749 \end{matrix}$$

$$= 0.1251$$

We have practiced the skill of finding areas under the standard normal curve for various intervals along the z-axis. This skill is important, since the probability that z lies in an interval is given by the area under the standard normal curve above that interval.

EXAMPLE   11

Let z be a random variable with a standard normal distribution. Find the probability that $-2 \leqslant z \leqslant 1.2$.

Figure 6-38 shows the area under the standard normal curve that corresponds to $P(-2 \leqslant z \leqslant 1.2)$.

FIGURE 6-38   Area Corresponding to $P(-2 \leqslant z \leqslant 1.2)$

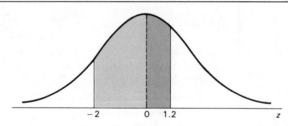

From the figure we see that

$$
\begin{aligned}
P(-2 \leqslant z \leqslant 1.2) &= P(-2 \leqslant z \leqslant 0) + P(0 \leqslant z \leqslant 1.2) \\
&= P(0 \leqslant z \leqslant 2) + P(0 \leqslant z \leqslant 1.2) \\
&= 0.4772 + 0.3849 \quad \text{(From Table 5, Appendix I)} \\
&= 0.8621
\end{aligned}
$$

_____ Exercise 14 _____

Let z be a random variable with a standard normal distribution. Find the probability that $1.41 \leqslant z \leqslant 2.18$.

a) Shade the area under the standard normal curve corresponding to $P(1.41 \leqslant z \leqslant 2.18)$.

a) **FIGURE 6-39** Area Corresponding to $P(1.41 \leqslant z \leqslant 2.18)$

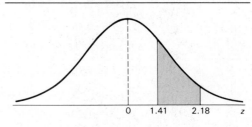

b) Find $P(0 \leqslant z \leqslant 2.18)$.

b) 0.4854

c) Find $P(0 \leqslant z \leqslant 1.41)$.

c) 0.4207

d) Use parts (b) and (c) to find $P(1.41 \leqslant z \leqslant 2.18)$.

d) $P(1.41 \leqslant z \leqslant 2.18) = P(0 \leqslant z \leqslant 2.18)$
$$- P(0 \leqslant z \leqslant 1.41)$$
$$= 0.4854 - 0.4207$$
$$= 0.0647$$

Section Problems 6.3

In problems 1–20 sketch the areas under the standard normal curve over the indicated intervals, and find the areas.

1. area between $z = 0$ and $z = 3.18$

2. area between $z = 0$ and $z = 2.92$

3. area between $z = 0$ and $z = -2.01$

4. area between $z = 0$ and $z = -1.93$

5. area between $z = -2.18$ and $z = 1.34$

6. area between $z = -1.4$ and $z = 2.03$

7. area between $z = 0.32$ and $z = 1.92$

8. area between $z = 1.42$ and $z = 2.17$

9. area between $z = -2.42$ and $z = -1.77$

10. area between $z = -1.98$ and $z = -0.03$

11. area to the right of $z = 0$

12. area to the left of $z = 0$

13. area to the right of $z = 1.52$

14. area to the right of $z = 0.15$

15. area to the left of $z = -1.32$

16. area to the left of $z = -0.47$

17. area to the right of $z = -1.22$

18. area to the right of $z = -2.17$

19. area to the left of $z = 0.45$

20. area to the left of $z = 0.72$

In problems 21–40 let z be a random variable with a standard normal distribution. Find the indicated probability, and shade the corresponding area under the standard normal curve.

21. $P(0 \leqslant z \leqslant 1.62)$

22. $P(0 \leqslant z \leqslant 0.54)$

23. $P(-0.82 \leq z \leq 0)$ 32. $P(z \geq 0)$

24. $P(-2.37 \leq z \leq 0)$ 33. $P(z \geq 1.35)$

25. $P(-0.45 \leq z \leq 2.73)$ 34. $P(z \geq 2.17)$

26. $P(1.73 \leq z \leq 3.12)$ 35. $P(z \leq -0.13)$

27. $P(-2.18 \leq z \leq -0.42)$ 36. $P(z \leq -2.15)$

28. $P(-1.78 \leq z \leq -1.23)$ 37. $P(z \geq -1.2)$

29. $P(-1.20 \leq z \leq 2.64)$ 38. $P(z \geq -1.5)$

30. $P(-2.20 \leq z \leq 1.04)$ 39. $P(z \leq 1.2)$

31. $P(z \leq 0)$ 40. $P(z \leq 3.2)$

SECTION 6.4 Areas Under Any Normal Curve

In many applied situations the original normal curve is not the standard normal curve. Generally, there will not be a table of areas available for the original normal curve. This does not mean that we cannot find the probability that a measurement x will fall in an interval from a to b. What we must do is *convert* original measurements x, a, and b to z values.

EXAMPLE 12

Let x have a normal distribution with $\mu = 10$ and $\sigma = 2$. Find the probability that an x value selected at random from this distribution is between 11 and 14. In symbols, find $P(11 \leq x \leq 14)$.

Since probabilities correspond to areas under the distribution curve, we want to find the area under the x-curve above the interval from $x = 11$ to $x = 14$. To do so, we will convert the x values to standard z values (*see* Figure 6-40) and then use Table 5 of Appendix I to find the corresponding area under the standard curve.

We use the formula

$$z = \frac{x - \mu}{\sigma}$$

to convert the given x interval to a z interval.

$$z_1 = \frac{11 - 10}{2} = 0.50 \qquad \text{(Use } x = 11, \mu = 10, \sigma = 2.\text{)}$$

$$z_2 = \frac{14 - 10}{2} = 2.00 \qquad \text{(Use } x = 14, \mu = 10, \sigma = 2.\text{)}$$

FIGURE 6-40   The Interval $11 \leqslant x \leqslant 14$ Corresponds
to the Interval $0.50 \leqslant z \leqslant 2.00$

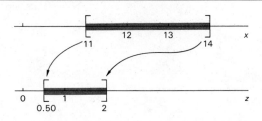

The corresponding areas under the x and z curves are shown in Figure 6-41.

FIGURE 6-41   Corresponding Areas Under the x-Curve and z-Curve

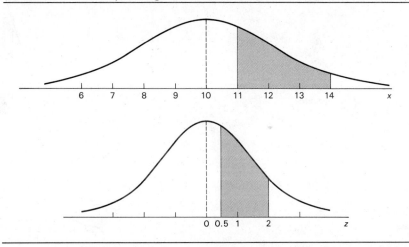

From Figure 6-41 we see that

$$P(11 \leqslant x \leqslant 14) = P(0.50 \leqslant z \leqslant 2.00)$$
$$= P(0 \leqslant z \leqslant 2.00) - P(0 \leqslant z \leqslant 0.50)$$
$$= 0.4772 - 0.1915 \quad \text{(From Table 5, Appendix I)}$$
$$= 0.2857$$

The probability is 0.2857 that an x value selected at random from a normal distribution with mean 10 and standard deviation 2 lies between 11 and 14.

Exercise 15

Given that x has a normal distribution with $\mu = 3$ and $\sigma = 0.5$, find $P(2.1 \leqslant x \leqslant 3.7)$ for an x value selected at random.

a) Convert the interval $2.1 \leqslant x \leqslant 3.7$ to a z interval. Do you expect the z values to be positive or negative?

a) Since 2.1 is below the mean 3, its corresponding z value should be negative. Since 3.7 is above the mean, its corresponding z value should be positive.

$$z_1 = \frac{x - \mu}{\sigma} = \frac{2.1 - 3}{0.5} = -1.80$$

$$z_2 = \frac{x - \mu}{\sigma} = \frac{3.7 - 3}{0.5} = 1.40$$

$$P(2.1 \leqslant x \leqslant 3.7) = P(-1.80 \leqslant z \leqslant 1.40)$$

b) Sketch the areas corresponding to the desired probabilities under the x-curve and under the z-curve.

b) **FIGURE 6-42**   Areas

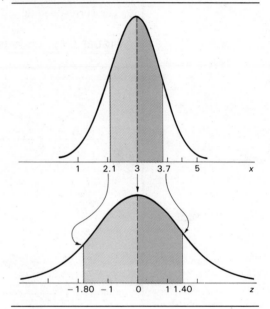

c) Find $P(2.1 \leqslant x \leqslant 3.7)$.

c) $P(2.1 \leqslant x \leqslant 3.7) = P(-1.80 \leqslant z \leqslant 1.40)$
$= P(0 \leqslant z \leqslant 1.80)$
$\quad\quad + P(0 \leqslant z \leqslant 1.40)$
$= 0.4641 + 0.4192$
$= 0.8833$

EXAMPLE   13

A factory has a machine that puts corn flakes into boxes. These boxes are advertised as having 20 oz of corn flakes. Assume that the distribution of weights is normal with $\mu = 20$ and $\sigma = 1.5$ oz. You pick a box at random from the production line. What is the probability that its weight will be between 19 and 21 oz?

The original normal curve of weights is not a standard normal curve because the mean is not zero and the standard deviation is not one. So we will convert our data to standard units.

Let z represent the standard unit weight of the corn flakes in the box. Next we convert the interval from 19 to 21 oz to standard z units.

$$z_1 = \frac{19 - 20}{1.5} = -0.67 \qquad z_2 = \frac{21 - 20}{1.5} = 0.67$$

FIGURE 6-43 Original Normal Curve and Standard Normal Curve

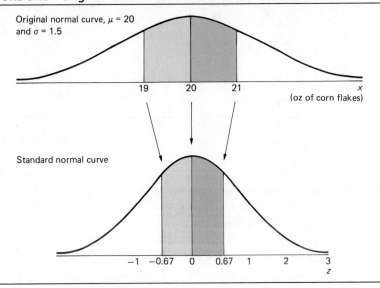

As shown in Figure 6-43, the area under the normal curve between 19 and 21 corresponds to the area under the standard normal curve between -0.67 and 0.67. So

$$P(19 \leq x \leq 21) = P(z_1 \leq z \leq z_2)$$
$$= P(-0.67 \leq z \leq 0.67)$$
$$= (\text{area from } -0.67 \text{ to } 0.67)$$
$$= (\text{area from } -0.67 \text{ to } 0) + (\text{area from } 0 \text{ to } 0.67)$$
$$= 0.2486 + 0.2486 = 0.4972$$

The probability that a corn flakes box picked at random will have from 19 to 21 oz of corn flakes is 0.4972, or about 50%.

_____ Exercise 16 _____

Early in Section 6.3 we talked about Sunshine Stereo turntables. The turntable life was normally distributed with a mean of 2.3 yr and a standard deviation of 0.4 yr. We wanted to know the probability that a turntable will break down during the guarantee period of 2 yr.

a) Let x represent the life of a turntable. The statement that the turntable breaks during the 2-yr guarantee period means the life is less than 2 yr, or $x \leqslant 2$. Convert this to a statement about z.

a)
$$z = \frac{x - \mu}{\sigma} = \frac{2 - 2.3}{0.4} = -0.75$$

So $x \leqslant 2$ means $z \leqslant -0.75$.

b) Indicate the area to be found in Figure 6-44. Does this area correspond to the probability that $z \leqslant -0.75$?

b)

FIGURE 6-44

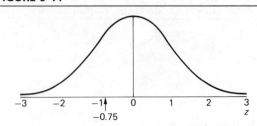

FIGURE 6-45 $z \leqslant -0.75$

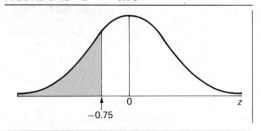

Yes, the shaded area does correspond to the probability that $z \leqslant -0.75$.

c) The shaded portion of Figure 6-44 corresponds to the probability that $z \leqslant -0.75$. To compute this probability we must do the subtraction of areas shown in Figure 6-46. Do that subtraction.

c) $P(z \leqslant -0.75) = P(z \leqslant 0) - P(-0.75 \leqslant z \leqslant 0)$
$$= \quad 0.5 \quad - \quad 0.2734$$
$$= 0.2266$$

FIGURE 6-46 $P(z \leqslant -0.75) = P(z \leqslant 0) - P(-0.75 \leqslant z \leqslant 0) = $ _____ $-$ _____ $= $ _____

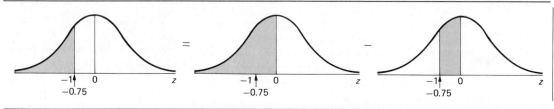

d) What is the probability that the turntable will break before the end of the guarantee period? [*Hint:* $P(x \leqslant 2) = P(z \leqslant -0.75)$.]

d) The probability is
$$P(x \leqslant 2) = P(z \leqslant -0.75) \simeq 0.23$$

This means the company will fix or replace about 23% of the turntables.

Sometimes we need to find z or x values that correspond to a given area under the normal curve. This situation arises when we want to specify a guarantee period so that a given percent of the total products produced by a company last at least as long as the duration of the guarantee period.

EXAMPLE 14

Magic Video Games, Inc., sells an expensive video computer games package. Because the games package is so expensive, the company wants to advertise an impressive guarantee for the life expectancy of their computer control system. The guarantee policy will refund full purchase price if the computer fails during the guarantee period. The research department has done tests which show that the mean life for the computer is 30 months, with standard deviation of four months. The computer life is normally distributed. How long can the guarantee period be if management does not want to refund the purchase price on more than 7% of the Magic Video Games packages?

Let us look at the distribution of lifetimes for the computer control system, and shade the portion of the distribution in which the computer lasts fewer months than the guarantee period. (*See* Figure 6-47.)

FIGURE 6-47   7% of the Computers Have a Lifetime Less Than the Guarantee Period

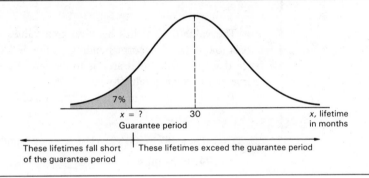

If a computer system lasts fewer months than the guarantee period, a full-price refund will have to be made. The lifetimes requiring a refund are in the shaded region in Figure 6-47, and this region represents 7% of the total area under the curve.

We can use Table 5 of Appendix I to find the z value so that 7% of the total area under the *standard* normal curve lies to the left of the z value. Then we convert the z value to its corresponding x value to find the guarantee period.

The z value with 7% of the area to the left of it is the negative of the z value with 50% − 7% = 43% of the area between 0 and z. (*See* Figure 6-48.)

FIGURE 6-48   Areas Under the Standard Normal Curve

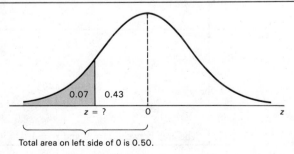

We find the number *closest* to 0.4300 in the area region of Table 5 and read the corresponding z value (*see* Table 6-3).

TABLE 6-3 Excerpt from Table 5 of Appendix I

z	. . .	0.4	.05	.06	.07		.08	.09
.								
.								
1.4	. . .	.4251	.4265	.4279	.4292	↑ .4300	.4306	.4319
.								
.								
.								

The value 0.4300 lies between area values 0.4292 and 0.4306, but it is closer to 0.4306. The corresponding z value is 1.48. Since we want the z value such that 7% of the total area is to the *left* of z, we use the symmetry of the curve and find the z value to be −1.48.

To translate this value back to an x value (in months) we use the formula

$x = \sigma z + \mu$

$\quad = 4(-1.48) + 30 \qquad$ (Use $\sigma = 4$ months and $\mu = 30$ months.)

$\quad = 24.08$ months

The company can guarantee the Magic Video Games package for $x = 24$ months. For this guarantee period, they expect to refund the purchase price of no more than 7% of the video games packages.

- **Comment:** When we use Table 5 to find a z value corresponding to a given area, we usually use the nearest area value rather than interpolation between values. However, when the area value given is exactly halfway between two area values of the table, we use the z value halfway between z values of the corresponding table areas. Table 6-4 shows an example in which the given

TABLE 6-4 Excerpt from Table 5 of Appendix I

z	. . .	.04	z = 1.645 is halfway between 1.64 and 1.65 ↓	.05	. . .
.					
.					
.					
1.6	. . .	.4495	↑	.4505	
		.4500 is exactly halfway between the area values .4495 and .4505			

The area between z = 0 and z = 1.645 is 0.4500.

area is halfway between two table areas. However, this interpolation convention is not always used, especially if the area is changing slowly, as it does in the tail ends of the distribution. When the z value corresponding to an area is larger than, say, 2, the standard convention is to use the z values corresponding to the next larger area. We see an example of this special case in Exercise 1 of the next chapter.

_____ Exercise 17 _____

Find the z value so that 3% of the area under the standard normal curve lies to the left of z.

a) Draw a standard normal curve and shade the region so that 3% of the area lies to the left of z.

a) **FIGURE 6-49**   **3% of the Total Area Lies to the Left of z**

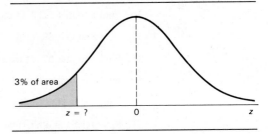

b) What portion of the area lies between this z value and 0?

b) Since half the area lies to the left of zero, then $50\% - 3\% = 47\%$ lies between the z value and 0.

c) Use Table 5 of Appendix I to find the z value so that 47% of the area under the standard normal curve is between 0 and z.

c) The area value of Table 5 nearest to 0.4700 is 0.4699. The corresponding z value is 1.88.

d) Use the symmetry of the standard normal curve to find the z value so that 3% of the area lies to the left of z. (That is, find the z value indicated in the figure of part (a).)

d) 3% of the area lies to the left of $z = -1.88$.

Section Problems 6.4

In problems 1–10 assume that x has a normal distribution, with the specified mean and standard deviation. Find the indicated probabilities.

1. $P(3 \leq x \leq 6)$; $\mu = 4$; $\sigma = 2$

2. $P(10 \leq x \leq 26)$; $\mu = 15$; $\sigma = 4$

3. $P(50 \leq x \leq 70)$; $\mu = 40$; $\sigma = 15$

4. $P(7 \leq x \leq 9)$; $\mu = 5$; $\sigma = 1.2$

5. $P(8 \leq x \leq 12); \mu = 15; \sigma = 3.2$ 8. $P(x \geq 120); \mu = 100; \sigma = 15$

6. $P(40 \leq x \leq 47); \mu = 50; \sigma = 15$ 9. $P(x \geq 90); \mu = 100; \sigma = 15$

7. $P(x \geq 30); \mu = 20; \sigma = 3.4$ 10. $P(x \geq 2); \mu = 3; \sigma = 0.25$

In problems 11–20, find the z value described and sketch the area described.

11. Find $z \geq 0$ so that 45% of the normal curve lies between 0 and z.

12. Find $z \geq 0$ so that 47.5% of the normal curve lies between 0 and z.

13. Find $z \leq 0$ so that 42% of the standard normal curve lies between z and 0.

14. Find $z \leq 0$ so that 33% of the standard normal curve lies between z and 0.

15. Find z so that 6% of the standard normal curve lies to the left of z.

16. Find z so that 5.2% of the standard normal curve lies to the left of z.

17. Find z so that 8% of the standard normal curve lies to the right of z.

18. Find z so that 5% of the standard normal curve lies to the right of z.

19. Find the z value so that 98% of the standard normal curve lies between $-z$ and z.

20. Find the z value so that 95% of the standard normal curve lies between $-z$ and z.

21. Researchers at a pharmaceutical company have found that the effective time duration of a safe dosage of a pain relief drug is normally distributed with mean 2 hr and standard deviation 0.3 hr. For a patient selected at random, what is the probability that the drug will be effective
    a) for 2 hours or less?
    b) for 1 hour or less?
    c) for 3 hours or more?

22. The ages of workers in the Illuminex plant are approximately normally distributed, with a mean of 45 years and a standard deviation of 12 years. A worker is stopped at random and asked to fill out a questionnaire. What is the probability that this worker is
    a) less than 30 years old?
    b) between 35 and 55 years old?
    c) more than 60 years old?

23. Ivy League Prep School claims that its seniors score well on the math portion of the Scholastic Aptitude Test (800 maximum score). The school brochure says that the scores of its students are normally distributed, with a mean score of 600 and a standard deviation of 57 points. What is the probability that an Ivy League Prep senior selected at random will have a math SAT score
    a) between 600 and 700?
    b) less than 450?
    c) more than 750?

24. Weights of the Pacific Yellowfin Tuna follow a normal distribution with mean weight 68 lb and standard deviation 12 lb. For a randomly caught Pacific Yellowfin Tuna, what is the probability that the weight is
    a) less than 50 lb?
    b) more than 80 lb?
    c) between 50 lb and 80 lb?

25. Police response time to an emergency call is the time difference between the time the call is first received by the dispatcher and the time a patrol car radios that it has arrived at the scene. Over a long period of time it has been determined that the police response time has a normal distribution with mean 8.4 minutes and standard deviation 1.7 minutes. For a randomly received emergency call, what is the probability that the response time will be
    a) between 5 and 10 minutes?
    b) less than 5 minutes?
    c) more than 10 minutes?

26. A single commercial jet engine uses an average of 718 gallons of jet fuel each hour it is in cruising position. The standard deviation of fuel consumption is 42 gallons per hour. When the jet is in cruising position, the fuel consumption follows a normal distribution. What is the probability that for a jet in cruising position, the fuel consumption for one hour for each engine is
    a) between 650 and 815 gallons?
    b) less than 600 gallons?
    c) more than 800 gallons?

27. At a ski area in Vermont the daytime high temperature is normally distributed during January, with mean of 22 °F and standard deviation of 10 °F. You are planning to ski there this January. What is the probability that you will encounter daytime highs of
    a) 42 °F or higher?
    b) 15 °F or lower?
    c) between 29 °F and 40 °F?

28. *Your Health* magazine did a study of college dorm food menus and weight gain of students. One of the less expensive meal plans has a lot of starch, and the study showed that the weight gain during the first three months using this plan was normally distributed, with mean 5 pounds and standard deviation 1.6 pounds. What is the probability that a student selected at random who is fed using this meal plan will gain
    a) between 2 and 6 pounds?
    b) 1 pound or less?
    c) 10 pounds or more?

29. Quality control for Speedie Typewriters, Inc., has done studies showing that the light-use model (150,000 words per year) has a mean life of 3.5 years and a standard deviation of 0.7 years. At the prescribed usage, how long should the guarantee period be if the company wishes to replace no more than 10% during the guarantee period?

30. The life of a Freeze Breeze electric fan is normally distributed, with mean 4 years and standard deviation 1.2 years. The manufacturer will replace any defective fan free of charge while it is under guarantee. For how many years should a Freeze Breeze fan be guaranteed if the manufacturer does not want to replace more than 5% of them? (Give your answer to the nearest month.)

31. Accrotime is a company that manufactures quartz crystal watches. Accrotime researchers have shown that the watches have an average life of 28 months before certain electronic components deteriorate, causing the watch to become unreliable. The standard deviation of watch lifetimes is 5 months, and the distribution of lifetimes is normal.
    a) If Accrotime guarantees a full refund on any defective watch for two years after purchase, what percent of total production will they expect to replace?
    b) If Accrotime does not want to make refunds on more than 12% of the watches they make, how long should the guarantee period be (to the nearest month)?

32. Quick Start Company makes 12-volt car batteries. After many years of product testing, they know that the average life of a Quick Start battery is normally distributed, with mean 45 months and standard deviation 8 months.
    a) If Quick Start guarantees a full refund on any battery that fails within the 36-month period after purchase, what percent of their batteries will they expect to replace?
    b) If Quick Start does not want to replace more than 10% of their batteries under the full refund guarantee policy, for how long should they guarantee the batteries (to the nearest month)?

33. A relay microchip in a telecommunications satellite has a life expectancy that follows a normal distribution with mean 90 months and standard deviation 3.7 months. When this computer-relay microchip malfunctions, the entire satellite is useless. A large London insurance company is going to insure the satellite for 50 million dollars. Assume that the only part of the satellite in question is the microchip. All other components will work indefinitely.
    a) For how many months should the satellite be insured to be 99% confident it will last beyond the insurance date?
    b) If the satellite is insured for 84 months, what is the probability that it will malfunction before the insurance coverage ends?

34. The Zinger is a modern sports car with an epoxy body. The manufacturer constructs Zinger bodies under pressure in a mold, and must wait a certain length of time to be sure enough resin bonding and hardening have occurred before removing the body from the mold. In the process of hardening, a very great number of individual chemical bonds are created. Chemical engineers have found that the hardening time required to solidify a bond is normally distributed, with mean $\mu = 28.7$ hours and standard deviation $\sigma = 6.4$ hours. What is the minimal length of time the body should stay in the mold if the manufacturer must be assured that at least 85% of the bonds have hardened?

35. A malaria prevention pill was developed to protect U.S. soldiers in the South Pacific during World War II. The pill had a number of mildly unpleasant side effects, so most soldiers wanted to take as few pills as possible. After extensive medical work using blood tests, it was found that for a single pill the duration of protection times was normally distributed, with mean $\mu = 72$ hours and standard deviation $\sigma = 8$ hours. If each soldier in a battalion was given a pill at breakfast mess (and ordered to take it), after how many hours should another pill be issued, so that
    a) fewer than 10% of the soldiers in the battalion were unprotected from malaria?
    b) fewer than 5% were unprotected from malaria?
    c) fewer than 1% were unprotected from malaria?

Summary

In this chapter we have examined graphs of normal distributions, standard units, z scores, and areas under the standard normal curve.

It is important to remember that a normal distribution with mean μ and standard deviation σ can be transformed into the standard normal distribution with mean zero and standard deviation one by the formula

$$z = \frac{x - \mu}{\sigma}$$

Values of z and Table 5 of Appendix I can be used to obtain the area under the standard normal curve from 0 to z (for positive z values). Other areas can be obtained from Table 5 and the symmetry of the normal curve. Areas under the normal curve represent probabilities that a z value will fall in the interval over which the area lies.

Given a probability associated with an interval on the standard normal curve, we can use Table 5 of Appendix I to obtain the associated z values. When x follows a normal distribution, we can find the x values associated with a given probability by first finding the corresponding z values and then converting to x values by using the formula

$$x = z\sigma + \mu$$

Important Words and Symbols

	Section		Section
Normal distributions	6.1	z value or z score	6.2
Normal curves	6.1	Standard normal distribution	
Upward cup and downward		($\mu = 0$ and $\sigma = 1$)	6.2
cup on normal curves	6.1	Raw score	6.2
Bell-shaped curves	6.1	Areas under the standard	
Symmetry of normal curves	6.1	normal curve	6.3
Standard units	6.2	Area under any normal curve	6.4

Chapter Review Problems

1. Given that z is the standard normal variable (with mean 0 and standard deviation 1) find
   a) $P(0 \leq z \leq 1.75)$
   b) $P(-1.29 \leq z \leq 0)$
   c) $P(1.03 \leq z \leq 1.21)$
   d) $P(z \geq 2.31)$
   e) $P(z \leq -1.96)$
   f) $P(z \leq 1.00)$

2. Given that z is the standard normal variable (with mean 0 and standard deviation 1) find
   a) $P(0 \leq z \leq 0.75)$
   b) $P(-1.5 \leq z \leq 0)$
   c) $P(-2.67 \leq z \leq -1.74)$
   d) $P(z \geq 1.56)$
   e) $P(z \leq -0.97)$
   f) $P(z \leq 2.01)$

3. Given that x is a normal variable with mean $\mu = 47$ and standard deviation $\sigma = 6.2$, find
   a) $P(x \leq 60)$
   b) $P(x \geq 50)$
   c) $P(50 \leq x \leq 60)$

4. Given that x is a normal variable with mean $\mu = 110$ and standard deviation $\sigma = 12$, find
   a) $P(x \le 120)$
   b) $P(x \ge 80)$
   c) $P(108 \le x \le 117)$

5. Find z so that 5% of the area under the standard normal curve lies to the right of z.

6. Find z so that 1% of the area under the standard normal curve lies to the left of z.

7. Find z so that 95% of the area under the standard normal curve lies between $-z$ and z.

8. Find z so that 99% of the area under the standard normal curve lies between $-z$ and z.

9. On a practical nursing licensing exam, the mean score is 79 and the standard deviation is 9 points.
   a) What is the standardized score of a student with a raw score of 87?
   b) What is the standardized score of a student with a raw score of 79?
   c) Assuming the scores follow a normal distribution, what is the probability that a score selected at random is above 85?

10. On an auto mechanic aptitude test, the mean score is 270 points and the standard deviation is 35 points.
    a) If a student has a standardized score of 1.9, how many points is that?
    b) If a student has a standardized score of -0.25, how many points is that?
    c) Assuming the scores follow a normal distribution, what is the probability that a student will get between 200 and 340 points?

11. In the town of Rockwood a survey found that the number of hours children watch TV per week is normally distributed with mean 10 hr and standard deviation 2 hr. If a child is chosen at random, what is the probability that he or she watches TV
    a) less than 4 hr per week?
    b) more than 12 hr per week?
    c) between 8 and 14 hr per week?

12. Future Electronics makes compact disc players. Their research department found that the life of the laser beam device is normally distributed with mean 5000 hr and standard deviation 450 hr.
    a) Find the probability that the laser beam device will wear out in 5000 hr or less.
    b) Future Electronics wants to place a guarantee on the players so that no more than 5% fail during the guarantee period. Since the laser pickup is the part most likely to wear out first, the guarantee period will be based on the life of the laser beam device. How many playing hours should the guarantee cover? (Round to the next playing hour.)

13. Express Courier Service has found that the delivery time for packages is normally distributed with mean 14 hr and standard deviation 2 hr.
    a) For a package selected at random, what is the probability that it will be delivered in 18 hr or less?
    b) What should be the guaranteed delivery time on all packages in order to be 95% sure that the package will be delivered before this time? (*Hint:* Note that 5% of the packages will be delivered at a time beyond the guarantee time period.)

14. The Customer Service Center in a large New York department store has determined that the amount of time spent with a customer about a complaint is normally distributed with mean 9.3 min and standard deviation 2.5 min. What is the probability that for a randomly

chosen customer with a complaint the amount of time spent resolving the complaint will be

a) less than 10 min?
b) more than 5 min?
c) between 8 and 15 min?

15. The Flight For Life emergency helicopter service is available for medical emergencies occurring from 15 to 90 miles from the hospital. Emergencies that occur closer to the hospital can be handled effectively by ambulance service. A long-term study of the service shows that the response time from receipt of the dispatch call to arrival at the scene of the emergency is normally distributed with mean 42 min and standard deviation 8 min. For a randomly received call, what is the probability that the response time will be

a) between 30 and 45 min?
b) less than 30 min?
c) more than 60 min?

16. The life of a Sunshine electric coffeepot is normally distributed with mean 4 yr and standard deviation 6 mo. The manufacturer will replace, free of charge, any Sunshine coffeepot that breaks down under guarantee. If the manufacturer does not want to replace more than 3% of the coffeepots, for how long should the guarantee be made (round to the nearest month)?

USING COMPUTERS

Professional statisticians in industry and research use computers to help them analyze and process statistical data. There are many computer programs available for statistics. Some commonly used statistical packages include Minitab, SAS, and SPSS. There are many others as well. Any of these statistical packages may be used with this text.

The authors of this text have also written a computer statistics package called ComputerStat. The programs in ComputerStat are available on disks with an accompanying manual that explains how to use the programs and gives additional computer applications. The disks are intended for the Apple IIe, IIc, or the IBM PC. Programs are written in the computer language BASIC and are designed for the beginning statistics student with no previous computer experience. ComputerStat is available as a supplement from the publisher of this text.

Problems in this section may be done using ComputerStat or other statistical computer programs.

1. The average earnings of a city government employee in October 1981 are given at right for a sample of 18 cities.

   In the ComputerStat menu of topics, select Probability Distributions and Central Limit Theorem. Then use program S7, Standardized Scores, to find the sample mean, sample standard deviation, and z values for the average salaries of these city government employees. Looking at the z values, which are above average? Which are below average? Which are within one standard deviation of the mean?

2. The standard normal probability distribution is very important in all of statistics. In Table 5 of Appendix I we have listed some standard normal probabilities. What if you wanted a more accurate table (that is, one with more significant digits displayed), with more entries? How could you use a computer to find probabilities in a standard normal table? The complete answer to this question is quite technical and requires mathematics beyond the scope of this text. However, the basic formulas are very accurate and can be used by

City	Average Earnings for October 1981
New York	1,854
Chicago	1,796
Los Angeles	2,216
Philadelphia	1,701
Houston	1,621
Detroit	2,392
Dallas	1,637
San Diego	1,917
Phoenix	1,729
Baltimore	1,453
San Antonio	1,282
Indianapolis	1,120
San Francisco	2,017
Memphis	1,479
Washington, D.C.	1,978
Milwaukee	1,647
San Jose	2,276
Cleveland	1,460

Source: United States Department of Commerce, Bureau of the Census. *Statistical Abstract of the United States.* 103d ed. Washington: GPO, 1982, 310.

anyone. These formulas can be found in the following reference:

Abramowitz and Stegun. *Handbook of Mathematical Functions*. National Bureau of Standards, 1968.

We suggest that interested readers consult these references.

Programs S8 and S9 in ComputerStat give normal probabilities and z values corresponding to the standard normal distribution. These computer programs are often more convenient to use than tables. The programs are used as subroutines in later computer programs for confidence intervals (Chapter 8) and hypothesis testing (Chapter 9).

a) In the ComputerStat menu of topics, select Probability Distributions and Central Limit Theorem. Then use program S8, Standard Normal Probability Distribution, to check your answers to problems 1–40 of Section 6.3. [*Hint:* To find the probability $P(0 \le z)$, use the computer to find the probability $P(0 \le z \le 10)$.] Explain why this method works and is appropriate for this type of computer application.

b) In the ComputerStat menu of topics, select Probability Distribution and Central Limit Theorem. Then use program S9, Inverse Normal Probability Distribution, to check your answers to problems 11–14 of Section 6.4.

7 Introduction to Sampling Distributions

"When we are not sure, we are alive."

Graham Green

The modern English author Graham Green implies that life and uncertainty are inseparable. Only those who are no longer living can escape chance happenings. Since life and therefore uncertainty are so important, it is necessary for us to learn all we can about these phenomena. In this chapter we will study how information from samples relates to information about populations. We cannot be certain that the information from a sample reflects corresponding information about the entire population, but we can describe likely differences. Study this chapter and the following material carefully. We believe your effort will be rewarded by helping you appreciate the joy and wonder of living in an uncertain universe.

SECTION 7.1   Sampling Distributions

Let us begin with some common statistical terms. Most of these have been discussed before, but this is a good time to review them.

From a statistical point of view a *population* can be thought of as a set of measurements (or counts), either existing or conceptual. We discussed populations at some length in Chapter 1. A *sample* is a subset of measurements from the population. For our purposes the most important samples are *random samples,* which were discussed in Section 2.1.

A *population parameter* is a numerical descriptive measure of a population. Examples of population parameters are the population mean μ, population variance σ^2, population standard deviation σ, and the population proportion of successes p in a binomial distribution.

A *statistic* is a numerical descriptive measure of a sample (usually a random sample). Examples of statistics are the sample mean $\bar{x}$, sample variance s^2, sample standard deviation s, and the sample estimate r/n for the proportion of successes in a binomial distribution. Notice that for a given population a specified parameter is a *fixed* quantity, while the statistic might vary depending on which sample has been selected.

Often we do not have access to all the measurements of an entire population, so we must use measurements from a sample instead. In such cases we will use a statistic (such as $\bar{x}$, s, or r/n) to make *inferences* about corresponding population parameters (e.g., μ, σ, or p). The principal types of inferences we will make are

a) to *estimate* the value of a population parameter or

b) to formulate a *decision* about the value of a population parameter.

In order to evaluate the reliability of our inferences, we will need to know the probability distribution for the statistic we are using. Such a probability distribution is called a *sampling distribution.* Perhaps an example will help clarify this discussion.

EXAMPLE 1

Pinedale, Wisconsin, is a rural community with a children's fishing pond. Posted rules say that all fish under six inches must be returned to the pond, only children under 12 years old may fish, and a limit of five fish may be kept per day. Susan is a college student who was hired by the community last summer to make sure the rules were obeyed and to see that the children were safe from accidents. The pond contains only rainbow trout, and has been well stocked for many years. Each child has no difficulty catching his or her limit of five trout.

As a project for her biometrics class Susan kept a record of the lengths (to the nearest inch) of all trout caught last summer. Hundreds of children visited the pond and caught their limit of five trout, so Susan has a lot of data. To make Table 7-1 Susan selected 100 children at random and listed the length of each of the five trout caught by a child in the sample. Then, for each child she listed the mean length of the five trout that child caught.

Now let us turn out attention to the following question: What is the average (mean) length of a trout taken from the Pinedale children's pond last summer? We can get an idea of the average length by looking at the far right column of Table 7-1. But just looking at 100 of the $\bar{x}$ values doesn't tell us much. Let's organize our $\bar{x}$ values into a frequency table. We used a class width of 0.65 to make Table 7.2.

TABLE 7-1 Length Measurements of Trout Caught by a Random Sample of 100 Children at the Pinedale Children's Pond

Sample	Length (to nearest inch)					$\bar{x}$ = Sample Mean
1	13	9	11	13	13	11.8
2	8	6	6	12	10	8.4
3	14	9	9	14	9	11.0
4	12	6	6	9	6	7.8
5	7	14	8	10	15	10.8
6	14	7	10	11	15	11.4
7	14	9	6	14	8	10.2
8	15	8	9	15	7	10.8
9	12	13	7	15	15	12.4
10	10	10	7	13	12	10.4
11	15	12	10	6	14	11.4
12	6	11	14	8	14	10.6
13	14	8	11	14	10	11.4
14	8	10	13	15	8	10.8
15	6	8	9	8	11	8.4
16	9	8	11	11	12	10.2
17	8	11	14	6	15	10.8
18	11	9	9	12	11	10.4
19	13	10	11	11	8	10.6
20	11	15	15	6	13	12.0
21	8	13	11	7	7	9.2
22	9	6	6	13	12	9.2
23	15	14	15	9	6	11.8
24	9	9	9	11	8	9.2
25	9	11	14	6	11	10.2
26	8	8	9	9	10	8.8
27	6	8	9	7	11	8.2
28	8	6	11	12	14	10.2
29	10	9	7	15	15	11.2
30	10	15	11	7	7	10.0
31	10	15	15	10	14	12.8
32	6	11	9	13	12	10.2
33	12	6	14	7	6	9.0
34	14	7	13	15	8	11.4
35	6	12	7	6	9	8.0

TABLE 7-1 *continued*

Sample	Length (to nearest inch)					$\bar{x}$ = Sample Mean
36	7	6	8	6	7	6.8
37	13	12	14	11	7	11.4
38	13	9	9	7	11	9.8
39	12	10	10	15	12	11.8
40	14	8	8	11	8	9.8
41	13	14	13	10	9	11.8
42	15	10	9	6	15	11.0
43	8	8	10	9	8	8.6
44	6	10	10	13	10	9.8
45	12	15	14	6	10	11.4
46	9	12	8	9	6	8.8
47	6	7	13	11	15	10.4
48	11	11	6	12	14	10.8
49	6	14	6	13	14	10.6
50	9	11	12	8	7	9.4
51	9	14	14	9	8	10.8
52	10	9	7	11	6	8.6
53	7	6	13	7	12	9.0
54	9	12	9	10	14	10.8
55	8	8	9	10	8	8.6
56	13	7	9	8	9	9.2
57	15	15	7	14	13	12.8
58	9	7	10	7	6	7.8
59	8	15	11	15	13	12.4
60	6	10	7	9	6	7.6
61	13	6	9	15	7	10.0
62	12	11	11	8	7	9.8
63	12	7	8	12	9	9.6
64	12	14	14	10	6	11.2
65	8	11	9	7	12	9.4
66	11	7	13	14	8	10.6
67	10	9	15	13	8	11.0
68	14	12	7	13	15	12.2
69	10	6	8	15	6	9.0
70	10	6	7	6	6	7.0

TABLE 7-1 *continued*

Sample	Length (to nearest inch)					$\bar{x}$ = Sample Mean
71	9	14	13	9	7	10.4
72	9	7	6	9	15	9.2
73	6	8	8	6	7	7.0
74	9	12	8	8	11	9.6
75	9	14	9	8	12	10.4
76	11	12	14	13	14	12.8
77	13	13	15	12	12	13.0
78	12	10	10	7	6	9.0
79	8	9	9	8	15	9.8
80	8	11	11	13	14	11.4
81	13	7	10	7	6	8.6
82	12	6	6	13	12	9.8
83	6	8	8	7	14	8.6
84	7	13	13	6	12	10.2
85	15	8	6	8	13	10.0
86	12	6	11	12	6	9.4
87	15	13	9	7	13	11.4
88	9	14	9	6	14	10.4
89	9	15	6	14	11	11.0
90	15	13	14	10	10	12.4
91	12	15	14	10	8	11.8
92	8	9	10	8	9	8.8
93	11	13	10	13	9	11.2
94	15	8	15	7	10	11.0
95	14	13	6	15	9	11.4
96	6	15	11	10	11	10.6
97	7	12	10	7	6	8.4
98	12	8	8	15	13	11.2
99	7	8	7	9	8	7.8
100	8	12	12	15	13	12.0

- **Comment:** Techniques of Section 2.3 dictate a class width of 0.7. However, this choice results in the tenth class being beyond the data. Consequently we shortened the class width slightly and also started the first class with a value slightly lower than the smallest data value.

TABLE 7-2 Frequency Table for 100 Values of $\bar{x}$

Class	Class Limits		f = Frequency	$f/100$ = Relative Frequency
	Lower	Upper		
1	6.8 –	7.44	3	0.03
2	7.45 –	8.09	5	0.05
3	8.10 –	8.74	9	0.09
4	8.75 –	9.39	12	0.12
5	9.40 –	10.04	14	0.14
6	10.05 –	10.69	17	0.17
7	10.70 –	11.34	16	0.16
8	11.35 –	11.99	14	0.14
9	12.00 –	12.64	6	0.06
10	12.65 –	13.29	4	0.04

The far right column of Table 7-2 contains relative frequencies $f/100$. Recall that the relative frequencies may be thought of as probabilities, so we effectively have a probability distribution. Since $\bar{x}$ represents the mean length of a trout (based on samples of five trout caught by each child), then we estimate the probability of $\bar{x}$ falling into each class by using the relative frequencies. Figure 7-1 is a relative frequency or probability distribution of the $\bar{x}$ values. For convenience we have labeled the horizontal axis of the histogram with the class limits.

FIGURE 7-1   Estimates of Probabilities of $\bar{x}$ Values

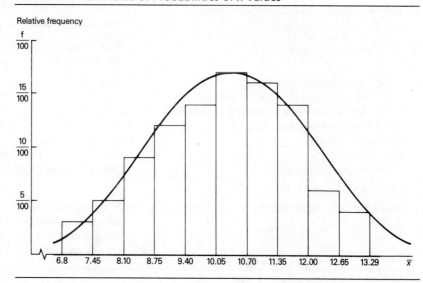

The bars of Figure 7-1 represent our estimated probabilities of $\bar{x}$ values based on the data of Table 7-1. The bell-shaped curve represents the theoretical probability distribution which would be obtained if the number of children (i.e., number of $\bar{x}$ values) were much larger.

Figure 7-1 represents a *probability sampling distribution* for the sample mean $\bar{x}$ of trout lengths based on random samples of size 5. We see that the distribution is mound-shaped and even somewhat bell-shaped. Irregularities are due to the small number of samples used (only 100 sample means) and the rather small sample size (five trout per child). These irregularities would become less obvious and even disappear if the sample of children became much larger, if we used a larger number of classes in Figure 7-1, and if the number of trout used in each sample became larger. In fact, the curve would eventually become a perfect bell-shaped curve. We will discuss this property at some length in the next section, which introduces the *Central Limit Theorem*.

There are other sampling distributions besides the $\bar{x}$ distribution. In the chapters ahead we will see that other statistics have different sampling distributions. However, the $\bar{x}$ sampling distribution is very important. It will serve us well in our inferential work in Chapters 8 and 9 on estimation and testing.

Let us summarize the information about sampling distributions in the following exercise.

_____ Exercise 1 _____

a) What is a population parameter? Give an example.

a) A population parameter is a numerical descriptive measure of a population. Examples are μ, σ, and p.

b) What is a sample statistic? Give an example.

b) A sample statistic or a statistic is a numerical descriptive measure of a sample. Examples are $\bar{x}$, s, and r/n.

c) What is a sampling distribution?

c) A sampling distribution is a probability distribution for the sample statistics we are using.

d) In Table 7-1 what makes up the members of the sample? What is the sample statistic corresponding to each sample? What is the sampling distribution? To which population parameter does this sampling distribution correspond?

d) There are 100 samples, each of which has five trout lengths. The first sample of five trout has lengths 13, 9, 11, 13, 13. The sample statistic is the sample mean $\bar{x} = 11.8$. The sampling distribution is shown in Figure 7-1. This sampling distribution relates to the population mean μ of all lengths of trout taken from the Pinedale children's pond (i.e., trout over six inches long).

e) Where will sampling distributions be used in our study of statistics?

e) Sampling distributions will be used for statistical inference. Chapter 8 will concentrate on the method of inference called estimation. Chapter 9 will concentrate on a method of inference called testing.

Section Problems 7.1

This is a good time to review several important concepts, some of which we have studied earlier. Please write out a careful but brief answer to each of the following questions.

1. What is a population? Give three examples.

2. What is a random sample from a population? (*Hint:* See Section 2.1.)

3. What is a population parameter? Give three examples.

4. What is a sample statistic? Give three examples.

5. What is the meaning of the term statistical inference? What types of inferences will we make about population parameters?

6. What is a sampling distribution?

7. How do frequency tables, relative frequencies, and histograms using relative frequencies help us understand sampling distributions?

8. How can relative frequencies be used to help us estimate probabilities occurring in sampling distributions?

9. How do sampling distributions help us make inferences about population parameters?

10. Give an example of a specific sampling distribution we studied in this section. Outline other possible examples of sampling distributions from areas such as business administration, economics, finance, psychology, political science, sociology, biology, medical science, sports, engineering, chemistry, linguistics, and so on.

SECTION 7.2   The Central Limit Theorem

In Section 7.1 we began a study of the distribution of $\bar{x}$ values where $\bar{x}$ was the (sample) mean length of five trout caught by children at the Pinedale children's fishing pond. Let's consider this example again in the light of a very important theorem of mathematical statistics.

- **Theorem 7.1** Let x be a random variable with a *normal distribution* whose mean is μ and standard deviation is σ. Let $\bar{x}$ be the sample mean corresponding to random samples of size n taken from the x distribution. Then the following are true:

  a) The $\bar{x}$ distribution is a *normal distribution*.
  b) The mean of the $\bar{x}$ distribution is μ.
  c) The standard deviation of the $\bar{x}$ distribution is $\sigma/\sqrt{n}$.

We conclude from Theorem 7.1 that when x has a normal distribution the $\bar{x}$ distribution will be normal *for any sample size n*. Furthermore, we can convert the $\bar{x}$ distribution to the standard normal z distribution using the following formulas.

$$\mu_{\bar{x}} = \mu$$

$$\sigma_{\bar{x}} = \frac{\sigma}{\sqrt{n}}$$

$$z = \frac{\bar{x} - \mu_{\bar{x}}}{\sigma_{\bar{x}}} = \frac{\bar{x} - \mu}{\sigma/\sqrt{n}} \quad \text{or} \quad \frac{\sqrt{n}(\bar{x} - \mu)}{\sigma}$$

where n is the sample size,
μ is the mean of the x distribution, and
σ is the standard deviation of the x distribution.

Theorem 7.1 is a wonderful theorem! It says that the $\bar{x}$ distribution will be *normal* provided the x distribution is normal. The sample size n could be 2, 3, or 4, or any (fixed) sample size we wish. Furthermore, the mean of the $\bar{x}$ distribution is μ (same as for the x distribution), but the standard deviation is $\sigma/\sqrt{n}$ (which is of course smaller than σ). The next example illustrates Theorem 7.1.

EXAMPLE 2

Suppose a team of biologists has been studying the Pinedale children's fishing pond. Let x represent the length of a single trout taken at random from the pond. This group of biologists has determined that x has a normal distribution with mean $\mu = 10.2$ in. and standard deviation $\sigma = 1.4$ in.

a) What is the probability that a *single trout* taken at random from the pond is between 8 and 12 in. long?

Since $\mu = 10.2$ and $\sigma = 1.4$ we use the methods of Chapter 6 to get

$$z = \frac{x - \mu}{\sigma} = \frac{x - 10.2}{1.4}$$

Therefore,

$$P(8 < x < 12) = P\left(\frac{8 - 10.2}{1.4} < \frac{x - 10.2}{1.4} < \frac{12 - 10.2}{1.4}\right)$$

$$= P(-1.57 < z < 1.29)$$

$$= 0.4418 + 0.4015 = 0.8433$$

Therefore, the probability is about 0.8433 that a *single* trout taken at random is between 8 and 12 in. long.

b) What is the probability that the *mean length* $\bar{x}$ of five trout taken at random is between 8 and 12 in.?

If we let $\mu_{\bar{x}}$ represent the mean of the $\bar{x}$ distribution, then Theorem 7.1 part (b) tells us that

$$\mu_{\bar{x}} = \mu = 10.2$$

If $\sigma_{\bar{x}}$ represents the standard deviation of the $\bar{x}$ distribution then Theorem 7.1 part (c) tells us that

$$\sigma_{\bar{x}} = \sigma/\sqrt{n} = 1.4/\sqrt{5} = 0.63$$

To create a standard normal z variable from $\bar{x}$ we subtract $\mu_{\bar{x}}$ and divide by $\sigma_{\bar{x}}$

$$z = \frac{\bar{x} - \mu_{\bar{x}}}{\sigma_{\bar{x}}} = \frac{\bar{x} - \mu}{\sigma/\sqrt{n}} = \frac{\bar{x} - 10.2}{0.63}$$

To standardize the interval $8 < \bar{x} < 12$ we use 8 and then 12 in place of $\bar{x}$ in the above formula for z.

$$8 < \bar{x} < 12$$

$$\frac{8 - 10.2}{0.63} < z < \frac{12 - 10.2}{0.63}$$

$$-3.49 < z < 2.86$$

Theorem 7.1 part (a) tells us that $\bar{x}$ has a normal distribution. Therefore,

$$P(8 < \bar{x} < 12) = P(-3.49 < z < 2.86)$$
$$= 0.4998 + 0.4979 = 0.9977$$

The probability is about 0.9977 that the mean length based on a sample size 5 is between 8 and 12 in.

c) Looking at the results of parts (a) and (b) we see that the probabilities (0.8433 and 0.9977) are quite different. Why is this the case?

According to Theorem 7.1 both x and $\bar{x}$ have a normal distribution and both have the same mean of 10.2 in. The difference is in the standard deviation for x and $\bar{x}$. The standard deviation of the x distribution is $\sigma = 1.4$. The standard deviation of the $\bar{x}$ distribution is

$$\sigma_{\bar{x}} = \sigma/\sqrt{n} = 1.4/\sqrt{5} = 0.63$$

The standard deviation of $\bar{x}$ is less than half the standard deviation of x. Figure 7-2 shows the distribution of x and $\bar{x}$.

Looking at Figure 7-2 (a) and (b) we see both curves aligned so that the mean of one curve is above the mean of the other. The interval from 8 to 12 is also aligned for each curve. It becomes clear that the smaller standard deviation of the $\bar{x}$ distribution has the effect of gathering together much more of the total probability into the region over its mean. Therefore, the region from 8 to 12 has a much higher probability for the $\bar{x}$ distribution.

Theorem 7.1 gives complete information about the $\bar{x}$ distribution provided the original x distribution is known to be normal. What happens if we don't have infor-

FIGURE 7-2   General Shapes of the *x* and *x̄* Distributions

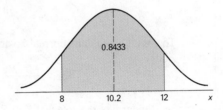

a) The *x* distribution with $\mu = 10.2$ and $\sigma = 1.4$

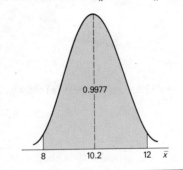

0.8433

8   10.2   12   *x*

b) The *x̄* distribution with $\mu_{\bar{x}} = 10.2$ and $\sigma_{\bar{x}} = 0.63$

0.9977

8   10.2   12   *x̄*

mation about the shape of the original *x* distribution? The *Central Limit Theorem* tells us what to expect.

> • **Theorem 7.2 The Central Limit Theorem:** If *x* possesses any distribution with mean μ and standard deviation σ, then the sample mean *x̄* based on a random sample of size *n* will have a distribution that approaches the distribution of a normal random variable with mean μ and standard deviation $\sigma/\sqrt{n}$ as *n* increases without limit.

The Central Limit Theorem is indeed surprising! It says that *x* can have *any* distribution whatever, but as the sample size gets larger and larger, the distribution of *x̄* will approach a *normal* distribution. From this relation, we begin to appreciate the scope and significance of the normal distribution.

In the Central Limit Theorem, the degree to which the distribution of *x̄* values fits a normal distribution depends on both the selected value of *n* and the original distribution of x values. A natural question is: How large should the sample size be if we want to apply the Central Limit Theorem? After a great deal of theoretical as well as empirical study statisticians agree that if *n* is 30 or larger, the *x̄* distribution will appear to be normal and the Central Limit Theorem will apply. However, this rule should not be blindly applied. If the x distribution is definitely not symmetric about its mean, then the $\bar{x}$ distribution will also display a lack of symmetry. In such a case, a sample size larger than 30 may be required to get a reasonable approximation to the normal.

In practice it is a good idea, when possible, to make a histogram of sample x values. If the histogram is approximately mound-shaped and if it is more or less symmetric, then we may be assured that for all practical purposes the $\bar{x}$ distribution will be well approximated by a normal distribution and the Central Limit Theorem will apply when the sample size is 30 or larger. The main thing to remember is that in almost all practical applications a sample size of 30 or more is adequate for the Central Limit Theorem to hold. However, in a few rare applications you may need a sample size larger than 30 to get good results.

Let's summarize this information for convenient reference.

For almost all x distributions, if we use a random sample of size 30 or larger, the $\bar{x}$ distribution will be approximately normal; and the larger the sample size becomes, the closer the $\bar{x}$ distribution gets to the normal. Furthermore, we may convert the $\bar{x}$ distribution to a standard normal distribution using the formulas shown below.

$$\mu_{\bar{x}} = \mu$$

$$\sigma_{\bar{x}} = \frac{\sigma}{\sqrt{n}}$$

$$z = \frac{\bar{x} - \mu_{\bar{x}}}{\sigma_{\bar{x}}} = \frac{\bar{x} - \mu}{\sigma/\sqrt{n}} \quad \text{or} \quad \frac{\sqrt{n}(\bar{x} - \mu)}{\sigma}$$

where n is the sample size ($n \geqslant 30$),
μ is the mean of the x distribution,
σ is the standard deviation of the x distribution.

_____ Exercise 2 _____

a) Suppose x has a *normal* distribution with mean $\mu = 18$ and standard deviation $\sigma = 3$. If we draw random samples of size 5 from the x distribution and $\bar{x}$ represents the sample mean, what can you say about the $\bar{x}$ distribution? How could you standardize the $\bar{x}$ distribution?

a) Since the x distribution is given to be *normal* then the $\bar{x}$ distribution will also be normal even though the sample size is much less than 30. The mean is $\mu_{\bar{x}} = \mu = 18$. The standard deviation is

$$\sigma_{\bar{x}} = \sigma/\sqrt{n} = 3/\sqrt{5} = 1.3$$

We could standardize $\bar{x}$ as follows

$$z = \frac{\bar{x} - \mu}{\sigma/\sqrt{n}}$$

$$z = \frac{\bar{x} - 18}{1.3}$$

b) Suppose we know that the x distribution has mean $\mu = 75$ and standard deviation $\sigma = 12$ but we have no information as to whether or not the x distribution is normal. If we draw samples of size 30 from the x distribution and $\bar{x}$ represents the sample mean, what can you say about the $\bar{x}$ distribution? How could you standardize the $\bar{x}$ distribution?

b) Since the sample size is large enough the $\bar{x}$ distribution will be approximately a normal distribution. The mean of the $\bar{x}$ distribution is

$$\mu_{\bar{x}} = \mu = 75$$

The standard deviation of the $\bar{x}$ distribution is

$$\sigma_{\bar{x}} = \sigma/\sqrt{n} = 12/\sqrt{30} = 2.2$$

We could standardize $\bar{x}$ as follows

$$z = \frac{\bar{x} - \mu}{\sigma/\sqrt{n}} = \frac{\bar{x} - 75}{2.2}$$

c) Suppose you did not know that x had a normal distribution. Would you be justified in saying that the $\bar{x}$ distribution is approximately normal if the sample size was $n = 8$?

c) No, the sample size should be 30 or larger if we don't know that x has a normal distribution.

Now let's look at an example that demonstrates the use of the Central Limit Theorem in a decision-making process.

EXAMPLE 3

A certain strain of bacteria occurs in all raw milk. Let x be the bacteria count per milliliter of milk. The health department has found that if the milk is not contaminated, then x has a distribution that is more or less mound-shaped and symmetric. The mean of the x distribution is $\mu = 2,500$ and the standard deviation is $\sigma = 300$. In a large commercial dairy the health inspector takes 42 random samples of the milk produced each day. At the end of the day the bacteria count in each of the 42 samples is averaged to obtain the sample mean bacteria count $\bar{x}$.

a) Assuming the milk is not contaminated, what is the distribution of $\bar{x}$?

The sample size is $n = 42$. Since this value exceeds 30, the Central Limit Theorem applies and we know $\bar{x}$ will be approximately normal with mean

$$\mu_{\bar{x}} = \mu = 2,500$$

and standard deviation

$$\sigma_{\bar{x}} = \sigma/\sqrt{n} = 300/\sqrt{42} = 46.3$$

b) Assuming the milk is not contaminated, what is the probability that the average bacteria count $\bar{x}$ for one day is between 2,350 and 2,650 bacteria per milliliter?

We convert the interval

$$2,350 \le \bar{x} \le 2,650$$

to a corresponding interval on the standard z-axis.

$$z = \frac{\bar{x} - \mu}{\sigma/\sqrt{n}} = \frac{\bar{x} - 2{,}500}{46.3}$$

$$\bar{x} = 2{,}350 \text{ converts to } z = \frac{2{,}350 - 2{,}500}{46.3} = -3.24$$

$$\bar{x} = 2{,}650 \text{ converts to } z = \frac{2{,}650 - 2{,}500}{46.3} = 3.24$$

Therefore,

$$
\begin{aligned}
P(2{,}350 \leq \bar{x} \leq 2{,}650) &= P(-3.24 \leq z \leq 3.24) \\
&= 2P(0 \leq z \leq 3.24) \\
&= 2(0.4994) \\
&= 0.9988
\end{aligned}
$$

The probability is 0.9988 that $\bar{x}$ is between 2,350 and 2,650.

c) At the end of each day the inspector must decide to accept or reject the accumulated milk that has been held in cold storage awaiting shipment. Suppose the 42 samples taken by the inspector have a mean bacteria count $\bar{x}$ that is *not* between 2,350 and 2,650. If you were the inspector, what would be your comment on this situation?

The probability that $\bar{x}$ is between 2,350 and 2,650 is very high. If the inspector finds that the average bacteria count for the 42 samples is not between 2,350 and 2,650 then it is reasonable to conclude that there is something wrong with the milk. If $\bar{x}$ is less than 2,350, you might suspect someone added chemicals to the milk to artificially reduce the bacteria count. If $\bar{x}$ is above 2,650 you might suspect some other kind of biological contamination.

_____ Exercise 3 _____

In mountain country, major highways sometimes use tunnels instead of long, winding roads over high passes. However, too many vehicles in a tunnel at the same time can cause a hazardous situation. Traffic engineers are studying a long tunnel in Colorado. If x represents the time for a vehicle to go through the tunnel, it is known that the x distribution has mean $\mu = 12.1$ minutes and standard deviation $\sigma = 3.8$ minutes under ordinary traffic conditions. From a histogram of x values, it was found that the x distribution is mound-shaped with some symmetry about the mean.

Engineers have calculated that *on the average* vehicles should spend from 11 to 13 minutes in the tunnel. If the time is less than 11 minutes traffic is moving too fast for safe travel in the tunnel. If the time is more than 13 minutes, there is a problem of bad air (too much carbon monoxide and other pollutants).

Under ordinary conditions, there are about 50 vehicles in the tunnel at one time. What is the probability that the mean time for 50 vehicles in the tunnel will be from 11 to 13 minutes?

We will answer this question in steps.

a) Let $\bar{x}$ represent the sample mean based on samples of size 50. Describe the $\bar{x}$ distribution.

a) From the Central Limit Theorem we expect the $\bar{x}$ distribution to be approximately normal with mean

$$\mu_{\bar{x}} = \mu_x = 12.1$$

and standard deviation

$$\sigma_{\bar{x}} = \frac{\sigma}{\sqrt{n}} = \frac{3.8}{\sqrt{50}} = 0.54$$

b) Find $P(11 < \bar{x} < 13)$.

b) We convert the interval

$$11 < \bar{x} < 13$$

to a standard z interval and use the standard normal probability table to find our answer. Since

$$z = \frac{\bar{x} - \mu}{\sigma/\sqrt{n}} = \frac{\bar{x} - 12.1}{0.54}$$

then $\bar{x} = 11$ converts to

$$z = \frac{11 - 12.1}{0.54} = -2.04$$

and $\bar{x} = 13$ converts to

$$z = \frac{13 - 12.1}{0.54} = 1.67$$

Therefore,

$$P(11 < \bar{x} < 13) = P(-2.04 < z < 1.67)$$
$$= 0.4793 + 0.4525$$
$$= 0.9318$$

c) Comment on your answer for part (b).

c) It would seem that about 93% of the time there should be no safety hazard for average traffic flow.

Section Problems 7.2

In these problems the word *average* refers to the arithmetic mean $\bar{x}$ or μ, as appropriate.

1. Suppose x has a distribution with $\mu = 15$ and $\sigma = 14$.
   a) If a random sample of size $n = 49$ is drawn, find $\mu_{\bar{x}}$, $\sigma_{\bar{x}}$, and $P(15 \leq \bar{x} \leq 17)$.
   b) If a random sample of size $n = 64$ is drawn, find $\mu_{\bar{x}}$, $\sigma_{\bar{x}}$, and $P(15 \leq \bar{x} \leq 17)$.
   c) Why should you expect the probability of part (b) to be higher than that of part (a)?
   [*Hint:* Consider the standard deviations in parts (a) and (b).]

2. Suppose x has a distribution with $\mu = 100$ and $\sigma = 48$.
   a) If a random sample of size $n = 81$ is drawn, find $\mu_{\bar{x}}$, $\sigma_{\bar{x}}$, and $P(92 \leq \bar{x} \leq 100)$.
   b) If a random sample of size $n = 121$ is drawn, find $\mu_{\bar{x}}$, $\sigma_{\bar{x}}$, and $P(92 \leq \bar{x} \leq 100)$.
   c) Again, comment on the differences in the probabilities in parts (a) and (b). Why do you expect the differences?

3. Suppose x has a distribution with $\mu = 25$ and $\sigma = 3.5$.
   a) If random samples of size $n = 9$ are selected, can we say anything about the $\bar{x}$ distribution of sample means?
   b) If the original x distribution is *normal*, can we say anything about the $\bar{x}$ distribution from samples of size $n = 9$? Find $P(23 \leq \bar{x} \leq 26)$.

4. Suppose x has a distribution with $\mu = 72$ and $\sigma = 8$.
   a) If random samples of size $n = 16$ are selected, can we say anything about the $\bar{x}$ distribution of sample means?
   b) If the original x distribution is *normal*, can we say anything about the $\bar{x}$ distribution of random samples of size 16? Find $P(68 \leq \bar{x} \leq 73)$.

5. Arthur is a night watchman in a large warehouse. During one complete round of the warehouse he must check in at 40 different checkpoints, all of which are about the same distance apart. At each checkpoint he inserts a key which tells a central computer the time he checked in. Let x be a random variable representing the length of time from one check-in to the next. While monitoring Arthur's check-in times for the past year the computer found the mean of the x distribution to be $\mu = 6.4$ min with standard deviation $\sigma = 1.5$ min.
   a) In one complete round of the warehouse there are 40 check-in time intervals. What is the probability that the average check-in time interval will be from 6 to 7 min?
   b) Answer part (a) for two complete rounds of the warehouse (i.e., use a sample of 80 check-in time intervals).
   c) If in two complete rounds of the warehouse the average check-in time interval is not between 6 and 7 min, do you think a second security guard should drop in for a look? Would this seem to indicate that Arthur is going too slowly or too fast for some unexplained reason? Assume Arthur begins a round at a time randomly selected by the computer in order to foil burglars looking for a time schedule.

6. The general manager of an amusement park wanted to estimate the length of time people spend in the park. After collecting data over several months it was found that the average amount of time spent by an individual in the park was $\mu = 3.4$ hr with standard deviation $\sigma = 1.1$ hr. It was not known if the distribution was normal, but it was mound-shaped.
   a) Thirty-five people just got off the city bus and went into the amusement park. What is the probability that the average time these people spend in the park is more than 3 hr?
   b) What is the probability that the average time these people spend in the park is from 3 to 3.5 hr?

7. We have all seen lights flicker momentarily during a storm. At these times the power went off for an instant, but then came on again. When a telecommunications computer loses power for an instant, it shuts down and then reboots itself. During this time important high-speed messages may be lost. After examining many such cases it was found that the reboot times had a mound-shaped distribution with mean $\mu = 20.3$ msec and standard deviation $\sigma = 6.2$ msec.
   a) What is the probability that an average time for 33 reboots will be less than 19 msec?

b) What is the probability that an average time for 33 reboots will be from 19 to 21 msec?

c) What is the probability that an average of 33 reboot times will be more than 22 msec?

8. Read-a-Lot Bookstores, Inc., claims that the mean gross of Read-a-Lot stores is $300,000 per year with a standard deviation of $72,000. If a random sample of 36 stores in the franchise is selected, what is the probability that the mean annual gross $\bar{x}$ for these stores is
   a) more than $300,500?
   b) less than $299,500?
   c) between $299,500 and $300,500?

9. The Oak Grove College financial aid office did a study showing that their students spend an average (mean) of $680 in the college bookstore on books and supplies per year. The standard deviation is $138. If a random sample of 36 students is surveyed, what is the probability that the mean amount spent for books and supplies is
   a) less than $600?
   b) more than $700?
   c) between $600 and $700?

10. Shoestring Travel published a guesthouse guide, recommending budget-priced lodging in major European cities. The average cost for one night's lodging for one person was $30, with a standard deviation of $3. What is the probability that the mean price for 49 guest houses selected at random is
    a) more than $31?
    b) less than $29?
    c) between $29 and $31?

11. The electric company believes that the mean kilowatts used by a family each month is 365 with standard deviation of 30 kW. To check their figures, a random sample of 40 families is to be used. If $\bar{x}$ is the mean number of kilowatts used per month and the electric company's estimates are correct, what is the probability that $\bar{x}$ is between 358 and 372 kilowatts? If $\bar{x} = 385$, do you think the company's figures $\mu = 365$ and $\sigma = 30$ might be wrong? Explain.

12. The manager of Mammon Savings and Loan has computed that the mean number of dollars borrowed for the purchase of an economy automobile is $4,473.18 with standard deviation $816.12. During the next few weeks 50 customers come in for an auto loan on an economy car. What is the probability that $\bar{x}$, the mean amount of money loaned for an automobile, is
    a) more than $4,600?
    b) less than $4,300?
    c) between $4,300 and $4,600?

13. The diameters of grapefruit in a certain orchard are *normally distributed* with mean 4.6 inches and standard deviation 1.3 inches. If a random sample of ten of these grapefruit are put in a bag and sold in a grocery store, what is the probability that the mean diameter $\bar{x}$ will be
    a) larger than 5 in.?
    b) between 4 in. and 5 in.?
    c) smaller than 4 in.?

14. At a solar energy research station in Arizona it was found that the number of British thermal units absorbed by a particular type of panel was *normally distributed* with a mean $\mu = 18.5$ Btu and standard deviation $\sigma = 4.7$ Btu.
    a) What is the probability that one such panel will absorb from 18 to 19 Btu?
    b) What is the probability that the average Btu absorbed by 15 such panels is from 18 to 19 Btu?
    c) Why was the probability of part (b) higher than that of part (a)?

15. Coal is carried from a mine in West Virginia to a power plant in New York in hopper cars on a long train. The automatic hopper car loader is set to put 36 tons of coal in each car. The actual weights of coal loaded into each car are *normally distributed* with mean $\mu = 36$ tons and standard deviation $\sigma = 0.8$ tons.
    a) What is the probability that one car chosen at random will have less than 35.5 tons of coal?
    b) What is the probability that 20 cars chosen at random will have a mean load weight $\bar{x}$ of less than 35.5 tons of coal?
    c) Suppose the weight of coal in one car was less than 35.5 tons. Would that fact make you suspect the loader had slipped out of adjustment? Suppose the weight of coal in 20 cars selected at random had an average $\bar{x}$ less than 35.5 tons. Would that fact make you suspect the loader had slipped out of adjustment? Why?

16. The scores on an aptitude test for finger dexterity are *normally distributed*, with mean 250 and standard deviation 65.
    a) What is the probability that a person selected at random scores between 240 and 270 on the test?
    b) The test is given to a random sample of seven people. What is the probability that the *mean* $\bar{x}$ of their scores is between 240 and 270?
    c) Compare your answers for parts (a) and (b). Was the probability in part (b) much higher? Why would you expect this?

17. The heights of 18-year-old men are known to be *normally distributed*, with mean 68 in. and standard deviation 3 in.
    a) What is the probability that an 18-year-old man selected at random is between 67 and 69 in. tall?
    b) If a random sample of nine 18-year-old men is selected, what is the probability that the mean height $\bar{x}$ is between 67 in. and 69 in.?
    c) Compare your answers for parts (a) and (b). Was the probability in part (b) much higher? Why would you expect this?

18. True Sound cassette tapes have playing times that are *normally distributed*, with mean 30 min and standard deviation 2.3 min.
    a) What is the probability that a tape selected at random will play for a time period x between 28 min and 33 min?
    b) What is the probability that four tapes selected at random will have a mean playing time $\bar{x}$ between 28 min and 33 min?
    c) Compare your answers for parts (a) and (b). Was the probability in part (b) much higher? Why would you expect this?

19. Nature Kiss Granola comes in a medium-size box. The weight of granola in the box is *normally distributed* with mean 14 oz and standard deviation 0.4 oz.
    a) What is the probability that the amount of granola in a box selected at random is between 13.9 oz and 14.2 oz?

b) What is the probability that the mean amount of granola $\bar{x}$ computed from a random sample of 12 boxes is between 13.9 oz and 14.2 oz?

c) Compare your answers for parts (a) and (b). Was the probability in part (b) much higher? Why would you expect this?

20. a) If we have a distribution of x values that is more or less mound shaped and somewhat symmetric, what is the size of sample needed in order to claim that the distribution of sample means $\bar{x}$ from random samples of that size is approximately normal?

b) If the original distribution of x values is known to be normal, do we need to make any restriction about sample size in order to claim that the distribution of sample means $\bar{x}$ taken from random samples of a given size is normal?

SECTION 7.3   Normal Approximation to the Binomial Distribution

The probability that a new vaccine will protect adults from cholera is known to be 0.85. It is administered to 300 adults who must enter an area where the disease is prevalent. What is the probability that more than 280 of these adults will be protected from cholera by the vaccine?

This question falls in the category of a binomial experiment with number of trials n equal to 300, the probability of success p equal to 0.85, and the number of successes r greater than 280. It is possible to use the formula for the binomial distribution to compute the probability that r is greater than 280. However, this approach would involve a number of tedious and long calculations. There is an easier way to do this problem, for under the conditions stated below the normal distribution can be used to approximate the binomial distribution.

Again let p be the probability of success and let q be the probability of failure in a single binomial trial. Let n be the number of trials in the binomial experiment. If n, p, and q are such that *both $np > 5$ and $nq > 5$*, then the normal probability distribution with $\mu = np$ and $\sigma = \sqrt{npq}$ will be a good approximation to the binomial distribution. As n gets larger, the approximation becomes better.

> • **Summary:** Consider the binomial distribution with
>
> $$n = \text{number of trials}$$
> $$r = \text{number of successes}$$
> $$p = \text{probability of success}$$
> $$q = \text{probability of failure} = 1 - p$$
>
> If $\qquad\qquad np > 5 \quad \text{and} \quad nq > 5$
>
> then r has a binomial distribution that is approximated by a *normal* distribution with
>
> $$\mu = np \quad \text{and} \quad \sigma = \sqrt{npq}$$

EXAMPLE   4

Consider binomial distributions where $p = 0.25$, $q = 0.75$, and the number of trials is first $n = 3$, then $n = 10$, $n = 25$, and finally $n = 50$. The authors used a programmable calculator to obtain the binomial distributions for the given values of p, q, and n. The results have been organized and graphed in Figures 7-3, 7-4, 7-5, and 7-6.

When $n = 3$ the outline of the histogram does not even begin to take the shape of a normal curve. But when $n = 10$, 25, or 50 it does begin to take a normal shape indicated by the dashed curve.

From a theoretical point of view, the histograms in Figures 7-4, 7-5, and 7-6 would have bars for all values of r from $r = 0$ to $r = n$. However, in the construction of these histograms the bars of height less than 0.001 unit have been omitted—i.e., in Examples 4 and 5 probabilities less than 0.001 have been rounded to 0.

FIGURE 7-3

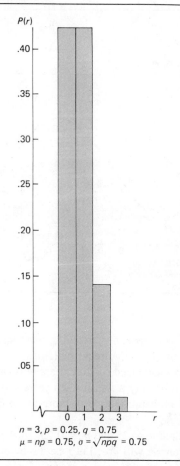

$n = 3, p = 0.25, q = 0.75$
$\mu = np = 0.75, \sigma = \sqrt{npq} = 0.75$

FIGURE 7-4

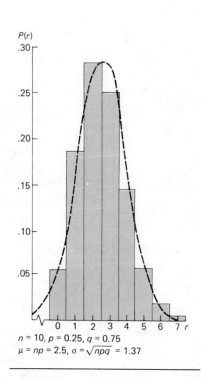

$n = 10, p = 0.25, q = 0.75$
$\mu = np = 2.5, \sigma = \sqrt{npq} = 1.37$

FIGURE 7-5

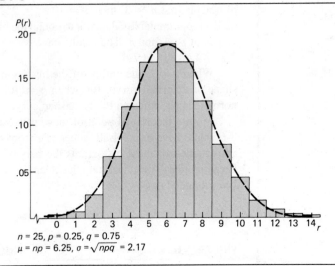

$n = 25$, $p = 0.25$, $q = 0.75$
$\mu = np = 6.25$, $\sigma = \sqrt{npq} = 2.17$

FIGURE 7-6

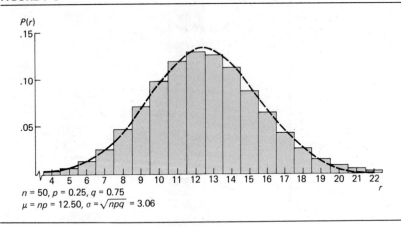

$n = 50$, $p = 0.25$, $q = 0.75$
$\mu = np = 12.50$, $\sigma = \sqrt{npq} = 3.06$

EXAMPLE 5

The owner of a new apartment building must install 25 water heaters. From past experience in other apartment buildings she knows that Quick Hot is a good brand. It is guaranteed for five years only, but from her past experience she knows the probability it will last ten years is 0.25.

What is the probability *eight* or more of the 25 water heaters will last at least ten years?

In this example $n = 25$ and $p = 0.25$, so Figure 7-5 represents the probability distribution we will use. Let r be the binomial random variable corresponding to the number of successes out of $n = 25$ trials. We want to find

$P(r \geq 8)$ by using the normal approximation. This probability is represented graphically (Figure 7-5) by the area of the bar over 8 and all bars to the right of the bar over 8.

Let x be a normal random variable corresponding to a normal distribution with $\mu = np = 25(0.25) = 6.25$ and $\sigma = \sqrt{npq} = \sqrt{25(0.25)(0.75)} \simeq 2.17$. This normal curve is represented by the dashed line in Figure 7-5. The area under the normal curve from $x = 7.5$ to the right is approximately the same as the area of the bars from the bar over $r = 8$ to the right. It is important to notice that we start with $x = 7.5$ because the bar over $r = 8$ really starts at $x = 7.5$.

Since the area of the bars and the area under the corresponding dashed (normal) curve are approximately equal, we conclude that $P(r \geq 8)$ is approximately equal to $P(x \geq 7.5)$.

When we convert $x = 7.5$ to standard units, we get

$$z = \frac{x - \mu}{\sigma} = \frac{7.5 - 6.25}{2.17} = 0.58$$

The probability we want is

$$P(x \geq 7.5) = P(z \geq 0.58) = 0.5 - P(0 \leq z \leq 0.58)$$
$$= 0.5 - 0.2190$$
$$= 0.2810$$

Using a programmable calculator and the binomial distribution, the authors computed that $p(r \geq 8) = 0.2735$. This means the probability is approximately 0.27 that eight or more water heaters will last at least ten years.

The error of approximation is the difference between the approximate value (0.2810) and the true value (0.2735). The error is only $0.2810 - 0.2735 = 0.0075$, which is negligible for most practical purposes.

We knew in advance that the normal approximation to the binomial probability would be good since $np = 25(0.25) = 6.25$ and $nq = 25(0.75) = 18.75$ are both greater than five. These are the conditions that assure us that the normal approximation will be sufficiently close to the binomial probability for most practical purposes.

Remember that when using the normal distribution to approximate the binomial, we are computing the areas under bars. The bar over r goes from $r - 0.5$ to $r + 0.5$. If r is a *left* endpoint of an interval, we *subtract* 0.5 to get the corresponding normal variable x. If r is a *right* endpoint of an interval, we *add* 0.5 to get the corresponding variable x. For instance, $P(6 \leq r \leq 10)$ where r is a binomial variable is approximated by $P(5.5 \leq x \leq 10.5)$ where x is the corresponding normal variable.

_____ Exercise 4 _____

From many years of observation a biologist knows the probability is only 0.65 that any given Arctic tern will survive the migration from its summer nesting area to its winter feeding grounds. A random sample of 500 Arctic terns were banded at their summer nesting area.

Use the normal approximation to the binomial and the following steps to find the probability that between 310 and 340 of the banded Arctic terns will survive the migration. Let r be the number of surviving terns.

a) To approximate $P(310 \leq r \leq 340)$ we use the normal curve with $\mu =$ _____ and $\sigma =$ _____ .

a) We use the normal curve with

$$\mu = np = 500(0.65) = 325$$
$$\sigma = \sqrt{npq} = \sqrt{500(0.65)(0.35)} = 10.67$$

b) $P(310 \leq r \leq 340)$ is approximately equal to $P(\underline{\hspace{0.7cm}} \leq x \leq \underline{\hspace{0.7cm}})$ where x is a variable from the normal distribution described in part (a).

b) Since 310 is the left endpoint, we subtract 0.5, and since 340 is the right endpoint, we add 0.5. Consequently,

$$P(310 \leq r \leq 340) \approx P(309.5 \leq x \leq 340.5)$$

c) Convert the condition $309.5 \leq x \leq 340.5$ to a condition in standard units.

c) Since $\mu = 325$ and $\sigma = 10.67$ the condition $309.5 \leq x \leq 340.5$ becomes

$$\frac{309.5 - 325}{10.67} \leq z \leq \frac{340.5 - 325}{10.67}$$

or

$$-1.45 \leq z \leq 1.45$$

d) $P(310 \leq r \leq 340) = P(309.5 \leq x \leq 340.5)$
$= (-1.45 \leq z \leq 1.45)$
$= \underline{\hspace{1cm}}$

d) $P(-1.45 \leq z \leq 1.45) = 2P(0 \leq z \leq 1.45)$
$= 2(0.4265)$
$= 0.8530$

which is approximately the probability we were seeking.

e) Will the normal distribution make a good approximation to the binomial for this problem? Explain your answer.

e) Since

$$np = 500(0.65) = 325 \text{ and}$$
$$nq = 500(0.35) = 175$$

are both greater than five, the normal will be a good approximation to the binomial.

Section Problems 7.3

Note: When we say *between a and b*, we mean every value from a to b *including a and b.*

1. A recent survey indicated that 63% of the adult population in Santa Fe watch the 10:00 P.M. news on TV. If this is true, what is the probability that in a random sample of 375 adults in Santa Fe
   a) 240 or more watch the 10:00 P.M. news?
   b) 225 or fewer watch the 10:00 P.M. news?
   c) between 220 and 250 watch the 10:00 P.M. news?

2. Joel is a door-to-door sales representative who makes a sale at only 8% of his house calls. If Joel makes 219 house calls today, what is the probability that
   a) he makes a sale at 12 or fewer houses?
   b) he makes a sale at 20 or more houses?
   c) he makes a sale at 15 to 25 houses?

3. At a large computer company 88% of all employees have a college degree. In a random sample of 410 employees, what is the probability that
   a) 370 or fewer have college degrees?
   b) 350 or more have college degrees?
   c) between 345 and 375 have college degrees?

4. One model of an imported automobile is known to have defective seat belts. An extensive study found that 17% of the seat belts were incorrectly installed at the factory. A car dealer has just received a shipment of 196 of this model car. What is the probability that in this shipment
   a) 22 or fewer have defective seat belts?
   b) 30 or more have defective seat belts?
   c) between 25 and 50 have defective seat belts?

5. It is estimated that 3.5% of the general population will live past their 90th birthday. In a graduating class of 753 high school seniors, what is the probability that
   a) 15 or more will live beyond their 90th birthday?
   b) 30 or more will live beyond their 90th birthday?
   c) between 25 and 35 will live beyond their 90th birthday?
   d) more than 40 will live beyond their 90th birthday?

6. The probability that an individual with a telephone has an unlisted phone number is 0.15. The Vote Now political action group is trying to get people to go to the polls, and so each district manager is to call all the households in his or her district during the week before elections. In a district with 416 households, assume all households have phones. What is the probability the district manager will find that
   a) 50 or fewer of the households in the district have unlisted phone numbers?
   b) 345 or more have listed numbers?
   c) between 40 and 80 have unlisted numbers?

7. Eureka Market Research Company conducts telephone interviews to determine product preference. They have found that the probability a person called at random agrees to answer a survey is 0.61. If 100 calls are made, what is the probability that
   a) 70 or more will respond to the survey?
   b) between 50 and 65 will respond?
   c) 55 or fewer will respond?

8. The probability of an adverse reaction to a flu shot is 0.02. If the shot is given to 1,000 people selected at random, what is the probability that
   a) 15 or fewer people will have an adverse reaction?
   b) 25 or more people will have an adverse reaction?
   c) between 20 and 30 people will have an adverse reaction?

9. From long experience, the registrar at Ocean View Community College has determined that the probability a student will change his or her schedule during the first week of classes is 0.47. If 3,000 students register, what is the probability that
   a) 1,400 or more will change their schedules?
   b) 1,320 or fewer will change their schedules?
   c) between 1,425 and 1,500 will change their schedules?

10. Canter Political Polling Service has found that the probability of a registered voter selected at random answering and returning a short questionnaire on a well-publicized issue is 0.31. If 1,000 questionnaires are sent to registered voters selected at random, what is the probability that
   a) 350 or more will be returned?
   b) 320 or fewer will be returned?
   c) between 315 and 355 will be returned?

11. Over the years it has been observed that of all the lawyers who take the state bar exam, only 60% pass. This year 850 lawyers are going to take the bar exam. What is the probability that
   a) 540 or more pass?
   b) 500 or fewer pass?
   c) between 485 and 525 pass?

12. It is known that 90% of all people who have a certain nervous disorder eventually recover from it. In the community hospital there are 75 people with this nervous disorder. What is the probability that
   a) 73 or more recover?
   b) between 60 and 70 recover?
   c) fewer than 65 recover?

13. The Country Boy Seed Company advertises that 80% of their tomato seeds will germinate. If 100 of these seeds are planted, what is the probability that
   a) 85 or more germinate?
   b) 68 or fewer germinate?
   c) between 69 and 84 germinate?

14. Suppose the probability is 0.65 that a thumbtack will land with its point up. If 5,000 tacks are dropped, what is the probability that
   a) 3,300 or more will land point up?
   b) 3,200 or fewer will land point up?
   c) between 3,150 and 3,350 will land point up?

15. In Summit County 35% of the registered voters are Democrats. What is the probability that in a random sample of 300 registered voters
   a) 100 or more are Democrats?
   b) between 100 and 120 are Democrats?

16. In Central City 40% of the families own their own homes. What is the probability that a random sample of 100 Central City families will contain
   a) between 45 and 55 homeowners?
   b) 56 or more homeowners?
   c) 44 or fewer homeowners?

Summary

Sampling distributions give us the basis for inferential statistics. By studying the distribution of sample statistics we can learn about a population parameter.

The central Limit Theorem describes the sampling distribution of sample means taken from samples of size *n*. It tells us that for increasing sample size *n*, the distribution of sample means $\bar{x}$ approaches a normal distribution with mean $\mu_{\bar{x}} = \mu$ and standard deviation $\sigma_{\bar{x}} = \sigma/\sqrt{n}$. The values of μ and σ are the population mean and standard deviation, respectively, of the original x distribution.

We concluded the chapter by studying how the normal distribution can be used to approximate the binomial distribution when both np and nq are greater than five. This approximation is especially convenient since the computation of binomial probabilities for large n is otherwise quite long.

Important Words and Symbols

	Section		Section
Population parameter	7.1	$\mu_{\bar{x}}$	7.2
Statistic	7.1	$\sigma_{\bar{x}}$	7.2
Sampling distribution	7.1	Normal approximation to the	
Central Limit Theorem	7.2	binomial distribution	7.3

Chapter Review Problems

1. Let x be a random variable representing the amount of sleep each adult in New York City got last night. Consider a sampling distribution of sample means $\bar{x}$.
   a) As the sample size becomes increasingly larger, what distribution does the $\bar{x}$ distribution approach?
   b) As the sample size becomes increasingly larger, what value will the mean $\mu_{\bar{x}}$ of the $\bar{x}$ distribution approach?
   c) What value will the standard deviation $\sigma_{\bar{x}}$ of the sampling distribution approach?
   d) How do the two $\bar{x}$ distributions for sample size $n = 50$ and $n = 100$ compare?

2. If x has a normal distribution with mean $\mu = 15$ and standard deviation $\sigma = 3$, describe the distribution of $\bar{x}$ values for sample size n, where $n = 4$, $n = 16$, $n = 100$. How do the $\bar{x}$ distributions compare for the various sample sizes?

3. For a binomial distribution with $n = 20$ trials and probability of success $p = 0.5$, we let $r =$ number of successes out of 20 trials.
   a) Explain why we can use a normal approximation to this binomial distribution.
   b) Use the normal approximation to estimate $P(6 \leq r \leq 12)$. How does this value compare to the corresponding probability obtained using the binomial table?

4. For a binomial distribution with $n = 16$ trials and probability of success $p = 0.5$, we let $r =$ number of successes out of 16 trials.
   a) Explain why we can use a normal approximation to this binomial distribution.
   b) Use the normal approximation to estimate $P(6 \leq r \leq 9)$. How does this value compare to the corresponding probability obtained using the binomial table?

5. The personnel office at a large electronics firm regularly schedules job interviews and maintains records of the interviews. From the past records, they have found that the length of a first interview is normally distributed with mean $\mu = 35$ min and standard deviation $\sigma = 7$ min.
   a) What is the probability that a first interview will last 40 min or longer?

b) Nine interviews are usually scheduled per day. What is the probability that the average length of time for the first interview for each of the nine individuals will be 40 min or longer?

6. A new muscle relaxant is available. Researchers of the firm developing the relaxant have done studies that indicate that the time lapse between administration of the drug and beginning effects of the drug is normally distributed with mean $\mu = 38$ min and standard deviation $\sigma = 5$ min.
   a) The drug is administered to one patient selected at random. What is the probability that the time it takes to go into effect is 35 min or less?
   b) The drug is administered to a random sample of ten patients. What is the probability that the average time before it is effective for all ten patients is 35 min or less?
   c) Comment on the differences of the results in parts (a) and (b).

7. In an Iowa community the average pay for nurses is $\mu = \$15.93$ per hour with standard deviation $\sigma = \$4.17$.
   a) For a random sample of 49 nurses, what is the probability that the average pay $\bar{x}$ is $15 or more?
   b) For a random sample of 36 nurses, what is the probability that the average pay is $15 or more?
   c) Compare the results of parts (a) and (b) and comment on any differences.

8. The test description accompanying a standardized English placement exam indicates that the average score is $\mu = 41$ with standard deviation $\sigma = 6$ for adults who have not taken any course work in ten years or more. Eastmore College has a degree program specifically designed for adults. The college administers the tests to all new students.
   a) For the Fall semester there are 64 new students who have not taken any course work in ten years or more. What is the probability that the average score for these students will be 40 or higher?
   b) For the Spring semester there are 81 new students who have not taken any course work in at least ten years. What is the probability that the average score for this group of students will be 40 or higher?
   c) Comment on any differences in the results of part (a) and part (b).

9. One environmental group did a study of recycling habits in a California community. They found that 70% of the aluminum cans sold in the area were recycled.
   a) If 400 cans were sold last month, what is the probability that 300 or more will be recycled?
   b) Of the 400 cans sold, what is the probability that between 260 and 300 will be recycled?

10. One study indicated that the probability that a hospital bill contains an error is 0.87. These errors might be as small as not charging for an aspirin or as large as charging for medical tests that were not performed on that individual. Of 500 bills handled by one insurance company in August, what is the probability that
   a) 450 or more have errors?
   b) from 425 to 445 have errors (including 425 and 445)?

11. The Aloha Taxi Company has found that the taxi fares are normally distributed with mean fare $8.27 and standard deviation $2.20.
   a) If a driver takes a taxi call at random, what is the probability that the fare will be less than $7.00? more than $10.00? between $7.00 and $10.00?
   b) If a driver takes ten calls at random, what is the probability that the mean fare of these ten calls is between $7.00 and $10.00?

c) Is the answer for part (b) considerably larger than the corresponding answer for part (a)? Why would you expect this?

12. A rare blood type is found in only 4% of the population. If 150 people are chosen at random, what is the probability that
   a) ten or more will have this blood type?
   b) between five and ten will have this blood type?

13. Assume that IQ scores are normally distributed with standard deviation of 15 points and mean of 100 points. If 100 people are chosen at random, what is the probability that the sample mean of IQ scores will not differ from the population mean by more than two points?

14. A large tank of fish from a hatchery is being delivered to a lake. The hatchery claims that the mean length of fish in the tank is 15 in. and the standard deviation is 2 in. A random sample of 36 fish is taken from the tank. Let $\bar{x}$ be the mean sample length of these fish. What is the probability that $\bar{x}$ is within 0.5 in. of the claimed population mean?

15. A company that makes light bulbs claims its bulbs have an average life of 750 hr with standard deviation of 20 hr. A random sample of 64 light bulbs is taken. Let $\bar{x}$ be the mean life of this sample.
   a) What is the probability that $\bar{x} \geq 750$?
   b) What is the probability that $745 \leq \bar{x} \leq 755$?

16. Camp Green has found the probability to be 0.07 that any one of their summer guests will get seriously ill. If Camp Green has 800 guests this summer, what is the probability that
   a) 61 or more will be seriously ill?
   b) 39 or fewer will be seriously ill?
   c) from 40 to 60 (including 40 and 60) will be seriously ill?

USING COMPUTERS

Professional statisticians in industry and research use computers to help them analyze and process statistical data. There are many computer programs available for statistics. Some commonly used statistical packages include Minitab, SAS, and SPSS. There are many others as well. Any of these statistical packages may be used with this text.

The authors of this text have also written a computer statistics package called ComputerStat. The programs in ComputerStat are available on disks with an accompanying manual that explains how to use the programs and gives additional computer applications. The disks are intended for the Apple IIe, IIc, or the IBM PC. Programs are written in the computer language BASIC and are designed for the beginning statistics student with no previous computer experience. ComputerStat is available as a supplement from the publisher of this text.

Problems in this section may be done using ComputerStat.

Part I: In-class project using calculators

1. Have a class discussion about how you could use the random number table (Table 2 of Appendix I) to get a random sample of the digits 0, 1, 2, 3, 4, 5, 6, 7, 8, and 9. In our sample we allow repetitions of the digits. Therefore 3, 6, 1, 6, and 8 would be a random sample of size $n = 5$ digits. The mean for this sample would be 4.80.

2. Have each student in class use the random number table to get a random sample of five digits from 0 to 9. Then have each student compute the sample mean of his or her sample of digits. Save these means for use in problems 3 and 4.

3. We assume the professor has a calculator with mean and standard deviation keys. Have each student read his or her mean (calculated in problem 2) to the professor. The professor enters each mean into the calculator and finds the grand sample mean $\bar{x}$ and sample standard deviation s of the means provided by the students.

4. Use the numbers $\bar{x}$ and s computed in problem 3 to create the intervals shown in Column I of Table 7-3. Again have each student read his or her mean (from problem 2) to the professor. Tally the sample means computed by the students in problem 2 to determine how many fall in each interval of Column II.

a) Examine Figure 6-7. Notice the areas under the normal curve in this figure. Compare the percents shown in this figure to those in Column III of Table 7-3 labeled Hypothetical Normal. Percents in the column labeled Hypothetical Normal represent percents of the total sample tally that would fall in each interval if the sampling distribution (based on samples of size $n = 5$) were in fact normal. How does your sample tally compare with the hypothetical normal? (*Hint:* Convert your sample tally to a percent of the total tally.)

b) Explain how this class project relates to the Central Limit Theorem. If we repeat problems 2, 3, and 4 with a large class of students and with a sample size n that is fairly large, would you say the sampling distribution should closely approximate the hypothetical normal? Why?

c) Using the frequency table, construct a histogram with six bars. Some bars may be of zero height depending on how your table turned out. Looking at your histogram, would you say it is approximately mound-shaped and symmetric? Does it seem to give the general outline of a normal curve?

TABLE 7-3 Frequency Table of Sample Means

I. Interval	II. Frequency	III. Hypothetical Normal
$\bar{x} - 3s$ to $\bar{x} - 2s$	Tally the sample	2 or 3%
$\bar{x} - 2s$ to $\bar{x} - s$	means computed	13 or 14%
$\bar{x} - s$ to $\bar{x}$	by the students	about 34%
$\bar{x}$ to $\bar{x} + s$	in problem 2 and	about 34%
$\bar{x} + s$ to $\bar{x} + 2s$	place here.	13 or 14%
$\bar{x} + 2s$ to $\bar{x} + 3s$		2 or 3%

Part II: Computer Demonstration

In the ComputerStat menu of topics, select Probability Distributions and Central Limit Theorem. Then use program S10, Central Limit Theorem Demonstration, to continue our study.

Computer program S10 is set up to essentially parallel the activities of problems 2, 3, and 4. However, the computer does the work much faster. You may use any sample size n from 2 to 50.

For each sample size you prescribe, the computer will use a random number generator to obtain a random sample of n real numbers from 0 to 9. It then computes the mean of these numbers. The process is repeated 100 times to get 100 sample means based on samples of size n.

Next the computer finds the grand mean $\bar{x}$ and sample standard deviation s for all 100 sample means. Then it constructs a frequency table similar to Table 7-3 of problem 4. Finally, the computer constructs a histogram from the frequency table.

5. Run S10 for a sample size $n = 2$
   a) Examine the frequency table. How close does the sample tally fit a hypothetical normal?
   b) Examine the histogram. Does the graph appear mound-shaped and symmetric? Does it seem to be approximately normal?
   c) Pay close attention to the standard deviation s and look at the histogram to see how "spread out" the graph is for this sample size n.

6. Repeat problem 5 for a sample of size $n = 5$.
7. Repeat problem 5 for a sample of size $n = 15$.
8. Repeat problem 5 for a sample of size $n = 30$.
9. In problems 5–8 the sample standard deviations s got smaller and smaller as n got larger. Also, the graph became less and less spread as n got larger. Furthermore, the entire graph seemed to strongly "peak out" over 4.5 as n got larger. Look again at the statement of the Central Limit Theorem. Why does the Central Limit Theorem predict all of the above behavior?
10. Notice that the hypothetical normal gives a good approximation even when the sample size is small. This is because the original distribution of numbers from zero to nine is symmetric about the value 4.5 when the numbers are drawn at random. Because of the symmetry of the original distribution, we got good results even for small samples. The Central Limit Theorem says the sampling distribution of means will become a normal distribution even if the original distribution was not symmetric. However, we usually need a sample size of $n \geq 30$ to get reasonably accurate results.

As a library assignment, interested readers are asked to give a report on the article "Understanding the Central Limit Theorem" by David A. Thomas in *The Mathematics Teacher*, vol 77, n 7, Oct. 1984, p. 452.

8 Estimation

> *We dance round in a ring and suppose,*
> *But the Secret sits in the middle and knows.*
>
> Robert Frost, "The Secret Sits"*

In Chapter 1 we said that statistics is the study of how to collect, organize, analyze, and interpret numerical data. That part concerned with analysis, interpretation, and forming conclusions about the source of the data is called *statistical inference*. Problems of statistical inference require us to draw a *sample* of observations from a larger *population*. A sample usually contains incomplete information, so in a sense we must "dance round in a ring and suppose." Nevertheless, conclusions about the population can be obtained from sample data by use of statistical estimates. This chapter will introduce you to several widely used methods of estimation.

SECTION 8.1   Estimating μ with Large Samples

An estimate of a population parameter given by a single number is called a *point estimate* of that parameter. It should be no great surprise that we use $\bar{x}$ (the sample mean) as a point estimate for μ (the population mean) and s (the sample standard deviation) as a point estimate for σ (the population standard deviation). In this section we will discuss estimates of μ and σ from large samples ($n \geq 30$).

Statistical theory and empirical results show that if a distribution is approximately mound-shaped and symmetric, then when the sample size is 30 or larger, we

---

*From *The Poetry of Robert Frost* edited by Edward Connery Lathem. Copyright 1942 by Robert Frost. Copyright © 1969 by Holt, Rinehart and Winston. Copyright © 1970 by Lesley Frost Ballantine. Reprinted by permission of Henry Holt and Company, Inc., and by Jonathan Cape Limited, Publishers.

are safe for most practical purposes if we estimate σ by s. The error resulting from taking the population standard deviation σ to be equal to the sample standard deviation s is negligible. However, when the sample size is less than 30, we will use special small sample methods, which we will study later in Section 8.2.

For large samples $n \geq 30$

$$\sigma \simeq s$$

is a good estimate for most practical purposes.

Using $\bar{x}$ to estimate μ is not quite so simple, even when we have a large sample size. The *error of estimate* is the magnitude of the difference between the point estimate and the true parameter value. Using $\bar{x}$ as a point estimate for μ, the error of estimate is the magnitude of $\bar{x} - \mu$. If we use absolute-value notation, we can indicate the error of estimate for $\bar{x}$ by the notation $|\bar{x} - \mu|$.

We cannot say exactly how close $\bar{x}$ is to μ when μ is unknown. Therefore, the exact error of estimate is unknown when the population parameter is unknown. Of course, μ is usually not known or there would be no need to estimate it. In this section we will use the language of probability to give us an idea of the size of the error of estimate when we use $\bar{x}$ as a point estimate of μ.

First we need to learn about *confidence levels*. The reliability of an estimate will be measured by the confidence level.

Suppose we want a confidence level of c (*see* Figure 8-1). Theoretically, you can choose c to be any value between 0 and 1, but usually c is equal to 0.90, 0.95, or 0.99. In any event the value z_c is the number such that the area under the standard normal curve falling between $-z_c$ and z_c is equal to c. The value z_c is called the *critical value* for a confidence level of c.

FIGURE 8-1 Confidence Level c and Corresponding Critical Value z_c Shown on the Standard Normal Curve

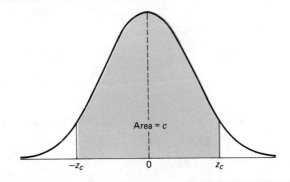

Area = c

$-z_c$ 0 z_c

Note: Area $= 2 \times$ table value associated with z_c.

The area under the normal curve from $-z_c$ to z_c is the probability that the standardized normal variable z lies in that interval. This means that

$$P(-z_c < z < z_c) = c$$

EXAMPLE   1

Let us use Table 5 in Appendix I to find a number $z_{0.95}$ so that 95% of the area under the standard normal curve lies between $-z_{0.95}$ and $z_{0.95}$. That is, we will find $z_{0.95}$ so that

$$P(-z_{0.95} < z < z_{0.95}) = 0.95$$

Table 5 gives the area under the normal curve from the mean 0 to any point z. The condition $P(-z_{0.95} < z < z_{0.95}) = 0.95$ is the same as the condition $2P(0 < z < z_{0.95}) = 0.95$ since the standard normal curve is symmetric about the mean 0. When we divide both sides of the last equation by 2 we get

$$P(0 < z < z_{0.95}) = \frac{0.95}{2} = 0.4750$$

Note that 0.4750 is an entry in Table 5. Table 8-1 is an excerpt from Table 5.

TABLE 8-1 Excerpt from Table 5 of Areas Under the Standard Normal Curve from 0 to z (Appendix I)

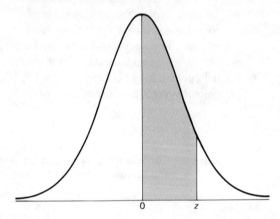

z	0.00	0.01	0.02	0.03	0.04	0.05	0.06	0.07...
0.0								
⋮								
⋮								
1.8	0.4641	0.4649	0.4656	0.4664	0.4671	0.4678	0.4686	0.4693
1.9	0.4713	0.4719	0.4726	0.4732	0.4738	0.4744	0.4750	0.4756

Area is 0.4750 for $z = 1.96$

We will use Table 8-1 to find $z_{0.95}$. From Table 8-1 we see that $z_{0.95} = 1.96$, so the probability is 0.95 that the standardized statistic z lies between -1.96 and 1.96. In symbols we have

$$P(-1.96 < z < 1.96) = 0.95$$

_____ Exercise 1 _____

a) Is it true that the condition

$$P(-z_{0.99} < x < z_{0.99}) = 0.99$$

is equivalent to the condition

$$2P(0 < z < z_{0.99}) = 0.99?$$

Why?

b) Use the information of part (a) and Table 5 in Appendix I to find the value of $z_{0.99}$.

a) It is true that the conditions are equivalent since the standard normal curve is symmetric about its mean 0.

b) To complete the computation, we divide both sides of the condition

$$2P(0 < z < z_{0.99}) = 0.99$$

by 2 and get the equivalent condition

$$P(0 < z < z_{0.99}) = \frac{0.99}{2} = 0.4950$$

We look up the area 0.4950 in Table 5 and then find the z value that produces that area. The value 0.4950 is not in the table; however, the values 0.4949 and 0.4951 are in the table. Even though 0.4950 is exactly halfway between the two values, the two values are so close together we use the higher value 0.4951. This gives us

$$z_{0.99} = 2.58$$

The results of Example 1 and Exercise 1 will be used a great deal in our later work. For convenience, Table 8-2 gives some levels of confidence and corresponding critical values z_c.

TABLE 8-2 Some Levels of Confidence and Their Corresponding Critical Values

Level of Confidence c	Critical Value z_c
0.80	1.28
0.85	1.44
0.90	1.645
0.95	1.96
0.99	2.58

An estimate is not very valuable unless we have some kind of measure of how "good" it is. Now that we have studied confidence levels and critical values, the language of probability can give us an idea of the size of the error of estimate caused by using the sample mean $\bar{x}$ as an estimate for the population mean.

Remember that $\bar{x}$ is a random variable. Each time we draw a sample of size n from a population, we can get a different value for $\bar{x}$. According to the Central Limit Theorem, if the sample size is large, then $\bar{x}$ has a distribution that is approximately normal with mean $\mu_{\bar{x}} = \mu$, the population mean we are trying to estimate. The standard deviation is $\sigma_{\bar{x}} = \sigma/\sqrt{n}$.

This information, together with our work on confidence levels, leads us (as shown in the optional derivation) to the probability statement

$$P\left(-z_c \frac{\sigma}{\sqrt{n}} < \bar{x} - \mu < z_c \frac{\sigma}{\sqrt{n}}\right) = c \tag{1}$$

Equation (1) uses the language of probability to give us an idea of the size of the error of estimate for the corresponding confidence level c. In words, equation (1) says that the probability is c that our point estimate $\bar{x}$ is within a distance $\pm z_c(\sigma/\sqrt{n})$ of the population mean μ. This relationship is shown in Figure 8-2.

In the following optional discussion, we derive formula (1). If you prefer, you may jump ahead to the summary about the error of estimate.

OPTIONAL DERIVATION OF EQUATION (1)

For a c confidence level we know

$$P(-z_c < z < z_c) = c \tag{2}$$

This statement gives us information about the size of z, but we want information about the size of $\bar{x} - \mu$. Is there a relationship between z and $\bar{x} - \mu$? The answer is yes since, by the Central Limit Theorem, $\bar{x}$ has a distribution that is approximately normal with mean μ and standard deviation $\sigma/\sqrt{n}$. We can convert $\bar{x}$ to a standard z score by using the formula

$$z = \frac{\bar{x} - \mu}{\sigma/\sqrt{n}} \tag{3}$$

Substituting this expression for z in equation (2) gives

$$P\left(-z_c < \frac{\bar{x} - \mu}{\sigma/\sqrt{n}} < z_c\right) = c \tag{4}$$

Multiplying all parts of the inequality in (4) by $\sigma/\sqrt{n}$ gives us

$$P\left(-z_c \frac{\sigma}{\sqrt{n}} < \bar{x} - \mu < z_c \frac{\sigma}{\sqrt{n}}\right) = c \tag{1}$$

Equation (1) is precisely the equation we set out to derive.

FIGURE 8-2   Distribution of Sample Means $\bar{x}$

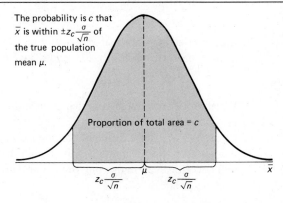

The probability is c that $\bar{x}$ is within $\pm z_c \dfrac{\sigma}{\sqrt{n}}$ of the true population mean μ.

Proportion of total area $= c$

$$z_c \frac{\sigma}{\sqrt{n}} \qquad \mu \qquad z_c \frac{\sigma}{\sqrt{n}}$$

- **Summary:**   The *error of estimate* using $\bar{x}$ as a point estimate for μ is $|\bar{x} - \mu|$. In most practical problems μ is unknown, so the error of estimate is also unknown. However, Equation (1) allows us to compute an *error tolerance E* which serves as a bound on the error of estimate. Using a $c\%$ level of confidence, we can say the point estimate $\bar{x}$ differs from the population mean μ by a maximal error tolerance of

$$E = z_c \frac{\sigma}{\sqrt{n}}$$

Since $\sigma \simeq s$ for large samples, we have

$$E \simeq z_c \frac{s}{\sqrt{n}} \qquad \text{when } n \geqslant 30 \qquad (5)$$

where E is the maximal error tolerance on the error of estimate for a given confidence level c (i.e., $|\bar{x} - \mu| < E$ with probability c); z_c is the critical value for the confidence level c (*see* Table 8-2); s is the sample standard deviation; n is the sample size.

Using Equations (1) and (5) we conclude

$$P(-E < \bar{x} - \mu < E) = c \qquad (6)$$

Equation (6) says that the probability is c that the difference between $\bar{x}$ and μ is no more than the maximal error tolerance E.

If we use a little algebra on the inequality

$$-E < \bar{x} - \mu < E \qquad (7)$$

we can rewrite it in the following mathematically equivalent way

$$\bar{x} - E < \mu < \bar{x} + E \tag{8}$$

Since (7) and (8) are mathematically equivalent, their probabilities are the same. Therefore, from (6), (7), and (8) we obtain

$$P(\bar{x} - E < \mu < \bar{x} + E) = c \tag{9}$$

Equation (9) says there is a chance of c that the population mean μ lies in the interval from $\bar{x} - E$ to $\bar{x} + E$. We call this interval a *c confidence interval for* μ. We will get a different confidence interval for each different sample that is taken. Some intervals will contain the population mean μ and others will not. However, in the long run the proportion of confidence intervals that contains μ is c.

- **Summary:** For large samples ($n \geq 30$) taken from a distribution that is approximately mound-shaped and symmetric, and for which the population standard deviation σ is unknown, a *c confidence interval for the population mean* μ is given by

c confidence interval for μ

$$\bar{x} - E < \mu < \bar{x} + E \tag{10}$$

where

$$\bar{x} = \text{sample mean}$$

$$E \simeq z_c \frac{s}{\sqrt{n}}$$

$$s = \text{sample standard deviation}$$

$$c = \text{confidence level } (0 < c < 1)$$

$$z_c = \text{critical value for confidence level } c$$

(*See* Table 8-2 for frequently used values.)

$$n = \text{sample size } (n \geq 30)$$

- **Important Comment:** What if you could assume

  i) you were drawing your samples from a *normal* population *and*
  ii) the population standard deviation σ was known to you?

Under these assumptions a c confidence interval for the population mean would be $\bar{x} - E < \mu < \bar{x} + E$ where

$$E = z_c \frac{\sigma}{\sqrt{n}}$$

and the condition that we must have a large sample ($n \geq 30$) could be dropped.

In most practical situations we don't know the population standard deviation σ. In these situations we use formula (10) with large samples to find a c confidence interval, or we use small samples ($n < 30$) and the methods of the next section to find a c confidence interval.

EXAMPLE 2

Julia enjoys jogging. She has been jogging over a period of several years, during which time her physical condition has remained constantly good. Usually she jogs 2 mi/day. During the past year Julia has sometimes recorded her times required to run 2 mi. She has a sample of 90 of these times. For these 90 times the mean was $\bar{x} = 15.60$ min and the standard deviation was $s = 1.80$ min. Let μ be the mean jogging time for the entire distribution of Julia's 2-mi running times (taken over the past year). Find a 0.95 confidence interval for μ.

The interval from $\bar{x} - E$ to $\bar{x} + E$ will be a 95% confidence interval for μ. In this case $c = 0.95$, so $z_c = 1.96$ (*see* Table 8-2). The sample size $n = 90$ is large enough that we may approximate σ as $s = 1.80$ min. Therefore,

$$E = z_c \frac{s}{\sqrt{n}}$$

$$E = 1.96 \left(\frac{1.80}{\sqrt{90}} \right)$$

$$E = 0.37$$

Using formula (10), the given value of $\bar{x}$, and our computed value for E, we get the 95% confidence interval for μ.

$$\bar{x} - E < \mu < \bar{x} + E$$

$$15.60 - 0.37 < \mu < 15.60 + 0.37$$

$$15.23 < \mu < 15.97$$

We conclude that there is a 95% chance that the population mean μ of jogging times for Julia is between 15.23 min and 15.97 min.

A few comments are in order about the general meaning of the term confidence interval. It is important to realize that the endpoints $\bar{x} \pm E$ are really statistical *variables*. Equation (9) says we have a chance c of obtaining a sample such that the interval, once it is computed, will contain the parameter μ. Of course, after the confidence interval is numerically fixed, it either does or does not contain μ. So the probability is 1 or 0 that the interval, when it is fixed, will contain μ. A nontrivial probability statement can be made only about variables, not constants. Therefore, Equation (9) really says that if we repeat the experiment many times and get lots of confidence intervals (for the same sample size), then the proportion of all intervals which will turn out to straddle the mean μ is c.

In Figure 8-3 the horizontal lines represent 0.90 confidence intervals for various samples of the same size from a distribution. Some of these intervals contain μ, others do not. Since the intervals are 0.90 confidence intervals, about 90% of all such intervals should contain μ. For each sample the interval goes from $\bar{x} - E$ to $\bar{x} + E$.

FIGURE 8-3 0.90 Confidence Intervals for Samples of the Same Size

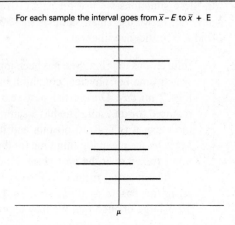

For each sample the interval goes from $\bar{x} - E$ to $\bar{x} + E$

μ

• **Comment:** Please see "Using Computers" at the end of this chapter for a computer demonstration of this discussion about confidence intervals.

_____ Exercise 2 _____

Walter usually meets Julia at the track. He prefers to jog 3 mi. While Julia kept her record he also kept one for his time required to jog 3 mi. For his 90 times the mean was $\bar{x} = 22.50$ min and the standard deviation was $s = 2.40$ min. Let μ be the mean jogging time for the entire distribution of Walter's 3-mi running times over the past several years. How can we find a 0.99 confidence interval for μ?

a) What is $z_{0.99}$? (*See* Table 8-2.)

b) Since the sample size is large, what can we use for σ?

c) What is the value of E?

d) What are the endpoints for a 0.99 confidence interval for μ?

a) $z_{0.99} = 2.58$

b) $\sigma \approx s = 2.40$

c) $E = z_c \dfrac{\sigma}{\sqrt{n}} \approx 2.58 \left(\dfrac{2.40}{\sqrt{90}} \right) = 0.65$

d) The endpoints are given by
$$\bar{x} - E \approx 22.50 - 0.65$$
$$= 21.85$$
$$\bar{x} + E \approx 22.50 + 0.65$$
$$= 23.15$$

_____ Exercise 3 _____

A large loan company specializes in making automobile loans. The board of directors wants to estimate the average amount loaned for cars during the past year. The company takes a random sample of 225 customer files for this period. The mean amount loaned for this sample of 225 loans is $\bar{x} = \$8{,}200$ and the standard deviation is $s = \$750$. Let μ be the mean of all car loans made over the past year.

Find a 0.95 confidence interval for μ.

Since $n = 225$ is a large sample we take $\sigma \approx s = 750$. From Table 8-2 we see $z_{0.95} = 1.96$. Then

$$E \approx z_c \frac{s}{\sqrt{n}} = 1.96 \left(\frac{750}{\sqrt{225}} \right) = 98$$

$$\bar{x} - E \approx 8{,}200 - 98 = \$8{,}102$$

$$\bar{x} + E \approx 8{,}200 + 98 = \$8{,}298$$

The interval from \$8,102 to \$8,298 is a 0.95 confidence interval for μ.

———————— Exercise 4 ————————

We have said that a sample of size 30 or larger is a large sample. In this section we indicated two important reasons why our methods require large samples.

What are these reasons?

Reason 1: Our methods require $\bar{x}$ to have approximately a normal distribution. We know from the Central Limit Theorem that this will be the case for large samples.

Reason 2: Unless we somehow know σ, our methods require us to approximate σ with the sample standard deviation s. This approximation will be good only if the sample size is large.

When we use samples to estimate the mean of a population, we generate a small error. However, samples are useful even when it is possible to survey the entire population because the use of a sample may yield savings of time or effort in collecting data.

Section Problems 8.1

In all problems in Chapter 8, the word average is taken to be the mean $\bar{x}$.

1. A random sample of 40 cups of coffee dispensed from an automatic vending machine showed that the mean amount of coffee the machine gave was $\bar{x} = 7.1$ oz with standard deviation $s = 0.3$ oz. Find a 90% confidence interval for the population mean of the amount of coffee dispensed.

2. The Junetag Company makes washing machines. In one phase of production it is necessary to estimate the drying time of enamel paint on metal sheeting. Analysis of a

random sample of 420 sheets shows that the average drying time is 56 hr with a standard deviation of 12 hr. Find a 99% confidence interval for the population mean drying time of the enamel.

3. A large car manufacturing company has hired a psychologist to examine the response time for a proposed tail light system. The response time is the length of time it takes a driver following a car to apply the brakes after the new tail light system is activated on the car in front. For a random sample of 283 subjects it was found that the average response time was 5.47 sec with sample standard deviation of 1.91 sec. Find a 90% confidence interval for the mean response time for the population of all drivers.

4. A sociologist is studying the length of courtship before marriage in a rural district near Kyoto, Japan. A random sample of 56 middle income families were interviewed. It was found that the average length of courtship was 3.4 yr with sample standard deviation 1.2 yr. Find an 85% confidence interval for the length of courtship for the population of all middle income families in this district.

5. The Roman Arches is an Italian restaurant. The manager wants to estimate the net income each day on the Monday through Friday lunch business. A random sample of 45 days gave an average net income of $522.17 from lunch time business. The sample standard deviation was $114.09. Find a 95% confidence interval for the daily net income from lunches Monday through Friday.

6. A random sample of 121 New York City policemen showed that they worked as off-duty policemen an average of 6.3 hr per week with standard deviation of 2.5 hr. Find a 0.90 confidence interval for the mean number of hours of off-duty work for the entire police department.

7. A random sample of 100 felony trials in Chicago shows the mean waiting time between arrest and trial is 173 days with standard deviation 28 days. Find a 0.99 confidence interval for the mean time interval between arrest and trial.

8. A local C.J. Nickel department store took a random sample of 49 charge accounts and found the average balance due was $63.19 with standard deviation $13.50. Find a 0.95 confidence interval for the average balance due in this store.

9. Ralph Smith took a random sample of 400 homes in El Paso and found that the homes received an average of 18.6 pieces of junk mail per week with a standard deviation of 5.2 pieces. Find a 0.95 confidence interval for the mean number of pieces of junk mail received per week by El Paso families.

10. Irv and Nancy are thinking about buying the Rockwood Motel located on Interstate 70. Before they make up their mind, they want to estimate the average number of vehicles that go by the motel each day in the summer. Fortunately, the highway department has been counting vehicles on I-70 near the motel. A random sample of 36 summer days shows an average of 16,000 cars per day with a standard deviation of 2,400 cars. Find a 0.90 confidence interval for the mean number of cars per summer day going past the Rockwood Motel.

11. The head of the hospital maternity ward wants to estimate the average number of days a patient stays in the ward. A random sample of 100 patients shows the average stay to be 5.2 days with standard deviation of 1.9 days. Find a 0.90 confidence interval for the mean number of days a patient stays in the ward.

12. The national program planner for the Girl Scouts wants to estimate the average number of girls per troop. A random sample of 81 troops shows an average (mean) of 14.8 girls

with standard deviation of 9.0 girls. Find a 0.99 confidence interval for the average (mean) number of girls per troop.

13. An anthropologist measured the heights of a random sample of 40 adult males native to a South Pacific atoll. The average height was 57.6 in. with a standard deviation of 3.7 in.
   a) Find a *c* confidence interval for the mean height of the adult males when *c* = 0.80, 0.90, 0.95, and 0.99.
   b) Find the length of each interval of part (a) and comment on how these lengths change as *c* increases.

14. In September a biological research team caught, weighed, and released a random sample of 54 chipmunks in Rocky Mountain National Park. The mean of the sample weights was $\bar{x}$ = 8.7 oz, with a standard deviation of *s* = 1.4 oz.
   a) Find a *c* confidence interval for the mean September weight of all Rocky Mountain chipmunks when *c* = 0.80, 0.90, 0.95, and 0.99.
   b) Find the length of each interval of part (a), and comment on how these lengths change as *c* increases.

15. In Roosevelt National Forest, the rangers took random samples of live aspen trees and measured the base circumference of each tree.
   a) The first sample had 30 trees with a mean circumference of $\bar{x}$ = 15.71 in. and a sample standard deviation of *s* = 4.63 in. Find a 95% confidence interval for the mean circumference of aspen trees from this data.
   b) The next sample had 90 trees with a mean of $\bar{x}$ = 15.58 in. and a sample standard deviation of *s* = 4.61 in. Again, find a 95% confidence interval from this data.
   c) The last sample had 300 trees with a mean of $\bar{x}$ = 15.59 in. and a sample standard deviation of *s* = 4.62 in. Again, find a 95% confidence interval from this data.
   d) Find the length of each interval of parts (a), (b), and (c). Comment on how these lengths change as the sample size increases.

16. Sarah works at the college library. She took random samples of overdue book cards and computed the number of days each book was overdue.
   a) The first sample had 36 book cards, and the mean number of days overdue was $\bar{x}$ = 20.3 days with standard deviation *s* = 9.6 days. Find a 95% confidence interval for the number of days that a book is overdue.
   b) The next sample had 64 book cards, and the mean number of days the books were overdue was $\bar{x}$ = 21.2 days with standard deviation *s* = 9.5 days. Find a 95% confidence interval for the number of days that a book is overdue.
   c) The last sample had 400 book cards, and the mean number of days the books were overdue was $\bar{x}$ = 20.9 days with standard deviation *s* = 9.6 days. Again find a 95% confidence interval for the number of days that a book is overdue.
   d) Find the length of each interval of parts (a), (b), and (c). Comment on how these lengths change as the sample size increases.

17. Pro Computer Services has computer monitoring of all outgoing and incoming telephone calls to record the length of time of the technical consulting calls. A random sample of 40 calls had the following lengths (to the nearest minute):

11	17	14	16	22	13	17	14	19	12
17	21	37	14	28	11	18	17	18	13
16	12	16	29	12	16	12	17	11	15
13	16	13	18	10	11	17	12	10	13

a) Use a calculator to verify that the sample mean length of calls is $\bar{x} = 15.95$ min and the sample standard deviation is $s = 5.43$ min.

b) Find a 90% confidence interval for the population mean of technical consulting telephone calls at Pro Computer Services.

18. A random sample of 40 full-time students at Ocean View College were asked to keep records of the number of hours they studied during midterm week. The responses (to the nearest hour) were:

3	6	12	15	21	26	2	7	11	17
24	27	4	8	14	16	22	1	5	13
15	24	9	11	15	21	7	10	17	9
10	18	8	14	15	12	11	10	12	13

a) Use a calculator to verify that the sample mean number of study hours for midterm week is $\bar{x} = 12.9$ hr and the sample standard deviation $s = 6.51$ hr.

b) Find a 95% confidence interval for the population mean study time for the entire population of Ocean View College students during midterm week.

SECTION 8.2   Estimating μ with Small Samples

For samples of size 30 or larger we can approximate the population standard deviation σ by s, the sample standard deviation. Then we can use the Central Limit Theorem to find bounds on the error of estimate and confidence intervals for μ.

There are many practical and important situations, however, where large samples are simply not available. Suppose an archaeologist discovers only seven fossil skeletons from a previously unknown species of miniature horse. Reconstructions of the skeletons of these seven miniature horses show their mean shoulder heights to be $\bar{x} = 46.1$ cm. Let μ be the mean shoulder height for this entire species of miniature horse. How can we find the maximal error of estimate $|\bar{x} - \mu|$? How can we find a confidence interval for μ? We will return to this problem later in this section.

To avoid the error involved in replacing σ by s—i.e., approximating σ by s—when the sample size is small (less than 30) we introduce a new variable called *Student's t variable*. The t variable and its corresponding distribution, called Student's t distribution, were discovered in 1908 by W.S. Gossett. He was employed as a statistician by a large Irish brewing company that frowned upon the publication of research by its employees, so Gossett published his research under the pseudonym Student. Gossett was the first to recognize the importance of developing statistical methods for obtaining reliable information from small samples. It might be more fitting to call this Gossett's t distribution; however, in the literature of mathematical statistics it is known as Student's t distribution.

The t variable is defined by the following formula.

$$t = \frac{\bar{x} - \mu}{\frac{s}{\sqrt{n}}} \qquad (11)$$

where $\bar{x}$ is the mean of a random sample of n measurements, μ is the population mean of the x distribution, and s is the sample standard deviation.

● **Comment:** You should note that our t variable is just like

$$z = \frac{\bar{x} - \mu}{\frac{\sigma}{\sqrt{n}}}$$

except that we replace σ with s. Unlike our methods for large samples, σ cannot be approximated by s when the sample size is less than 30 and we cannot use the normal distribution. However, we will be using the same methods as in Section 8.1 to find the maximal error of estimate and to find confidence intervals, but we use the Student's t distribution.

If lots of random samples of size n are drawn, then we get lots of t values from equation (11). These t values can be organized into a frequency table and a histogram can be drawn, thereby giving us an idea of the shape of the t distribution (for a given n).

Fortunately all this work is not necessary because mathematical theories can be used to obtain a formula for the t distribution. However, it is important to observe that these theories say the shape of the t distribution depends only on n provided the basic variable x has a normal distribution. So *when we use the t distribution we will assume that the x distribution is normal*.

Table 6 in Appendix I gives values of the variable t corresponding to what we call the number of *degrees of freedom*, abbreviated *d.f.* For the methods used in this section the number of degrees of freedom is given by the formula

$$d.f. = n - 1 \qquad (12)$$

where $d.f.$ stands for the degrees of freedom and n is the sample size being used.

Each choice for $d.f.$ gives a different t distribution. However, for $d.f.$ larger than about 30, the t distribution and the standard normal z distribution are almost the same.

The graph of a t distribution is always symmetrical about its mean, which (as for the z distribution) is 0. The main observable difference between a t distribution and the standard normal z distribution is that a t distribution has somewhat thicker tails.

Figure 8-4 shows a standard normal z distribution and Student's t distribution with $d.f. = 3$ and $d.f. = 5$.

FIGURE 8-4   A Standard Normal Distribution and Student's *t* distribution with
d.f. = 3 and *d.f.* = 5

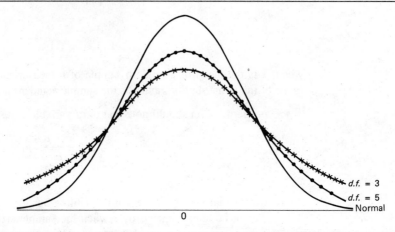

Table 6 of Appendix I gives various *t* values for different degrees of freedom
d.f. We will use this table to find critical values t_c for a *c* confidence level. In other
words, we want to find t_c so that an area equal to *c* under the *t* distribution for a
given number of degrees of freedom falls between $-t_c$ and t_c. In the language of
probability we want to find t_c so that

$$P(-t_c < t < t_c) = c$$

This probability corresponds to the area shaded in Figure 8-5.

Table 6 has been arranged so that *c* is one of the column headings, and the
degrees of freedom *d.f.* are the row headings. To find t_c for any specific *c*, we find
the column headed by that *c* value and read down until we reach the row headed by
the appropriate number of degrees of freedom *d.f.* (You will notice two other column
headings: α' and α''. We will use these later, but for the time being, ignore them.)

FIGURE 8-5   Area Under the *t* Curve Between $-t_c$ and t_c

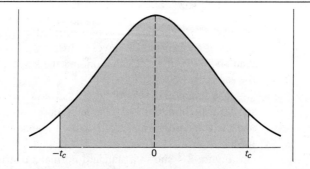

EXAMPLE 3 Use Table 8-3 (an excerpt from Table 6) to find the critical value t_c for a 0.99 confidence level for a t distribution with sample size $n = 5$.

TABLE 8-3 Student's t Distribution (Excerpt from Table 6, Appendix I)

c	0.90	0.95	0.98	0.99
α'	—	—	—	—
α''	—	—	—	—
$d.f.$				
$\vdots$				
3	2.353	3.182	4.541	5.841
4	2.132	2.776	3.747	4.604←
$\vdots$				
7	1.895	2.365	2.998	3.499
8	1.860	2.306	2.896	3.355

Source: Table 6, Appendix I, is taken from Table III of Fisher and Yates: *Statistical Tables for Biological, Agricultural and Medical Research,* published by Longman Group Ltd., London (previously published by Oliver & Boyd Ltd., Edinburgh) and by permission of the authors and publishers.

a) First we find the column with c heading 0.99. This is the last column.

b) Next we compute the number of degrees of freedom: $d.f. = n - 1 = 5 - 1 = 4$.

c) We read down the column under the heading $c = 0.99$ until we reach the row headed by 4 (under $d.f.$). The entry is 4.604. Therefore, $t_{0.99} = 4.604$.

_____ Exercise 5 _____

Use Table 8-3 to find t_c for a 0.90 confidence level of a t distribution with sample size $n = 9$.

a) We find the column headed by $c =$ _____. This is the _____ (first, second, third, fourth) column.

a) $c = 0.90$. This is the first column.

b) The degrees of freedom is given by

$$d.f. = n - 1 = \underline{\quad}$$

b) $d.f. = n - 1 = 9 - 1 = 8$

c) Read down the column found in part (a) until you reach the entry in the row headed by $d.f. = 8$. The value of $t_{0.90}$ is _____ for a sample size of 9.

c) $t_{0.90} = 1.860$ for a sample size $n = 9$.

d) Find t_c for a 0.95 confidence level of a t distribution with sample size $n = 9$.

d) $t_{0.95} = 2.306$ for a sample of size $n = 9$.

In Section 8.1 we found bounds $\pm E$ on the error of estimate for a c confidence level. Using the same basic approach, we arrive at the conclusion that

$$E = t_c \frac{s}{\sqrt{n}}$$

is the maximal error of estimate for a c confidence level with small samples (i.e., $|\bar{x} - \mu| < E$ with probability c). The analogue of equation (1) in Section 8.1 is

$$P\left(-t_c \frac{s}{\sqrt{n}} < \bar{x} - \mu < t_c \frac{s}{\sqrt{n}}\right) = c \qquad (13)$$

- **Comment:** Comparing equation (13) with equation (1) in Section 8.1, it becomes evident that we are using the same basic method on the t distribution that we did on the z distribution.

Likewise, for small samples from normal populations equation (9) of Section 8.1 becomes

$$P(\bar{x} - E < \mu < \bar{x} + E) = c \qquad (14)$$

where $E = t_c (s/\sqrt{n})$. Let us organize what we have been doing in a convenient summary.

- **Summary:** For small samples ($n < 30$) taken from a normal population where σ is unknown, a c confidence interval for the population mean μ is as follows

c confidence interval for μ (small sample)

$$\bar{x} - E < \mu < \bar{x} + E \qquad (15)$$

where

$\bar{x}$ = sample mean

$$E = t_c \frac{s}{\sqrt{n}}$$

c = confidence level ($0 < c < 1$)

t_c = critical value for confidence level c,

and degrees of freedom $d.f. = n - 1$

taken from t distribution

n = sample size (since this formula applies

to small samples, $n < 30$)

s = sample standard deviation

- **Comment:** In our applications of Student's t distribution we have made the basic assumption that x has a normal distribution. However, the same methods apply even if x is only approximately normal. In fact the main requirement for using the Student's t distribution is that the distribution of x values be reasonably symmetric and mound-shaped. If this is the case, then the methods we employ with the t distribution can be considered valid for most practical applications.

EXAMPLE 4

Let's return to our archaeologist and the newly discovered (but extinct) species of miniature horse. There are only seven known existing skeletons with shoulder heights (in centimeters) 45.3, 47.1, 44.2, 46.8, 46.5, 45.5, and 47.6. For this sample data the mean is $\bar{x} = 46.14$ and the sample standard deviation is $s = 1.19$. Let μ be the mean shoulder height (in centimeters) for this entire species of miniature horse and assume the population is approximately normal.

Find a 99% confidence interval for μ, the mean shoulder height of the entire population of such horses. In this case $n = 7$, so $d.f. = n - 1 = 7 - 1 = 6$. For $c = 0.99$ Table 6 gives $t_{0.99} = 3.707$ (for $d.f. = 6$). The sample standard deviation is $s = 1.19$.

$$E = t_c \frac{s}{\sqrt{n}} = (3.707)\frac{1.19}{\sqrt{7}} = 1.67$$

The 99% confidence interval is

$$\bar{x} - E < \mu < \bar{x} + E$$
$$46.14 - 1.67 < \mu < 46.14 + 1.67$$
$$44.5 < \mu < 47.8$$

——————————— Exercise 6 ———————————

A company has a new process for manufacturing large artificial sapphires. The production of each gem is expensive, so the number available for examination is limited. In a trial run 12 sapphires are produced. The mean weight for these 12 gems is $\bar{x} = 6.75$ carats and the sample standard deviation is $s = 0.33$ carats. Let μ be the mean weight for the distribution of all sapphires produced by the new process.

a) What is $d.f.$ for this setting?

a) $d.f. = n - 1$ where n is the sample size. Since $n = 12$, $d.f. = 12 - 1 = 11$.

b) Use Table 6 to find $t_{0.95}$.

b) Using Table 6 with $d.f. = 11$ and $c = 0.95$, we find $t_{0.95} = 2.201$.

c) Find E.

c) $E = t_{0.95}\frac{s}{\sqrt{n}} = (2.201)\frac{0.33}{\sqrt{12}} = 0.21$

d) Find a 95% confidence interval for μ.

d) $\bar{x} - E < \mu < \bar{x} + E$

$6.75 - 0.21 < \mu < 6.75 + 0.21$

$6.54 < \mu < 6.96$

e) What assumption about the distribution of all sapphires had to be made to obtain these answers?

e) The population of artificial sapphire weights is approximately normal.

We have several formulas for confidence intervals for the population mean μ. How do we choose an appropriate one? We need to look at sample size, the distribution of the original population, as well as whether or not the population standard deviation σ is known. There are essentially four cases for which we have the tools to find c confidence intervals for the mean μ.

- **Summary**

Confidence Intervals for the Mean

Large Sample Cases

$n \geq 30$

1. If σ is not known, then a c confidence interval for μ is

$$\bar{x} - z_c \frac{s}{\sqrt{n}} < \mu < \bar{x} + z_c \frac{s}{\sqrt{n}}$$

2. If σ is known, then a c confidence interval for μ is

$$\bar{x} - z_c \frac{\sigma}{\sqrt{n}} < \mu < \bar{x} + z_c \frac{\sigma}{\sqrt{n}}$$

Small Sample Case

$n < 30$

3. If the population is approximately normal and σ is not known then a c confidence interval for μ is

$$\bar{x} - t_c \frac{s}{\sqrt{n}} < \mu < \bar{x} + t_c \frac{s}{\sqrt{n}}$$

For *any* Sample Size

4. If the population *is normal* and σ is known, then for any sample size (large or small) a c confidence interval for μ is

$$\bar{x} - z_c \frac{\sigma}{\sqrt{n}} < \mu < \bar{x} + z_c \frac{\sigma}{\sqrt{n}}$$

Section Problems 8.2

In all the following problems assume that the population of x values has an approximately normal distribution.

1. Use Table 6 in Appendix I to find t_c for a 0.95 confidence level when the sample size is 18.

2. Use Table 6 to find t_c for a 0.99 confidence level when the sample size is 4.

3. Use Table 6 to find t_c for a 0.90 confidence level when the sample size is 22.

4. Symptoms of a new flu virus have been determined from 18 sufferers to have a mean duration time of 9.7 days with standard deviation of 4.8 days. Find a 0.95 confidence interval for the mean duration time of these flu symptoms.

5. Four boxes of Purr Lots Cat Food are found to weigh 12.7, 11.6, 11.2, and 12.4 ounces. Find a 0.99 confidence interval for the mean weight of all such boxes of cat food. (*Hint:* First find the sample mean $\bar{x}$ and sample standard deviation s of the weight of food in the boxes.)

6. A biologist has found the average weight of 12 randomly selected adult mud turtles to be 8.7 lb with standard deviation 3.6 lb. Find a 0.90 confidence interval for the average weight of all such turtles.

7. A park ranger has timed the mating calls of 22 bull elk. The mean duration of these calls was 14.6 sec with standard deviation 2.8 sec. Find a 0.95 confidence interval for the mean duration of mating calls of the bull elk.

8. The Aloha Taxi Cab Company of Honolulu, Hawaii, wants to estimate the mean life of tires on its cabs. A random sample of 15 tires from the cabs are found to have a life of 18,280 mi with standard deviation 1,310 mi. Find a 0.90 confidence interval for the mean life of all such tires on Aloha Taxi Cabs.

9. The average cholesterol level for a random sample of 26 adult women from Dairyhaven, Wisconsin, is 263 units (mg of cholesterol per dl of blood). The sample standard deviation is 43 units. Find a 0.99 confidence interval for the mean cholesterol level for all adult women of Dairyhaven.

10. A Superball is dropped from a fixed height and the height of the bounce is measured. The coefficient of restitution is the height of the bounce divided by the original height from which the ball is dropped. Five bounces of the ball resulted in the following measurements of the coefficient of restitution: 0.87, 0.79, 0.85, 0.82, and 0.80. Find a 0.95 confidence interval for the mean coefficient of restitution of the Superball. (*Hint:* First find the sample mean $\bar{x}$ and sample standard deviation s of the restitution coefficients.)

11. André is head waiter at one of the famous gourmet restaurants in San Francisco. The Internal Revenue Service is doing an audit on his tax return this year. In particular, the IRS wants to know the average amount André gets for a tip. In an effort to satisfy the IRS, André took a random sample of eight credit card receipts, each of which indicated his tip. The results were:

| $10.00 | $11.93 | $15.70 | $ 9.10 |
| $12.75 | $11.15 | $14.50 | $13.65 |

a) Use a calculator to verify that the sample mean is $12.35 and the sample standard deviation is $2.25.

b) Find a 90% confidence interval for the population mean of tips received by André.

12. A random sample of records of ten night shifts at the Emergency Medical Center gave the following information about the number of people admitted each night:

18   23   15   19   21   19   24   22   23   22

a) Use a calculator to verify that the sample mean is 20.6 and the sample standard deviation is 2.8.

b) Find a 95% confidence interval for the mean number of people admitted to the Emergency Medical Center during the night shift.

13. In wine making, acidity of the grape is a crucial factor. A ph range of 3.1 to 3.6 is considered very acceptable. A random sample of 12 bunches of ripe grapes was taken from a California vineyard. For each bunch of grapes the acidity as measured by ph level was found to be:

3.2   3.7   3.4   3.6   3.5   3.1
3.5   3.3   3.6   3.1   3.2   3.4

a) Use a calculator to verify that the sample mean acidity is 3.38 and the sample standard deviation is 0.20.

b) Find a 99% confidence interval for the mean acidity of the entire harvest of grapes from this vineyard.

14. Certain rivers in Labrador are known to have phenomenally large brook trout. A biologist using catch-and-release tagging of trout obtained the following information about annual weight gain of seven trout (in pounds). The results are:

0.73   1.05   0.81   1.12   0.92   1.17   0.98

a) Use a calculator to verify that the sample mean of the annual weight gain is 0.97 lb and the sample standard deviation is 0.16 lb.

b) Find a 90% confidence interval for the mean annual weight gain of brook trout.

15. Error of depth perception is very important in dental work. A certain aptitude test asks subjects to estimate the distances between a fixed object and a second object of variable position. A random sample of 14 subjects gave the following information about errors in a particular depth perception test (units in millimeters):

1.1   1.5   0.9   2.1   1.4   1.7   0.8
1.3   1.8   1.1   1.6   1.9   1.2   1.6

a) Use a calculator to verify that the sample mean of the above data is 1.43 mm and the sample standard deviation is 0.38 mm.

b) Find a 90% confidence interval for the population mean of errors for this depth perception test.

16. At a large industrial computer center a computer failure is extremely expensive. The longer the computer is down, the greater the expense becomes. For the past six computer

downs, the support service contractor required the following times (in hours) to bring the computer back up:

4.1   3.8   5.2   4.4   5.1   3.9

a) Use a calculator to verify that the sample mean for computer downs is 4.42 hr and the sample standard deviation is 0.60 hr.
b) Find a 90% confidence interval for the population mean of all such computer down times.

17. A stockbroker took a random sample of 18 clients and found the number of stock trans-actions (buying or selling a single stock) for each client over the last year. The results were:

21   15   17   9   28   19   22   16   15
14   13   16   15   26   10   18   21   14

a) Use a calculator to verify that the sample mean is 17.17 transactions with sample standard deviation 5.00 transactions.
b) Find a 95% confidence interval for the population mean of all such stock transactions each year.

18. Shoplifting has been a problem in a large men's clothing store. Using special security measures to monitor shoplifting, it was found that there were attempts to shoplift the following dollar values of merchandise each week for the past nine weeks:

$356 $285 $310 $375 $290
$325 $331 $342 $335

a) Use a calculator to verify that the sample mean is $327.67 with sample standard deviation $29.31.
b) Find a 90% confidence interval for the population mean.

SECTION 8.3   Estimating *p* in the Binomial Distribution

The binomial distribution is completely determined by the number of trials n and the probability p of success in a single trial. For most experiments the number of trials is chosen in advance. Then the distribution is completely determined by p. In this section we will consider the problem of estimating p under the assumption that n has already been selected.

Again we are employing what are called large sample methods. We will assume the normal curve is a good approximation to the binomial distribution and, when necessary, we will use sample estimates for the standard deviation. Empirical studies have shown that these methods are quite good provided that *both*

$$np > 5 \quad \text{and} \quad nq > 5 \quad \text{where } q = 1 - p$$

Let r be the number of successes out of n trials in a binomial experiment. We will take the proportion of successes r/n as our point estimate for p. For example, suppose 800 students are selected at random from a student body of 20,000 students and they are each given shots to prevent a certain type of flu. These 800 students are then exposed to the flu, and 600 of them do not get the flu. What is the probability p that the shot will be successful for any single student selected at random from the entire population of 20,000 students? We estimate p for the entire student body by computing r/n from the sample of 800 students. The value r/n is 600/800 or 0.75. The value 0.75 is then the estimate for p.

The difference between the actual value of p and the estimate r/n is the size of our error caused by using r/n as a point estimate for p. The magnitude of $(r/n) - p$ is called the *error of estimate* for r/n as a point estimate for p. In absolute value notation the error of estimate is $|(r/n) - p|$.

To compute the bounds for the error of estimate we need some information about the distribution of r/n values for different samples of the same size n. It turns out that for large samples the distribution of r/n values is well approximated by a *normal curve with*

$$\mu = p \quad \text{and} \quad \sigma = \sqrt{pq/n}$$

Since the distribution of r/n is approximately normal, we use features of the standard normal distribution to find the bounds for the difference $(r/n) - p$. Recall z_c is the number such that an area equal to c under the standard normal curve falls between $-z_c$ and z_c. Then, in terms of the language of probability,

$$P\left(-z_c \sqrt{\frac{pq}{n}} < \frac{r}{n} - p < z_c \sqrt{\frac{pq}{n}} \right) = c \qquad (16)$$

Equation (16) says the chance is c that the numerical difference between r/n and p is between $-z_c \sqrt{pq/n}$ and $z_c \sqrt{pq/n}$. With the c confidence level our estimate r/n differs from p by no more than

$$E = z_c \sqrt{pq/n}$$

As in Section 8.1, we call E the maximal error tolerance on the error of estimate $|(r/n) - p|$ for a confidence level c.

OPTIONAL

Derivation of Equation (16)

First we need to show that r/n has a distribution that is approximately normal with $\mu = p$ and $\sigma = \sqrt{pq/n}$. From Section 7.3 we know that for sufficiently large n the binomial distribution can be approximated by a normal distribution with mean $\mu = np$ and standard deviation

$\sigma = \sqrt{npq}$. If r is the number of successes out of n trials of a binomial experiment, then r is a binomial random variable with a binomial distribution. When we convert r to standard z units, we obtain

$$z = \frac{r - \mu}{\sigma} = \frac{r - np}{\sqrt{npq}}$$

For sufficiently large n, r will be approximately normally distributed, so z will be too.

If we divide both numerator and denominator of the last expression by n, the value of z will not change.

$$z = \frac{\dfrac{r - np}{n}}{\dfrac{\sqrt{npq}}{n}}$$

When we simplify we find

$$z = \frac{\dfrac{r}{n} - p}{\sqrt{\dfrac{pq}{n}}} \tag{17}$$

The last equation tells us the r/n distribution is approximated by a normal curve with $\mu = p$ and $\sigma = \sqrt{pq/n}$.

The probability is c that z lies in the interval between $-z_c$ and z_c since an area equal to c under the standard normal curve lies between $-z_c$ and z_c. Using the language of probability we write

$$P(-z_c < z < z_c) = c$$

From equation (17) we know that

$$z = \frac{\dfrac{r}{n} - p}{\sqrt{\dfrac{pq}{n}}}$$

If we put this expression for z into the previous equation, we obtain

$$P\left(-z_c < \frac{\dfrac{r}{n} - p}{\sqrt{\dfrac{pq}{n}}} < z_c \right) = c$$

If we multiply all parts of the inequality by $\sqrt{pq/n}$ we obtain the equivalent statement

$$P\left(-z_c \sqrt{\frac{pq}{n}} < \frac{r}{n} - p < z_c \sqrt{\frac{pq}{n}} \right) = c \tag{16}$$

To find a c confidence interval for p we will use E in place of the expression $z_c\sqrt{pq/n}$ in equation (16). Then we get

$$P\left(-E < \frac{r}{n} - p < E\right) = c \qquad (18)$$

Some algebraic manipulation produces the mathematically equivalent statement

$$P\left(\frac{r}{n} - E < p < \frac{r}{n} + E\right) = c \qquad (19)$$

Equation (19) says the probability is c that p lies in the interval from $(r/n) - E$ to $(r/n) + E$. Therefore, the interval from $(r/n) - E$ to $(r/n) + E$ is the c confidence interval for p that we wanted to find.

There is one technical difficulty in computing the c confidence interval for p. The expression $E = z_c\sqrt{pq/n}$ requires that we know the values of p and q. In most situations, we will not know the actual values of p or q, so we will use our point estimates

$$p \simeq \frac{r}{n} \quad \text{and} \quad q = 1 - p \simeq 1 - \frac{r}{n}$$

to estimate E. These estimates are safe for most practical purposes since we are dealing with large sample theory ($np > 5$ and $nq > 5$).

For convenient reference, we'll summarize the information about c confidence intervals for p, the probability of success.

• **Summary:** Consider a binomial distribution where n = number of trials, r = number of successes out of the n trials, p = probability of success on each trial, and q = probability of failure on each trial.

If n, p, and q are such that

$$np > 5 \quad \text{and} \quad nq > 5$$

then a c confidence interval for p is

$$\frac{r}{n} - E < p < \frac{r}{n} + E$$

where

$$E \simeq z_c \sqrt{\frac{\left(\frac{r}{n}\right)\left(1 - \frac{r}{n}\right)}{n}}$$

z_c = critical value for confidence level c taken from a normal distribution (*see* Table 8-2)

EXAMPLE 5

Let's return to our flu shot experiment. Suppose 800 students were selected at random from a student body of 20,000 and given shots to prevent a certain type of flu. All 800 students were exposed to the flu and 600 of them did not get the flu. Let p represent the probability that the shot will be successful for any single student selected at random from the entire population of 20,000. Let q be the probability that the shot is not successful.

a) What is the number of trials n? What is the value of r?

Since each of the 800 students receiving the shot may be thought of as a trial, then $n = 800$, and $r = 600$ is the number of successful trials.

b) What are the point estimates for p and q?

We estimate p by

$$\frac{r}{n} = \frac{600}{800} = 0.75$$

We estimate q by

$$1 - \frac{r}{n} = 1 - 0.75 = 0.25$$

c) Would it seem that the number of trials is large enough to justify a normal approximation to the binomial?

Since $n = 800$, $p \simeq 0.75$, and $q \simeq 0.25$, then

$$np \simeq (800)(0.75) = 600 > 5 \quad \text{and} \quad nq \simeq (800)(0.25) = 200 > 5$$

A normal approximation is certainly justified.

d) Find a 99% confidence interval for p.

$$z_{99} = 2.58 \; (\textit{see} \text{ Table 8-2})$$

$$E \simeq z_{99} \sqrt{\frac{\left(\dfrac{r}{n}\right)\left(1 - \dfrac{r}{n}\right)}{n}}$$

$$\simeq 2.58 \sqrt{\frac{(0.75)(0.25)}{800}}$$

$$\simeq 0.0395$$

The 99% confidence interval is then

$$\frac{r}{n} - E < p < \frac{r}{n} + E$$

$$0.75 - 0.0395 < p < 0.75 + 0.0395$$

$$0.71 < p < 0.79$$

_____ Exercise 7 _____

A random sample of 188 books purchased at a local bookstore showed that 66 of the books were murder mysteries. Let p represent the proportion of books sold by this store that are murder mysteries.

a) What is a point estimate for p?

a) $p \approx \dfrac{r}{n} = \dfrac{66}{188} = 0.35$

b) Find a 90% confidence interval for p.

b) $E = z_c \sqrt{\dfrac{\left(\dfrac{r}{n}\right)\left(1 - \dfrac{r}{n}\right)}{n}}$

$= 1.645 \sqrt{\dfrac{(0.35)(1 - 0.35)}{188}}$

$= 0.0572$

The confidence interval is

$$\frac{r}{n} - E < p < \frac{r}{n} + E$$

$$0.35 - 0.0572 < p < 0.35 + 0.0572$$

$$0.29 < p < 0.41$$

c) What is the meaning of the confidence interval you just computed?

c) If we had computed the interval for many different sets of 188 books, we would find that about 90% of the intervals actually contained p, the population proportion of mysteries. Consequently, we can be 90% sure that our interval contains the unknown value p.

d) To compute the confidence interval, we used a normal approximation. Does this seem justified?

d) $n = 188$

$p \approx 0.35$ and $q \approx 0.65$

Since $np \approx 65.8 > 5$ and $nq \approx 122.2 > 5$ the approximation is justified.

It is interesting to note that our point estimate r/n and the confidence interval for p do not depend on the size of the population. In our bookstore example it made no difference how many books the store sold. On the other hand, the size of the sample does affect the accuracy of a statistical estimate. In the next section we will study the effect of sample size on the reliability of our estimate.

Section Problems 8.3

1. Anita is doing a research project on criminal justice and the state prison system. In her research the term "reformed criminal" means a convicted criminal who is not convicted of another crime for at least five years after release from prison. To estimate the proportion

of criminals who become reformed criminals, Anita took a random sample of 316 criminal files of convicts who had been released from prison at least five years ago. After careful follow-up studies she found that 145 qualified as reformed criminals.

a) Let p represent the proportion of criminals who become reformed out of the population of all criminals in the state prison system. Find a point estimate for p.

b) Find a 95% confidence interval for p. Give a brief interpretation of the meaning of the confidence interval you have found.

c) Do you think that the condition $np > 5$ and $nq > 5$ are valid in this problem? Why is this an important consideration?

2. The past few years a nursery has kept records of blue spruce trees they have replanted. Over this period it was found that 421 out of 518 blue spruce trees survived replanting.

a) Let p be the proportion of blue spruce trees that survive replanting in the population of all such trees replanted by this nursery. Find a point estimate for p.

b) Find a 99% confidence interval for p. Give a brief interpretation of the meaning of your confidence interval.

c) Do you think the conditions $np > 5$ and $nq > 5$ are satisfied in this problem? Why is this important?

3. The manager of the dairy section of a large supermarket took a random sample of 250 egg cartons and found that 40 cartons had at least one broken egg.

a) Let p be the proportion of egg cartons with at least one broken egg out of the population of all egg cartons stocked by this store. Find a point estimate for p.

b) Find a 90% confidence interval for p. Give a brief interpretation of the meaning of your confidence interval.

c) Do you think that the conditions $np > 5$ and $nq > 5$ will be satisfied? Why is this important?

4. In a large midwestern city, a random sample of 395 homes on the resale market showed that 32 of the homes were on the market because of a mortgage default.

a) Let p be the proportion of all homes on the market due to mortgage default. Find a point estimate for p.

b) Find a 95% confidence interval for p. Give a brief interpretation of the meaning of your confidence interval.

c) Do you think the conditions $np > 5$ and $nq > 5$ will be satisfied? Why is this important?

5. A CPA is auditing the accounts of a large interstate banking system. Out of a random sample of 152 accounts it was found that 19 had transaction errors.

a) Let p be the proportion of all such accounts with transaction errors. Find a point estimate for p.

b) Find a 99% confidence interval for p.

6. Out of a random sample of 2,466 customers making airline reservations, it was found that 1,903 requested a nonsmoking section.

a) Let p be the proportion of all airline customers requesting nonsmoking seating. Find a point estimate for p.

b) Find a 99% confidence interval for p.

7. A new fast food franchise specializes in chicken wings. A random sample of 110 of the new restaurants showed that 67 went out of business in three years.

a) Let p be the proportion of all such franchises that go out of business within three years. Find a point estimate for p.

b) Find a 90% confidence interval for p.

8. At a sales conference for executives, Ellen noticed that 27 of the executives were women out of a random observation of 53 executives at the conference.
   a) Let p be the proportion of women executives at the conference. Find a point estimate for p.
   b) Find a 95% confidence interval for p.

9. At Children's Hospital, a random sample of 300 surgery cases showed that in 231 of the cases, a parent requested permission to stay in the room with the child all night on the night after surgery.
   a) Let p represent the proportion of all such requests. Find a point estimate for p.
   b) Find a 95% confidence interval for p.

10. A random sample of 1,800 registered voters showed that 792 favor Senator Wolf for re-election.
    a) Let p be the proportion of all registered voters who favor Senator Wolf. Find a point estimate for p.
    b) Find a 0.90 confidence interval for p.

11. The Book Worm Book Club sends each of its members a book once a month. The members then have the option of returning the book without cost within ten days. A random sample of 360 members showed that 216 had returned at least one book in the last year.
    a) Let p be the proportion of all members who returned at least one book last year. Find a point estimate for p.
    b) Find a 0.95 confidence interval for p.

12. Air Connect airlines found that 88 out of a random sample of 121 passengers purchased round trip tickets.
    a) Let p be the proportion of all Air Connect passengers who purchase round trip tickets. Find a point estimate for p.
    b) Find a 0.95 confidence interval for p.

13. The U.S. Department of the Interior is checking cattle on the Windgate open range in Montana. A random sample of 900 cattle shows that 54 are undernourished.
    a) Let p represent the proportion of undernourished cattle on the Windgate range. Find a point estimate for p.
    b) Find a 0.99 confidence interval for p.

14. The Highway Patrol did a safety spot check of 256 cars on Arizona highways and found that 18 cars did not pass.
    a) Let p be the probability that a single car selected at random on the Arizona highways does not pass a safety check. Find a point estimate for p.
    b) Find a 0.90 confidence interval for p.

15. A labor union took a random sample of 85 workers at the Fleetfoot Shoe Factory and found that 41 favored a strike for shorter hours and higher wages.
    a) Let p be the probability that a randomly selected worker at Fleetfoot Shoe Factory favors the strike. Find a point estimate for p.
    b) Find a 0.99 confidence interval for p.

16. The Tender Turkey Farm hatches turkey eggs. A random sample of 100 eggs hatched 84 chicks.
    a) Let p be the probability that an egg selected at random at Tender Turkey Farm hatches. Find a point estimate for p.
    b) Find a 0.90 confidence interval for p.

17. The Postmaster General in Atlanta, Georgia found that 44 out of a random sample of 400 packages had insufficient postage.
    a) Let p be the probability that a package selected at random has insufficient postage. Find a point estimate for p.
    b) Find a 0.99 confidence interval for p.

18. A random sample of 683 fish taken from a popular fishing lake in Maine showed that 217 were yellow perch.
    a) Let p be the proportion of fish taken from the lake that are yellow perch. Find a point estimate for p.
    b) Find a 0.95 confidence interval for p.

19. An anthropologist, specializing in Southwest Indian culture, is studying a large prehistoric communal dwelling in northern Arizona. A random sample of 127 individual family dwellings showed signs that 19 belonged to the Sun Clan.
    a) Let p be the probability that a dwelling selected at random was a dwelling of a Sun Clan member. Find a point estimate for p.
    b) Find a 0.90 confidence interval for p.

SECTION 8.4   Choosing the Sample Size

In the design stages of statistical research projects it is a good idea to decide in advance upon the confidence level you wish to use and to select the *maximum* error of estimate E you want for your project. How you choose to make these decisions depends on the requirements of the project and the practical nature of the problem. Whatever specifications you make, the next step is to determine the sample size. In this section we will assume the distribution of sample means $\bar{x}$ is approximately normal, and when necessary we will approximate σ by the sample standard deviation s. These methods are technically justifiable since our samples will be of size at least 30.

Let's say that at a confidence level of c, we want our point estimate $\bar{x}$ for μ to be in error either way by less than some quantity E. In other words, E is the maximum error of estimate we can tolerate. Using the language of probability this means we want the following to be true:

$$P(-E < \bar{x} - \mu < E) = c \qquad (20)$$

This is essentially the same as equation (1) of Section 8.1. Let's compare them.

$$P(-E < \bar{x} - \mu < E) = c \qquad (20)$$

$$\downarrow \qquad \downarrow \qquad \downarrow$$

$$P\left(-z_c \frac{\sigma}{\sqrt{n}} < \bar{x} - \mu < z_c \frac{\sigma}{\sqrt{n}}\right) = c \qquad (1)$$

From this comparison, we see that we want E to be

$$E = z_c \frac{\sigma}{\sqrt{n}}$$

Solving this equation for *n*, we get

$$n = \left(\frac{z_c \sigma}{E}\right)^2 \qquad (21)$$

To compute *n* from equation (21) we must know the value of σ. If the value of σ is not previously known, we do a preliminary sampling to approximate it. For most practical purposes a preliminary sample of size 30 or larger will give a sample standard deviation *s* which we may use to approximate σ.

EXAMPLE 6

A wildlife study is designed to find the mean weight of salmon caught by an Alaskan fishing company. As a preliminary study a random sample of 50 freshly caught salmon are weighed. The sample standard deviation of the weights of these 50 fish is $s = 2.15$ lb. How large a sample should be taken to be 99% confident that the sample mean $\bar{x}$ is within 0.20 lb of the true mean weight μ?

In this problem $z_{0.99} = 2.58$ (*see* Table 8-2) and $E = 0.20$. The preliminary study of 50 fish is large enough to permit a good approximation of σ by $s = 2.15$. Therefore, equation (21) becomes

$$n = \left(\frac{z_c \sigma}{E}\right)^2 \simeq \left(\frac{(2.58)(2.15)}{0.20}\right)^2 = 769.2$$

In determining sample size, any fraction value of *n* is always rounded to the *next higher whole number.* We conclude that a sample size of 770 will be enough to satisfy the specifications. Of course, a sample size larger than 770 will also work.

EXAMPLE 7

A certain company makes light fixtures on an assembly line. An efficiency expert wants to determine the mean time it takes an employee to assemble the switch on one of these fixtures. A preliminary study used a random sample of 45 observations and found that the sample standard deviation was $s = 78$ sec. How many more observations are necessary for the efficiency expert to be 95% sure that the point estimate $\bar{x}$ will be "off" from the true mean μ by at most 15 sec?

In this example we approximate σ by $s = 78$. We use $z_{0.95} = 1.96$ (*see* Table 8-2). The maximum error of estimate is specified to be $E = 15$ sec. Equation (21) gives us

$$n = \left(\frac{z_c \sigma}{E}\right)^2 = \left(\frac{(1.96)(78)}{15}\right)^2 = 103.9$$

The efficiency expert should use a sample of minimum size 104. Since the preliminary study has 45 observations, an additional $104 - 45 = 59$ observations are necessary.

_____ Exercise 8 _____

A large state college has over 1,800 faculty members. The dean of faculty wants to estimate the average teaching experience (in years) of the faculty members. A preliminary random sample of 60 faculty members yields a sample standard deviation of $s = 3.4$ yr. The dean wants to be 99% confident that the sample mean $\bar{x}$ does not differ from the population mean by more than half a year. How large a sample should be used? Let's answer this question in parts.

a) What can we use to approximate σ? Why can we do this?

a) $s = 3.4$ yr is a good approximation because a preliminary sample of 60 is fairly large.

b) What is $z_{0.99}$? (*Hint: See* Table 8-2.)

b) $z_{0.99} = 2.58$

c) What is E for this problem?

c) $E = 0.5$ yr

d) Which is the correct formula for n:

$$\left(\frac{z_c\sigma}{n}\right)^2,\ \left(\frac{z_c\sigma}{E}\right)^2,\ \text{or } \left(\frac{z_cE}{\sigma}\right)^2$$

d) $n = \left(\dfrac{z_c\sigma}{E}\right)^2$

e) Use the formula for n to find the minimum sample size. Should your answer be rounded up or down to a whole number?

e) $n = \left(\dfrac{(2.58)(3.4)}{0.5}\right)^2 = (17.54)^2 = 307.8$

Always round n up to the next whole number. Our final answer $n = 308$ is the minimum size.

(If you omitted the binomial distribution, omit the rest of this section.)

Next we will determine the minimum sample size when we use r/n as a point estimate for p in a binomial distribution. We will use the methods of normal approximation (large samples) discussed in Section 8.3. Suppose for a confidence level c we want the estimate r/n for p to be in error either way by less than some quantity E. Using the language of probability, this means we want the following to be true.

$$P\left(-E < \frac{r}{n} - p < E\right) = c \tag{22}$$

Let's compare this with equation (16) of Section 8.3. For convenience, they both are written together:

$$P\left(-E < \frac{r}{n} - p < E\right) = c \tag{22}$$

$$\downarrow$$

$$P\left(-z_c\sqrt{\frac{pq}{n}} < \frac{r}{n} - p < z_c\sqrt{\frac{pq}{n}}\right) = c \tag{16}$$

The comparison of the two equations gives a formula for E:

$$E = z_c\sqrt{\frac{pq}{n}}$$

Solving the last equation for n we get

$$n = pq\left(\frac{z_c}{E}\right)^2$$

Since $q = 1 - p$, our equation for n can be written

$$n = p(1 - p)\left(\frac{z_c}{E}\right)^2 \tag{23}$$

Equation (23) cannot be used unless we already have a preliminary estimate for p. To get around this difficulty we will use the equation $p(1 - p) = \frac{1}{4} - (p - \frac{1}{2})^2$. This is an algebraic identity which you are asked to verify in an optional exercise at the end of this chapter. In this exercise you are also asked to use a little logical deduction to show that the maximum possible value of $p(1 - p)$ is $\frac{1}{4}$. Therefore, *when we have no preliminary estimate for p we use the formula*

$$n = \frac{1}{4}\left(\frac{z_c}{E}\right)^2 \tag{24}$$

Since formula (24) may make the sample size unnecessarily large, we can say the probability is *at least* (and possibly more than) c that the point estimate r/n for p will be in error either way by less than the quantity E.

EXAMPLE 8

A company is in the business of selling wholesale popcorn to grocery stores. The company buys directly from farmers. A buyer for the company is examining a large amount of corn from a certain farmer. Before the purchase is made, the buyer wants to estimate p, the probability that a kernel will pop.

Suppose a random sample of n kernels is taken and r of these kernels pop. The buyer wants to be 95% sure that the point estimate r/n for p will be in error either way by less than 0.01.

a) If no preliminary study is made to estimate p, how large a sample should the buyer use?

In this case we use formula (24) with $z_{0.95} = 1.96$ (*see* Table 8-2) and $E = 0.01$.

$$n = \frac{1}{4}\left(\frac{z_c}{E}\right)^2 = \frac{1}{4}\left(\frac{1.96}{0.01}\right)^2 = 0.25(38{,}416) = 9{,}604$$

We would need a sample of $n = 9{,}604$ kernels.

b) A preliminary study showed that p was approximately 0.86. If the buyer uses the results of the preliminary study, how large a sample should be used?

In this case we use formula (23) with $p \simeq 0.86$. Again from Table 8-2, $z_{0.95} = 1.96$, and from the problem, $E = 0.01$.

$$n = p(1 - p)\left(\frac{z_c}{E}\right)^2 = (0.86)(0.14)\left(\frac{1.96}{0.01}\right)^2$$

$$= 4{,}625.29$$

The sample size should be at least $n = 4{,}626$ kernels. This sample is less than half the sample size necessary without the preliminary study.

_____ Exercise 9 _____

In Indianapolis the department of public health wants to estimate the proportion of children (grades 1–8) who require corrective lenses for their vision. A random sample of n children is taken and r of these children are found to require corrective lenses. Let p be the true proportion of children requiring corrective lenses in Indianapolis. The health department wants to be 99% sure that the point estimate r/n for p will be in error either way by less than 0.03.

a) If no preliminary study is made to estimate p, how large a sample should the health department use? Let's answer this question in parts.

i) Which formula shall we use: (23) or (24)?

ii) What is the value of E, and what is the value of z_c in this problem?

iii) What is n?

a)

i) We use equation (24) since we do not have an estimate for p.

ii) $E = 0.03$ and $z_{0.99} = 2.58$ (*see* Table 8-2)

iii) $n = \dfrac{1}{4}\left(\dfrac{z_c}{E}\right)^2 = \dfrac{1}{4}\left(\dfrac{2.58}{0.03}\right)^2 = 1{,}849$

So without a preliminary study to find p, we will need a sample size of at least $n = 1{,}849$ children.

b) A preliminary random sample of 100 children indicates that 23 require corrective lenses. Using the results of this preliminary study, how large a sample should the health department use? Again let's answer this question in parts.

i) Which formula should we use: (23) or (24)?

ii) What is E and what is z_c for this problem?

iii) What approximate value shall we use for p?

iv) What is the value of n?

b)

i) We use equation (23) since we have an estimate of p from a preliminary study.

ii) $E = 0.03$ and $z_{0.99} = 2.58$

iii) $p \simeq 0.23$

iv) $n = p(1 - p)\left(\dfrac{z_c}{E}\right)^2 = (0.23)(0.77)\left(\dfrac{2.58}{0.03}\right)^2$

$= 1{,}309.83$

Therefore, the sample size should be at least 1,310 children.

Section Problems 8.4

1. A sociologist is studying marriage customs in a rural community in Denmark. A random sample of 35 women who have been married was used to determine the age of the woman at the time of her first marriage. The sample standard deviation of these ages was 2.3 years. The sociologist wants to estimate the population mean age of a woman at the time of her first marriage. How many *more* women should be included in the sample to be 95% confident that the sample mean $\bar{x}$ of ages is within 0.25 yr of the population mean μ?

2. An automobile manufacturer used a random sample of 50 cars of a certain model to estimate the miles per gallon (mpg) this model car gets in highway driving. The sample standard deviation was found to be 5.7 mpg. How many *more* cars should be included in the sample if we are to be 90% sure the sample mean $\bar{x}$ mpg is within 1 mpg of the population mean μ of all cars of this model?

3. Gordon is thinking of buying a combination hardware/sporting goods store in Montana. However, daily cash flow is an important consideration. A random sample of 40 business days from the past year was taken from the store records. For each day the net income was determined. The sample standard deviation was found to be $57.19. How many *more* business days should be included in the sample to be 85% confident that the sample mean of daily net incomes is within $10 of the population mean μ of daily net incomes?

4. When customers phone an airline to make reservations they usually find it irritating if they are kept on hold for a long time. In an effort to determine how long phone customers are kept on hold, one airline took a random sample of 167 phone calls and determined the length of time (in minutes) each was kept on hold. The sample standard deviation was 3.8 minutes. How many more phone customers should be included in the sample to be 99% sure that the sample mean $\bar{x}$ of hold times is within 30 sec of the population mean μ of hold times?

5. Pyrometric cones are used in the brick-making and pottery industries. Each cone is a small piece of specially engineered ceramic designed to melt at a set temperature. The pyrometric cone is placed in the firing kiln with bricks or pottery. When the cone melts we know a certain temperature has been reached. A random sample of 100 pyrometric cones that are supposed to melt at 1250 °F are being tested. The sample standard deviation of melting temperatures was 38.5 °F. How many more such cones should be tested to be 90% sure that the sample mean of melting temperatures is within 5 °F of the population mean for melting temperatures?

* 6. At many gasoline stations customers have the option of paying cash and receiving a discount. The question is: What proportion of customers take advantage of this option?
   a) Let p be the proportion of customers who take advantage of the cash discount option. If no preliminary study is made to estimate p, how large a sample of customers is necessary to be 90% sure that a point estimate r/n will be within a distance of 0.08 of p?
   b) A preliminary study of 76 customers showed 19 used the cash discount option. How

---

*Omit if you omitted the binomial distribution.

many *more* customers should be included in the sample to be 90% sure that a point estimate r/n will be within a distance of 0.08 of p?

* 7. In a particular neighborhood of a large city it is thought that most crimes are drug-related. Let p be the proportion of crimes in this neighborhood that are in fact drug-related.
   a) If no preliminary study is made to estimate p, how large a sample is necessary to be 99% sure that a point estimate r/n will be within a distance of 0.05 from p?
   b) In a preliminary study a random sample of 116 crime cases showed 103 were drug-related. How many *more* crime cases should be included in the sample to be 99% sure that a point estimate r/n will be within a distance of 0.05 from p?

* 8. The national council of small businesses is interested in the proportion of small businesses that declared Chapter 11 bankruptcy last year. Since there are so many small businesses, the national council intends to estimate the proportion from a random sample. Let p be the proportion of small businesses that declared Chapter 11 bankruptcy last year.
   a) If no preliminary sample is taken to estimate p, how large a sample is necessary to be 95% sure that a point estimate r/n will be within a distance of 0.10 from p?
   b) In a preliminary random sample of 38 small businesses, it was found that six had declared Chapter 11 bankruptcy. How many *more* small businesses should be included in the sample to be 95% sure that a point estimate r/n will be within a distance of 0.10 from p?

* 9. A ponderosa pine forest in Colorado has a pine beetle infestation. The beetles bore into a tree and carry a fungus which ultimately kills the tree. Let p be the proportion of trees in the forest that are infested.
   a) If no preliminary sample is taken to estimate p, how large a sample is necessary to be 85% sure that a point estimate r/n will be within a distance of 0.06 from p?
   b) A preliminary study of 58 trees showed that 19 were infested. How many *more* trees should be included in the sample to be 85% sure that a point estimate r/n will be within a distance of 0.06 from p?

*10. Jim is doing a research project in political science to determine the proportion p of voters in his district who favor capital punishment.
   a) If no preliminary sample of voters is taken to estimate p, how large a sample is necessary to be 99% sure that a point estimate r/n will be within a distance of 0.01 from p?
   b) Jim did a preliminary study in which he found that in a random sample of 932 voters, 36 favored capital punishment. How many *more* voters should be included in the sample to be 99% sure that a point estimate r/n will be within a distance of 0.01 from p?

11. Super Goo Ice Cream Parlor wants to estimate the average number of scoops of ice cream sold each day. A random sample of 36 days shows the standard deviation to be 53.8 dips per day. How many more days should be included in the sample to be 95% confident that the sample mean $\bar{x}$ is within five dips of the population mean μ?

12. Big Pill Pharmacy wants to estimate the average amount a customer pays for a prescription. A preliminary random sample of 150 customers shows the sample standard deviation to be $3.62. How many *more* customers should be included in a random sample to be 99% confident that the sample mean $\bar{x}$ is within $0.50 of the population mean cost μ?

*13. Gassum Heater Company has been known to make defective heaters. Let p be the proportion of defective heaters made by the company.
   a) If no preliminary sample is made to estimate p, how large a sample is necessary to be 95% sure that a point estimate r/n will be within a distance of 0.05 from p?
   b) A preliminary study shows p is approximately 0.12. Answer part (a) using the results of the preliminary study.

*14. Union officials want to estimate the percent p of workers in the Big Bend Metalworks who favor a strike. The union wants to be 90% sure that its estimate for p is within 2.5% of the true percent p who favor a strike.
   a) If no accurate estimate for p is available, how large a sample of workers is necessary?
   b) A preliminary random sample of 100 workers shows p is approximately 35%. How many *more* workers should be included in this sample?

15. A random sample of 40 Salt Lake City school teachers shows the standard deviation for the number of years teaching experience is 5.3. How many more teachers should be included in the sample to be 95% sure the sample mean $\bar{x}$ of years teaching experience is within six months of the population mean μ?

*16. The Country Boy Seed Company wants to estimate the proportion p of its hay seeds that will germinate.
   a) If no reliable estimate for p is known, how large a random sample of hay seeds should be used if Country Boy wants to be 99% confident their estimate is within a distance of 0.01 of the population value of p?
   b) If preliminary studies show p is approximately 0.86, use this information to answer part (a).

*17. a) Show that $p(1 - p) = 1/4 - (p - 1/2)^2$.
   b) Why is $p(1 - p)$ never greater than $1/4$?

18. At the site of a nuclear waste spill, cleanup crews were working to bring the mean radiation level down to 16 microcuries, with a maximum error tolerance of $\pm$ 0.5 microcuries. A random sample of 30 readings around the spill area gave a sample standard deviation of $s = 1.75$ microcuries. How many more sample readings should be taken to be 99% confident that the sample mean $\bar{x}$ is within ± 0.5 microcuries of the population mean radiation μ?

19. Aerodynamic engineers want to know the mean shear force on the wings of an FX5000 fighter jet as the jet comes out of a 3,000-foot vertical dive. A random sample of FX5000 jets flown by Air Force fighter pilots was used for 37 test flights. The sample standard deviation of wing shear forces was $s = 948$ foot-pounds. The engineers want to be 95% confident that the sample mean $\bar{x}$ is within 250 foot-pounds of the population mean μ. How many more flight tests should be made?

*20. Linda Silbers is a social scientist studying voter opinion about a city bond proposal for a light-rail mass transit system in Denver.
   a) If Linda wants to be 90% sure that her sample estimate of the proportion of voters who favor the bond is within 5% of the population percent p who favor the bond issue, how large a sample should she use?
   b) If a preliminary study showed that p is approximately 73%, how large a sample is required?

Summary

We have studied point estimates and interval estimates. For point estimates we found E, the maximal error of estimate, and for interval estimates we found the interval end points. In each case E or end points were determined by four things: the confidence level c, the sample estimate for μ or p, the sample standard deviation, and the sample size n.

For large samples ($n \geq 30$) we used the normal distribution, the Central Limit Theorem, and sometimes the normal approximation to the binomial distribution. For small samples ($n < 30$) we made use of Student's t distribution. In applications to choose the sample size we found formulas for n so that with probability c the estimate for μ or p is in error by less than a preassigned number E.

Important Words and Symbols

	Section		*Section*
Error of estimate $\lvert x - \mu \rvert$	8.1	c confidence interval	8.1
Confidence level c	8.1	Small samples, $n < 30$	8.2
Critical values z_c	8.1	Student's t variable	8.2
Point estimate for μ	8.1	Critical values t_c	8.2
E, the maximal		Degrees of freedom ($d.f.$)	8.2
error of estimate	8.1	Point estimate of p	8.3
Interval estimate for μ	8.1	Interval estimate for p	8.3
Large samples, $n \geq 30$	8.1	Sample size n	8.4

Chapter Review Problems

Categorize each problem according to (a) parameter being estimated, proportion p, or mean μ, and (b) large sample or small sample. Then solve the problem.

1. In your own words carefully explain the meaning of the following terms: point estimate, critical value, maximal error of estimate, confidence level, confidence interval, large samples, small samples.

2. Anystate Auto Insurance Company took a random sample of 370 insurance claims paid out during a one-year period. The average claim paid was $750 with a standard deviation of $150. Find 0.90 and 0.99 confidence intervals for the mean claim payment.

3. Zapa Zapa Zoo fraternity wants to estimate the quarter grade average of its members. A random sample of eight brothers shows their average (mean) to be 2.09 with sample standard deviation 0.63. Find a 0.95 confidence interval for the mean grade average. (Assume the grades follow a normal distribution.)

4. Fisher State College is thinking of going from the quarter to the semester system. A random sample of 169 students shows 100 are in favor of the change. Find a 0.90 confidence interval for the proportion of students at Fisher who favor the change.

5. At the annual Madison Square Garden Persian Cat Show a random sample of 50 show

cats had a mean weight of 12.8 lb with standard deviation 3.6 lb. Find a 0.95 confidence interval for the mean weight of Persian show cats.

6. How large a sample is needed in problem 5 if we wish to be 99% confident that the sample mean is within 0.5 lb of the population mean?

7. The president of the Clipper Steamship Company wants to estimate the percent of his ships that run more than two days behind schedule. A random sample of 121 ship voyages shows ten ships were more than two days late. Find a 0.90 confidence interval for the percent of late ships.

8. How large a sample is needed in problem 7 if we wish to be 95% confident that the sample percent of late ships is within 3% of the population percent of late ships? (*Hint:* Use $p \simeq 0.083$ as a preliminary estimate.)

9. Country Boy Seed Company wants to estimate the yield from their corn seed on prime Iowa farmland. A random sample of 20 one-acre plots of prime Iowa farmland gave an average yield of 170 bushels per acre with standard deviation 6.4 bushels. Find a 0.99 confidence interval for the true mean yield per acre on prime Iowa farmland.

10. A random sample of 49 medical doctors in Los Angeles showed they worked an average of 53.1 hr/wk with standard deviation 7.2 hr. Find a 0.95 confidence interval for the mean number of hours worked by all Los Angeles doctors.

11. How large a sample is needed in problem 10 if we want to be 99% confident that our point estimate is within 1 hr of the population mean?

12. Ms. Green of Quicksell Realty Company wants to estimate the proportion of four-bedroom homes on the private residential market in Boston. A random sample of 63 private homes on the housing market shows 24 of them are four-bedroom. Find a 0.85 confidence interval for the proportion of four-bedroom homes on the market in Boston.

13. How large a sample is needed in problem 12 if we want to be 90% sure our point estimate for the proportion is within 0.05 of the population proportion? If no preliminary estimate for the proportion is available, how large a sample is necessary?

14. A random sample of six Big Blow cigars has an average nicotine content of 138.4 mg and a sample standard deviation of 25.7 mg. Find a 0.95 confidence interval for the mean nicotine content of all Big Blow cigars.

15. The Department of Public Health wants to estimate the proportion of child deaths in the age bracket 1–6 that occur because of accidental poisoning. A random sample of 100 children's death certificates (age 1–6) indicates 11 died because of accidental poisoning. Find a 0.99 confidence interval for the proportion of children's deaths due to poisoning.

16. How large a sample is needed in problem 15 if we want to be 99% sure the point estimate for the proportion is within 0.02 of the population proportion? If no preliminary estimate were available for the proportion, how large a sample would be necessary?

17. Six beagles selected at random lived 15, 12, 20, 17, 8, and 16 years. Find a 0.90 confidence interval for the mean life span of beagles. (Assume the population follows a normal distribution.)

18. A physics lab instructor timed a random sample of 30 physics students completing a required experiment. She found the average time to be 43 min with a sample standard deviation of 16 min. Find a 0.90 confidence interval for the mean time.

19. In problem 18 how large a sample is required to be 99% sure the point estimate is within 5 min of the population mean time?

20. The Virginia State Patrol took a random sample of 225 accident reports from their files and found that 90 involved alcohol as the underlying cause. Find a 0.99 confidence interval for the proportion of Virginia auto accidents due to alcohol.

21. How large a sample is needed in problem 20 to be 99% confident the point estimate for the proportion will be within 0.05 of the population proportion? If no preliminary estimate for the proportion were available, how large a sample would be necessary?

22. A random sample of 16 days showed the Dodge City Police Department received an average of 223 phone calls per day with a sample standard deviation of 31 calls. Find a 0.95 confidence interval for the mean number of phone calls received each day.

23. The U.S. Department of Consumer Protection is investigating the Willie's quarter-pound hamburger. In order to estimate the precooked weight of the meat patty a random sample of 100 meat patties is obtained. The mean weight is 3.91 oz with a sample standard deviation of 0.50 oz. Find a 0.99 confidence interval for the mean weight of all the Willie's quarter-pound meat patties.

24. How large a sample is required in problem 23 if the Department of Consumer Protection wants to be 99% confident the sample estimate for the mean is within 0.1 oz of the population mean weight?

25. How large a sample is required in problem 23 if the Department of Weights and Measures wants to be only 90% confident the sample estimate for the mean is within 0.1 oz of the population mean weight?

26. The Summa is a new sports car. An extended transmission warranty is available for $1,500 that will replace or repair the transmission if it fails during the first 65,000 miles. The normal warranty period is for 24,000 miles. A random sample of eight Summas were driven under a variety of conditions until the transmission failed. The mileage at the time of transmission failure was recorded. These data are (in miles):

75,436	80,212	42,316	25,415
61,335	50,720	52,318	66,143

a) Use a calculator to verify that the sample mean mileage at the time of transmission failure is $\bar{x} = 56,737$ miles with standard deviation $s = 17,949$ miles.
b) Use these values to find a 90% confidence interval for mileage at the time of transmission failure.

27. During a television miniseries what is the average length of time between commercial breaks? A random sample of 20 such periods were selected from miniseries that were aired on commercial television stations last year. The times between commercial breaks were (to the nearest minute):

5	7	8	14	13
10	9	8	11	12
14	11	9	10	6
8	12	5	11	8

a) Use a calculator to verify that the sample mean of the times between commercial breaks is $\bar{x} = 9.55$ min with standard deviation $s = 2.72$ min.
b) Find a 95% confidence interval for the mean length of time between commercial breaks.

USING COMPUTERS

Professional statisticians in industry and research use computers to help them analyze and process statistical data. There are many computer programs available for statistics. Some commonly used statistical packages include Minitab, SAS, and SPSS. There are many others as well. Any of these statistical packages may be used with this text.

The authors of this text have also written a computer statistics package called ComputerStat. The programs in ComputerStat are available on disks with an accompanying manual that explains how to use the programs and gives additional computer applications. The disks are intended for the Apple IIe, IIc, or the IBM PC. Programs are written in the computer language BASIC and are designed for the beginning statistics student with no previous computer experience. ComputerStat is available as a supplement from the publisher of this text.

Problems in this section may be done using ComputerStat or other statistical computer programs.

1. Suppose you select 30 numbers at random on the line segment from 0 to 1. Then you find the sample mean $\bar{x}$ and sample standard deviation s of the numbers you selected at random. Using the material we learned in this chapter we realize that we have a large sample and a 90% confidence interval would be given by

$$\bar{x} - 1.645 \frac{s}{\sqrt{30}} < \mu < \bar{x} + 1.645 \frac{s}{\sqrt{30}}$$

where μ is the population mean of all numbers selected at random from 0 to 1.

However, in this special case mathematical theories can be used to show that for the *population* of all random numbers taken from the interval 0 to 1 the population mean is just $\mu = 0.5$. Since in this special case we actually know $\mu = 0.5$, we can use this information to do a little checking up on our theory of confidence intervals. What does it mean to say we have a 90% confidence interval for the population mean μ? It means that if we construct lots and lots of confidence intervals (from lots and lots of samples) then in the long run, about 90% of these intervals will in fact contain μ. Figure 8-3 illustrated this property.

Now let's use the computer to check up on this statement. Notice that we cannot use the computer to actually *prove* we have a 90% confidence interval. However, we can use the computer to give a numerical demonstration that we have a 90% confidence interval.

In the ComputerStat menu of topics, select Confidence Intervals. Then select S11 Confidence Interval Demonstration. This computer program selects 20 random samples each of size N (you select N from 30 to 500). Each sample consists of N random numbers from 0 to 1. Then the computer finds 90% confidence intervals for each sample and keeps a running tally of how many intervals actually contain the population mean $\mu = 0.5$. As we know from theory, about 90% of the intervals should contain μ. It is interesting to see how our theory shapes up against actual calculations.

a) Run S11 for a sample size of $N = 30$. The computer will find 20 confidence intervals for $\mu = 0.5$ based on samples of size 30. Look at the displayed confidence intervals and their graphs. Look at the tally and percent of intervals that contain $\mu = 0.5$. Look at the length of each interval on the graph. Now use the program option to find more confidence intervals of size $N = 30$. Keep on finding intervals until you have 100 of them. In the long run what percent of the intervals seem to contain $\mu = 0.5$? How is this predicted by what we have learned in this chapter?

b) Repeat part (a) for a sample size of $N = 60$. Pay close attention to the lengths of the intervals graphed. Are these intervals longer or shorter than those of part (a)? Explain why you should expect them to be longer or shorter. We have a larger sample size than that of part (a) but we still only get about 90% of the intervals containing $\mu = 0.5$. Again explain why we expect only 90% to contain μ and not a higher percent.

c) Run S11 for a sample size of $N = 500$. Just compute 20 intervals and no more (this will take about 90 seconds). Again compare the graphs of these intervals with those of parts (a) and (b). Which are the longest? The shortest? Why do we expect these results? A sample size of 500 is a very large sample, but we still get only about 90% of the intervals containing $\mu = 0.5$. Explain why we expect this to be the case.

2. Cryptanalysis, the science of breaking codes, makes extensive use of language patterns. The frequency of various letter combinations is an important part of the study. A letter combination consisting of a single letter is a monograph, while combinations consisting of two letters are called digraphs, and those with three letters are called trigraphs. In the English language the most frequent digraph is the letter combination TH.

The characteristic rate of a letter combination is a measurement of its rate of occurrence. To compute the characteristic rate, count the number of occurrences of a given letter combination and divide by the number of letters in the text. For instance, to estimate the characteristic rate of the digraph TH, you could select a newspaper text and pick a random starting place. From that place mark off 2,000 letters and count the number of times that TH occurs. Then divide the number of occurrences by 2,000.

The characteristic rate of a digraph can vary slightly depending on the style of the author. So to estimate an overall characteristic frequency you want to consider several samples of newspaper text by different authors. Suppose you did this with a random sample of 15 articles and found the characteristic rate of the digraph TH in the articles. The results follow.

0.0275	0.0290	0.0315	0.0265	0.0255
0.0280	0.0230	0.0295	0.0250	0.0265
0.0240	0.0295	0.0300	0.0275	0.0265

In the ComputerStat menu of topics, select Confidence Intervals. Then use program S12, Confidence Intervals for a Population Mean *MU*, to solve the following problems.

a) Find a 95% confidence interval for the mean characteristic rate of the digraph TH.

b) Repeat part (a) for a 90% confidence interval. Note: you do not need to enter the data again; simply use the option to rerun the program with data you have already entered.

c) Repeat part (a) for an 80% confidence interval.

d) Repeat part (a) for a 70% confidence interval.

e) Repeat part (a) for a 60% confidence interval.

f) For each confidence interval in parts (a)–(e) compute the length of the given interval. Do you notice a relation between the confidence level and the length of the interval?

A good reference for cryptanalysis is a book by Sinkov:

Sinkov, Abraham, *Elementary Cryptanalysis,* New York: Random House, 1968

In the book other common digraphs and trigraphs are given.

3. There must be nurses on duty in hospitals around the clock. Therefore, many nurses work various shifts. Tasto, Colligan *et al.* did a study of the health consequences of shift work. Their results are published in the following government document:

United States Department of Health, Education, and Welfare, NIOSH Technical Report. Tasto, Colligan, *et al., Health Consequences of Shift Work.* Washington: GPO, 1978. 25.

They used a large random sample of nurses on various shifts in 12 hospitals. Part of the report concerns the number of sick days nurses on various shifts take. A random sample of 315 day-shift nurses showed that 62 took no sick days during a six-month period. During that same period a random sample of 309 nurses on rotating duty showed that 51 took no sick days.

In the ComputerStat menu of topics, select Confidence Intervals. Then use program S13, Confidence Intervals for the Probability of Success in a Binomial Distribution, to solve the following problems.

a) We wish to estimate the proportion p of rotating-shift nurses who take no sick days in a six-month period. Find a $c\%$ confidence interval for p when $c = 98, 90, 85, 75, 60$. Notice that you enter the data only once and rerun the program with the same data for the different c values.

b) For each confidence interval in part (a) compute the length of the interval. Do you notice a relation between confidence level and length of the interval?

c) We wish to estimate the proportion p of day-shift nurses who take no sick days in a six month period. Find a $c\%$ confidence interval for p when $c = 99, 95, 80, 70, 60$. Is there a relation between confidence level and length of these intervals?

9 Hypothesis Testing

> *"Would you tell me, please, which way I ought to go from here?"*
> *"That depends a good deal on where you want to get to," said the Cat.*
> *"I don't much care where—" said Alice.*
> *"Then it doesn't matter which way you go," said the Cat.*
>
> Lewis Carroll, Alice's Adventures in Wonderland

Charles Dodgson was an English mathematician who loved to write children's stories in his free time. The above dialogue between Alice and the Cheshire Cat occurs in the masterpiece *Alice's Adventures in Wonderland* written by Dodgson under the pen name Lewis Carroll. These lines relate to our study of hypothesis testing. Statistical tests cannot answer all of life's questions. They cannot always tell us "where to go"; but after this decision is made on other grounds, they can help us find the best way to get there.

SECTION 9.1   Introduction to Hypothesis Testing

In Chapter 1 we emphasized the fact that a statistician's most important job is to draw inferences about populations based on samples taken from the population. Most statistical inference centers around the parameters of a population (usually the mean, or probability of success in a binomial trial). Methods for drawing inferences about parameters are of two types: Either we make decisions concerning the value of the parameter, or we actually estimate the value of the parameter. When we estimate the value (or location) of a parameter we are using methods of estimation as studied in Chapter 8. Decisions concerning the value of a parameter are obtained by hypothesis testing, the topic we shall study in this chapter.

Students often ask which method should be used on a particular problem—that is, should the parameter be estimated or should we test a hypothesis involving the parameter? The answer lies in the practical nature of the problem and the questions posed about it. Some people prefer to test theories concerning the parameters. Others prefer to express their inferences as estimates. Both estimation and hypothesis testing are found extensively in the literature of statistical applications.

A *statistical hypothesis,* or a *hypothesis,* is an assumption about one or more population parameters of a probability distribution. The hypothesis will be accepted or rejected on the basis of information taken from a sample of the distribution. Hypothesis *testing* is the procedure whereby we decide whether to accept or reject a hypothesis.

Sometimes we formulate a statistical hypothesis for the sole purpose of trying to reject it. If we want to decide whether one method of doing something is better than another, we might formulate a hypothesis that says there is no difference in the methods. The term *null hypothesis* is used for any hypothesis which is set up primarily for the purpose of seeing whether it can be rejected.

If the null hypothesis cannot be rejected, we do not claim that it is true beyond all doubt; we just claim that under our method of testing, it cannot be rejected. Any hypothesis which differs from the null hypothesis is called an *alternate hypothesis.* An alternate hypothesis is constructed in such a way that it is the one to be accepted when the null hypothesis must be rejected. The *null hypothesis is denoted by* H_0 and the *alternate hypothesis is denoted by* H_1.

EXAMPLE 1

A car dealer advertises that its new subcompact models get 47 mpg. Let μ be the mean of the mileage distribution for these cars. You assume the dealer will not underrate the car, but you suspect the mileage might be overrated.

a) What shall we use for H_0?
   We want to see if the dealer's claim $\mu = 47$ can be rejected. Therefore, our null hypothesis is simply that $\mu = 47$. We denote the null hypothesis as

$$H_0: \mu = 47$$

b) What shall we use for H_1?
   From experience with this dealer we have every reason to believe that the advertised mileage is not too low. If μ is not 47, we are sure it is less than 47. Therefore, the alternate hypothesis is

$$H_1: \mu < 47$$

_____ Exercise 1 _____

A company manufactures ball bearings for precision machines. The average diameter of a certain type of ball bearing should be 6.0 mm. To check that the average diameter is correct the company decides to formulate a statistical test.

a) What should be used for H_0? (*Hint:* What are they trying to test?)

a) If μ is the mean diameter of the ball bearings, the company wants to test $\mu = 6.0$ mm. Therefore, $H_0: \mu = 6.0$.

b) What should be used for H_1 (*Hint:* An error either way, too small or too large, would be serious.)

b) An error either way could occur and it would be serious. Therefore, H_1: $\mu \neq 6.0$ (μ is either smaller than or larger than 6.0).

_____ Exercise 2 _____

A package delivery service claims it takes an average of 12 days to send a package from New York to San Francisco. An independent consumer agency is doing a study to test the truth of this claim. Several complaints have led the agency to suspect that the delivery time is longer than 12 days.

a) What should be used for the null hypothesis?

a) The claim $\mu = 12$ days is in question, so we take H_0: $\mu = 12$.

b) Assuming the delivery service does not underrate itself, what should be used for the alternate hypothesis?

b) If the delivery service does not underrate itself, then the only reasonable alternate hypothesis is H_1: $\mu > 12$.

If we *reject the null hypothesis when it is* in fact *true*, we have made an error that is called a *type I error.* On the other hand, if we *accept the null hypothesis when it is* in fact *false*, we have made an error that is called a *type II error.*

Table 9-1 indicates how these errors occur.

TABLE 9-1 Type I and Type II Errors

	Our Decision	
Truth of H_0	And if we accept H_0	And if we reject H_0
If H_0 is true	correct decision; no error	type I error
If H_0 is false	type II error	correct decision; no error

In order for tests of hypotheses to be good they must be designed to minimize possible errors of decision. (Often we do not know if an error has been made and therefore we can only talk about the probability of making an error.) Usually for a given sample size an attempt to reduce the probability of one type of error results in an increase in the probability of the other type of error. In practical applications one type of error may be more serious than another. In such a case careful attention is given to the more serious error. If we increase the sample size, it is possible to reduce both types of errors, but increasing the sample size may not be possible.

The probability with which we are willing to risk a type I error is called the *level of significance* of a test. The level of significance is denoted by the Greek letter α (pronounced *alpha*). In good statistical practice α is specified in advance before any samples are drawn so that results will not influence the choice for the level of significance.

The probability of making a type II error is denoted by the Greek letter β (pronounced *beta*). Methods of hypothesis testing require us to choose α and β

values to be as small as possible. In elementary statistical applications we usually choose α first.

The quantity $1 - \beta$ is called the *power* of the test and represents the probability of rejecting H_0 when it is in fact false. For a given level of significance, how much power can we expect from a test? The actual value of the power is usually difficult (and sometimes impossible) to obtain, since it requires us to know the H_1 distribution. However, we can make the following general comments:

1. The power of a statistical test increases as the level of significance α increases. A test performed at the $\alpha = 0.05$ level has more power than one at $\alpha = 0.01$. This means that the less stringent we make our significance level α, the more likely we will reject the null hypothesis when it is false. Using a larger value of α will increase the power, but it will also increase the probability of a type I error. In spite of this fact, most business executives, administrators, social scientists, and scientists use *small* α values. This choice reflects the conservative nature of administrators and scientists, who are usually more willing to make an error by failing to reject a claim (i.e., H_0) than to make an error by accepting another claim (i.e., H_1) when it is false.

2. The power of a statistical test increases as the sample size n increases. Therefore, the probability of correctly rejecting H_0 when it is false is better when the sample size is larger. Of course, the probability of correctly accepting the null hypothesis when it is true is also larger when the sample size is larger. The difficulty is that sometimes it may be too costly, or perhaps even impossible, to get a larger sample.

3. Some statistical tests are designed to be more powerful than others. The techniques described in Chapters 9, 10, and 11 are called parametric tests; those discussed in Chapter 12 are called nonparametric tests. Most parametric tests require special and stringent assumptions about the nature of the data. However, the nonparametric tests make very few restrictive assumptions about the data. In general, parametric tests are more powerful than nonparametric tests because we can employ more theoretical assumptions with parametric tests. However, nonparametric tests are extremely important (although less powerful) when certain theoretical assumptions cannot be made.

TABLE 9-2 Probabilities Associated with a Statistical Test

	Our Decision	
Truth of H_0	If we accept H_0 as True	If we reject H_0 as false
H_0 is true	correct decision with corresponding probability $1 - \alpha$	type I error with corresponding probability α called the *level of significance* of the test
H_0 is false	type II error with corresponding probability β	correct decision with corresponding probability $1 - \beta$ called the *power* of the test

- **Comment:** Since the calculation of the probability of a type II error is treated in advanced statistics courses, we will restrict our attention to the probability of a type I error.

In most statistical applications the level of significance is specified to be $\alpha = 0.05$ or $\alpha = 0.01$, although other values can be used. If $\alpha = 0.05$, then we say we are using a 5% level of significance; this means there are about five chances out of 100 that we will reject H_0 when H_0 should be accepted. Or put another way, if $\alpha = 0.05$, then our *confidence level* is $1 - \alpha$, or 0.95, that our test has led us to the correct conclusion H_0 is true when, in fact, it is true.

If a hypothesis cannot be rejected on the basis of a statistical test, we say we *accept* the hypothesis. However, the acceptance of a hypothesis does not mean that the hypothesis is true beyond all doubt. It simply means that our methods of testing indicate the hypothesis cannot be rejected on the basis of the sample data. This is why some statisticians prefer *not to reject H_0* rather than *to accept* it. For instance, suppose we accept a null hypothesis H_0: $\mu = 8$. Furthermore, suppose H_0 is false and μ is really equal to 8.001. Since 8.001 lies so close to 8, it could be extremely difficult to detect the difference without very large samples. Consequently, when a test concludes that we should accept H_0, many statisticians will elect simply *not to reject H_0* and follow up the test by estimating the desired parameter using a confidence interval. The confidence interval will give the statistician a range of possible values for the parameter.

A decision to accept or reject the null hypothesis is made on the basis of sample data. How is this done? Let's look at Example 2.

EXAMPLE 2

Statement of Problem: The St. Louis Zoo wishes to obtain eggs of a rare Mississippi river turtle. The zoo will hatch the eggs and raise the turtles as an exhibit of a rare and endangered species. Carol Wright is the staff biologist at the zoo who has been given the job of finding the eggs to be hatched. Turtles of the area bury their eggs in a nest in sandbanks along the river. Then the nest is abandoned and the eggs hatch by themselves. A number of different species of turtles live in the region, and eggs from each species look much alike. Past research has shown that lengths of turtle eggs are normally distributed, and lengths of the rare turtle eggs have population mean $\mu = 7.50$ cm with standard deviation $\sigma = 1.5$ cm. The rare turtle egg is the only one with mean length 7.50 cm. The mean lengths of eggs from each of the other species are longer than 7.50 cm. The eggs of all the local turtle species have the same standard deviation $\sigma = 1.5$ cm for length.

After searching for some time, Carol found a nest with 36 eggs. The way the nest was constructed makes her suspect that it was made by the rare turtle. The mean length of this collection of 36 eggs is $\bar{x} = 7.74$ cm. Since $\bar{x} = 7.74$ cm is longer than the population mean $\mu = 7.50$ cm of the rare turtle, Carol is a little worried that the eggs may come from a species that lays larger eggs.

Let's use a statistical test to help the biologist make a decision. Pay close attention to this example; it contains principal features common to most statistical tests.

I. *Summary of Known Facts:* In any testing problem it is a good idea to make a short summary of known facts before we start to construct the test.

a) The distribution of lengths of eggs from the rare turtle is *normal,* with population mean $\mu = 7.50$ cm and standard deviation $\sigma = 1.5$ cm.

b) The nest contains $n = 36$ eggs with an observed mean $\bar{x} = 7.74$ cm.

c) Since the population is *normal* and μ and σ are known, Theorem 7.1 tells us that $\bar{x}$ is also normally distributed. The mean and standard deviation of the $\bar{x}$ distribution are

$$\mu_{\bar{x}} = \mu = 7.50 \text{ cm}$$

$$\sigma_{\bar{x}} = \frac{\sigma}{\sqrt{n}} = \frac{1.5}{\sqrt{36}} = 0.25 \text{ cm}$$

II. *Establishing H_0 and H_1:* The null hypothesis H_0 is set up for the primary purpose of seeing whether or not it can be rejected. The alternate hypothesis H_1 is the one to be accepted when the null hypothesis must be rejected.

Let μ be the mean length of the population distribution from which our sample of 36 eggs is drawn. Our biologist suspects that the nest contains the rare turtle eggs; therefore, we will use the null hypothesis

$$H_0: \mu = 7.50$$

since the rare turtle eggs are known to have mean length $\mu = 7.50$ cm. However, all other local species lay eggs with a longer population mean. Therefore, the alternate hypothesis is

$$H_1: \mu > 7.50$$

III. *Choosing the Level of Significance α:* The null hypothesis says that the eggs in the nest are from the rare turtle. A type I error means we reject the nest as being formed by the rare turtle when it was in fact formed by the rare turtle. A type II error means we accept the nest as coming from the rare turtle when it really came from a common species.

A type I error could be serious; we don't want to reject the eggs if they are from the rare turtle. A type II error is not too serious. If the eggs are from a common species, the turtles can be released after they hatch. Naturally the zoo doesn't want to incubate the wrong eggs, but it would not be too serious if they did.

Although other levels of significance may be used, most researchers use either $\alpha = 0.05$ or $\alpha = 0.01$. Our biologist has agreed to use a level of significance $\alpha = 0.05$. This means she is willing to risk a type I error with a probability of $\alpha = 0.05$.

IV. *Geometric Model for the Test:* What decision procedure shall we use to test our hypothesis? In a way, the logic of our decision process is similar to that used in a typical courtroom setting, but our methods are mathematical rather than legal. In a court setting, the person charged with a crime is initially considered innocent (null hypothesis). If evidence (data) presented in court can sufficiently discredit the person's innocence, he or she is then judged to be guilty (alternate hypothesis). In a similar way, we will consider the null hypothesis to be *true* until there is enough data (mathematical evidence) to discredit it at the $\alpha = 0.05$ level of significance. Figure 9-1 shows the geometrical model for a test with

$$H_0: \mu = 7.50$$
$$H_1: \mu > 7.50$$

FIGURE 9-1 Geometrical Model for a Right-Tail Test, $\alpha = 0.05$

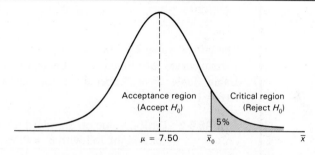

Let us examine the details of Figure 9-1.

a) Figure 9-1 shows an $\bar{x}$ distribution. The $\bar{x}$ values represent mean lengths of eggs based on samples of size $n = 36$. From conditions given in the problem, we know $\bar{x}$ follows a normal distribution.

b) Initially we assume that the null hypothesis is true. Therefore, the mean of the $\bar{x}$ distribution is $\mu = 7.50$ cm.

c) The *critical region* is a right tail of the normal curve. This tail has an area $\alpha = 0.05$. Our observed value of $\bar{x}$ in this case is $\bar{x} = 7.74$ cm, computed from the lengths of the 36 eggs in the discovered nest. If it falls under the critical region, we say there is enough evidence to discredit the null hypothesis, and we reject H_0 at the $\alpha = 0.05$ level of significance. If our observed value $\bar{x} = 7.74$ falls under the acceptance region, we conclude that the evidence is not strong enough to reject H_0, so in this case we "accept" H_0.

    The critical or rejection region is always determined from the alternate hypothesis. Since $H_1: \mu > 7.50$ claims μ is greater than 7.50, then observed $\bar{x}$ values sufficiently far to the *right* of $\mu = 7.50$ discredit the null hypothesis and support the alternate hypothesis. This circumstance is why our critical region is on the right. In future examples, we will also encounter left-tail and two-tail tests.

d) The value $\bar{x}_0$ is called the *critical value* for the test. It is the separation point between the acceptance region and the critical region. If our observed value $\bar{x} = 7.74$ falls to the left of $\bar{x}_0$, we accept H_0, and if it falls to the right of $\bar{x}_0$, we reject H_0. Once we know $\bar{x}_0$, we can quickly finish the test.

V. *Finding the Critical Value and Critical Region:* In the next section you will be given formulas for finding the critical value(s) $\bar{x}_0$ for a test. The use of formulas saves time and energy; however, in this section we will do examples showing you why the formulas work.

From the statement of the problem and our previous work in Chapter 7, we know that if H_0 is assumed to be true, then $\bar{x}$ has a normal distribution with mean $\mu_{\bar{x}} = 7.50$ and standard deviation

$$\sigma_{\bar{x}} = \frac{\sigma}{\sqrt{n}} = \frac{1.5}{\sqrt{36}} = 0.25$$

The critical value $\bar{x}_0$ is the cutoff point for a 5% area in the right tail of this normal curve. (*See* Figure 9-2.) Let us first examine a standard normal curve. Using Table 5 of Appendix I for the standard normal curve we find $P(z \geq 1.645) = 0.05$.

FIGURE 9-2 Critical Value on Standard Normal Curve
for a Right Tail with $\alpha = 0.05$ is 1.645

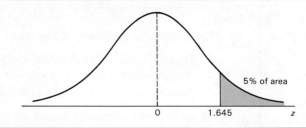

We wish to find the $\bar{x}_0$ value corresponding to $z = 1.645$ so that 5% of the area under the $\bar{x}$ distribution lies to the right of $\bar{x}_0$ (*see* Figure 9-1). The formula to convert $\bar{x}_0$ to standard units is

$$z = \frac{\bar{x}_0 - \mu}{\sigma/\sqrt{n}}$$

By rearranging the formula, we get

$$\bar{x}_0 = \mu + z\frac{\sigma}{\sqrt{n}}$$

Since $\mu = 7.50$, $\sigma/\sqrt{n} = 0.25$, and $z = 1.645$, for an $\alpha = 0.05$ right-tail test,

$$\bar{x}_0 = \mu + z\frac{\sigma}{\sqrt{n}}$$

$$= 7.50 + 1.645(0.25) = 7.91$$

Now we have all the information we need to complete the test.

VI. *Conclusion:* Earlier we pointed out a similarity between the logic of a statistical test and proceedings in a court of law. We are now at the point where all the (sample) evidence has been presented, and we are awaiting a decision.

Who makes the decision? In a court, the jury or judge would make the decision, but in our statistical test, it is Mother Nature who makes the final decision. The observed value $\bar{x} = 7.74$ cm of lengths of turtle eggs taken from the turtle nest lies in the acceptance region (*see* Figure 9-3).

We conclude that there is not enough evidence to discredit the null hypothesis at the $\alpha = 0.05$ level of significance. Therefore, we accept H_0: $\mu = 7.50$ and conclude that the nest was made by the rare endangered species of turtle. It is important to remember that we do not claim to have *proven* the eggs are from the rare turtle; all we say is that there is not enough evidence to reject H_0, and therefore we "accept" it.

FIGURE 9-3   The Observed Value $\bar{x} = 7.74$ Lies in the Acceptance Region

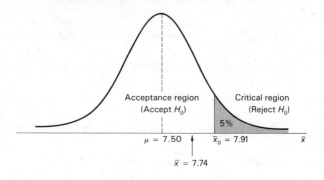

VII. *Meaning of α and β*: Remember that β is the probability of accepting H_0 when it is false. In our setting, β is the probability that the zoo incubates and hatches eggs from a common turtle species. To calculate β requires knowledge of the H_1 distribution. Since there are several species of common turtle in the area and we don't know which species made the nest, we really don't know the H_1 distribution. However, a type II error is not too serious because if the wrong eggs are hatched, there is little harm done.

The level of significance α is the probability of rejecting H_0 when it is true. This is the probability of saying the eggs are not from the rare turtle species when they in fact are. It would be a serious mistake to reject the nest of the rare turtle. To guard against such a mistake, we have taken a relatively small α value of 0.05.

In this section we have introduced you to all the essential ingredients of hypothesis testing. The turtle example contains all the important, basic ideas. For convenience let's summarize the main points:

A statistical test is a package of four basic ingredients:

1. H_0, the null hypothesis.
2. H_1, the alternate hypothesis.

3. A critical value $\bar{x}_0$ and critical region that depends on H_0, H_1, and the level of significance α.
4. An estimate of the parameter in question obtained from a sample. (In our example the sample consisted of measurements of lengths of 36 turtle eggs and the estimate for the mean μ was $\bar{x} = 7.74$.)

Once we have the four basic ingredients for a statistical test our strategy is simple. If the estimate of the parameter falls inside the critical region, then we reject H_0. If the estimate for the parameter falls outside the critical region—that is, falls in the acceptance region—we accept H_0 as true or, more specifically, we *do not* reject H_0. In either case the probability of rejecting H_0 when it is really true is the level of significance α, and the probability of correctly accepting H_0 when it is in fact true is the level of confidence, which is $1 - \alpha$.

_____ Exercise 3 _____

Suppose the biologist found a nest with 36 eggs for which $\bar{x} = 7.78$.

Would you accept or reject H_0 at the 5% level of significance? Explain.

The critical region is still such that we reject H_0 if $\bar{x}$ is larger than 7.91. Since our sample mean $\bar{x}$ is 7.78, we accept H_0 again. That is to say, we do not reject H_0.

_____ Exercise 4 _____

Suppose that the biologist found another nest with 64 eggs. The observed mean length of the 64 eggs is $\bar{x} = 7.90$ cm, and the length of rare turtle eggs is normally distributed with mean $\mu = 7.50$ cm and standard deviation $\sigma = 1.5$ cm. The other turtle species have eggs with longer mean length but with the same standard deviation as the length of turtle eggs from the rare turtle.

a) We wish to test the claim that the nest was made by the rare turtle. What are the null and alternate hypotheses?

a) H_0: $\mu = 7.50$
   H_1: $\mu > 7.50$

b) Describe the $\bar{x}$ distribution for the mean length of turtle eggs. According to H_0, what is the value of $\mu_{\bar{x}}$? What is the value of $\sigma_{\bar{x}}$?

b) By Theorem 7.1, $\bar{x}$ has a normal distribution because x has a normal distribution. $\mu_{\bar{x}} = 7.50$ and

$$\sigma_{\bar{x}} = \frac{\sigma}{\sqrt{n}} = \frac{1.5}{\sqrt{64}} = 0.19$$

c) By Table 5 of Appendix I, 5% of the area under the standard normal curve lies to the right of $z = 1.645$. Use this information to find the critical value $\bar{x}_0$ so that 5% of the area under the $\bar{x}$ curve lies to the right of $\bar{x}_0$. Use the formula

$$\bar{x}_0 = \mu_{\bar{x}} + z\sigma_{\bar{x}}$$

c) **FIGURE 9-4   5% Right Tail of z-curve**

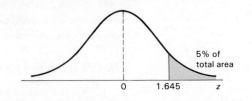

5% of total area

0   1.645   z

$$\bar{x}_0 = \mu_{\bar{x}} + z\sigma_{\bar{x}} = 7.50 + (1.645)(0.19) = 7.81$$

d) Sketch the critical region of the $\bar{x}$ distribution. Does the sample mean $\bar{x} = 7.90$, computed from the nest of 64 eggs, fall in the critical region?

d) **FIGURE 9-5   Critical Region**

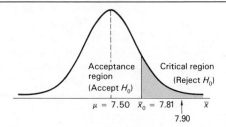

Acceptance region (Accept H_0)   Critical region (Reject H_0)

$\mu = 7.50$ $\bar{x}_0 = 7.81$ $\bar{x}$
7.90

e) Do we accept or reject H_0 based on the sample of 64 eggs?

e) Since $\bar{x} = 7.90$ falls in the critical region, we reject H_0 and accept H_1 at the 0.05 level of significance. The eggs seem to be from a common turtle.

In this section we have introduced the main ingredients of hypothesis testing. In the next section we will show you how to calculate critical values for other values of α and for two-tail and left-tail tests. For now concentrate on identifying suitable hypotheses, determining α, and looking for sample information that will determine the outcome of the test. Don't try to calculate the critical value $\bar{x}_0$ yet. Just concentrate on understanding the machinery of a statistical test.

Section Problems 9.1

In all problems in Chapter 9, *average* is taken to be the mean, $\bar{x}$.

1. In your own words carefully and completely answer each of the following questions:
   a) What is a statistical hypothesis?
   b) What is a null hypothesis H_0?
   c) What is an alternate hypothesis H_1?
   d) What type of errors are associated with hypothesis testing?
   e) What is the level of significance α of a test? What is the power of a test?
   f) What is a critical region? What is an acceptance region? What is a critical value?
   g) What are the basic features common to all statistical tests?

2. In a statistical test we have a choice of a left-tail critical region, right-tail critical region, or two-tail critical region. Is it the null hypothesis or alternate hypothesis that is used to determine which type of critical region is used?

3. If we accept the null hypothesis, does that mean we have *proven* it to be true beyond *all* doubt? Explain your answer.

4. If we reject the null hypothesis, does that mean we have *proven* it to be false beyond *all* doubt? Explain your answer.

5. Participate in a class discussion in which you compare similarities of statistical hypothesis testing with legal methods used in a courtroom setting.

   a) What corresponds to the null hypothesis? the alternate hypothesis?

   b) If someone is found innocent, the court has not usually *proven* that person to be innocent beyond *all* doubt. It simply means that the evidence against the person was not adequate to find him or her guilty. Compare this situation with your answer to problem 3.

   c) If someone is found guilty, the court has not usually *proven* that the person is guilty beyond *all* doubt. It simply means that the evidence against the person was adequate to discredit his or her position as innocent. Compare this situation with your answer to problem 4.

   d) In a courtroom setting the working hypothesis throughout the trial is that the person charged is innocent. In hypothesis testing, which hypothesis is used as the working hypothesis throughout the test: the null hypothesis or the alternate hypothesis?

   e) In a courtroom setting the final decision as to whether the person charged is innocent or guilty is made at the end of the trial, usually by a jury of impartial people. In hypothesis testing, the final decision to accept or reject the null hypothesis is made at the end of the test by using information from an (impartial) random sample. Discuss these similarities between hypothesis testing and a courtroom setting.

   f) We hope you are able to use this discussion to increase your understanding of statistical testing by comparing it with something that is a well-known part of our American way of life. However, all analogies have weak points. It is important not to take the analogy between statistical hypothesis testing and legal courtroom methods too far. For instance, the judge does not set a level of significance and tell the jury to determine a verdict that is wrong only 5% or 10% of the time. Discuss some of these weak points in the analogy between the courtroom setting and hypothesis testing.

6. A fast food franchise in Portland has launched a massive advertising campaign to increase sales. Before the campaign, the average net income per week for each of the franchise outlets was $27,490.

   a) Suppose we want to set up a statistical test to challenge the statement that the average net income per week is *still* $27,490. What should the null hypothesis be?

   b) It is generally accepted that the advertising campaign could only increase sales, and not hurt sales. What should the alternate hypothesis be?

   c) We now have two of the four basic ingredients of a statistical test. What are the others? In the next section we will show you how to complete such a test on your own.

7. An astronomer has been receiving radio signals from a planet. These signals have an average frequency of 750 cycles per second. More signals are being received from the planet and the astronomer wants a statistical test to determine if the average frequency has changed.

   a) What should we use for the null hypothesis?

   b) If the average frequency has changed we don't know if it has gone up or down. What should we use for the alternate hypothesis?

   c) We now have two of the four basic ingredients of a statistical test. What are the others? In the next section we will show you how to complete such a test.

8. The Diet Club claims its overweight members lose an average of 4 lb per week. A consumer group wants to test this claim.

   a) What should we use for the null hypothesis?

   b) Assuming the club does not underrate its effectiveness, what should we use for the alternate hypothesis?

c) We now have two of the four basic ingredients of a statistical test. What are the others? In the next section we will show you how to complete such a test.

9. The Illuminex Company claims that the 40W light bulbs they sell have an average life of 885 hr. Having used several Illuminex bulbs, you suspect the claim is overrated. Let μ be the mean of the lifetime distribution of their 40W bulbs.
   a) We want to set up a statistical test to challenge the statement $\mu = 885$. What should we use for the null hypothesis?
   b) Assuming Illuminex does not underrate their product, what shall we use for the alternate hypothesis?
   c) We now have two of the four basic ingredients of a statistical test. What are the others? In the next section we will show you how to complete such a test.

10. *Comprehensive News Magazine* did a survey of salaries for women in management in the Cincinnati area last year. They found the mean salary to be $\mu = \$46,300$. The Women Ahead group believes that the average salary for women in management has increased since that survey.
    a) To do a statistical test, what should the null hypothesis be?
    b) What should the alternate hypothesis be?
    c) We now have two of the four basic ingredients of a statistical test. What are the others? In the next section we will show you how to complete such a test.

SECTION 9.2

In most instances of hypothesis testing the alternate hypothesis is the condition that the mean μ is less than ($<$), greater than ($>$), or not equal to ($\neq$) some specified value. In Section 9.1 we saw an example of each. The types of critical regions used with these kinds of alternate hypotheses are shown in Figures 9-6, 9-7, and 9-8.

Figures 9-6 and 9-7 correspond to what are called *one-tail tests*. A one-tail test is a statistical test for which the critical region is located in the left tail *or* in the right tail of the distribution (but not both tails).

Figure 9-8 corresponds to what is called a *two-tail test*. A two-tail test is a statistical test for which the critical region is located in *both* tails of the distribution.

FIGURE 9-6 If the Alternate Hypothesis Is H_1: $\mu < k$, Use a Single Critical Region on the *Left* Side

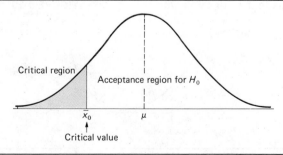

Critical region

Acceptance region for H_0

x_0

μ

↑
Critical value

FIGURE 9-7   If the Alternate Hypothesis Is H_1: $\mu > k$, Use a Single Critical Region on The *Right* Side

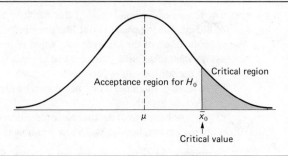

Critical region

Acceptance region for H_0

μ $\bar{x}_0$

Critical value

FIGURE 9-8   If the Alternate Hypothesis Is H_1: $\mu \neq k$, Use *Two* Critical Regions

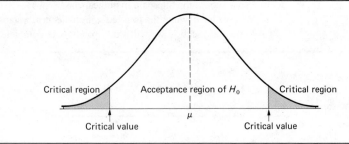

Critical region   Acceptance region of H_0 Critical region

μ

Critical value   Critical value

────────── Exercise 5 ──────────

a) Suppose an alternate hypothesis is

$$H_1: \mu > \text{some value}$$

Let $\bar{x}_0$ be the critical value. Do we have a one-tail test or a two-tail test? Shade the critical region in Figure 9-9.

a) One-tail test. Since H_1: $\mu >$ some value, we shade the right tail (Figure 9-10).

FIGURE 9-9   Shade the Critical Region

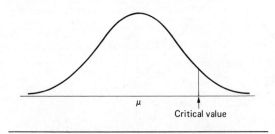

μ

Critical value

FIGURE 9-10   Completion of Figure 9-9

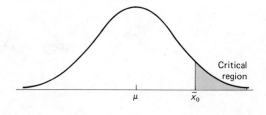

Critical region

μ $\bar{x}_0$

b) Suppose an alternate hypothesis is

$$H_1: \mu \neq \text{some value}$$

There are two critical values. Is this a one-tail test or a two-tail test? Shade the critical region(s) in Figure 9-11.

b) Two-tail test. Since $H_1: \mu \neq$ some specific value, we shade both tails (Figure 9-12).

FIGURE 9-11   Shade the Critical Regions

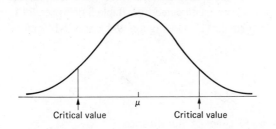

FIGURE 9-12   Completion of Figure 9-11

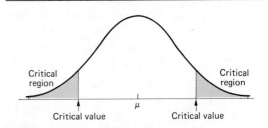

The type of symbol in the alternate hypothesis tells us what kind of critical region to use. For convenience in later work the discussion about critical regions is summarized in Table 9-3.

TABLE 9-3 Types of Critical Regions Used in Hypothesis Testing

Alternate Hypothesis H_1 About μ and Any Constant k	$H_1: \mu < k$	$H_1: \mu > k$	$H_1: \mu \neq k$
Type of critical region to be used	One tail on the left side	One tail on the right side	Two tails, one on each side

The *critical value* separates the critical region (or rejection region) from the acceptance region. When testing a hypothesis involving a *mean,* we compute the critical value $\bar{x}_0$ from formula (1) of Section 9.1. This formula is

$$\bar{x}_0 = \mu + z\frac{\sigma}{\sqrt{n}} \tag{1}$$

In the formula for $\bar{x}_0$ we obtain μ from the null hypothesis H_0; σ is the standard deviation of the x distribution implied by H_0; n is the sample size being used; and z is the critical value of a standard normal curve corresponding to the type of test being used and the level of significance.

The methods used in this section will require $\bar{x}$ to be approximately normally distributed, and we will approximate σ by the sample standard deviation s when necessary. To ensure that our methods yield accurate results, we will *require* our *sample size to be 30 or larger* ($n \geq 30$). Tests involving small samples will be presented in Section 9.4.

What values of z shall we use in formula (1)? The values of z are determined by our choice of the level of significance α. In theory we could use any value of α between 0 and 1. However, it is customary in most practical work to use either $\alpha = 0.05$ or $\alpha = 0.01$.

Figures 9-13 and 9-14 show the z values for different values of α and various one- or two-tailed tests. These values of z were computed from Table 5 of Appendix I.

The information contained in formula (1) and Figures 9-13 and 9-14 can be summed up as in Table 9-4.

FIGURE-13   Critical *z* Values for a 5% Level of Significance

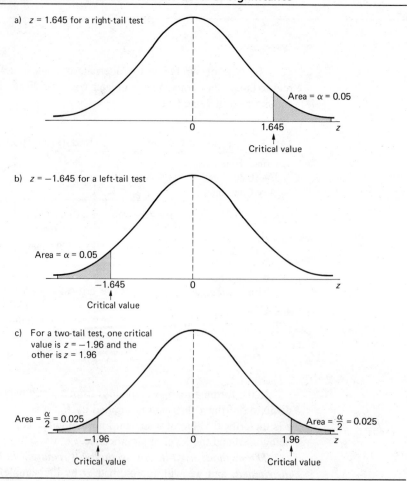

a) $z = 1.645$ for a right-tail test

Area = α = 0.05

0   1.645   z

Critical value

b) $z = -1.645$ for a left-tail test

Area = α = 0.05

−1.645   0   z

Critical value

c)   For a two-tail test, one critical value is $z = -1.96$ and the other is $z = 1.96$

Area = $\frac{\alpha}{2}$ = 0.025

Area = $\frac{\alpha}{2}$ = 0.025

−1.96   0   1.96   z

Critical value   Critical value

FIGURE 9-14   Critical z Values for a 1% Level of Significance

a) $z = 2.33$ for a right-tail test

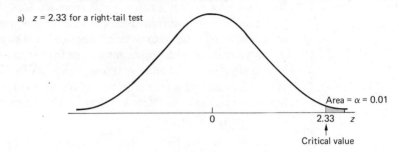

b) $z = -2.33$ for a left-tail test

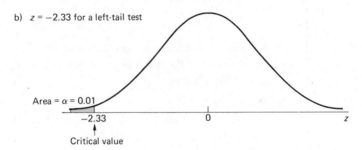

c) For a two-tail test one critical value is $z = -2.58$ and the other is $z = 2.58$

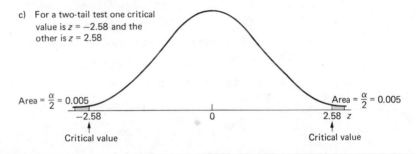

TABLE 9-4 Critical Values for Tests Involving a Mean (Large Samples)

Level of Significance	$\alpha = 0.05$	$\alpha = 0.01$
Critical value for a right-tail test	$\bar{x}_0 = \mu + 1.645 \dfrac{\sigma}{\sqrt{n}}$	$\bar{x}_0 = \mu + 2.33 \dfrac{\sigma}{\sqrt{n}}$
Critical value for a left-tail test	$\bar{x}_0 = \mu - 1.645 \dfrac{\sigma}{\sqrt{n}}$	$\bar{x}_0 = \mu - 2.33 \dfrac{\sigma}{\sqrt{n}}$
Critical values for a two-tail test	$\bar{x}_0 = \mu + 1.96 \dfrac{\sigma}{\sqrt{n}}$ and $\bar{x}_0 = \mu - 1.96 \dfrac{\sigma}{\sqrt{n}}$	$\bar{x}_0 = \mu + 2.58 \dfrac{\sigma}{\sqrt{n}}$ and $\bar{x}_0 = \mu - 2.58 \dfrac{\sigma}{\sqrt{n}}$

EXAMPLE 3

A research meteorologist has been studying wind patterns over the Pacific Ocean. Based on these studies a new route is proposed for commercial airlines going from San Francisco to Honolulu. The new route is intended to take advantage of existing wind patterns to reduce flying time. It is known that for the old route the distribution of flying times for a large four-engine jet has mean $\mu = 5.25$ hr with standard deviation $\sigma = 0.6$ hr. Thirty-six flights on the new route have yielded a mean flying time of $\bar{x} = 4.85$ hr. Does this indicate that the average flying time for the new route is less than 5.25 hr? Use a 5% level of significance.

a) What is H_0?

The average time for the old route is $\mu = 5.25$ hr. So we will set up the null hypothesis

$$H_0: \mu = 5.25$$

(In words we are saying that the mean flying time on the new route is the same as that of the old route.) We will see if H_0 can be rejected at the 5% level of significance.

b) What is H_1?

We want to see if the average flying time for the new route is less than 5.25, so the alternate hypothesis is

$$H_1: \mu < 5.25$$

c) What type of critical region must be used?

Since the $<$ symbol is used in the alternate hypothesis, Table 9-3 tells us to use a one-tail test on the left.

d) What critical value $\bar{x}_0$ must we use for a 5% level of significance? Table 9-4 tells us to use

$$\bar{x}_0 = \mu - 1.645 \frac{\sigma}{\sqrt{n}}$$

In this example $\sigma = 0.60$, $n = 36$, and $\mu = 5.25$ (use μ and σ from H_0). Therefore,

$$\bar{x}_0 = 5.25 - 1.645 \left(\frac{0.6}{\sqrt{36}} \right) = 5.09$$

e) Shall we accept or reject H_0 at $\alpha = 0.05$?

Let's look at the critical region and the acceptance region illustrated in Figure 9-15.

When $\alpha = 0.05$, $\bar{x}_0 = 5.09$. Since the average time for the 36 flights on the new route is $\bar{x} = 4.85$ hr, we see that our estimate $\bar{x} = 4.85$ is in the rejection region of Figure 9-15. So we reject H_0 and accept H_1 at the 0.05 level. This means that we accept, or at least do not reject, the alternate hypothesis that the flying time on the new route is less than that on the old route.

FIGURE 9-15   Critical Region for $\alpha = 0.05$

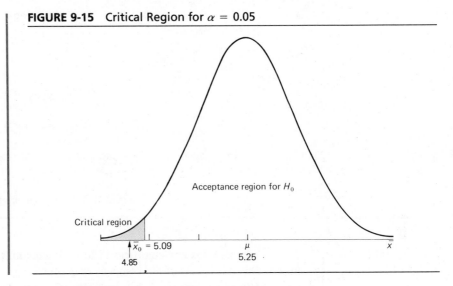

Acceptance region for H_0

Critical region

$\bar{x}_0 = 5.09$

4.85

μ

5.25

$\bar{x}$

_____ Exercise 6 _____

Pam likes to practice golf at a driving range. From thousands of shots at the range she knows that her average distance is about 225 yd. A friend asked Pam to experiment with a ball made of a new kind of material to see if it will increase her distance. Using the new type ball for 100 shots, she found that her average distance was 236 yd with a standard deviation of 25 yd. Let us see if we can reject the hypothesis H_0: $\mu = 225$ (the new ball goes no farther on the average than the old ball). We will use the 1% level of significance.

a) From the statement of the problem, what is a reasonable choice for H_1?

H_1: $\mu < 225$ H_1: $\mu \neq 225$ H_1: $\mu > 225$

b) What kind of test should we use: left-tail test, right-tail test, two-tail test? (*Hint: See* Table 9-3.)

c) What critical value $\bar{x}_0$ should we use? (*Hint: See* Table 9-4.) Draw a sketch showing the critical region and the acceptance region.

a) A reasonable alternate hypothesis is that the new ball increases the average driving distance. Therefore, we choose H_1: $\mu > 225$.

b) Since the alternate hypothesis is H_1: $\mu > 225$, we use a right-tail test.

c) From Table 9-4, with $\alpha = 0.01$ and a right-tail test, we see that the critical value should be

$$\bar{x}_0 = \mu + 2.33 \frac{\sigma}{\sqrt{n}}$$

In our case we use $\mu = 225$ (from H_0). Since $n = 100$ is a large sample, we can substitute $s = 25$ (the sample standard deviation) for σ. Therefore,

$$\bar{x}_0 \approx 225 + 2.33 \frac{25}{\sqrt{100}} \approx 231$$

The critical region is shown in Figure 9-16.

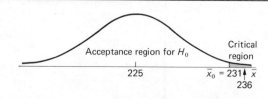

FIGURE 9-16 Critical Region for $\alpha = 0.01$

Acceptance region for H_0

Critical region

225

$\bar{x}_0 = 231$ $\bar{x}$

236

d) Does the value $\bar{x} = 236$ fall in the critical region or the acceptance region? Should we accept or reject H_0?

d) The mean of Pam's 100 shots with the new ball is $\bar{x} = 236$, and since this falls in the critical region, we reject H_0 and conclude that the new ball has increased Pam's average driving distance.

EXAMPLE   4

A large company has branch offices in several major cities of the world. From time to time it is necessary for company employees to move their families from one city to another. From long experience the company knows that its employees move on the average of once every 8.50 yr. However, trends for the past few years have led people to think a change might have occurred. To determine if such a change has occurred, a random sample was taken of 49 employees (from the entire company). The employees were asked to provide either the number of years since the company asked them to move or the number of years employed by the company if they had never been asked to move. For this sample of 49 employees the mean time was $\bar{x} = 7.91$ yr with sample standard deviation 3.62 yr.

Let us see if we can reject the hypothesis H_0: $\mu = 8.50$ at the $\alpha = 0.05$ level of significance. Since we have no way of knowing if the average moving time (μ) has increased or decreased, the alternate hypothesis will simply be H_1: $\mu \neq 8.50$.

a) Should we use a right-tail, left-tail, or two-tail test?

Table 9-3 and the alternate hypothesis H_1: $\mu \neq 8.50$ indicate a two-tail test.

b) What are the critical values for a two-tail test? What is the critical region?

In our example we use Table 9-4 with $\mu = 8.50$, $n = 49$, and $\sigma \simeq 3.62$. (Since the sample size of 49 is large we can replace σ with the sample standard deviation 3.62.) From Table 9-4 we obtain

$$\bar{x}_0 = \mu + 1.96 \frac{\sigma}{\sqrt{n}} \simeq 8.50 + 1.96 \left(\frac{3.62}{\sqrt{49}} \right) = 9.51$$

$$\bar{x}_0 = \mu - 1.96 \frac{\sigma}{\sqrt{n}} \simeq 8.50 - 1.96 \left(\frac{3.62}{\sqrt{49}} \right) = 7.49$$

c) Can we reject H_0 at a 5% level of significance? (*See* Figure 9-17.)

FIGURE 9-17   Critical Region for $\alpha = 0.05$

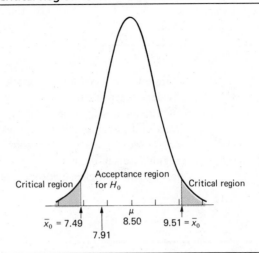

Critical region   Acceptance region   Critical region
                       for H_0

$\bar{x}_0 = 7.49$ μ $9.51 = \bar{x}_0$
                 8.50

         7.91

Our sample mean $\bar{x} = 7.91$ is in the acceptance region so we cannot reject H_0 at the 5% level of significance. It appears that employees still move on the average of once every 8.50 yr.

_____ Exercise 7 _____

A machine makes twist-off caps for bottles. The machine is adjusted to make caps of diameter 1.85 cm. Production records show that when the machine is so adjusted it will make caps with mean diameter 1.85 cm and with standard deviation $\sigma = 0.05$ cm. During production an inspector checks the diameters of caps to see if the machine has slipped out of adjustment. A random sample of 64 caps is taken. If the mean diameter for this sample is $\bar{x} = 1.89$, does this indicate the machine has slipped out of adjustment and the average diameter of caps is no longer $\mu = 1.85$ cm? (Use a 5% level of significance.)

a) What is the null hypothesis?

    $H_0: \mu = 0.05$ $H_0: \mu = 1.85$ $H_0: \mu = 1.90$

a) $H_0: \mu = 1.85$

b) We want to test the null hypothesis against the hypothesis that the mean diameter is *not* 1.85. Therefore, the alternate hypothesis is:

    $H_1: \mu < 1.85$ $H_1: \mu \neq 1.85$ $H_1: \mu > 1.85$

b) $H_1: \mu \neq 1.85$

c) Should we use a one- or two-tail test? (*Hint: See* Table 9-3 and consider H_1.)

c) Two-tail test

d) What are the critical values (*Hint: See* Table 9-4.) and what is the critical region for this test?

d) $\bar{x}_0 = \mu + 1.96 \dfrac{\sigma}{\sqrt{n}} = 1.85 + 1.96\left(\dfrac{0.05}{\sqrt{64}}\right) = 1.86$

    $\bar{x}_0 = \mu - 1.96 \dfrac{\sigma}{\sqrt{n}} = 1.85 - 1.96\left(\dfrac{0.05}{\sqrt{64}}\right) = 1.84$

e) Does the test indicate the machine needs adjustment at the 5% level of significance?

e) The sample mean $\bar{x} = 1.89$ is in the critical region. At the 5% level of significance we must reject H_0, so the test indicates the machine needs adjustment.

- **Comment:** In a sense the sample estimate of the parameter represents the evidence favoring rejection of H_0. For this reason statisticians say the results of a random sample are *statistically significant* if the estimated parameter falls in the critical region of a test. In Exercise 7 the random sample of 64 caps gave a parameter estimate $\bar{x} = 1.89$. Since 1.89 was in the critical region, we say the results of the sample were statistically significant—that is, there was sufficient evidence to reject H_0 at the specified level of significance.

Section Problems 9.2

For each of the following problems,
- a) State the null and alternate hypotheses.
- b) Find the critical value or values.
- c) Sketch the critical region and the acceptance region.
- d) On the sketch of part (c) show the location of $\bar{x}$, the observed sample mean.
- e) Decide whether you should accept or reject the null hypothesis.

1. Judy Povich is a fashion design artist who designs the display windows in front of a large clothing store in New York City. Electronic counters at the entrances total the number of people entering the store each business day. Before Judy was hired by the store the mean number of people entering the store each day was 3,218. However, since Judy has started working, it is thought that this number has increased. A random sample of 42 business days after Judy began work gave an average $\bar{x} = 3,492$ people entering the store each day. The sample standard deviation was $s = 287$ people. Does this indicate that the average number of people entering the store each day has increased? Use a 1% level of significance.

2. A new state-run job assistance program helps the unemployed find work. In one community where there is a job assistance program, a sociologist is studying unemployment. This sociologist found that before the job assistance program the mean length of time it took an unemployed person to find work was 57 days. A random sample of 33 people using the new job assistance program showed that it took an average of 48 days to find work. The sample standard deviation was 14 days. Does this indicate that the average length of time to find a job is now less than 57 days? Use a 5% level of significance.

3. A large waterfront warehouse installed a new security system a few years ago. Under the old system the warehouse managers estimated that they were losing an average of $678 worth of merchandise to thieves each week. A random sample of 30 weekly records under the current system showed that they were still losing an average of $\bar{x} = \$650$ worth of merchandise each week. The sample standard deviation was $s = \$93$. Does this

indicate that the average loss each week is different (either more or less) than the previous $678 per week? Use a 5% level of significance.

4. The college basketball coach has discovered a new tactical playing strategy which he uses in the second half of a game. Long-term records of practice games played without the strategy indicate that the team scores an average of 46.3 points in the second half. However, for a random sample of 30 recent practice games using the new second-half strategy the sample mean score in the second half was 50 points. The sample standard deviation was 12.7 points. Does this indicate that the mean score in the second half is more than 46.3 points using the new playing strategy? Use a 1% level of significance.

5. Maureen is a cocktail hostess in a very exclusive private club. The Internal Revenue Service is auditing her tax return this year. Maureen claims that her average tip last year was $4.75. To support this claim, she sent the IRS a random sample of 52 credit card receipts showing her bar tips. When the IRS got the receipts, they computed the sample average and found it to be $\bar{x}$ = $5.25 with sample standard deviation s = $1.15. Do these receipts indicate that the average tip Maureen received last year was more than $4.75? Use a 1% level of significance.

6. Dental associations recommend that the time lapse between routine dental checkups should average six months. A random sample of 36 patient records at one dental clinic showed the average time between routine checkups to be 9.2 months with standard deviation 2 months. Do the sample data indicate that patients wait longer than recommended between dental checkups? Use a 1% level of significance.

7. The Smoky Bear Trucking Company claims that the average weight of a fully loaded moving van is 12,000 lb. The highway patrol decides to check this claim. A random sample of 30 Smoky Bear moving vans shows that the average weight is 12,100 lb with a standard deviation of 800 lb. Does this indicate the average weight of a Smoky Bear moving van is more than 12,000 lb? (Use a 5% level of significance.)

8. The Magic Dragon Cigarette Company claims that their cigarettes contain an average of only 10 mg of tar. A random sample of 100 Magic Dragon cigarettes shows the average tar content per cigarette to be 11.5 mg with a standard deviation of 4.5 mg. Does this indicate that the average tar content of Magic Dragon cigarettes exceeds 10 mg? (Use a 5% level of significance.)

9. Jerry is doing a project for his sociology class in which he tests the claim that the Pleasant View housing project contains family units of average size 3.3 people (the national average). A random sample of 64 families from the Pleasant View project shows that there is an average of 4.3 people per family unit with a standard deviation of 1.3. Does this indicate that the average size of a family unit in Pleasant View is different from the national average of 3.3? (Use a 5% level of significance.)

10. The Mammon Savings and Loan Company claims that the average amount of money on deposit in a savings account in their bank is $5,200 with a standard deviation of $670. Suppose a random sample of 49 accounts shows the average amount on deposit to be $4,975. Does this indicate that the average amount on deposit per account is not $5,200? (Use a 1% level of significance.)

11. The manufacturer of a compact car claims that it averages 38.6 mpg of gasoline. To test this claim 30 cars were randomly selected and driven under normal conditions. The results for these cars gave an average of 31.5 mpg with a standard deviation of 9.3. Does this indicate that the manufacturer's claim is too high? (Use a 1% level of significance.)

12. A pharmaceutical company makes tranquilizers that are claimed to have a mean effective period of 2.8 hr. Researchers in a hospital used the drug on a random sample of 81 patients and found the mean effective period to be 2.5 hr with a standard deviation of 0.4 hr. Does this indicate that the company's claim is too high? (Use a 1% level of significance.)

13. An Air Force base mess hall has received a shipment of 10,000 gallon-size cans of cherries. The supplier claims that the average amount of liquid is 0.25 gal per can. A government inspector took a random sample of 100 cans and found the average liquid content to be 0.32 gal per can with a standard deviation of 0.10. Does this indicate that the supplier's claim is too low? (Use a 5% level of significance.)

14. Robert Jeffries is an industrial psychologist working for an auto manufacturing company in Detroit. His job is to invent ways to speed up production without increasing cost. After examining the production line, he thinks he can reorder the door assembly sequence and reduce assembly time. Under the old routine, the mean time for a door assembly was 8.50 min. Under Bob's new assembly method, a random sample of 47 door jobs showed that the doors were done in a mean time of $\bar{x} = 7.19$ min with sample standard deviation $s = 0.86$ min. Does this sample indicate that Bob's new method has reduced the mean assembly time? (Use a 5% level of significance.)

15. Walter Gleason is an astronomer who has been studying radio signals from the planet Jupiter. Over a long period of time, the planet's electromagnetic field has been sending low-frequency signals with a mean frequency of 14.2 megahertz. A recent space module that went by Jupiter recorded a major volcanic eruption that may have covered a large surface on the planet. For the past several months, Walter has been measuring Jupiter's radio signals on a random time schedule. A group of 75 measurements gave a mean of $\bar{x} = 14.9$ megahertz with sample standard deviation $s = 0.9$ megahertz. Using a 1% level of significance, can we say that the mean radio frequency of these signals has changed (either up or down)?

16. Julia Ching is a botanist hired by the city of Tulsa to study the problem of finding a better summer watering plan for the many acres of lawn in the city parks. Julia believes that less frequent but longer watering is better, because it gives the grass a better root system. The water board says it has been using an average of 137,000 gal of water a day for city parks, and Julia cannot go above this average. After Julia's new watering plan was put into effect, a random sample of 40 summer watering days showed that the mean amount of water being used per day was $\bar{x} = 132,119$ gal with standard deviation $s = 6,410$ gal. Using a 1% level of significance, test the claim that Julia's system uses less water on the average than the amount being used with the old watering system.

SECTION 9.3

Tests Involving the Difference of Two Means (Independent Samples)

Many practical applications of statistics involve a comparison of two population means. In this section we will study distributions and statistical tests which arise from the difference of sample means taken from two different populations. First let's look at an example to focus our attention on the type of problem that calls for a test involving the difference of means.

EXAMPLE 5

A teacher wishes to compare the effectiveness of two teaching methods. Students are randomly divided into two groups: the first group is taught by method 1, the second group by method 2. At the end of the course a comprehensive exam is given to all students. The mean score for group 1 is $\bar{x}_1$ and the mean score for group 2 is $\bar{x}_2$. How can we decide if the difference between these means is due just to chance, or if the difference is statistically significant?

The first step toward resolving problems as posed in Example 5 is to find a probability distribution of the differences of the sample means. The probability distributions described in this section require the two original distributions to be *independent* of each other.

• **Definition:** We say that two sampling distributions are *independent* if there is no relation whatsoever between specific values of the two distributions.

EXAMPLE 6

Are the distributions of the means from each group of students in Example 5 independent? Explain.

Because the students were *randomly* divided into two groups, it is reasonable to say the $\bar{x}_1$ distribution is independent of the $\bar{x}_2$ distribution.

EXAMPLE 7

Suppose a shoe manufacturer claims that for the general population of adult United States citizens the average length of the left foot is larger than the average length of the right foot. To study this claim the manufacturer gathers data in this fashion: Sixty adult United States citizens are drawn at random, and for these 60 people both their left and right feet are measured. Let $\bar{x}_1$ be the mean length of the left feet and $\bar{x}_2$ be the mean length of the right feet.

Are the $\bar{x}_1$ and $\bar{x}_2$ distributions independent for this method of collecting data?

In this method there is only one random sample of people drawn and both the left and right feet are measured from this sample. The length of a person's left foot is usually related to the length of the right foot, so in this case the $\bar{x}_1$ and $\bar{x}_2$ distributions are not independent.

_____ Exercise 8 _____

Suppose the shoe manufacturer of Example 7 gathers data in the following way: Sixty adult United States citizens are drawn at random and their left feet are measured; then another 60 adult United States citizens are drawn at random and their right feet are measured. Again $\bar{x}_1$ is the mean of the left foot measurements and $\bar{x}_2$ is the mean of the right foot measurements.

Are the $\bar{x}_1$ and $\bar{x}_2$ distributions independent for this method of collecting data?

For this method of gathering data two random samples are drawn: one for the left foot measurements and one for the right foot measurements. The first sample is not related to the second sample. The $\bar{x}_1$ and $\bar{x}_2$ distributions are independent.

In this section we will use distributions that arise from a difference of means. How do we obtain such distributions? If we have two statistical variables $\bar{x}_1$ and $\bar{x}_2$, each with its own distribution, we take independent random samples of size n_1 from the $\bar{x}_1$ distribution and size n_2 from the $\bar{x}_2$ distribution. Then we can compute the respective means $\bar{x}_1$ and $\bar{x}_2$. Consider the difference $\bar{x}_1 - \bar{x}_2$. This represents a difference of means. If we repeat the sampling process over and over, we will come up with lots of $\bar{x}_1 - \bar{x}_2$ values. These values can be arranged in a frequency table, and we can make a histogram for the distribution of $\bar{x}_1 - \bar{x}_2$ values. This would give us an experimental idea of the theoretical distribution of $\bar{x}_1 - \bar{x}_2$.

Fortunately, it is not necessary to carry out this lengthy process for each example. The results have already been worked out mathematically. The next theorem presents the main results.

- **Theorem 9.1:** Let x_1 have a normal distribution with mean μ_1 and standard deviation σ_1. Let x_2 have a normal distribution with mean μ_2 and standard deviation σ_2. If we take independent random samples of size n_1 from the x_1 distribution and of size n_2 from the x_2 distribution, then the variable $\bar{x}_1 - \bar{x}_2$ has

  1. a normal distribution,
  2. mean $\mu_1 - \mu_2$, and
  3. standard deviation

$$\sqrt{\frac{\sigma_1^2}{n_1} + \frac{\sigma_2^2}{n_2}}$$

- **Comment:** The theorem requires that x_1 and x_2 have normal distributions. However, if both n_1 and n_2 are 30 or larger, then for most practical applications the Central Limit Theorem assures us that $\bar{x}_1$ and $\bar{x}_2$ are approximately normally distributed. In this case the conclusions of the theorem are again valid even if the original x_1 and x_2 distributions were not normal.

When testing the difference of means it is customary to use the null hypothesis

$$H_0: \mu_1 - \mu_2 = 0$$

or, equivalently,

$$H_0: \mu_1 = \mu_2$$

As mentioned in Section 9.1 the null hypothesis is set up to see if it can be rejected. When testing the difference of means we first set up the hypothesis H_0 that there is no difference. The alternate hypothesis could then be any of the ones listed in Table 9-5. The alternate hypothesis and consequent type of test used depend on the particular problem. Note μ_1 is always listed first.

The critical values and critical regions are much the same as those used before except that the standard deviation of the $\bar{x}_1 - \bar{x}_2$ distribution is obtained from Theorem 9.1 and the null hypothesis stipulating that $\mu_1 - \mu_2 = 0$. As a convenient reference, critical values for a 5% and a 1% level of significance are given in Table 9-6.

TABLE 9-5 Alternate Hypotheses and Type of Test: Difference of Two Means

H_1			Type of Test
$H_1: \mu_1 - \mu_2 < 0$	or equivalently	$H_1: \mu_1 < \mu_2$	Left-tail test
$H_1: \mu_1 - \mu_2 > 0$	or equivalently	$H_1: \mu_1 > \mu_2$	Right-tail test
$H_1: \mu_1 - \mu_2 \neq 0$	or equivalently	$H_1: \mu_1 \neq \mu_2$	Two-tail test

TABLE 9-6 Critical Values for Tests Involving a Difference of Two Means (Large Samples)

Level of Significance	$\alpha = 0.05$	$\alpha = 0.01$
Critical value for a right-tail test	$1.645 \sqrt{\dfrac{\sigma_1^2}{n_1} + \dfrac{\sigma_2^2}{n_2}}$	$2.33 \sqrt{\dfrac{\sigma_1^2}{n_1} + \dfrac{\sigma_2^2}{n_2}}$
Critical value for a left-tail test	$-1.645 \sqrt{\dfrac{\sigma_1^2}{n_1} + \dfrac{\sigma_2^2}{n_2}}$	$-2.33 \sqrt{\dfrac{\sigma_1^2}{n_1} + \dfrac{\sigma_2^2}{n_2}}$
Critical values for a two-tail test	$1.96 \sqrt{\dfrac{\sigma_1^2}{n_1} + \dfrac{\sigma_2^2}{n_2}}$ and $-1.96 \sqrt{\dfrac{\sigma_1^2}{n_1} + \dfrac{\sigma_2^2}{n_2}}$	$2.58 \sqrt{\dfrac{\sigma_1^2}{n_1} + \dfrac{\sigma_2^2}{n_2}}$ and $-2.58 \sqrt{\dfrac{\sigma_1^2}{n_1} + \dfrac{\sigma_2^2}{n_2}}$

EXAMPLE 8

A consumer study group is testing camp stoves. To test the heating capacity of a stove they measure the time required to bring 2 qt of water from 50 °F to boiling (at sea level).

Two competing models are under consideration. Thirty-six stoves of each model were tested and the following results were obtained.

Model 1:   mean time $\bar{x}_1$ = 11.4 min; standard deviation s_1 = 2.5 min

Model 2:   mean time $\bar{x}_2$ = 9.9 min; standard deviation s_2 = 3.0 min

Is there any difference between the performances of these two models? (Use a 5% level of significance.)

a) The problem is whether the observed difference between sample means is significant or not at the 0.05 level. Let μ_1 and μ_2 be the mean of the distribution of times for models 1 and 2, respectively. We set up the null hypothesis to say there is no difference

$$H_0: \mu_1 = \mu_2$$

The alternate hypothesis says there is a difference

$$H_1: \mu_1 \neq \mu_2$$

b) Since the alternate hypothesis is $H_1: \mu_1 \neq \mu_2$, we use a two-tail test (*see* Table 9-5).

c) Since α = 0.05, Table 9-6 gives the critical values for the test. In this problem we are given

$$\bar{x}_1 = 11.4 \qquad s_1 = 2.5 \quad \text{and} \quad n_1 = 36$$

$$\bar{x}_2 = \ \ 9.9 \qquad s_2 = 3.0 \quad \text{and} \quad n_2 = 36$$

The sample size of 36 for both n_1 and n_2 is large enough to permit us to replace σ_1 with s_1 and σ_2 with s_2. Therefore, from Table 9-6 we have

$$\text{critical values} = \pm 1.96 \sqrt{\frac{\sigma_1^2}{n_1} + \frac{\sigma_2^2}{n_2}} \approx \pm 1.96 \sqrt{\frac{(2.5)^2}{36} + \frac{(3.0)^2}{36}}$$

$$= \pm 1.96(0.65) = \pm 1.3 \text{ min}$$

Table 9-6 indicates that for a two-tail test the critical values are then 1.3 and -1.3 (*see* Figure 9-18).

FIGURE 9-18   Critical Values for a Two-Tail Test

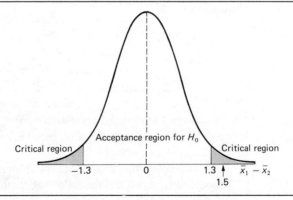

d) Since $\bar{x}_1 = 11.4$ and $\bar{x}_2 = 9.9$, then

$$\bar{x}_1 - \bar{x}_2 = 11.4 - 9.9 = 1.5$$

is in the critical region. Therefore, at the 5% level of significance we must reject H_0 and accept H_1 that the mean times for the two stoves to boil water are statistically different.

—————————— Exercise 9 ——————————

Is the difference between the sample means of Example 8 significant at the 0.01 level? (When $\alpha = 0.01$, should we accept or reject H_0?)

a) From part (c) of Example 8, we see that

$$\sqrt{\frac{\sigma_1^2}{n_1} + \frac{\sigma_2^2}{n_2}} = \underline{\hspace{2cm}}$$

a) 0.65, as before.

b) Use Table 9-6 to find the critical values for a two-tail test when $\alpha = 0.01$.

b) critical value $= 2.58 \sqrt{\dfrac{\sigma_1^2}{n_1} + \dfrac{\sigma_2^2}{n_2}}$

$$= 2.58(0.65) = 1.7$$

Therefore, the critical values are -1.7 and 1.7.

c) Is the difference between the sample means significant at the 0.01 level? Should we accept or reject H_0? Shade in the critical regions in Figure 9-19 to help make your decisions.

c) At the 0.01 level the difference between the sample means

$$\bar{x}_1 - \bar{x}_2 = 11.4 - 9.9 = 1.5$$

is in the acceptance region. Therefore, we accept H_0 and conclude at the 0.01 level that the difference is *not* statistically significant. (*See* Figure 9-20.)

FIGURE 9-19 Shade the Critical Regions

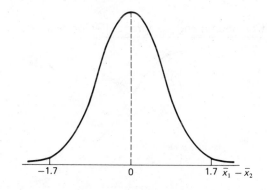

FIGURE 9-20 Completion of Figure 9-19

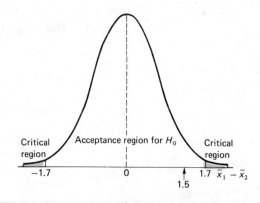

_____ Exercise 10 _____

Let us return to Example 5 at the beginning of this section. A teacher wishes to compare the effectiveness of two teaching methods. Students are randomly divided into two groups: the first group is taught by method 1, the second group by method 2. At the end of the course a comprehensive exam is given to all students.

The first group consists of $n_1 = 49$ students with a mean score of $\bar{x}_1 = 74.8$ points and standard deviation $s_1 = 14$ points. The second group has $n_2 = 50$ students with a mean score of $\bar{x}_2 = 81.3$ points and standard deviation $s_2 = 15$ points.

The teacher claims that the second method will increase the mean score on the comprehensive exam. Is this claim justified at the 5% level of significance?

Let μ_1 and μ_2 be the mean score of the distribution of all scores using method 1 and method 2, respectively.

a) Which is the null hypothesis?

$$H_0: \mu_1 = \mu_2 \qquad H_0: \mu_1 \neq \mu_2$$

$$H_0: \mu_1 < \mu_2 \qquad H_0: \mu_1 > \mu_2$$

a) $H_0: \mu_1 = \mu_2$

b) To examine the validity of the teacher's claim, what will we use for the alternate hypothesis?

$$H_1: \mu_1 \neq \mu_2 \qquad H_1: \mu_1 > \mu_2 \qquad H_1: \mu_1 < \mu_2$$

b) $H_1: \mu_1 < \mu_2$ (the second method gives a higher average score)

c) Shall we use a right-tail, left-tail, or two-tail test?

c) Since the alternate hypothesis is $H_1: \mu_1 < \mu_2$, Table 9-5 tells us to use a left-tail test.

d) Use Table 9-6 to find the critical value(s) when $\sigma_1 \approx 14$, $n_1 = 49$, $\sigma_2 \approx 15$, and $n_2 = 50$. Draw a diagram showing the acceptance region and the critical region of the $\bar{x}_1 - \bar{x}_2$ distribution.

d) critical value $= -1.645 \sqrt{\dfrac{(14)^2}{49} + \dfrac{(15)^2}{50}}$

$$= -1.65 \sqrt{4.00 + 4.50}$$

$$= -1.645(2.9) = -4.8$$

Figure 9-21 shows the critical region.

FIGURE 9-21 Critical Region and Acceptance Region of $\bar{x}_1 - \bar{x}_2$

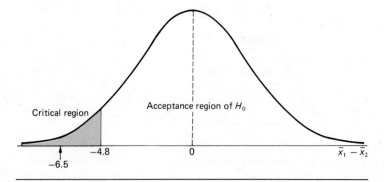

e) Since $\bar{x}_1 = 74.8$ and $\bar{x}_2 = 81.3$, is $\bar{x}_1 - \bar{x}_2$ in the acceptance region or the critical region? At the 5% level of significance is the claim that method 2 is better unjustified?

e) $\bar{x}_1 - \bar{x}_2 = 74.8 - 81.3 = -6.5$ is in the critical region. Therefore, we reject H_0 and conclude that at the 5% level of significance the claim that method 2 is better is justified.

Section Problems 9.3

For each problem do the following:

a) State the null and alternate hypotheses.
b) Find the critical value or values.
c) Sketch the critical region and the acceptance region.
d) On the sketch of part (c) show the location of $\bar{x}_1 - \bar{x}_2$, the observed sample difference of means.
e) Decide whether you should accept or reject the null hypothesis.

1. The personnel manager of a large manufacturing plant suspects a difference in the mean amount of sick leave taken by workers on the day shift compared to the night shift. It is suspected that the night workers take more sick leave on the average. A random sample of 38 day workers had a mean $\bar{x}_1 = 10.6$ days sick leave last year; the standard deviation was $s_1 = 3.3$ days. A random sample of 41 night workers had a mean $\bar{x}_2 = 12.9$ days sick leave last year with standard deviation $s_2 = 4.5$ days. At the 5% level of significance, can we say the night workers take more sick leave on the average?

2. Tracie is doing a research project for her political science class regarding opinions about a proposed bill before the state legislature. Each person interviewed indicated his or her opinion on a continuous scale from 0 (strongly disagree) to 10 (strongly agree). A random sample of 53 college students gave a mean opinion rating of $\bar{x}_1 = 4.7$ with standard deviation $s_1 = 2.1$. A random sample of 46 community people who are not students had an opinion rating with mean $\bar{x}_2 = 5.1$ and standard deviation $s_2 = 2.5$. At the 1% level of significance, is there a difference either way between student opinion and community opinion about this bill?

3. A new model compact car is being tested for gasoline consumption. A random sample of 40 such cars were tested at sea level and found to get a mean of $\bar{x}_1 = 36.2$ mpg with standard deviation $s_1 = 4.8$ mpg. Then another random sample of 35 cars were tested one mile above sea level in Denver. The mean for the second group was $\bar{x}_2 = 31.5$ mpg with standard deviation $s_2 = 5.5$ mpg. Using a 1% level of significance test the claim that these cars burn more gasoline at one mile above sea level.

4. Do students who study more get higher grades? A random sample of 73 freshmen who got a grade point average of 3.5 to 4.0 (i.e., B+ to A) last semester were interviewed and asked the number of hours they study each week night. For this group the sample mean was $\bar{x}_1 = 2.8$ hr with standard deviation $s_1 = 0.6$ hr. A second random sample of 84 freshmen who got a grade point average of 2.0 to 2.5 (i.e., C to C+) were asked the same question. For this group the sample mean was $\bar{x}_2 = 1.5$ hr with standard deviation $s_2 = 0.9$ hr. Use a 1% level of significance to test the claim that students in the 3.5 to 4.0 grade average range study longer.

5. A random sample of 68 upper income divorced men were asked how long their marriage lasted before the divorce. The sample mean was $\bar{x}_1 = 8.1$ yr with standard deviation $s_1 = 2.2$ yr. Another random sample of 51 middle income divorced men were asked the same question. For this group the sample mean was $\bar{x}_2 = 7.8$ yr with standard deviation $s_2 = 2.6$ yr. At the 5% level of significance, is there a difference either way between the mean length of marriage for these income groups?

6. A chemist has invented a new preservative for cut flowers and wants to test its effectiveness against the leading commercial preservative. He took two random samples of 100 freshly cut carnations each. One group of flowers was set in vases containing the new preservative, and the other group was set in vases containing the commercial preservative. Flowers in the new preservative began to wilt after an average of $\bar{x}_1 = 75$ hr with a standard deviation $s_1 = 15$ hr. In the commercial preservative the average was $\bar{x}_2 = 71$ hr with a standard deviation $s_2 = 10$ hr. Is the new preservative more effective? (Use a 5% level of significance.)

7. A large company has been hiring graduates from two secretarial schools. The company efficiency expert gave a typing test to a random sample of 30 new graduates from one school and found the graduates' mean score to be $\bar{x}_1 = 68$ words per minute with standard deviation $s_1 = 16$. Another random sample of 30 new graduates from the other school was tested. The mean score of the second group was $\bar{x}_2 = 71$ words per minute with standard deviation $s_2 = 12$. At the 5% level of significance, can we say there is a significant difference between the average scores?

8. Professor Adams taught the same course for two terms. Except for negligible differences, the two courses were the same except that one met at 7:00 A.M. and the other at 11:00 A.M. The average score for the 49 students in the 7:00 A.M. class was $\bar{x}_1 = 73.2$ with standard deviation $s_1 = 8.1$. For the 11:00 A.M. class there were 36 students with average score $\bar{x}_2 = 78.1$ and standard deviation $s_2 = 10.0$. Does this indicate the average score for the 11:00 A.M. class is significantly higher? (Use a 5% level of significance.)

9. A large shipping company has a fleet of 90 almost identical cargo ships. A fluid dynamics engineer claims that a certain modification of the front hull will allow the ships to glide better and faster at the same engine speed than the present design does. The fleet is randomly divided into two groups of 60 and 30 ships. The group of 30 ships have their front hulls modified. When the engines are at 3/4 speed, the mean speed in a calm sea for the unmodified ships is $\bar{x}_1 = 25.6$ knots per hour with standard deviation $s_1 = 4.3$. The mean speed for the modified ships is $\bar{x}_2 = 29.0$ knots per hour and standard deviation $s_2 = 5.0$. Do these data support the engineer's claim at the 1% level of significance?

10. At Camp Grosso a group of 112 overweight soldiers "volunteered" to go on diets. The 112 soldiers were randomly divided into two groups and both groups dieted for six weeks. The group using diet A had 51 soldiers. The average weight loss for this group was $\bar{x}_1 = 19.5$ lb with standard deviation $s_1 = 9.0$ lb. The other group used diet B with an average weight loss of $\bar{x}_2 = 15.7$ lb and standard deviation $s_2 = 8.1$ lb. Is there a difference between the average weight loss with these two diets? (Use a 1% level of significance.)

11. The Sparkle Toothpaste Company, Inc. claims that after one year of brushing with Sparkle your children will have fewer cavities than if they had brushed with brand X. The *Consumer's Friend* magazine tested this claim on two random samples of 100 children each. Group 1 used Sparkle and had an average of $\bar{x}_1 = 3.3$ cavities with standard deviation $s_1 = 1.9$. Group 2 used brand X and had an average of $\bar{x}_2 = 3.6$ cavities with standard deviation $s_2 = 2.0$. Is the Sparkle claim justified at the 5% level of significance?

12. The English teachers in a certain state school system claim their average salary is less than the average salary of the physical education coaches. To test this claim a random sample of 50 coaches showed their average salary was $\bar{x}_1 = \$29,500$ with standard deviation $s_1 = \$800$. A random sample of 40 English teachers had an average salary of $\bar{x}_2 = \$29,300$ with standard deviation $s_2 = \$1,000$. Is the claim of the English teachers justified at the 5% level of significance?

13. Inner city residents claim that their average grocery costs are higher than the average grocery costs of people living in the suburbs. To test this claim the price of a quart of milk, a pound of bread, a dozen medium eggs, and a pound of butter was obtained from a random sample of 30 suburban stores. The average cost was $\bar{x}_1 = \$5.20$ with standard deviation $s_1 = \$0.30$. For a random sample of 36 inner city stores the same items cost an average of $\bar{x}_2 = \$5.54$ with standard deviation $s_2 = \$0.50$. Is the claim justified at the 5% level of significance?

14. A psychologist is testing a memory-enhancement drug. The effects of the drug wear off completely after about ten hours. However, while the drug is effective, it is thought to greatly improve the accuracy of the subject's memory A random sample of 75 students volunteered to participate in a test of the drug. All students were given 30 min to memorize a long passage of written material that they had not seen before. The students were randomly assigned to two groups. Group I had $n_1 = 37$ students who attempted to memorize the material without any drugs. Group II had $n_2 = 38$ students who took the drug and then attempted to memorize the material. All students took an exam testing their memory of the newly learned material. For Group I, the mean score was $\bar{x}_1 = 116$ with sample standard deviation $s_1 = 32$. For Group II, the mean was $\bar{x}_2 = 125$ with standard deviation $s_2 = 51$. Using an 0.01 level of significance, test the claim that the memory enhancement drug tended to improve memory.

15. Two neighbors, Ben and Jim, have solar hot water heaters that are much the same except that Jim's uses water in the collector coils, whereas Ben's uses a light-weight oil (the oil does not freeze in winter) which heats water in coils of a secondary tank. Ben and Jim are wondering if there is a difference in the heat-collecting capacity of the two solar systems. Both collected data on the average length of time it took their systems to raise the temperature of 80 gal of water from 50 °F to 95 °F. Ben took a random sample of $n_1 = 41$ readings and found his average time was $\bar{x}_1 = 7.12$ hr with standard deviation $s_1 = 1.88$ hr. Jim took a random sample of $n_2 = 39$ readings and found his average time was $\bar{x}_2 = 7.73$ hr with standard deviation $s_2 = 1.62$ hr. Using a 5% level of significance, test the claim that Jim's solar system is slower than Ben's system.

SECTION 9.4   Tests Involving Small Samples

Sometimes it is not practical or even possible to obtain large samples. Cost, available time, and other factors may require us to work with *small samples*. For our purposes we will say a sample of size less than 30 is a small sample. In Section 8.2 we found that the t distribution is suitable for small samples in which the sampled population has a normal distribution. However, it can be shown that our applications of the t distribution are still appropriate for populations which are not normal, but possess a "mound-shaped" probability distribution. Because such distributions commonly occur in nature, the t distribution is very useful in applied work.

When working with small samples we will use essentially the same methods we used for large samples. *The main difference is that to find critical values, we use the t distribution instead of the z distribution.*

In Section 8.3 we saw

$$t = \frac{\bar{x} - \mu}{\dfrac{s}{\sqrt{n}}}$$

Let $\bar{x}_0$ be the critical value to be used for a test. We solve the last equation for $\bar{x}_0$ and obtain

$$\bar{x}_0 = \mu + t\frac{s}{\sqrt{n}} \tag{1}$$

The value of μ is obtained from the null hypothesis H_0, s is the sample standard deviation, and n is the sample size. The value of t is obtained from the level of significance, the degrees of freedom (*d.f.*), and Table 6 (*t* distribution) of Appendix I.

In Table 6 you see the column headings c, α', and α''. (Up to now we have ignored the α' and α'' headings.) In this table c represents the level of confidence; α' is the significance level of a one-tail test (α' is the area to the right of t or, equivalently, to the left of $-t$); and α'' is the significance level for a two-tail test (α'' is the area beyond $-t$ and t, so in fact $\alpha'' = 2\alpha'$). Figure 9-22 illustrates these different values.

FIGURE 9-22   Meaning of α' and α'' in Table 6 of Appendix I

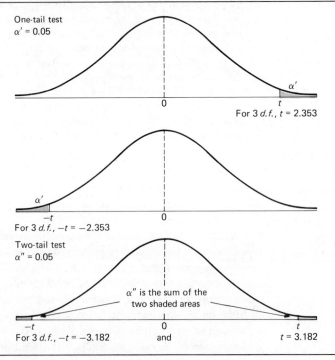

One-tail test
$\alpha' = 0.05$

For 3 *d.f.*, $t = 2.353$

For 3 *d.f.*, $-t = -2.353$

Two-tail test
$\alpha'' = 0.05$

α'' is the sum of the two shaded areas

For 3 *d.f.*, $-t = -3.182$ and $t = 3.182$

EXAMPLE 9

Use Table 6 to find a number t such that for a t curve with 10 degrees of freedom

a) 1% of the area under the curve lies to the right of t.

The desired t value is in the column under $\alpha' = 0.01$ and the row headed by 10 $d.f.$ The value of t is $t = 2.764$. (*See* Figure 9-23.)

b) 1% of the area under the curve lies beyond $-t$ and t.

In this case the desired t value is in the column under $\alpha'' = 0.01$ and the row headed by 10 $d.f.$ So $t = 3.169$ and $-t = -3.169$. (*See* Figure 9-24.)

FIGURE 9-23 t value for $\alpha' = 0.01$ and $d.f. = 10$

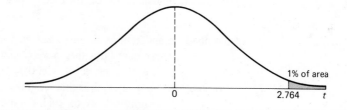

FIGURE 9-24 t values for $\alpha'' = 0.01$ and $d.f. = 10$

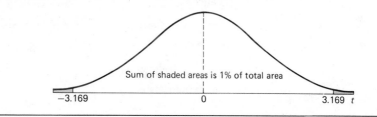

———— Exercise 11 ————

Use Table 6 of Appendix I to find a number t such that for a t curve with 7 degrees of freedom

a) 5% of the area under the curve lies to the left of $-t$ (*see* Figure 9-25).

FIGURE 9-25 t distribution with $d.f. = 7$

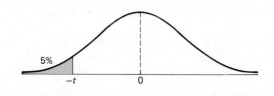

a) In this case we use $\alpha' = 0.05$. We read down the $d.f.$ column until we get to 7, and then we go to the right until we are in the column under $\alpha' = 0.05$. This brings us to $t = 1.895$. By the symmetry of the curve, $-t = -1.895$.

b) 5% of the area under the curve lies beyond $-t$ and t (*see* Figure 9-26).

FIGURE 9-26

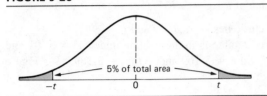

5% of total area

$-t$ 0 t

b) This problem calls for us to use $\alpha'' = 0.05$. We read down the $d.f.$ column until we reach $d.f. = 7$, then we go to the right until we are in the column headed by $\alpha'' = 0.05$. This brings us to $t = 2.365$. Therefore, 5% of the area under the curve (with $d.f. = 7$) lies beyond -2.365 and 2.365.

The next example demonstrates the method of testing a mean using small samples.

EXAMPLE   10

A company manufactures large rocket engines used to project satellites into space. The government buys the rockets, and the contract specifies that these engines are to use an average of 5,500 lb of rocket fuel the first 15 sec of operation. The company claims their engines fit specifications. To test the claim an inspector randomly selects six such engines from the warehouse. These six engines are fired 15 sec each and the fuel consumption for each engine is measured. For all six engines the mean fuel consumption is $\bar{x} = 5,690$ pounds and the standard deviation is $s = 250$ lb. Is the claim justified at the 5% level of significance?

a) We want to see if we can reject the hypothesis $\mu = 5,500$ lb, so the null hypothesis is

$$H_0: \mu = 5,500$$

A substantial difference either way from 5,500 lb could be important, so the alternate hypothesis is

$$H_1: \mu \neq 5,500$$

b) Since the alternate hypothesis is $H_1: \mu \neq 5,500$, we will use a two-tail test (*see* Table 9-3).

c) The critical values are obtained from formula (1)

$$\bar{x}_0 = \mu + t\frac{s}{\sqrt{n}}$$

with $\mu = 5,500$, $s = 250$, and $n = 6$. To get t we use Table 6. Since $n = 6$ and $d.f. = n - 1 = 6 - 1 = 5$, look down the $d.f.$ column until you get to 5, then go to the right until you are in the $\alpha'' = 0.05$ column (use α'' for two-tail tests). The corresponding value of t is $t = 2.571$. For a two-tail test we use both $-t = -2.571$ and $t = 2.571$. Therefore, the critical values are

$$\bar{x}_0 = \mu + t\frac{s}{\sqrt{n}} = 5,500 + (2.571)\left(\frac{250}{\sqrt{6}}\right) = 5,762.4$$

and

$$\bar{x}_0 = \mu - t\frac{s}{\sqrt{n}} = 5,500 - (2.571)\left(\frac{250}{\sqrt{6}}\right) = 5,237.6$$

The critical regions are shown in Figure 9-27.

FIGURE 9-27   Critical Regions for $\alpha'' = 0.05$

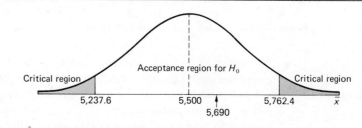

d) The experimental value of $\bar{x}$ is 5,690, which falls in the acceptance region of H_0. Therefore, we do not reject H_0. The data do not present sufficient evidence to indicate that the average fuel consumption for the first 15 sec of operation is different from $\mu = 5,500$.

───────────── Exercise 12 ─────────────

Suppose in Example 10 a random sample of eight engines was used and the mean fuel consumption was $\bar{x} = 5,880$ lb (for the first 15 sec). Again the standard deviation was 250 lb. Use a 1% level of significance to test the claim that the average fuel consumption *exceeds* 5,500 lb.

a) If we use H_0: $\mu = 5,500$, what should we use for H_1?

   H_1: $\mu < 5,500$ H_1: $\mu \neq 5,500$

   H_1: $\mu > 5,500$

a) H_1: $\mu > 5,500$ because we want to test the claim that the average fuel consumption *exceeds* 5,500 lb.

b) Should we use a left-, right-, or two-tail test? (*Hint:* Look at H_1 and Table 9-3.)

b) Right-tail test.

c) In the formula for the critical value, what should we use for μ? for s? for n? for t? Compute the critical value and draw a diagram to indicate the critical region.

c) $\mu = 5,500$; $s = 250$; $n = 8$

   Since $n = 8$, the $d.f. = 8 - 1 = 7$. Table 6 gives $t = 2.998$ with $d.f. = 7$ and $\alpha = 0.01$. We obtain the critical value from formula (1)

   $$\bar{x}_0 = \mu + t\frac{s}{\sqrt{n}} = 5,500 + (2.998)\left(\frac{250}{\sqrt{8}}\right) \approx 5,765$$

   Figure 9-28 shows the critical region.

FIGURE 9-28   Critical Region for $\alpha' = 0.01$

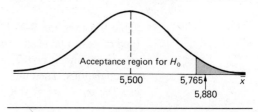

d) Shall we accept or reject H_0?

d) The experimental value $\bar{x} = 5{,}880$ is in the critical region, so we reject H_0 and conclude that the average fuel consumption exceeds 5,500 lb during the first 15 seconds of operation.

Statistical methods involving small samples and the difference between two means are much like the methods we studied in Section 9.3.

Independent random samples of size n_1 and n_2, respectively, are drawn from two populations which possess means μ_1 and μ_2. Again *we assume the parent populations have normal distributions, and we also assume the standard deviations σ_1 and σ_2 for the two populations are equal.* The condition $\sigma_1 = \sigma_2$ may seem quite restrictive. However, in a great many practical applications this condition is satisfied. Furthermore, our methods still apply even if the standard deviations are known to be only approximately equal.

Suppose we draw two independent random samples, one from the x_1 population and one from the x_2 population. Say the sample from the x_1 population is of size n_1 and has sample standard deviation s_1. Likewise for the x_2 population, the sample size is n_2 and the sample standard deviation is s_2. We estimate the common standard deviation for the two populations by using a pooled variance of the s_1^2 and s_2^2 values. The best estimate of the common variance of the x_1 and x_2 populations is given by the formula

$$s^2 = \frac{(n_1 - 1)s_1^2 + (n_2 - 1)s_2^2}{n_1 + n_2 - 2} \tag{2}$$

The best estimate of the common standard deviation is then

$$s = \sqrt{\frac{(n_1 - 1)s_1^2 + (n_2 - 1)s_2^2}{n_1 + n_2 - 2}}$$

The null hypothesis is usually set up to see if it can be rejected. When testing the difference of means, we first set up the hypothesis H_0 that there is no difference. That is, we take the null hypothesis to be

$$H_0\!: \mu_1 - \mu_2 = 0$$

or, equivalently,

$$H_0\!: \mu_1 = \mu_2$$

Table 9-5 (Section 9-3) lists the possibilities for the alternate hypothesis and the type of test to be used with the alternate hypothesis.

Let $\bar{x}_1$ and $\bar{x}_2$ be the sample means for our two random samples from the x_1 and x_2 populations. Then under the assumption that $\sigma_1 = \sigma_2$ and under the null hypothesis $H_0\!: \mu_1 - \mu_2 = 0$ it is possible to show that

$$t = \frac{\bar{x}_1 - \bar{x}_2}{s\sqrt{\dfrac{1}{n_1} + \dfrac{1}{n_2}}} \tag{4}$$

has a t distribution with degrees of freedom

$$d.f. = n_1 + n_2 - 2$$

Critical values and critical regions are obtained by solving (4) for $\bar{x}_1 - \bar{x}_2$. The critical values are given by

$$\text{critical values} = ts\sqrt{\frac{1}{n_1} + \frac{1}{n_2}} \tag{5}$$

where n_1 and n_2 are the sample sizes used, s is obtained from formula (3), and t is obtained from Table 6 by using the degrees of freedom $d.f. = n_1 + n_2 - 2$ and the level of significance.

Example 11 demonstrates the method of testing a difference of means using small samples.

EXAMPLE 11

Two competing headache remedies claim to give fast-acting relief. An experiment was performed to compare the mean lengths of time required for bodily absorption of brand A and brand B headache remedies.

Twelve people were randomly selected and given an oral dosage of brand A. Another 12 were randomly selected and given an equal dosage of brand B. The length of time in minutes for the drugs to reach a specified level in the blood was recorded. The means, standard deviations, and sizes of the two samples follow:

$$Brand\ A:\quad \bar{x}_1 = 20.1; \quad s_1 = 8.7; \quad n_1 = 12$$
$$Brand\ B:\quad \bar{x}_2 = 18.9; \quad s_2 = 7.5; \quad n_2 = 12$$

Past experience with the drug composition of the two remedies permits researchers to assume the standard deviations of the two time distributions are approximately equal. Let us use a 5% level of significance to test the claim that there is no difference in the mean time required for bodily absorption.

a) The null hypothesis is

$$H_0: \mu_1 = \mu_2$$

Since we have no prior knowledge about which brand is faster, the alternate hypothesis will be simply

$$H_1: \mu_1 \neq \mu_2$$

b) Since the alternate hypothesis is $H_1: \mu_1 \neq \mu_2$, we use a two-tail test.

c) From formula (5) we know that the critical values for a two-tail test are

$$\text{critical values} = \pm ts \sqrt{\frac{1}{n_1} + \frac{1}{n_2}}$$

In our example, $n_1 = n_2 = 12$ and formula (3) gives s to be

$$s = \sqrt{\frac{(n_1 - 1)s_1^2 + (n_2 - 1)s_2^2}{n_1 + n_2 - 2}}$$

$$= \sqrt{\frac{(12 - 1)(8.7)^2 + (12 - 1)(7.5)^2}{12 + 12 - 2}}$$

$$= \sqrt{\frac{(11)(75.69) + (11)(56.25)}{22}}$$

$$= \sqrt{\frac{832.59 + 618.75}{22}}$$

$$= \sqrt{65.97}$$

$$\approx 8.12$$

Table 6 with $d.f. = 12 + 12 - 2 = 22$ and $\alpha'' = 0.05$ gives $t = 2.074$. Therefore, the *critical values* are (*see* Figure 9-29)

$$\pm ts \sqrt{\frac{1}{n_1} + \frac{1}{n_2}} = \pm (2.074)(8.12) \sqrt{\frac{1}{12} + \frac{1}{12}}$$

$$\approx \pm (2.074)(8.12)(0.408)$$

$$\approx \pm 6.9$$

d) Since $\bar{x}_1 = 20.1$ and $\bar{x}_2 = 18.9$, their difference is

$$\bar{x}_1 - \bar{x}_2 = 20.1 - 18.9 = 1.2$$

This difference is in the acceptance region. We conclude that the data do not contain sufficient evidence to reject H_0. Therefore, at the 5% level of significance we conclude that there is no difference in the mean times.

FIGURE 9-29 Critical Region for $\alpha'' = 0.05$

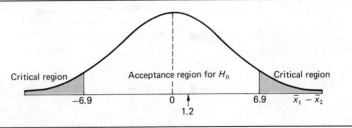

_____ Exercise 13 _____

Suppose the experiment with headache remedies (Example 11) yielded sample means, standard deviations, and sample sizes as follows:

$$\text{Brand A: } \bar{x}_1 = 20.1; \quad s_1 = 8.7; \quad n_1 = 12$$

$$\text{Brand B: } \bar{x}_2 = 11.2; \quad s_2 = 7.5; \quad n_2 = 12$$

Brand B claims to be faster. Is this claim justified at the 1% level of significance? (Use the following steps to obtain the answer.)

a) Choose H_0 from the following:

$$H_0: \mu_1 = \mu_2 \qquad H_0: \mu_1 \neq \mu_2$$

$$H_0: \mu_1 < \mu_2 \qquad H_0: \mu_1 > \mu_2$$

b) What would we use for H_1?

$$H_1: \mu_1 = \mu_2 \qquad H_1: \mu_1 < \mu_2$$

$$H_1: \mu_1 > \mu_2 \qquad H_1: \mu_1 \neq \mu_2$$

c) Should we use a right-, left-, or two-tail test? (*Hint:* Consider H_1 and see Table 9-5 of Section 9.3.)

d) In formula (5) for the critical value, what is the value of n_1? n_2? t? s? (*Hint: See* part (c) of Example 11.) Compute the critical value(s) and draw a diagram showing the critical regions and the acceptance region.

a) $H_0: \mu_1 = \mu_2$

b) $H_1: \mu_1 > \mu_2$. This says the mean time for brand B is less than the mean time for brand A.

c) Right-tail test.

d) $n_1 = 12$; $n_2 = 12$; $s = 8.12$ (This s value is the same as the s value computed in Example 11, part (c), since the n_1, n_2, s_1, and s_2 values of this exercise are the same as those of Example 11.)

To compute t we use $d.f. = n_1 + n_2 - 2 = 12 + 12 - 2 = 22$ and $\alpha' = 0.01$ (we use α' since this is a one-tail test). Table 6 gives $t = 2.508$. Therefore, the critical value is

$$\text{critical value} = ts \sqrt{\frac{1}{n_1} + \frac{1}{n_2}}$$

$$= (2.508)(8.12) \sqrt{\frac{1}{12} + \frac{1}{12}}$$

$$\approx (2.508)(8.12)(0.408)$$

$$\approx 8.3$$

Figure 9-30 shows the critical region.

FIGURE 9-30   Critical Region for $\alpha' = 0.01$

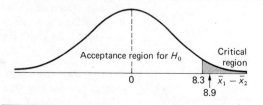

e) At the 1% level should we accept or reject H_0?

e) $\bar{x}_1 = 20.1$ and $\bar{x}_2 = 11.2$

Therefore, $\bar{x}_1 - \bar{x}_2 = 20.1 - 11.2 = 8.9$.

This value is in the critical region. We reject H_0 at the 1% level and conclude that brand B works faster.

Section Problems 9.4

1. Use Table 6 of Appendix I with 8 *d.f.* to find a number *t* such that 5% of the area under the corresponding Student's *t* distribution lies to the left of *t*.

2. Use Table 6 of Appendix I with 12 *d.f.* to find a number *t* such that 1% of the area under the corresponding Student's *t* distribution lies to the right of *t*.

3. Use Table 6 of Appendix I with 23 *d.f.* to find a number *t* such that 1% of the area under the corresponding Student's *t* distribution lies outside the interval from $-t$ to *t*.

4. Use Table 6 of Appendix I with 17 *d.f.* to find a number *t* such that 5% of the area under the corresponding Student's *t* distribution lies to the left of *t*.

5. Use Table 6 of Appendix I with 11 *d.f.* to find a number *t* such that 5% of the area under the corresponding Student's *t* distribution lies outside the interval from $-t$ to *t*.

6. Use Table 6 of Appendix I with 28 *d.f.* to find a number *t* such that 1% of the area under the corresponding Student's *t* distribution lies to the right of *t*.

For each of the problems do the following:

a) State the null and alternate hypotheses.
b) Find the critical value or values.
c) Sketch the critical region and acceptance region.
d) On the sketch of part (c) show the location of $\bar{x}$ or $\bar{x}_1 - \bar{x}_2$ according to the type of problem and information given.
e) Decide whether you should accept or reject the null hypothesis.

7. The Association of Greek Students (those belonging to a fraternity or sorority) claim that their grade point average is higher than the 2.68 grade average for the entire student body. To test this claim a random sample of 15 Greek students was used. Their grade point average was $\bar{x} = 2.81$ with sample standard deviation $s = 0.92$. Is the claim justified at the 5% level of significance?

8. The owner of a private fishing pond says that the average length of fish in the pond is 15 in. A sample of 23 fish are caught from various locations in the pond. The average length of these fish is $\bar{x} = 10.7$ in. with sample standard deviation $s = 4.8$ in. Do the data support the claim that the average fish length is less than 15 in.? (Use a 1% level of significance.)

9. The Dog Days Lawn Service advertises that it will completely maintain your lawn at an average cost per customer of $35.00 a month. A random sample of 18 Dog Days cus-

tomers shows the average cost to be $\bar{x}$ = $39.50 a month with sample standard deviation s = $8.10. Do the data support the claim that the average cost per month is more than $35.00? (Use a 5% level of significance.)

10. At a certain manufacturing plant, the general manager in charge of production is always promoted from the position of assistant manager. You are now assistant manager and your boss says the average waiting time for promotion is 2.5 yr. A random sample of six previous assistant managers had to wait the following times (in years) for promotion:

    3.5   2.75   2.5   2.25   2.75   3.0

    a) Use a calculator to verify that the sample mean is 2.79 yr and the sample standard deviation is 0.43 yr.
    b) Using a 5% level of significance, test the claim that the waiting period is longer than 2.5 yr.

11. An overnight package delivery service has a promotional discount rate in effect this week only. For several years the mean weight of a package delivered by this company has been 10.7 oz. However, a random sample of 12 packages mailed this week gave the following weights in ounces:

    12.1   15.3   9.5   10.5   14.2   8.8
    10.6   11.4   13.7   15.0   9.5   11.1

    a) Use a calculator to verify that the sample mean is 11.81 oz and the sample standard deviation is 2.24 oz.
    b) Use a 1% level of significance to test the claim that the packages are averaging more than 10.7 oz during the discount week.

12. The entrance onto a major bridge in New York City was engineered to accommodate a mean of 3,800 vehicles each hour. However, a random sample of nine observations gave the following counts per hour.

    4,115   3,960   3,680   4,075   4,253
    4,180   3,997   3,759   4,178

    a) Use a calculator to verify that the sample mean is 4,021.9 vehicles and the sample standard deviation is 195.2 vehicles.
    b) Use a 5% level of significance to test the claim that the mean number of vehicles per hour is more than 3,800.

13. One of the sports car magazines has a very prestigious and wealthy group of readers. The magazine editor says the average annual income of his subscribers is $95,000. Because advertisements in this magazine target the rich, the magazine charges a lot for each ad. A potential advertiser asked the magazine for a random sample of ten names of readers. After checking out these names through credit reference bureaus, it was found that the ten subscribers had the following annual incomes (in thousands of dollars)

    73   81   65   42   57
    110   93   52   70   96

a) Use a calculator to verify that the sample mean is 73.9 and the sample standard deviation is 21.3.

b) Use a 1% level of significance to test the claim that the mean annual income of subscribers is less than $95,000.

14. The airport has made improvements in its baggage claim system. It used to take passengers a mean of 20 min to get their luggage. A random sample of 14 passengers received their luggage in the following times (in minutes):

17	13	19	22	15	20	18
15	21	20	18	22	21	19

a) Use a calculator to verify that the sample mean is 18.6 min and the sample standard deviation is 2.8 min.

b) Use a 1% level of significance to test if the mean time required for passengers to receive their baggage is different (either shorter or longer) from 20 min.

15. The manufacturer of a new drug claims that the product causes a faster white cell buildup in people who have an internal infection than the drug on the market. The drug currently used has an average buildup factor of 9.2. When the new drug was used on four hospital patients, the buildup factors were 10.9, 9.5, 8.7, and 11.2. Use a 1% level of significance to test the manufacturer's claim.

16. Irv is doing a psychology study of susceptibility of people to perceptual illusions. He claims airplane pilots are less susceptible than the general population. To test the claim a random sample of 11 airplane pilots judged the length of a line in a large illusionary figure. For each judgment the magnitude of the deviation from the actual length was recorded. The mean of these magnitudes was $\bar{x}_1 = 76$ mm with sample standard deviation $s_1 = 9$ mm. A random sample of 15 nonpilots were then asked to judge the length, and the magnitudes of their deviations were recorded. For the nonpilots the mean of the magnitudes was $\bar{x}_2 = 82$ mm with standard deviation $s_2 = 7$ mm. Do the data support the claim at the 1% level of significance?

17. Linda is an Alpine botanist who thinks she has discovered a new species of wildflower. The only morphological difference from that of a known species is the petal length. A random sample of 12 flowers of the known species has a mean petal length $\bar{x}_1 = 9.3$ mm and sample standard deviation $s_1 = 1.1$ mm. A random sample of 15 "new species" flowers has mean petal length $\bar{x}_2 = 11.9$ mm with sample standard deviation $s_2 = 1.9$ mm. Linda claims that the mean petal lengths of the two types of flowers are different. Is this claim justified at the 1% level of significance?

18. To test the claim that fluoride helps prevent cavities, 16 children had their teeth cleaned and coated with a fluoride emulsion. Another 16 children had their teeth cleaned, but received no fluoride coating. The children without fluoride coating had a mean number of cavities $\bar{x}_1 = 2.4$ with sample standard deviation $s_1 = 0.8$. For the children with the fluoride coating the mean was $\bar{x}_2 = 1.7$ with $s_2 = 1.0$. Is the claim justified at the 5% level of significance?

19. A large electric power plant uses ocean water for its cooling system (and returns the water to the ocean). A random sample of ten temperature readings showed the changes in water temperature to be (in °F):

6	8	4	5	10	3	9	11	7	9

a) For these temperatures verify that the mean is $\bar{x}_1 = 7.2$ °F and the sample standard deviation is $s_1 = 2.7$ °F.

A new generator was added to the plant, and environmentalists fear that the average change in water temperature has increased. A random sample of 12 temperature readings showed the changes in water temperatures now to be (in °F):

9	11	15	12	7	12
10	13	8	11	14	8

b) For these temperatures verify that the sample mean is $\bar{x}_2 = 10.8$ °F with sample standard deviation $s_2 = 2.5$ °F.

c) Use a 5% level of significance to test the claim that the mean change in water temperature has increased since the addition of the new generator.

20. Paramedics in a large city say they need more staff at night because they get more emergency calls then. A random sample of seven days showed that the paramedics received the following numbers of emergency calls during the day:

65	54	81	67	75	83	79

a) Verify that the sample mean is $\bar{x}_1 = 72$ calls with sample standard deviation $s_1 = 10.47$ calls.

A random sample of seven nights showed that the paramedics received the following numbers of emergency calls on those nights.

72	81	85	80	88	82	56

b) Verify that the sample mean for night calls is $\bar{x}_2 = 77.7$ with sample standard deviation $s_2 = 10.78$.

c) Use a 1% level of significance to test the claim that there is a difference (either more or fewer) between the mean number of calls during the day and the mean number of calls during the night.

21. The fire department tested two types of fire extinguishers on gasoline fires. They set ten gallons of gasoline ablaze and then timed how long it took to extinguish the blaze using one extinguisher. This process was repeated six times using a random sample of type 1 extinguishers. The lengths of time (in minutes) to put out the blaze each time were:

10	16	15	12	9	11

a) Verify that the sample mean of these data is $\bar{x}_1 = 12.2$ min and the standard deviation is $s_1 = 2.8$ min.

Six more blazes of the same type were started. This time a random sample of type 2 extinguishers were used. The lengths of time (in minutes) to put out the blaze each time were:

18	17	14	16	10	15

b) Verify that the sample mean of these data is $\bar{x}_2 = 15.0$ min and the standard deviation is $s_2 = 2.8$ min.

c) Use a 1% level of significance to test the claim that it takes longer to put out the gasoline fire using a type 2 extinguisher.

SECTION 9.5   Tests Involving Paired Differences (Dependent Samples)

Many statistical applications use *paired data* samples to draw conclusions about the difference between two population means. Data *pairs* occur very naturally in "before and after" situations, where the *same* object or item is measured both before and after a treatment. Applied problems in social science, natural science, and business administration frequently involve a study of matching pairs. Psychological studies of identical twins; biological studies of plant growth on plots of land matched for soil type, moisture, and sun; and business studies on sales of matched inventories are examples of paired data studies.

When working with paired data, it is very important to have a definite and uniform method of creating data pairs that clearly utilizes a natural matching of characteristics. The next example and exercise demonstrate this feature.

EXAMPLE   12

In Section 9.3, we discussed the claim of a shoe manufacturer that among the general population of adults in the United States, the average length of the left foot is larger than that of the right. To compare the average length of the left foot with that of the right, we can take a random sample of 15 U.S. adults and measure the length of the left foot and then the length of the right foot for each person in the sample. Is there a natural way of pairing the measurements? How many pairs will we have?

In this case, we can pair each left foot measurement with the same person's right foot measurement. The person serves as the "matching link" between the two distributions. We will have 15 pairs of measurements.

Exercise 14

A psychologist has developed a series of exercises called the Instrumental Enrichment (IE) program, which he claims to be useful in overcoming cognitive deficiencies in mentally retarded children. To test the program, extensive statistical tests are being conducted. In one simple test, a random sample of ten-year-old students with IQ scores below 80 was selected. An IQ test was given to these students before they spent two years in an IE program, and an IQ test was given to the same students after the program.

a) On what basis can you pair the IQ scores?

a) Take the "before and after" IQ scores of each individual student.

b) If there were 20 students in the sample, how many data pairs would you have?

b) Twenty data pairs. Note that there would be 40 IQ scores, but only 20 pairs.

To compare two populations we cannot always employ paired data tests, but when we can, what are the advantages? In Sections 9.3 and 9.4 we used independent samples to test differences between population means. With independent samples we may inadvertently introduce either extraneous or uncontrollable factors into our sample measurements. Using matched or paired data can often reduce this danger, simply because the matched or paired data have essentially the *same* characteristics except for the *one* characteristic that is being measured. Furthermore, it can be shown that pairing data has the theoretical effect of reducing measurement variability (i.e., variance), which increases the accuracy of statistical conclusions.

When we wish to compare the means of two samples, the first item to be determined is whether the sampling distributions are independent (as in Sections 9.3 and 9.4), or if there is a natural pairing between the data in the two samples. Again, data pairs are created from "before and after" situations, or from matching data by using studies of the same object, or by a process of taking measurements of closely matched items.

If data come from independent samples, we use the techniques in Section 9.3 or 9.4 to test the difference of means of the two samples. Recall that in such a case, we calculate the means of both samples and look at the distribution $\bar{x}_1 - \bar{x}_2$. When testing *paired* data, we take the difference d of the data pairs *first* and look at the mean difference $\bar{d}$. We use a t test on $\bar{d}$. Theorem 9.2 provides the basis for our work with paired data.

- **Theorem 9.2:**   Consider a random sample of n data pairs. Suppose the differences d between the first and second members of each data pair are (approximately) normally distributed with population mean μ_d. Then the t values

$$t = \frac{\bar{d} - \mu_d}{s_d / \sqrt{n}}$$

where $\bar{d}$ is the sample mean of the d values and

$$s_d = \sqrt{\frac{\Sigma(d - \bar{d})^2}{n - 1}}$$

is the sample standard deviation of the d values, follow a Student's t-distribution with degrees of freedom $d.f. = n - 1$.

When testing the difference $\bar{d}$ of paired data values, the null hypothesis is that there is no difference among the pairs. That is, the mean of the differences μ_d is zero.

$$H_0: \mu_d = 0$$

The alternate hypothesis depends on the problem and can be

$$H_1: \mu_d < 0 \text{ (left tail)} \quad \text{or} \quad H_1: \mu_d > 0 \text{ (right-tail)} \quad \text{or} \quad H_1: \mu_d \neq 0 \text{ (two-tail)}$$

To find the critical value $\bar{d}_0$ of the $\bar{d}$ distribution, we use Theorem 9.2 and the null hypothesis.

$$t = \frac{\bar{d} - \mu_d}{s_d/\sqrt{n}} \quad \text{(from Theorem 9.2)}$$

Solve for $\bar{d}$:
$$\bar{d} = t\left(\frac{s_d}{\sqrt{n}}\right) + \mu_d$$

$$\bar{d}_0 = t\left(\frac{s_d}{\sqrt{n}}\right)$$

The critical value $\bar{d}_0$ is computed under the assumption that the null hypothesis $H_0: \mu_d = 0$ is true. Recall that the critical values are always computed under the (at least temporary) assumption that H_0 is valid.

EXAMPLE 13

A team of heart surgeons at Saint Ann's Hospital knows that many patients who undergo corrective heart surgery have a dangerous buildup of anxiety before their scheduled operations. The staff psychiatrist at the hospital has started a new counseling program, intended to reduce this anxiety. A test of anxiety is given to patients who know they must undergo heart surgery. Then each patient participates in a series of counseling sessions with the staff psychiatrist. At the end of the counseling sessions, each patient is retested to determine anxiety level. The first three columns of Table 9-7 indicate the results for a random sample of nine patients. (The other three columns will be explained later.) Higher scores mean higher levels of anxiety.

TABLE 9-7 Anxiety Levels of Heart Patients

Patient	B Score before Counseling	A Score after Counseling	$d = B - A$ Difference	$d - \bar{d}$	$(d - \bar{d})^2$
Jan	121	76	45	11.67	136.19
Tom	93	93	0	−33.33	1,110.89
Diane	105	64	41	7.67	58.83
Barbara	115	117	− 2	−35.33	1,248.21
Mike	130	82	48	14.67	215.21
Bill	98	80	18	−15.33	235.01
Frank	142	79	63	29.67	880.31
Carol	118	67	51	17.67	312.23
Alice	125	89	36	2.67	7.13
			$\Sigma d = 300$		$\Sigma(d - \bar{d})^2 = 4{,}204.01$

From the given data, can we conclude that the counseling sessions reduce anxiety? Use a 0.01 level of significance.

Before we answer this question, let us notice two important points: (1) we have a *random sample* of nine patients, and (2) we have a *pair* of measurements taken on the same patient before and after counseling sessions.

In the solution of paired difference problems we will use the following notation:

n = number of pairs in the sample

d = difference between first and second measurement of a pair

$\bar{d} = \dfrac{\Sigma d}{n}$ the sample mean of d values

$s_d = \sqrt{\dfrac{\Sigma(d - \bar{d})^2}{n - 1}}$ the sample standard deviation of d values

In our problem the sample size is $n = 9$ pairs (i.e., patients), and the d values are found in the fourth column of Table 9-7. The sum of the fourth column in Table 9-7 gives $d = 300$. Therefore,

$$\bar{d} = \frac{\Sigma d}{n} = \frac{300}{9} = 33.33$$

Furthermore, in the last column we have $\Sigma(d - \bar{d})^2 = 4{,}204.01$, so

$$s_d = \sqrt{\frac{\Sigma(d - \bar{d})^2}{n - 1}} = \sqrt{\frac{4{,}204.01}{9 - 1}} = 22.92$$

In our problem we want to test the claim that the counseling sessions reduce anxiety. This means that the differences $d = B - A$ should tend to be positive, and the population mean of difference values μ_d should also be positive. Therefore, we have

$$H_0\text{: } \mu_d = 0$$

$$H_1\text{: } \mu_d > 0$$

By Theorem 9.2 and the null hypothesis

$$\bar{d}_0 = t\left(\frac{s_d}{\sqrt{n}}\right)$$

Since the alternate hypothesis is H_1: $\mu_d > 0$, we have a right-tail test.

$$s_d = 22.92$$
$$n = 9$$
$$d.f. = n - 1 = 9 - 1 = 8$$
$$t = 2.896 \quad \text{for a right-tail test with level of significance } 0.01$$
$$(\alpha' = 0.01 \text{ in Table 6 of Appendix I})$$

Then

$$\bar{d}_0 = 2.896\left(\frac{22.92}{\sqrt{9}}\right)$$

$$\bar{d}_0 = 22.13 \quad \text{(critical value)}$$

FIGURE 9-31 Critical Region and Acceptance Region

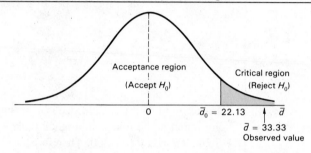

The observed value of $\bar{d}$ is 33.33, which is in the critical region shown in Figure 9-31. Therefore, we reject H_0: $\mu_d = 0$, and conclude that $\mu_d > 0$. The counseling sessions do reduce anxiety (at the 0.01 level of significance).

The problem we have just solved was a paired difference problem of the "before and after" type. The next exercise demonstrates a paired difference problem of the "matched pair" type.

_____ Exercise 15 _____

In a program intended to promote early development of preschool children's reading ability, the participating children spend two hours each day in a room well supplied with "educational" toys such as alphabet blocks, puzzles, ABC readers, lettered coloring books, and so forth. To evaluate the effectiveness of this experience on reading ability, it was decided to compare the children in the program to a control group of children who spent two hours a day in a "noneducational" toy room. It was anticipated that IQ differences and home environment might be uncontrollable factors unless identical twins could be used. Therefore, five pairs of identical twin boys of preschool age were randomly selected. From each pair, one member was randomly selected to participate in the experimental (i.e., educational toy room) group and the other in the control (i.e., noneducational toy room) group. For each twin, the data item recorded is the age in months when the child began reading at the primary level (Table 9-8).

TABLE 9-8 Reading Ages for Identical Twins

Twin Pair	Experimental Group B = Reading Age in Months	Control Group A = Reading Age in Months	Difference $d = B - A$	$d - \bar{d}$	$(d - \bar{d})^2$
1	58	60			
2	61	64			
3	52	52			
4	60	65			
5	71	75			

a) Compute the entries in the $d = B - A$ column.

b) Compute $\bar{d}$ and the entries in the $d - \bar{d}$ column.

c) Compute the entries in the $(d - \bar{d})^2$ column and compute $\Sigma(d - \bar{d})^2$.

a-c)

Pair	$d = B - A$	$d - \bar{d}$	$(d - \bar{d})^2$
1	−2	0.80	0.64
2	−3	−0.20	0.04
3	0	−2.8	7.84
4	−5	−2.2	4.84
5	−4	−1.2	1.44

$$\Sigma d = -14; \, \bar{d} = \frac{-14}{5} = -2.8$$

$$\Sigma(d - \bar{d})^2 = 14.80$$

d) Compute s_d.

d) $s_d = \sqrt{\dfrac{14.80}{5 - 1}} = 1.92$

e) What is the null hypothesis?

e) $H_0: \mu_d = 0$

f) To test the claim that the experimental group learned to read at an earlier age, what would you use for the alternate hypothesis?

f) $H_1: \mu_d < 0$

g) Using $\alpha = 0.05$, what is your critical value $\bar{d}_0$?

g) $\bar{d}_0 = t\left(\dfrac{s_d}{\sqrt{n}}\right) = -2.132\left(\dfrac{1.92}{\sqrt{5}}\right) = -1.83$

Note: d.f. $= n - 1 = 5 - 1 = 4$, so Table 6 of Appendix I gives $t = -2.132$ for a left-tail test with $\alpha' = 0.05$.

h) Sketch the acceptance and critical regions for H_0. When $\alpha = 0.05$, would we accept or reject H_0?

h) Figure 9-32 shows the critical region.

FIGURE 9-32 Critical Region

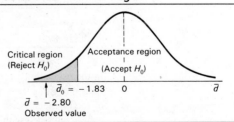

We reject H_0.

i) What conclusion can we draw regarding the experience in the educational toy room?

i) Since we reject H_0 at the 0.05 level, we conclude that the group using the educational toy room learned to read at an earlier age.

Section Problems 9.5

For each problem do the following:

a) State the null and alternate hypotheses.
b) Find the critical value or values.
c) Sketch the critical region and acceptance region.
d) On the sketch of part (c) show the location of the sample value of $\bar{d}$.
e) Decide whether you should accept or reject the null hypothesis.

1. The western United States has a number of four-lane interstate highways that cut through long tracts of wilderness. To prevent car accidents with wild animals, the highways are bordered on both sides with 12-foot-high woven wire fences. Although the fences prevent accidents, they also disturb the winter migration pattern of many animals. To compensate for this disturbance, the highways have frequent wilderness underpasses designed for exclusive use by deer, elk, and other animals.

    In Colorado there is a large group of deer that spend their summer months in a region on one side of a highway, and survive the winter months in a lower region on the other side. To determine if the highway has disturbed deer migration to the winter feeding area, the following data were gathered on a random sample of ten wilderness districts in the winter feeding area. Row B represents the average January deer count in a five-year period before the highway was built, and row A represents the average January deer count for a five-year period after the highway was built. The highway department claims that the January population has not changed. Test this against the claim that the January population has dropped. Use a 1% level of significance. Units used in Table 9-9 are hundreds of deer.

TABLE 9-9 Deer Count in Winter Feeding Area (Hundreds)

Wilderness District	1	2	3	4	5	6	7	8	9	10
B: Before highway	10.3	7.2	12.9	5.8	17.4	9.9	20.5	16.2	18.9	11.6
A: After highway	9.1	8.4	11.6	5.8	16.1	10.2	22.7	14.3	21.8	9.9

2. In the psychology lab students are studying what they call Rat Olympics. First, a rat is trained to run a maze with ten major turns and is rewarded only at the end with 20 grams of food pellets. Later, the rat is permitted to run the maze again but this time is immediately rewarded with two grams of food for each correct turn. No reward is given for a wrong turn. Six rats were randomly chosen to participate in the Rat Olympics. In Table 9-10, row B gives the number of seconds for a trained rat to run the maze correctly and

get a single reward of 20 grams of food at the end. Row A gives the number of seconds for the rat to correctly run the maze while being rewarded immediately for each correct turn. Using a 5% level of significance, test the claim that immediate rewards tend to shorten the time needed to correctly run the maze.

TABLE 9-10 Time to Run Maze (seconds)

Rat	1	2	3	4	5	6
B: Time for reward at end	35.7	21.2	53.9	44.0	39.5	47.1
A: Time for immediate reward	30.2	19.9	46.1	44.7	31.0	42.6

3. Many mothers say they can recognize the cries of their own babies and that they can distinguish between a cry of pain and a hunger cry. A psychologist studied this phenomenon using the following experiment. A random sample of mothers listened to a tape-recorded set of five cries from different babies, one of which was their own. They had to decide which was their baby. Each mother heard 20 such sets, in which ten were cries of hungry babies and ten were cries produced by a slight pin prick on a foot. The results are shown below in Table 9-11, where row B is the correct number of identifications (out of ten) for a hunger cry, and row A is the correct number of identifications (out of ten) for a pain cry. The psychologist claims that the mothers are more successful in picking out their own babies when a hunger cry is involved, since the mothers have more experience with that situation. Test this claim at the 1% level of significance.

TABLE 9-11 Correct Number of Identifications

Mother	1	2	3	4	5	6	7	8
B: Hunger cry	6	6	6	5	3	7	9	2
A: Pain cry	5	4	8	3	2	6	6	5

4. A systems specialist has studied the work flow of clerks all doing the same inventory work. Based on this study, she designed a new work-flow layout for the inventory system. To compare average production for the old and new methods, a random sample of six clerks was used. The average production rate (number of inventory items processed) for each clerk was measured both before and after the new system was introduced. The results are shown in Table 9-12. Test the claim that the new system speeds up the work rate (use $\alpha = 0.05$).

TABLE 9-12 Production Rate

Clerk	1	2	3	4	5	6
B: Rate of old system	110	100	97	85	117	101
A: Rate of new system	118	112	115	83	125	109

5. At Hinsdale College, students are employed by the college library to reshelve books. The students claim that a 15-minute coffee break every hour would relieve boredom and increase the average number of books reshelved per hour. To test the students' claim, the following data were gathered on a random sample of seven student employees. Over a period of one week, each student worked a normal schedule (three hours per day) without coffee breaks. Then, during the next week, the students were allowed coffee breaks. Unknown to the students, a member of the library staff secretly recorded the average number of books shelved per hour and computed the averages with and without the coffee break system. The results are given in Table 9-13. Test the claim that the students shelve more books when they have a coffee break. (Use 0.05 level of significance.)

TABLE 9-13 Number of Books Shelved

Student	1	2	3	4	5	6	7
B: Average w/o coffee break	75	63	82	53	79	96	73
A: Average with coffee break	90	72	95	50	85	110	89

6. The Hawaiian Planters Association is thinking of planting a new type of sugar cane which is supposed to yield pulp with higher sugar content. To test the new type of sugar cane against the one now in use, they chose an acre in eight different localities where soil and general climatic conditions vary. Each acre was divided into two equal parts so that the soil and general conditions were essentially identical for each half acre. The old type of sugar cane was planted in one of the halves, and the new cane was planted in the other half acre. After harvest, the average sugar content of the pulp was determined (Table 9-14). Use a 1% level of significance to test the claim that the new cane has higher sugar yield.

TABLE 9-14 Sugar Yield
(grams/liter of pulp)

Plot	1	2	3	4	5	6	7	8
B: Old cane	320	185	270	199	243	277	296	310
A: New cane	375	180	291	215	221	305	320	351

7. Fremont Fisheries of Maine specialize in commercial cod fishing. A food expert claims that a new packaging and fast-freeze process will improve the flavor of their product. To test this claim, Fremont Fisheries took a random sample of ten cod. Each fish was cut down the middle along the backbone from head to tail. One side was frozen and packaged by the old method; the other side was frozen and packaged by the new process. After uniform thawing and uniform cooking, a panel of taste experts rated each half on a score from 1 (poor) to 10 (excellent). The average scores given by the panel members were as shown in Table 9-15. Do the data support the claim that the new packaging and fast-freeze method improves flavor? (Use $\alpha = 0.01$.)

TABLE 9-15 Average Scores Given by Taste Experts

Cod	1	2	3	4	5	6	7	8	9	10
B: Old packaging method	3.5	7.5	5.2	9.4	8.1	4.4	6.8	7.7	9.2	8.6
A: New packaging method	3.1	9.8	6.4	9.7	9.2	3.1	7.4	8.3	9.7	9.4

8. To test the wearing quality of two brands of auto tires, one tire of each brand was placed on each of six test cars. The position was determined by using a random number table. The amount of wear (in thousandths of an inch) was determined for each car after six months of use (*see* Table 9-16). Do the data give sufficient evidence to conclude that the two tire brands show unequal wear? (Use $\alpha = 0.05$ level of significance.)

TABLE 9-16 Wear in Thousandths of an Inch

Car	1	2	3	4	5	6
Soapstone	132	71	90	37	93	107
Bigyear	140	74	110	36	105	119

9. An antismoking program has been using a motion picture film to depict the dangers of cigarette smoking to a person's health. To determine the effect of the film on attitudes of smokers about smoking, a random sample of seven cigarette smokers was selected. Each person filled out a questionnaire before and after viewing the film. (Higher scores in Table 9-17 indicate a more positive attitude about smoking.) Do the data indicate that cigarette smokers' attitudes about smoking become less positive after viewing the film? (Use $\alpha = 0.05$ level of significance.)

TABLE 9-17 Smoker Attitude Score

Person	1	2	3	4	5	6	7
B: Before film	31	27	25	36	34	29	38
A: After film	19	30	10	20	28	29	19

10. The college ski coach has discovered a new edge modification for skis used in the slalom races. It is thought that the new edges will give more control and reduce the time needed to run the race. To test the new edges, a random sample of five students ran the slalom twice, once with "old" edges and again with the "new" edges. The resulting times were (in seconds) as shown in Table 9-18. Do the data indicate a change in times (either faster or slower) between old or new ski edges? (Use $\alpha = 0.01$ level of significance.)

TABLE 9-18 Time in Seconds for the Slalom

Student	1	2	3	4	5
Old edges	35.7	38.6	32.5	40.7	33.2
New edges	34.1	39.0	31.0	38.9	30.8

11. A new law has been passed giving city police greater powers in apprehending suspected criminals. For six neighborhoods, the numbers of reported crimes one year before and one year after the new law are shown in Table 9-19.

TABLE 9-19 Reported Crimes

Neighborhood	1	2	3	4	5	6
B: Before new law	18	35	44	28	22	37
A: After new law	21	23	30	19	24	29

Use a 1% level of significance to test the claim that one year after the new law, the number of reported crimes has dropped.

12. During the national presidential campaign Professor Adams asked each of his political science students to do the following project. Take a random sample of your friends. Before the upcoming national TV debate, ask each friend to rate the Republican candidate on a scale from 0 (very negative) to 10 (very positive). Then, after your friends have watched the TV debate, have them rate the candidate again on the same scale. Mary did the project using a random sample of nine of her friends. The results are shown in Table 9-20.

TABLE 9-20 Candidate Rating Before and After TV Debate

Friend	1	2	3	4	5	6	7	8	9
B: Rating before debate	6	9	1	5	0	7	3	6	2
A: Rating after debate	8	7	1	7	3	8	5	4	3

Use a 5% level of significance to test the claim that the ratings before and after the debate were different (either more or less positive).

13. Late-night truck drivers sometimes take an over-the-counter nonprescription drug to keep them from falling asleep. The main ingredient is caffeine, but too much caffeine may not be good for a person's health. A random sample of eight truck drivers agreed to have their pulse rate (per minute) measured one half-hour before and one half-hour after taking such a drug. The results are shown in Table 9-21.

TABLE 9-21 Pulse Rate Before and After Drug

Driver	1	2	3	4	5	6	7	8
B: Before	68	75	110	96	72	80	73	67
A: After	68	83	110	94	71	85	70	69

Use a 1% level of significance to test the claim that pulse rate per minute was different (either higher or lower) after taking the drug.

14. Five members of the college track team in Denver (elevation 5,280 ft) went up to Leadville (elevation 10,152 ft) for a track meet. The times in minutes for these team members to run two miles at each location are shown in Table 9-22.

TABLE 9-22 Time to Run Two Miles at Different Elevations

Team Member	1	2	3	4	5
B: Time in Denver	10.3	9.8	11.4	9.7	9.2
A: Time in Leadville	11.5	10.6	11.0	10.8	10.1

Use a 5% level of significance to test the claim that the times were longer at the higher elevation.

15. The manager of a sporting goods store offered a bonus commission to his salespeople when they sold more goods. A new manager dropped the bonus system. For a random sample of six salespeople the monthly sales (in thousands of dollars) are shown with and without the bonus system in Table 9-23.

TABLE 9-23 Monthly Sales in Thousands of Dollars

Salesperson	1	2	3	4	5	6
B: With bonus	2.4	3.1	5.7	4.4	5.2	5.6
A: Without bonus	2.8	2.7	5.9	3.5	4.6	4.5

Use a 1% level of significance to test the claim that the monthly sales dropped when the bonus system was discontinued.

16. The computation formula for computing the standard deviation of data values x (see Section 3.3) can be used to compute the standard deviation of d values. We simply use d in place of x.

$$s_d = \sqrt{\frac{SS_d}{n-1}}$$

where

$$SS_d = \Sigma d^2 - \frac{(\Sigma d)^2}{n}$$

To use this formula, make the following column heads in the computation table:

B	A	$d = B - A$	d^2

and sum the d and d^2 columns. You will find this formula easy to use, but sensitive to rounding, so be sure to carry several digits in the intermediate steps.

a) Use this formula to compute s_d for the data in Table 9-7 (Example 13).

b) Compare your result with that obtained in Example 13, $s_d = 22.92$.

Tests Involving Proportions and Differences of Proportions

Many situations arise that call for tests of proportions or percentages rather than means. For example, a welfare office claims that the proportion of incomplete applications they receive is now 47%. The office is using this claim to justify a request for two more staff members whose main duty will be to help applicants complete the forms properly. The funding agency wants to test the claim that 47% of the applications are incomplete.

How can we make such a test? In this section we will study tests involving proportions (i.e., percentages or proportions). In principle, such tests are the same as those in Sections 9.2 and 9.3. The main difference is that we are working with a distribution of proportions instead of means.

Throughout this section we will assume that the situations we are dealing with satisfy the conditions underlying the binomial distribution. In particular, we will let r be a binomial random variable. This means r is the number of successes out of n independent binomial trials (for the definition of binomial trial see Section 5.1). We will use r/n as our estimate for p, the probability of success on each trial. The letter q again represents the probability of failure on each trial, and so $q = 1 - p$. We also assume the samples are large (i.e., $np > 5$ and $nq > 5$).

For large samples the distribution of r/n values is well approximated by a *normal curve* with mean μ and standard deviation σ as follows:

$$\mu = p \qquad \sigma = \sqrt{\frac{pq}{n}}$$

The construction of critical regions using a left-, right-, or two-tail test is essentially the same as we did in Section 9.2. The only change is that we use different formulas for the critical values. These values are given in Table 9-24 for easy reference. We have boxed the optional derivation of one of the table entries. The other entries are derived in a similar manner.

TABLE 9-24 Critical Values p_0 for Tests Involving a Proportion (Large Samples)

Level of Significance	$\alpha = 0.05$	$\alpha = 0.01$
Critical value for a right-tail test	$p_0 = p + 1.645\sqrt{\dfrac{pq}{n}}$	$p_0 = p + 2.33\sqrt{\dfrac{pq}{n}}$
Critical value for a left-tail test	$p_0 = p - 1.645\sqrt{\dfrac{pq}{n}}$	$p_0 = p - 2.33\sqrt{\dfrac{pq}{n}}$
Critical values for a two-tail test	$p_0 = p + 1.96\sqrt{\dfrac{pq}{n}}$	$p_0 = p + 2.58\sqrt{\dfrac{pq}{n}}$
	$p_0 = p - 1.96\sqrt{\dfrac{pq}{n}}$	$p_0 = p - 2.58\sqrt{\dfrac{pq}{n}}$

where p = probability of success,
$q = 1 - p$, or probability of failure,
n = sample size.

As before, the value of p to be used in Table 9-24 comes from the null hypothesis. From this value of p we can find the value of $q = 1 - p$. The type of test (left-, right-, or two-tail) comes from the alternate hypothesis.

OPTIONAL

Suppose we want to find the *critical value for a 5% level of significance with a right-tail test*. In this section we will use large samples so r/n is approximately normally distributed with mean $\mu = p$ and standard deviation $\sigma = \sqrt{pq/n}$. To convert from r/n to standard z units we use

$$z = \frac{\frac{r}{n} - \mu}{\sigma} = \frac{\frac{r}{n} - p}{\sqrt{\frac{pq}{n}}} \tag{6}$$

Using Table 5, we find that (*see* Figure 9-33)

$$P(z \geq 1.645) = 0.05$$

FIGURE 9-33 5% of Area Under Standard Normal Curve

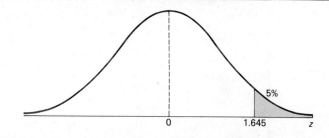

For a standard normal distribution $z = 1.645$ is the critical value for a 5% level of significance on a right-tail test. When we translate from this value of z to a corresponding value of r/n we obtain the critical value p_0 of the r/n distribution. To do this translation we use 1.645 in place of z and solve equation (6) for r/n. We obtain

$$\frac{r}{n} = p + 1.645 \sqrt{\frac{pq}{n}} \tag{7}$$

The right-hand side of equation (7) is our critical value p_0 for a right-tail test with a 5% level of significance. Thus we have

$$p_0 = p + 1.645 \sqrt{\frac{pq}{n}}$$

EXAMPLE 14

A team of eye surgeons has developed a new technique for a risky eye operation to restore the sight of people blinded from a certain disease. Under the old method it is known that only 30% of the patients who undergo this operation recover their eyesight.

Suppose that surgeons in various hospitals have performed a total of 225 operations using the new method and that 88 have been successful (the patients fully recovered their sight). Can we justify the claim that the new method is better than the old one? (Use a 1% level of significance.)

a) Let p be the probability that a patient fully recovers his or her eyesight. The null hypothesis is that p is still 0.30, even for the new method of operation. Therefore,

$$H_0: p = 0.30$$

b) The alternate hypothesis is that the new method has improved the patient's chances for eyesight recovery. Therefore,

$$H_1: p > 0.30$$

c) Since the alternate hypothesis is $H_1: p > 0.30$, we use a right-tail test (*see* Table 9-3 with μ replaced by p).

d) From Table 9-24 we see that the critical value to be used for a right-tail test with $\alpha = 0.01$ is

$$p_0 = p + 2.33 \sqrt{\frac{pq}{n}}$$

In our case the sample size is $n = 225$. From the null hypothesis, $p = 0.30$, and so $q = 1 - p = 1 - 0.30 = 0.70$. Therefore, (*see* Figure 9-34)

$$p_0 = 0.30 + 2.33 \sqrt{\frac{0.30(0.70)}{225}}$$

$$= 0.30 + 2.33 \sqrt{\frac{0.21}{225}}$$

$$= 0.30 + 2.33 \sqrt{0.0009}$$

$$\simeq 0.30 + 2.33(0.03)$$

$$\simeq 0.30 + 0.07 \simeq 0.37$$

e) Since $r = 88$, our sample estimate for p is

$$\frac{r}{n} = \frac{88}{225} = 0.39$$

The sample value $r/n = 0.39$ falls in the critical region; therefore, we reject H_0. At the 1% level of significance we conclude that the new method seems to have improved a patient's chances for eyesight recovery.

FIGURE 9-34   Critical Region for $\alpha = 0.01$

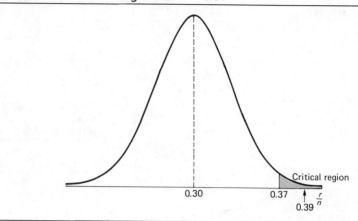

_____ Exercise 16 _____

A botanist has produced a new variety of hybrid wheat that is better able to withstand drought that other varieties. The botanist knows that for the parent plants the probability of seed germination is 80%. The probability of seed germination for the hybrid variety is unknown, but the botanist claims that it is 80%. To test this claim 400 seeds from the hybrid plant are tested and it is found that 312 germinated. Use a 5% level of significance to test the claim that the probability of germination for the hybrid is 80%.

a) Let p be the probability that a hybrid seed will germinate. What is the null hypothesis?

H_0: $p > 0.80$ H_0: $p = 0.80$ H_0: $p \neq 0.80$

b) Because we have no prior knowledge about germination probability for the hybrid plant, what would be a good choice for the alternate hypothesis?

c) Should our test be a one-tail or two-tail test? (*Hint:* Use the alternate hypothesis and Table 9-3.)

d) What is the sample size n for this problem? Based on our choice for H_0, what value should we use for p when computing critical values? Since $q = 1 - p$, what value should we use for q? What critical values should we use? (*Hint:* See Table 9-24.) Draw a diagram showing the acceptance region and the critical region for H_0.

a) H_0: $p = 0.80$

b) H_1: $p \neq 0.80$

c) Two-tail test

d) $n = 400$; $p = 0.80$;

$q = 1 - p = 1 - 0.80 = 0.20$

Therefore, from Table 9-24 with $\alpha = 0.05$ we obtain

$$p_0 = p + 1.96 \sqrt{\frac{pq}{n}}$$

$$= 0.80 + 1.96 \sqrt{\frac{(0.80)(0.20)}{400}}$$

$$= 0.80 + 1.96 \sqrt{0.0004}$$

$$\approx 0.80 + 1.96(0.02)$$

$$\approx 0.80 + 0.04$$

$$\approx 0.84$$

and $\qquad p_0 = p - 1.96 \sqrt{\dfrac{pq}{n}}$

$$= 0.80 - 1.96 \sqrt{\dfrac{(0.80)(0.20)}{400}}$$

$$\simeq 0.80 - 0.04$$

$$\simeq 0.76$$

Figure 9-35 shows the critical region.

FIGURE 9-35 **Critical Region for $\alpha = 0.05$**

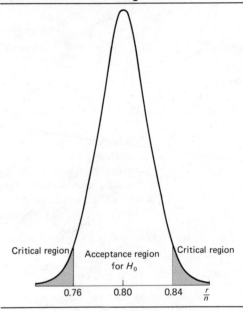

Critical region Acceptance region for H_0 Critical region

0.76   0.80   0.84 $\dfrac{r}{n}$

e) Since our number of successful germinations is $r = 312$ out of $n = 400$ seeds, what is the value of the estimate r/n for p? Does this estimate fall in the acceptance region or the critical region? Do we accept or reject H_0?

e) Our estimate for p is

$$\frac{r}{n} = \frac{312}{400} = 0.78$$

Since this value lies in the acceptance region of H_0, we accept the null hypothesis that 80% of the hybrid seeds germinate.

The construction of critical regions for tests involving a difference of proportions is essentially the same as we used in Section 9.3 for the difference of means. The main change is that we use a probability distribution of proportions rather than one of means.

Suppose we draw independent random samples of size n_1 and n_2, respectively, from two binomial distributions. Let r_1 be the number of successes in the sample of size n_1 and r_2 be the number of successes in the sample of size n_2. Let the probability of success on each trial be p_1 for the first binomial distribution and let it be p_2 for the second binomial distribution.

For large values of n_1 and n_2 the distribution of the differences

$$\frac{r_1}{n_1} - \frac{r_2}{n_2}$$

is closely approximated by a *normal distribution* with mean μ and standard deviation σ as shown

$$\mu = p_1 - p_2$$

$$\sigma = \sqrt{\frac{p_1 q_1}{n_1} + \frac{p_2 q_2}{n_2}}$$

where $q_1 = 1 - p_1$, and $q_2 = 1 - p_2$.

For most practical problems involving a comparison of two binomial populations the experimenters will want to test the null hypothesis that $p_1 = p_2$. Consequently, this is the only type of test we shall consider. Since the values of p_1 and p_2 are unknown and since specific values are not assumed under the hypothesis $p_1 = p_2$, they must be approximated by sample estimates. Under the condition $p_1 = p_2$ the best estimate for their common value is the total number of successes $(r_1 + r_2)$ divided by the total number of trials $(n_1 + n_2)$. If we denote this value by $\hat{p}$ (read "p hat") then

$$\hat{p} = \frac{r_1 + r_2}{n_2 + n_2} \tag{8}$$

Formula (8) gives the best estimate $\hat{p}$ for p_1 and p_2 under the assumption that $p_1 = p_2$.

Now using the normal approximation for p_1 and p_2, Table 5, and the methods outlined earlier in this section for critical values involving a single proportion, it follows that the critical values to be used for the differences of proportions are those in Table 9-25.

TABLE 9-25 Critical Values for Tests Involving a Difference of Two Proportions (Large Samples)

Level of Significance	$\alpha = 0.05$	$\alpha = 0.01$
Critical value for a right-tail test	$1.645 \sqrt{\dfrac{\hat{p}\hat{q}}{n_1} + \dfrac{\hat{p}\hat{q}}{n_2}}$	$2.33 \sqrt{\dfrac{\hat{p}\hat{q}}{n_1} + \dfrac{\hat{p}\hat{q}}{n_2}}$
Critical value for a left-tail test	$-1.645 \sqrt{\dfrac{\hat{p}\hat{q}}{n_1} + \dfrac{\hat{p}\hat{q}}{n_2}}$	$-2.33 \sqrt{\dfrac{\hat{p}\hat{q}}{n_1} + \dfrac{\hat{p}\hat{q}}{n_2}}$
Critical values for a two-tail test	$1.96 \sqrt{\dfrac{\hat{p}\hat{q}}{n_1} + \dfrac{\hat{p}\hat{q}}{n_2}}$ and $-1.96 \sqrt{\dfrac{\hat{p}\hat{q}}{n_1} + \dfrac{\hat{p}\hat{q}}{n_2}}$	$2.58 \sqrt{\dfrac{\hat{p}\hat{q}}{n_1} + \dfrac{\hat{p}\hat{q}}{n_2}}$ and $-2.58 \sqrt{\dfrac{\hat{p}\hat{q}}{n_1} + \dfrac{\hat{p}\hat{q}}{n_2}}$

where $\hat{p} = \dfrac{r_1 + r_2}{n_1 + n_2}$ and $\hat{q} = 1 - \hat{p}$

• **Comment:** For most practical applications the sample sizes n_1 and n_2 will be considered large samples if the four quantities

$$n_1\hat{p} \qquad n_1\hat{q} \qquad n_2\hat{p} \qquad n_2\hat{q}$$

are larger than 5 (*see* Section 7.3, Normal Approximation to the Binomial Distribution).

EXAMPLE 15

The Macek County Clerk wishes to improve voter registration. One method under consideration is to send reminders in the mail to all citizens in the county who are eligible to register. As a pilot study to determine if this method will actually improve voter registration, a random sample of 1,250 potential voters was taken. Then this sample was randomly divided into two groups.

Group 1: There were 625 people in this group. No reminders to register were sent to them. The number of potential voters from this group who registered was 295.

Group 2: This group also contained 625 people. Reminders were sent in the mail to each member in the group, and the number who registered to vote was 350.

The county clerk claims that the proportion of people to register was significantly greater in group 2. On the basis of this claim the clerk recommends that the project be funded for the entire population of Macek County. Use a 5% level of significance to test the claim that the proportion of potential voters who registered was greater in group 2, the group that received reminders.

a) Let p_1 be the proportion of voters who registered from group 1 and let p_2 be the proportion registered from group 2. The null hypothesis is that there is no difference in proportions, so

$$H_0: p_1 = p_2$$

The alternate hypothesis is that the proportion of voters was greater in the group which received reminders.

$$H_1: p_1 < p_2$$

b) Use a left-tail test (*see* Table 9-5 with μ replaced by p).

c) What critical value should we use?

In our setting the number of successes is the number of registered voters in the sample. For the first sample this is $r_1 = 295$ and for the second sample is $r_2 = 350$. Since $n_1 = n_2 = 625$, then formula (8) gives

$$\hat{p} = \frac{r_1 + r_2}{n_1 + n_2} = \frac{295 + 350}{625 + 625} \approx 0.52$$

Then

$$\hat{q} = 1 - \hat{p} = 1 - 0.52 = 0.48$$

FIGURE 9-36 Critical Region for $\alpha = 0.05$

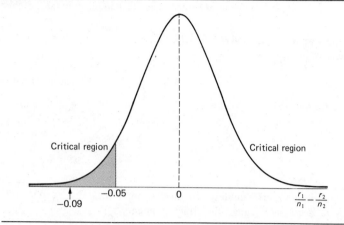

Table 9-25 tells us to use the critical value (*see* Figure 9-36)

$$-1.645 \sqrt{\frac{\hat{p}\hat{q}}{n_1} + \frac{\hat{p}\hat{q}}{n_2}} \approx -1.645 \sqrt{\frac{(0.52)(0.48)}{625} + \frac{(0.52)(0.48)}{625}}$$

$$\approx -1.645 \sqrt{0.0004 + 0.0004}$$

$$\approx -1.645 \sqrt{0.0008}$$

$$\approx -1.645(0.028)$$

$$\approx -0.05$$

d) Shall we accept or reject H_0?
   Since

$$\frac{r_1}{n_1} = \frac{295}{625} = 0.47 \quad \text{and} \quad \frac{r_2}{n_2} = \frac{350}{625} = 0.56$$

then

$$\frac{r_1}{n_1} - \frac{r_2}{n_2} = 0.47 - 0.56 = -0.09$$

is in the critical region. We reject H_0 and conclude that the county clerk's claim is valid at the 0.05 level of significance.

_____ Exercise 17 _____

In Example 15 about voter registration suppose a random sample of 1,100 potential voters was randomly divided into two groups.

Group 1: 500 potential voters; no registration reminders sent; 248 registered to vote
Group 2: 600 potential voters; registration reminders sent; 332 registered to vote

Do these data support the claim that the proportion of voters to register was greater in the group that received reminders than in the group that did not? Use a 1% level of significance.

a) What should we use for H_0 and H_1? Should we use a left-, right- or two-tail test?

a) As before,

$$H_0: p_1 = p_2 \quad \text{and} \quad H_1: p_1 < p_2$$

We use a left-tail test.

b) What is $\hat{p}$ for this problem? What is $\hat{q}$?

b) $n_1 = 500$ and $r_1 = 248$; $n_2 = 600$ and $r_2 = 332$

$$\hat{p} = \frac{r_1 + r_2}{n_1 + n_2} = \frac{248 + 332}{500 + 600} \approx 0.53$$

$$\hat{q} = 1 - \hat{p} = 1 - 0.53 = 0.47$$

c) What is the critical value for this test? Draw a diagram showing the acceptance region, the critical region, and the critical value.

c) $-2.33 \sqrt{\dfrac{\hat{p}\hat{q}}{n_1} + \dfrac{\hat{p}\hat{q}}{n_2}}$

$$= -2.33 \sqrt{\frac{(0.53)(0.47)}{500} + \frac{(0.53)(0.47)}{600}}$$

$$\approx -2.33 \sqrt{0.0005 + 0.0004}$$

$$\approx -2.33 \sqrt{0.0009}$$

$$\approx -2.33(0.03)$$

$$\approx -0.07$$

Figure 9-37 shows the critical region.

FIGURE 9-37 Critical Region for $\alpha = 0.01$

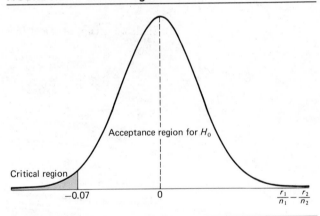

d) Shall we accept or reject H_0? Do the data support the claim that the reminders increased the proportion of registered voters?

d) $\dfrac{r_1}{n_1} - \dfrac{r_2}{n_2} = \dfrac{248}{500} - \dfrac{332}{600} = 0.496 - 0.553 = -0.057$

Since -0.057 is in the acceptance region, we cannot reject H_0. At the 1% level of significance the data do not support the claim that the reminders increased the proportion of registered voters.

Section Problems 9.6

For each problem do the following:

a) State the null and alternate hypotheses.

b) Find the critical value or values.

c) Sketch the critical region and acceptance region.

d) On the sketch of part (c) show the location of the sample value r/n or (r_1/n_1) — (r_2/n_2) as appropriate for the given information.

e) Decide whether you should accept or reject the null hypothesis.

1. Sometimes people who are required to take a medication develop a chemical response that makes the medication ineffective. A random sample of 163 people taking insulin showed that 11 had developed blood chemicals acting against the insulin. The pharmaceutical company manufacturing this brand of insulin says that only 3% of the people who use the insulin will develop such chemical reactions. Using a 1% level of significance, test the claim that the actual proportion of people who develop such inhibiting chemicals is higher than 3%.

2. A city council member gave a speech in which she said that 18% of all private homes in the city had been undervalued by the county tax assessor's office. In a follow-up story the local newspaper reported that it had taken a random sample of 91 private homes. Using professional realtors to evaluate the property, and checking against county tax records, it was found that 14 of the homes had been undervalued. Do these data indicate that the proportion of private homes that are undervalued by the county tax assessor is different from 18%? Use a 5% level of significance.

3. Internal Revenue Service employees who process income tax returns say that 10% of all tax returns contain arithmetic errors in excess of $1,200. A random sample of 817 income tax returns showed that 86 of them contain such errors. Do these data indicate that the proportion of income tax returns with arithmetic errors in excess of $1,200 is different from 10%? Use a 1% level of significance.

4. You are interviewing with a large company for a position as Director of Personnel. The company says that 75% of its employees have been with the company for three or more years. However, you spend a day walking around meeting employees and you find that 22 out of a random sample of 37 have been with the company three or more years. Do these data indicate that less than 75% of the employees have been with the company three or more years? Use a 5% level of significance.

5. The board of real estate developers claims that 55% of all voters will vote for a bond issue to construct a massive new water project. A random sample of 215 voters was taken and 96 said that they would vote for the new water project. Does this indicate that less than 55% of the voters favor such a project? Use a 1% level of significance.

6. At a national convention for travel agents, Ellen heard a report that 27% of the people who reserve a trip more than ten months in advance cancel the trip. After the convention, Ellen did a study for her travel agency. From a random sample of 78 reservations made more than ten months in advance, 23 were cancelled. Do these data seem to indicate that the percent of cancellations made at her agency is different from the percent indicated at the conference? Use a 1% level of significance.

7. The American Southpaw Association claims that the proportion of left-handed people in *Who's Who* is higher than the overall national proportion, which is one out of 21. A random sample of 180 people listed in the current *Who's Who* showed that 12 were left-handed. Is the claim justified at the 5% level of significance?

8. A washing machine manufacturer says that 85% of its washers last five years before repairs are necessary. A random sample of 100 washers showed that 73 of them lasted five years before repair. Is the manufacturer's claim too high? Use a 1% level of significance.

9. A real estate broker says that 68% of all retired couples prefer a condominium to a single-family house. A random sample of 75 retired couples showed that 61 actually preferred a condominium to single-unit living. Use a 5% level of significance to test the claim that the real estate broker's figure should be higher.

10. Professor Levings gave 58 A's and B's to a class of 125 English-101 students. The next term Professor Hardy gave 45 A's and B's to a class of 115 English-101 students. Use a 5% level of significance to test the claim that Professor Levings gives a higher proportion of A's and B's.

11. The manager of the student cafeteria is trying to decide which of two types of vending machines to install. Each machine is tested 200 times, and machine A fails to work 16 times while machine B fails to work 24 times. The manager claims machine A is better. Is this claim justified at the 5% level of significance?

12. Two different vaccines have been developed to prevent a cattle disease. Vaccine A prevented the disease in 300 out of a random sample of 350 cattle exposed to the disease, and vaccine B prevented the disease in 372 out of another random sample of 400 cattle exposed to the disease. Use a 1% level of significance to test the claim that there is a difference in the effectiveness of the two vaccines.

13. A dentist claims that 15% of all appointments are cancelled. In one month 15 of the dentist's 136 appointments were cancelled. Do the data indicate that the percentage of cancellations is different than his claim? Use a 5% level of significance.

14. A test of 80 youths and 120 adults showed that 18 of the youths and 10 of the adults were careless drivers. Use a 1% level of significance to test the claim that the youth percentage of careless drivers is higher than the adult percentage.

15. An army psychologist is studying a possible relation between accident rate on the base and the incidence of a full moon. A random sample of 420 base personnel were interviewed immediately after periods of a full moon, and 34 reported some kind of accident during this period. Another random sample of 485 base personnel were interviewed during other periods when there was no full moon, and 35 reported some kind of accident. Use a 1% level of significance to test the claim that there is a difference (either way) between the accident rate during a full moon and the accident rate at other times.

16. The Police Department in Los Angeles claims that the incidence of violent crime is up compared to last year, when 19.7% of all crimes were classified as violent crimes. A member of the city council took a random sample of 877 crimes reported in L.A. police records this year and found that 178 were classified by the police as violent crimes. Using a 5% level of significance, test the claim that the incidence of violent crimes has increased.

17. A random sample of 530 union bricklayers showed that 40 were unemployed. A random sample of 640 nonunion bricklayers showed that 60 were unemployed. Do these data

indicate that the proportion of unemployed bricklayers is greater for the nonunion people? Use a 5% level of significance.

18. A random sample of 378 airline passengers was taken one year ago and it was found that 178 requested nonsmoking seating. Recently a random sample of 516 airline passengers showed that 320 requested nonsmoking seating. Do these data indicate that the proportion of airline passengers requesting nonsmoking seating has increased? Use a 5% level of significance.

19. Sometimes porpoises are accidentally caught in tuna nets and killed. The United States government requires all commercial tuna boats to use nets that have escape chutes for porpoises. An observer on board a foreign tuna boat that does not use nets with escape chutes said that out of 1,279 fish killed, 44 were porpoises. An observer on a U.S. tuna boat said that out of 1,642 fish killed, 10 were porpoises. Do these data indicate that the proportion of porpoises killed is lower on the U.S. tuna boat? Use a 1% level of significance.

20. A forest in Oregon has an infestation of spruce moths. In an effort to control the moth one area has been regularly sprayed from airplanes. In this area a random sample of 495 spruce trees showed that 81 had been killed by moths. A second nearby area receives no treatment. In this area a random sample of 518 spruce trees showed that 92 had been killed by the moth. Do these data indicate that the proportion of spruce trees killed by the moth is different for these areas? Use a 1% level of significance.

21. A large shipment of produce contains two kinds of pears: the Bartlett pear and the LeConte pear. Both types have been harvested and stored about the same time. To determine if there is a difference in percentages of bad fruit for the two varieties a random sample of 850 Bartlett pears and 850 LeConte pears was selected. It was found that 36 Bartlett and 54 LeConte pears had gone bad. Do these data indicate a difference between percentages of bad fruit for these two varieties of pears? Use a 1% level of significance.

SECTION 9.7 The *P* Value in Hypothesis Testing

The level of significance α in a test of hypothesis is the probability of making a type I error; that is, α is the probability of rejecting the null hypothesis when it is true.

The experimenter sets α before beginning the statistical test. Then this level of significance is used to determine whether or not H_0 is rejected. However, as we shall see in the next example, the same sample data may lead to two different test conclusions depending on the level of significance used.

EXAMPLE 16

Viva, a credit card company, lowered their annual interest rate by 1%. Past records of accounts before the rate change showed the average outstanding balance on credit card accounts to be $\mu = \$576$. The managers believe that reducing interest rates spurs customers to greater credit card use, which will result in higher outstanding balances. To test this claim, a random sample of 36 accounts were examined six months after the interest rate reduction. The average outstanding balance was $\bar{x} = \$615$ with standard deviation $s = \$120$. Test the managers' claim at the 0.05 and 0.01 levels of significance.

In both cases we use the same null and alternate hypotheses.

$$H_0\!: \mu = 576$$

$$H_1\!: \mu > 576$$

This is a right-tail test, and Figure 9-38 shows the test conclusion for the two levels of significance.

FIGURE 9-38 Test Conclusion

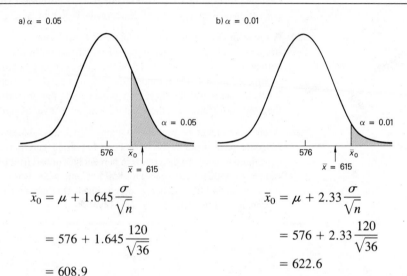

a) $\alpha = 0.05$

$\alpha = 0.05$

576   $\bar{x}_0$

$\bar{x} = 615$

b) $\alpha = 0.01$

$\alpha = 0.01$

576   $\bar{x}_0$

$\bar{x} = 615$

$$\bar{x}_0 = \mu + 1.645 \frac{\sigma}{\sqrt{n}}$$

$$= 576 + 1.645 \frac{120}{\sqrt{36}}$$

$$= 608.9$$

$$\bar{x}_0 = \mu + 2.33 \frac{\sigma}{\sqrt{n}}$$

$$= 576 + 2.33 \frac{120}{\sqrt{36}}$$

$$= 622.6$$

Reject H_0 and accept the managers' claim that the average outstanding balance is higher.

Accept H_0. There is not enough evidence to conclude that the average outstanding balance is higher.

As we see in Example 16, smaller levels of significance move the critical value for the rejection region farther from the mean μ. A natural question arises: What is the smallest level of significance at which the sample data will tell us to reject H_0? The answer is the *P value* (or probability of chance) associated with the observed sample statistic.

For the distribution described by the null hypothesis, the *P value* is the smallest level of significance for which the observed sample statistic tells us to reject H_0.

How do we go about computing the P value? We need to express the P value in terms of probability. Let's use the model of a right-tail test of the mean. For a right-tail test of the mean, the P value is simply the probability that the sample mean from any random sample of suitable size will be greater than or equal to the observed sample mean $\bar{x}$. In symbols we have

$$P \text{ value} = P(\bar{x} \text{ computed from any random sample} \geq \text{observed } \bar{x})$$
$$\text{for right-tail test of } \mu$$

P values are the areas in the tail or tails of a probability distribution beyond the observed sample statistic. Figure 9-39 shows the P values for right-, left-, and two-tail tests of the mean.

FIGURE 9-39 *P* Values for Different Types of Tests of the Mean

Hypothesis	Type of Test	P value

The shaded area is the P value

$H_0: \mu = k$
$H_1: \mu > k$

Right-tail test

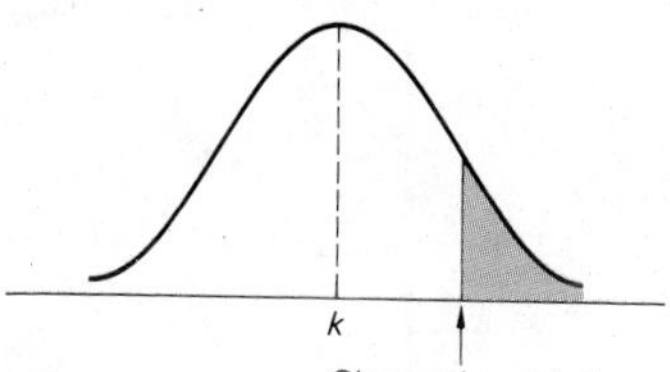

k
Observed sample $\bar{x}$

The shaded area is the P value

$H_0: \mu = k$
$H_1: \mu < k$

Left-tail test

Observed sample $\bar{x}$
k

P value is sum of areas in two tails

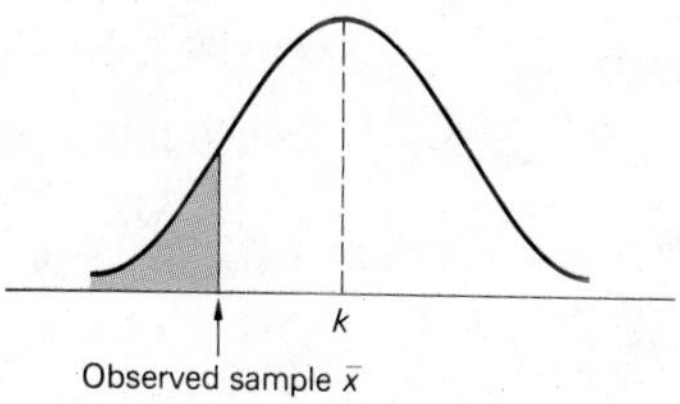

$H_0: \mu = k$
$H_1: \mu \neq k$

Two-tail test

k
k plus or minus $|\bar{x} - k|$
where $\bar{x}$ is the observed sample mean

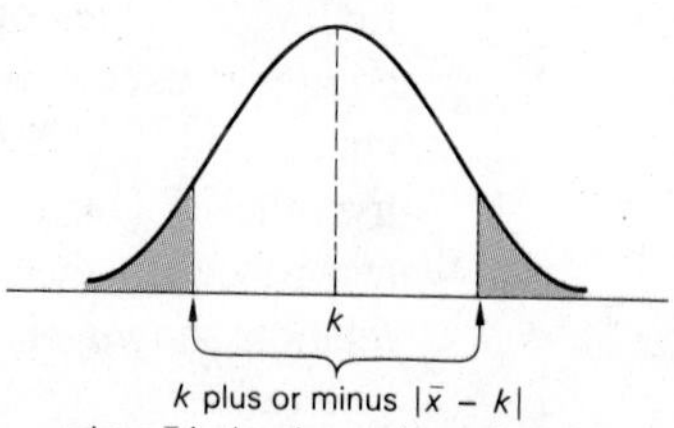

EXAMPLE 17

Compute the P value for the hypothesis test of Example 16. Recall that the problem involved the claim that the outstanding balance on Viva credit cards increased after the annual interest decreased. We had

$$H_0: \mu = 576$$

$$H_1: \mu > 576$$

Observed sample data: $n = 36, \bar{x} = \$615, s = \$120$

The P value is simply the area to the right of the observed sample mean $\bar{x} = 615$ under the normal curve with mean $\mu = 576$. (*See* Figure 9-40.)

FIGURE 9-40 *P* Value

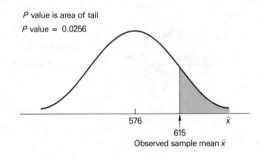

P value is area of tail
P value $= 0.0256$

576
615
Observed sample mean $\bar{x}$

$\bar{x}$

To find the P value, we convert $\bar{x}$ to z using the formula

$$z = \frac{\bar{x} - \mu}{\sigma/\sqrt{n}} = \frac{615 - 576}{120/\sqrt{36}} = 1.95$$

The P value $= P(\bar{x} \geq 615) = P(z \geq 1.95) = 0.5 - 0.4744 = 0.0256$

Since this is the smallest level of significance for which the sample data tells us to reject H_0, we would reject H_0 for any $\alpha \geq 0.0256$. For $\alpha < 0.0256$, we would accept H_0.

As we have seen, P values are areas in the tail or tails of a probability distribution beyond the observed sample statistic. For normal distributions, we found such areas in Chapter 6. However, for other distributions, our tables are not well-designed for finding P values other than those that happen to be 0.01 or 0.05. However, most computer packages that do statistical computations will give you the P value associated with the observed sample value and the distribution described by the null hypothesis. Our task then becomes one of interpreting the results of the statistical test based on the specified level of significance and the computed P value. The relation shown between the level of significance and the P value in Example 17 generalizes.

Let α be the specified level of significance in a statistical test.

> If $\alpha \geq P$ value, reject H_0.
> If $\alpha < P$ value, accept H_0.

See Figure 9-41.

FIGURE 9-41   Values of α For Which We Reject or Accept H_0

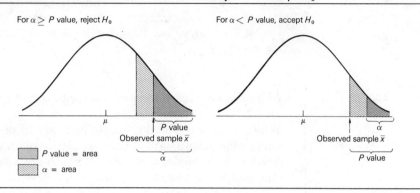

- **Comment:** Looking at the critical region for a one-tail test we see that the decision to reject H_0 if $\alpha \geq P$ value is a correct decision. Why would we also reject H_0 if we were dealing with a two-tail test? For a two-tail test using level of significance α we put an area of $\alpha/2$ in the left-tail critical region and an area of $\alpha/2$ in the right-tail critical region. Therefore, the decision procedure would be to reject H_0 if $(\alpha/2) \geq (P$ value/2$)$. However, this is logically the same as saying we should reject H_0 if $\alpha \geq P$ value. Consequently, we will reject H_0 whenever $\alpha \geq P$ value. This criterion is true of a one-tail test or a two-tail test.

The advantage of knowing the P value is that we know *all* levels of significance for which the observed sample statistic tells us to reject H_0. Many research journals require authors to include the P value of the observed sample statistic. Then readers will have more information and will know the test conclusion for any preset level of significance.

Again, let us caution you to establish the level of significance α before doing the hypothesis test. The level of significance reflects the probability level at which you are willing to risk a type I error. Also, the accuracy and reliability of your measurement instruments might affect your choice of α. So select α first, then use a computer to find the P value of your test statistic, and finally draw the appropriate conclusion.

Since you will usually find P values by using statistical software packages on a computer, the next example and exercise show you how to conclude your test using the P value. We follow the process outlined below.

> ● **Hypothesis Testing Using *P* Values from a Computer**
>
> 1. Establish the level of significance α of the test.
> 2. Determine the null and alternate hypotheses, H_0 and H_1.
> 3. Enter information about the observed sample into the computer and look for the *P* value of the observed sample statistic in the computer output.
> 4. Compare your level of significance with the *P* value.
>   If $\alpha \geq P$ value, reject H_0
>   If $\alpha < P$ value, accept H_0 (or, if you prefer, do not reject H_0)

EXAMPLE 18

Whitehall Construction company figures the cost of a project based on an average idle time of 68 minutes per worker per 8-hour shift. Idle time does not include lunch or rest breaks, but can be caused by delays in material delivery, delays in the completion of a prior stage of the project, worker injury, inadequate crew size due to absenteeism, and so forth. A new work schedule has just been completed by the construction foreman. It is hoped that the new work schedule will reduce the average idle time. To determine if the new schedule is having the desired effect, a random sample of 49 workers were observed during an 8-hour shift. The average idle time was $\bar{x} = 65$ min with standard deviation $s = 12$ min. Use a 5% level of significance. Did the new schedule reduce average idle time? Use the *P* value of the sample statistic to make the decision. The level of significance is $\alpha = 0.05$.

$$H_0: \mu = 68 \text{ min}$$

$$H_1: \mu < 68 \text{ min}$$

Observed sample data: $\bar{x} = 65$ min, $s = 12$ min, $n = 49$

A computer run will give you the *P* value based on the sample data.

$$P \text{ value} = 0.0401$$

Since $\alpha = 0.05$ and *P* value $= 0.0401$, we see that

$$\alpha \geq P \text{ value}$$

and we reject H_0. The new work schedule has reduced average idle time. Although the reduction seems to be slight, if Whitehall had several hundred workers, the total reduced idle time could be important.

_____ Exercise 18 _____

Compare the given *P* value and the level of significance α to conclude each test.

a) Given

$$H_0: \mu = 30$$

$$H_1: \mu \neq 30$$

$$\alpha = 0.01$$

$$P \text{ value} = 0.0213$$

Do we reject or accept H_0?

b) Given

$$H_0: \mu_1 = \mu_2$$

$$H_1: \alpha_1 < \mu_2$$

$$\alpha = 0.05$$

$$P \text{ value} = 0.0316$$

Do we reject or accept H_0?

c) Given

$$H_0: p = 0.15$$

$$H_1: p < 0.15$$

$$\alpha = 0.05$$

$$P \text{ value} = 0.0171$$

Do we reject or accept H_0?

d) Given

$$H_0: p_1 = p_2$$

$$H_1: p_1 \neq p_2$$

$$\alpha = 0.01$$

$$P \text{ value} = 0.0321$$

Do we reject or accept H_0?

a) Since $\alpha = 0.01$ and P value $= 0.0213$ we see $\alpha < P$ value, so we accept H_0.

b) Since $\alpha = 0.05$ and P value $= 0.0316$ we see $\alpha \geq P$ value, so we reject H_0.

c) Since $\alpha = 0.05$ and P value $= 0.0171$, we see that $\alpha \geq P$ value, so we reject H_0.

d) Since $\alpha = 0.01$ and P value $= 0.0321$, we see that $\alpha < P$ value, so we accept H_0.

- **Summary**

  1. The P value (probability of chance) is the area of the sampling distribution that lies beyond the observed sample statistic. It tells us the probability that a sample statistic will be more extreme than the observed sample statistic. We compute the P values as follows:
     a) When H_1 indicates a right-tail test,

     $P \text{ value} = $ area to the right of the observed sample statistic

     $P \text{ value} = P(\text{sample statistic} > \text{observed sample statistic})$

b) When H_1 indicates a left-tail test,

P value = area to the left of the observed sample statistic

P value = P(sample statistic < observed sample statistic)

c) When H_1 indicates a two-tail test,

P value = sum of the areas in the two tails

i) In the case that the observed sample statistic falls in the right half of a symmetric curve then

P value = $2P$(sample statistic > observed sample statistic)

ii) In the case that the observed sample statistic falls in the left half of a symmetric curve then

P value = $2P$(sample statistic < observed sample statistic)

2. To conclude the test, we compare the level of significance α with the P value.
a) If α is *greater than or equal to* the P value, reject H_0.
b) If α is *less than* the P value, accept H_0.

Section Problems 9.7

1. Given H_0: $\mu = 3$, H_1: $\mu > 3$, and P value = 0.0312
   a) do we accept or reject H_0 at a 1% level of significance?
   b) do we accept or reject H_0 at a 5% level of significance?

2. Given H_0: $p = 0.3$, H_1: $p < 0.3$, and P value = 0.0532
   a) do we accept or reject H_0 at a 1% level of significance?
   b) do we accept or reject H_0 at a 5% level of significance?

3. Given H_0: $\mu_1 = \mu_2$, H_1: $\mu_1 \neq \mu_2$, and P value = 0.0093
   a) do we accept or reject H_0 at a 1% level of significance?
   b) do we accept or reject H_0 at a 5% level of significance?

4. Given H_0: $\mu = 5$, H_1: $\mu < 5$, and P value = 0.0473
   a) do we accept or reject H_0 at the 1% level of significance?
   b) do we accept or reject H_0 at the 5% level of significance?

5. Given H_0: $p = 0.7$, H_1: $p > 0.7$, and P value = 0.0262
   a) do we accept or reject H_0 at a 1% level of significance?
   b) do we accept or reject H_0 at a 5% level of significance?

6. Given H_0: $\mu = 4.5$, H_1: $\mu > 4.5$, and P value = 0.0673
   a) do we accept or reject H_0 at a 1% level of significance?
   b) do we accept or reject H_0 at a 5% level of significance?

7. Given H_0: $\mu = 5$ and H_1: $\mu > 5$, and observed sample mean $\bar{x} = 8$, sketch a picture of the $\bar{x}$ distribution and show the location of the area corresponding to the P value.

8. Given H_0: $\mu = 3$ and H_1: $\mu > 3$, and observed sample mean $\bar{x} = 6$, sketch a picture of the $\bar{x}$ distribution and show the location of the area corresponding to the P value.

9. Given H_0: $\mu = 5$ and H_1: $\mu < 5$, and observed sample mean $\bar{x} = 3$, sketch a picture of the $\bar{x}$ distribution and show the location of the area corresponding to the P value.

10. Given H_0: $\mu = 12$ and H_1: $\mu < 12$, and observed sample mean $\bar{x} = 7$, sketch a picture of the $\bar{x}$ distribution and show the location of the area corresponding to the P value.

11. Given H_0: $\mu = 10$ and H_1: $\mu \neq 10$, and observed sample mean $\bar{x} = 12$, sketch a picture of the $\bar{x}$ distribution and show the location of the area corresponding to the P value.

12. Given H_0: $\mu = 6$ and H_1: $\mu \neq 6$, and observed sample mean $\bar{x} = 8$, sketch a picture of the $\bar{x}$ distribution and show the location of the area corresponding to the P value.

13. Given H_0: $\mu = 10$ and H_1: $\mu \neq 10$, and observed sample mean $\bar{x} = 8$, sketch a picture of the $\bar{x}$ distribution and show the location of the area corresponding to the P value.

14. Given H_0: $\mu = 12$ and H_1: $\mu \neq 12$, and observed sample mean $\bar{x} = 9$, sketch a picture of the $\bar{x}$ distribution and show the location of the area corresponding to the P value.

Summary

We have encountered two types of errors in hypothesis testing: a type I error occurs when a true null hypothesis is rejected, and a type II error occurs when a false null hypothesis is accepted. Because type II errors are difficult to control and beyond the scope of this book, we formulate all our statistical tests in such a way that the probability α of a type I error can be controlled.

The null hypothesis H_0 is set up to see if it can be rejected. The hypothesis which will be accepted if H_0 must be rejected is H_1, called the alternate hypothesis. The probability of a type I error, α, is called the level of significance and should be chosen prior to sampling. In testing hypotheses the following procedure should be used:

1. State the null hypothesis H_0.
2. State the alternate hypothesis H_1.
3. Determine the critical value and the critical region.
4. If the estimated value of the parameter falls in the critical region, reject H_0. Otherwise, accept it.

We used the preceding method of hypothesis testing to conduct tests involving a single mean (large or small sample), a difference of means (independent or dependent samples), a single proportion, and a difference of proportions. The same process can be used with other distributions to conduct tests of hypothesis with other parameters.

We concluded the chapter with a section on P values. Most statistical computer software generate the P value associated with the sample statistic and the null hypothesis. The P value gives the smallest level of significance for which we can reject H_0.

Important Words and Symbols

	Section		Section
Hypothesis	9.1	Critical region	9.1
Hypothesis testing	9.1	Acceptance region	9.1
Null hypothesis H_0	9.1	One-tail tests	9.2
Alternate hypothesis H_1	9.1	Two-tail tests	9.2
Type I error	9.1	Independent distributions	9.3
Type II error	9.1	*d.f.* for testing with small samples	9.4
α, the level of significance of a test and the probability of a type I error	9.1	α' and α'' with respect to the t distribution	9.4
β, the probability of a type II error	9.1	Pooled variance	9.4
		Data pairs	9.5
Confidence level $(1 - \alpha)$	9.1	Critical value or values for testing proportions	9.6
Power of a test $(1 - \beta)$	9.1		
Critical value or values for testing means	9.1	P value	9.7

Chapter Review Problems

Before you solve each problem, first categorize it by answering the following questions:

a) Are we testing a single mean, a difference of means, a single proportion, or a difference of proportions?

b) Are we using a large sample or a small sample?

c) If we are testing a difference of means, are the samples independent, or are the data paired?

d) Then solve each problem by

    i) stating the null and alternate hypotheses.

    ii) finding critical value or values.

    iii) sketching the critical region and acceptance region.

    iv) making your decision to accept or reject the null hypothesis from the sample information.

1. The Big Break Moving Company claims that a typical family moves an average of once every 5.2 yr. In a random sample of 100 families the average length of time between moves was $\bar{x} = 5.0$ yr with a standard deviation of 1.8 yr. Does this indicate that the average moving time is different from 5.2 yr? (Use a 5% level of significance.)

2. A random sample of 400 families in the city of Minneapolis showed that 355 owned a dog. The city council claims that 83% of the families own a dog. Do the data indicate that the actual percentage of families owning a dog is different from 83%? (Use a 5% level of significance.)

3. Two competing companies make TV picture tubes. A random sample of 100 Nippon tubes showed that the average lifetime was $\bar{x}_1 = 6.3$ yr with standard deviation 0.8 yr.

Another random sample of 100 Sunnyvale tubes showed that the average lifetime was $\bar{x}_2 = 6.2$ yr with standard deviation 0.5 yr. Do the data indicate that the Nippon tubes last longer? (Use a 1% level of significance.)

4. Professor Jennings claims that only 35% of the students at Flora College work while attending school. Dean Renata thinks the professor has underestimated the number of students with part-time or full-time jobs. A random sample of 81 students shows that 39 have jobs. Do the data indicate that more than 35% of the students have jobs? (Use a 5% level of significance.)

5. The Toylot Company makes an electric train with a motor that they claim will draw an average of only 0.8 amperes (A) under a normal load. A sample of nine motors was tested and it was found that the mean current was $\bar{x} = 1.6$ A with a sample standard deviation of $s = 0.41$ A. Do the data indicate the Toylot claim of 0.8 A is too low? (Use a 1% level of significance.)

6. The Quickie Cake Mix Company has decided to market gingerbread mixes. They have two recipes. Recipe A is easier to make than recipe B, so they prefer to use recipe A unless a higher proportion of people favor recipe B. Of a random sample of 62 people given a taste of gingerbread from recipe A, 35 said they liked it well enough to buy the mix. The members of another random sample of 62 people were given a taste from recipe B, and 41 said they liked it well enough to buy the mix. Use a 1% level of significance to test the claim that a higher proportion of people might purchase the recipe B mix.

7. The highway department is testing two types of reflecting paint for concrete bridge end pillars. The two kinds of paint are alike in every respect except that one is red and the other yellow. The red paint is applied to 12 bridges and the yellow paint is applied to 12 bridges. After a period of one year, reflectometer readings were made on all these bridge end pillars. (A higher reading means better visibility.) For the red paint the mean reflectometer reading was $\bar{x}_1 = 9.4$ with standard deviation $s_1 = 2.1$. For the yellow paint the mean was $\bar{x}_2 = 6.8$ with standard deviation $s_2 = 2.0$. Based on these data can we conclude that the yellow paint has less visibility after one year? (Use a 1% level of significance.)

8. A hospital reported that the normal death rate for patients with extensive burns (more than 40% of skin area) has been significantly reduced by the use of new fluid plasma compresses. Before the new treatment the mortality rate for extensive burn patients was about 60%. Using the new compresses the hospital found that only 40 out of 90 patients with extensive burns died. Use a 1% level of significance to test the claim that the mortality rate has dropped.

9. The Fleetfoot Shoe Company claims that the average yearly salary of its workers is $19,800. The union believes that the average salary is much less. A random sample of 64 employees shows their average salary to be $19,400 with a standard deviation of $800. Use a 5% level of significance to test the company's claim.

10. A random sample of 100 recorded deaths in the United States during the past year gave an average life span of 71.2 yr with standard deviation 9.3 yr. Does this indicate that the average life span today is more than 70 yr? Use a 1% level of significance.

11. The student council is thinking about discontinuing the student poetry magazine because only 20% of the students read it. A vote was taken and it was decided to continue the magazine if more than 20% of the students are known to read it. A random sample of 256 students showed that 77 of them had read the last issue. Use a 5% level of significance to determine whether the magazine should be continued.

12. A comparison is made between two bus lines to determine if arrival times of their regular buses from Denver to Durango are off schedule by the same amount of time. For 81 randomly selected runs, bus line A was observed to be off schedule an average time of 53 min with standard deviation of 19 min. For 100 randomly selected runs, bus line B was observed to be off schedule an average of 62 min with standard deviation 15 min. Do the data indicate a significant difference in off-schedule times? Use a 5% level of significance.

13. The Nero Match Company sells matchboxes that are supposed to have an average of 40 matches per box. A random sample of 94 Nero matchboxes shows the average number of matches per box to be 43.1 with standard deviation 6. Using a 1% level of significance, can you say that the average number of matches per box is more than 40?

14. A study is made of residents in Phoenix and its suburbs concerning the proportion of residents that subscribe to *Sporting News*. A random sample of 88 urban residents showed that 12 subscribed, and a random sample of 97 suburban residents showed that 18 subscribed. Does this indicate that a higher proportion of suburban residents subscribe to *Sporting News?* (Use a 5% level of significance.)

15. An independent rating service is trying to determine which of two hamburger stands has quicker service. Over a period of 16 randomly selected times the average waiting period at Queen Burger is 4.8 min with standard deviation 2.0 min. The average waiting period at McGregor over a period of 14 randomly selected times is 5.2 min with standard deviation 1.8 min. Using a 5% level of significance, can we say there is a difference in the average waiting time at Queen Burger and McGregor?

16. A random sample of 200 of the state's registered voters shows 86 favor Dermont for senator. Dermont claims to have 51% of the voters. Use a 5% level of significance to determine if the data indicate that the true proportion is less than 51%.

17. Professor Gillespie claims that a college algebra student can increase his or her test score if provided with sample problems in advance of the exam. To test this claim, Professor Gillespie gave an exam but no advance sample problems to a random sample of 10 algebra students. The average score on this exam was 73.1 with standard deviation 6.9. She gave another random sample of 12 algebra students the same exam but also gave the students sample problems in advance. The students' average score was 79.4 with standard deviation 8.0.
    a) Using a 1% level of significance, can we say that Professor Gillespie's claim is justified?
    b) Answer part (a) using a 5% level of significance.

18. The percent of adults living in Rocky Valley who are college graduates is estimated to be 41%. A random sample of 49 adults shows that 13 are college graduates. Does this indicate that the percentage of college graduates is lower than 41? (Use a 5% level of significance.)

19. A machine in the student lounge dispenses coffee. The average cup of coffee is supposed to contain 7.0 oz. Eight cups of coffee from this machine show the average content to be 7.3 oz with a standard deviation of 0.5 oz. Do you think the machine has slipped out of adjustment and the average amount of coffee per cup is different from 7 oz? Use a 5% level of significance.

20. Two types of flu vaccine are being tested to determine if there is a difference in the proportion of adult patient reactions. Twelve out of a random sample of 196 adults given

vaccine A had a reaction. Seven out of another random sample of 196 adults given vaccine B had a reaction. Use a 1% level of significance to test the claim that a higher proportion of adults have a reaction to vaccine A than vaccine B.

21. Six sets of identical twins were randomly selected from a population of identical twins. One child was taken at random from each pair to form an experimental group. These children participated in a program designed to promote creative thinking. The other child from each pair was part of the control group that did not participate in the program to promote creative thinking. At the end of the program, a creative problem-solving test was given with the results shown in Table 9-26.

TABLE 9-26 Scores on Creative Problem-Solving Test

Twin Pair	A	B	C	D	E	F
Experimental group	53	35	12	25	33	47
Control group	39	21	5	18	21	42

Higher scores indicate better performance in creative problem solving. Do the data support the claim that the program of the experimental group did promote creative problem solving? ($\alpha = 0.01$.)

22. A marketing consultant was hired to visit a random sample of five sporting goods stores across the state of California. Each store was part of a large franchise of sporting goods stores. The consultant taught the managers of each store better ways to advertise and display their goods. The net sales for one month before and one month after the consultant's visit were recorded for each store (in thousands of dollars) (Table 9-27).

TABLE 9-27 Net Sales (in Thousands of Dollars)

Store	1	2	3	4	5
Before visit	57.1	94.6	49.2	77.4	43.2
After visit	63.5	101.8	57.8	81.2	41.9

Do the data indicate that the net sales improved? (Use $\alpha = 0.05$.)

23. The manufacturer of a sports car claims that the fuel injection system lasts 48 months before it needs to be replaced. A consumer group tests this claim by surveying a random sample of 10 owners who had the fuel injection system replaced. The ages of the car at the time of replacement were (in months):

29 42 49 58 53
46 30 51 42 62

a) Use your calculator to verify that the mean age of a car when the fuel injection system fails is $\bar{x} = 46.2$ mo with standard deviation $s = 10.85$ mo.
b) Test the claim that the fuel injection system lasts less than 48 months before needing replacement. Use a 5% level of significance.

24. Leeville has two high schools; one in the older part of town and the other in the new section of town. The school board wants to determine if student performance at the two

schools is different. To make the comparison, they decided to compare composite scores on the ACT college admissions test for seniors seeking admission to college. A random sample of ten ACT composite scores from each school was used. The ACT composite scores follow.

Old school $\bar{x}_1$:	20	24	27	26	21	16	29	14	8	20
New school $\bar{x}_2$:	12	28	20	20	18	27	24	7	19	24

a) Use a calculator to verify that $\bar{x}_1 = 20.5$ and $s_1 = 6.47$ while $\bar{x}_2 = 19.9$ and $s_2 = 6.52$.

b) Test the claim that there is a difference in the ACT test scores from the two schools at the 5% level of significance.

25. Given $H_0: \mu = 5$, $H_1: \mu \neq 5$, and P value 0.0213
   a) do we accept or reject H_0 at a 1% level of significance?
   b) do we accept or reject H_0 at a 5% level of significance?

26. Given $H_0: p = 0.5$ and $H_1: p > 0.5$, and P value $= 0.0023$
   a) do we accept or reject H_0 at a 1% level of significance?
   b) do we accept or reject H_0 at a 5% level of significance?

27. Given $H_0: \mu = 2$ and $H_1: \mu < 2$, and observed sample mean $\bar{x} = 1.3$, sketch a picture of the $\bar{x}$ distribution and show the location of the area corresponding to the P value.

28. Given $H_0: \mu = 6$ and $H_1: \mu \neq 6$, and observed sample mean $\bar{x} = 2$, sketch a picture of the $\bar{x}$ distribution and show the location of the areas corresponding to the P value.

USING COMPUTERS

Professional statisticians in industry and research use computers to help them analyze and process statistical data. There are many computer programs available for statistics. Some commonly used statistical packages include Minitab, SAS, and SPSS. There are many others as well. Any of these statistical packages may be used with this text.

The authors of this text have also written a computer statistics package called ComputerStat. The programs in ComputerStat are available on disks with an accompanying manual that explains how to use the programs and gives additional computer applications. The disks are intended for the Apple IIe, IIc, or the IBM PC. Programs are written in the computer language BASIC and are designed for the beginning statistics student with no previous computer experience. ComputerStat is available as a supplement from the publisher of this text.

Problems in this section may be done using ComputerStat or other statistical computer programs.

Note on P Values

In ComputerStat most of the programs are written to reflect the methods presented in hypothesis testing. In such cases we use a specified value for the level of significance to find the critical value and critical region. We accept or reject H_0 based on whether or not the sample statistic falls in the acceptance or critical region. This is the same procedure used in most hypothesis testing problems and examples in the text. However, all ComputerStat programs for statistical testing also give the P value for the test. The P value method for accepting or rejecting the null hypothesis is commonly used in statistical software packages. Students should be familiar with the use of P values. (*See* Section 9.7.)

People who do shift work must often adjust their eating habits, sleep habits, exercise habits, family contact, social life, and overall life-style to accommodate their job. Extensive rearrangement of a person's habits and life-style can sometimes result in tension, anxiety, and overall health problems. An extensive study of the health consequences of shift work can be found in the following publication. Interested readers are referred to this report:

United States Department of Health, Education, and Welfare, NIOSH Technical Report. Tasto, Colligan, *et al., Health Consequences of Shift Work.* Washington: GPO, 1978.

1. In an effort to study mood levels of nurses working in large hospitals an opinion scale was used. Opinion ratings ranged from 0 = no feeling of tension and anxiety to a rating of 4 = extensive feelings of tension and anxiety. The scale was continuous so a nurse could mark any number between 0 and 4. Suppose a random sample of 35 nurses on the day shift of a very large hospital gave the ratings shown in Table 9-28 of their feelings of tension and anxiety at the end of the day shift.

   a) In the ComputerStat menu of topics, select Hypothesis Testing. Then use program S14,

TABLE 9-28 Data for Day-Shift Nurses

3.50	3.75	2.33	2.16	3.50	0.80	1.25
1.33	2.67	2.50	1.50	0.75	0.00	0.67
4.00	3.75	3.50	3.25	2.40	3.50	2.75
3.50	2.67	2.80	2.33	3.50	3.80	2.75
1.50	1.33	0.00	2.25	1.75	0.50	1.75

Testing a Single Population Mean, to complete parts (b)–(g).

b) On a scale of 0 to 4 a moderate (i.e. medium) level of tension and anxiety is 2. Use the null hypothesis that the population mean tension level for nurses after the day shift is 2.

c) Use the alternate hypothesis that the population mean tension level is different (either higher or lower) than 2.

d) Use a 5% level of significance.

e) Next enter the sample size and sample data.

f) Have the computer find the critical region and conclude the test. What is the *P* value? Compare the *P* value with the level of significance. Do you think we should accept or reject the null hypothesis?

g) Rerun the program with the same data (you do not need to re-enter the data). Use the same null hypothesis, but use a right-tail test with level of significance 0.10. Do we accept or reject the null hypothesis? Look at the P value. What is the smallest level of significance that will result in a rejection of the null hypothesis using a right-tail test?

2. Suppose a random sample of 33 nurses on the night shift were also asked to give their opinion about feelings of tension and anxiety at the end of the night shift. They used the same rating scale as described in problem 1. The ratings they gave are in Table 9-29.

TABLE 9-29 Data for Night-Shift Nurses

3.50	3.75	3.50	3.10	3.20	3.33	1.75
2.75	2.50	2.75	3.20	3.75	4.00	2.00
1.00	0.00	1.80	2.50	3.50	3.00	2.60
3.10	2.75	4.00	2.90	1.75	2.20	3.50
1.00	2.50	0.80	3.70	2.60		

a) Explain why the samples of problems 1 and 2 are independent, or at least why it is reasonable to assume they are independent. We want to compare mean tension levels of day- and night-shift nurses. In the ComputerStat menu of topics select Hypothesis Testing. Then use program S15, Testing a Difference of Means (Independent Samples), to complete parts (b)–(f).

b) In this problem let $MU1$ = population mean tension level for the day-shift nurses and let $MU2$ = population mean tension level for the night-shift nurses. What is the null hypothesis?

c) If we want to test the claim that there is a difference (either way) between tension levels of day-shift and night-shift nurses, which alternate hypothesis do we use? Tell the computer to use a 10% level of significance.

d) Using $N1$ = 35 as the first sample size of day-shift nurses and $N2$ = 33 as the second sample size of night-shift nurses, enter the sample data.

e) Have the computer find the critical region and complete the test. What is the *P* value? Compare the *P* value with the level of significance. Shall we accept or reject the null hypothesis? What is the smallest level of significance that will result in a rejection of the null hypothesis in this two-tail test?

f) Rerun the program with the same data (you do not need to re-enter the data). Use the same null hypothesis, but use a left-tail test with a level of significance 10%. Do we accept or reject the null hypothesis? What is the smallest level of significance at which we can say the night-shift nurses show a higher average level of tension?

3. Suppose a random sample of eight nurses were changed from the day shift to a rotating shift of some night work and some day work. For each of these nurses the information shown in Table 9-30 was recorded about feelings of tension and anxiety at the end of a day shift and also at the end of a night shift.

a) Explain why the sample data for the day shift cannot be thought of as independent of the

TABLE 9-30 Data for Nurses Working Both Shifts

Nurse	1	2	3	4	5	6	7	8
Day shift (B)	1.5	3	2	3	2	2	1	2
Night shift (A)	3.5	2	4	4	3.5	2	3.5	3

sample data for the night shift. In the ComputerStat menu of topics select Hypothesis Testing. Then use program S16, Paired Difference Test (Dependent Samples), to complete parts (b)–(d).

b) Let us say that *A* is the random variable representing tension levels of night nurses and *B* is the random variable representing tension levels of day nurses. If we want to test the claim that nurses have a higher level of tension after the night shift, what would we use for the null hypothesis? What would we use for the alternate hypothesis? Choose the appropriate hypothesis and enter your choice on the computer. Use a 2% level of significance.

c) Let the computer find the critical region and complete the test. Shall we accept or reject the claim that after a night shift nurses express more feelings of tension on the average than they do after a day shift?

d) What is the smallest level of significance at which this data will allow us to accept the claim that after a night shift nurses express more feelings of tension?

Data for Problems 4 and 5 are taken from the following source:

Miller, Miller, and Schneider. *American National Election Studies Data Sourcebook 1952–1978.* Cambridge: Harvard University Press, 1980. 250, 261.

4. In a study of support of the American political system the reference by Miller, Miller, and Schneider reports that the following statement was given to a large random sample of adults:

People like me don't have any say about what the government does.

a) In the ComputerStat menu of topics, select Hypothesis Testing. Then use program S17, Testing a Proportion, to solve the following problem.

For 1978 Miller, Miller, and Schneider reported that out of 2,291 people interviewed 1,214 disagreed with the statement. Let *p* represent the proportion of all adults in the United States who disagreed with the statement (in 1978). Use the null hypothesis that $p = 0.50$ and the alternate hypothesis that $p > 0.50$ to test the claim that more than half the population in 1978 believed that they had a say in what government does. Use a 1% level of significance.

b) Recall that the level of significance is the probability of a type I error; that is, the probability that we are *wrong* in rejecting the null hypothesis and accepting the alternate hypothesis. Using the data provided, what is the smallest level of significance at which we can accept the claim that more than 50% of the American people believed they had a say in what the government does?

5. In a study similar to that of problem 4, Miller, Miller, and Schneider gave the following statement to a large random sample of American adults:

(Political) Parties are only interested in people's votes, but not in their opinions.

For the year 1970 it was reported that 1,487 people were asked about the statement. Of those asked, 803 agreed with the statement. For the year 1978 it was reported that 2,278 people were asked about the statement and 1,412 agreed with it.

In the ComputerStat menu of topics, select Hypothesis Testing. Then use program S18,

Testing a Difference of Proportions, to complete parts (a)–(d).

a) Let $P1$ = proportion of all American adults who agreed with the statement in 1970 and let $P2$ = proportion of all American adults who agreed with the statement in 1978.

   i) What value will you use for $N1$ = number of trials in the first distribution?

   ii) What value will you use for $R1$ = number of successes in the first distribution?

   iii) What value will you use for $N2$ = number of trials in the second distribution?

   iv) What value will you use for $R2$ = number of successes in the second distribution?

b) What statement will you use for the null hypothesis? If we want to test the claim that a higher proportion of American adults agreed with the statement in 1978 than agreed with the statement in 1970, what should the alternate hypothesis be?

c) Use a 2% level of significance and the results of the computer run to complete the test. What is the critical region? Should we accept or reject the claim that a higher proportion of American adults agreed with the statement in 1978 than in 1970?

d) What is the smallest level of significance at which we can accept the claim that a higher proportion of the American adults agreed with the statement in 1978 than agreed in 1970?

10 Regression and Correlation

A mathematician, like a poet or painter, is a maker of patterns. If his patterns are more permanent than theirs it is because they are made with ideas. . . . . A mathematician has no material to work with but ideas, and so his patterns are likely to last longer.

G. H. Hardy

In business administration, economics, social science, and natural science we often encounter statistical variables that seem to follow a pattern. In this chapter we will study mathematical "patterns" that originate from linear correlations between statistical variables.

SECTION 10.1   Introduction to Paired Data and Scatter Diagrams

Statistical applications often involve several random variables. In this chapter we will study the techniques for dealing with data associated with two variables.

In some problems variables are studied simultaneously to see how they are interrelated. These are called *correlation* problems. The word correlation literally means "related together" or "co-relation."

For other problems there may be one variable of particular interest. In this case the other variables are used to *predict* how the first variable will behave under given conditions. Problems of this type are called *regression* problems. The methods of regression literally predict the value of one variable by going back to (or regressing to) the values of another related variable.

The study of correlation and regression usually begins with a table and/or a graph of *paired data values*. Example 1 shows what we mean.

EXAMPLE 1   Jan is doing a project for her botany class. She wonders if there is a connection between the average weight of watermelons a vine produces and the root depth of the vine. Jan suspects that vines with deeper roots have a better water supply and also larger average melons. From a large watermelon field 30 vines are chosen at random. At the end of eight weeks the watermelons are removed from each vine, and the average weight of watermelons from each vine is determined. Then each plant is carefully dug up and its root depth (or length) is measured. Table 10-1 shows the results.

TABLE 10-1 Results of a Botany Experiment

| | x = Root Depth (in.) y = Mean Wt of Watermelon (lb) |

x	y	x	y	x	y
26.7	20.3	17.5	4.3	9.1	4.5
14.0	4.8	13.1	8.7	17.5	11.1
18.0	9.0	16.5	9.1	20.5	12.3
10.5	9.0	28.4	17.1	4.5	3.1
26.1	10.7	23.9	9.7	23.4	20.8
21.5	17.4	27.0	13.1	27.0	15.2
7.0	8.2	10.2	2.1	24.6	16.1
26.0	7.9	13.1	11.0	15.8	5.7
13.9	3.5	19.1	16.0	27.0	19.3
19.3	8.7	22.6	12.9	21.3	4.2

a) For each plant there is an x,y pair of data values. If we plot the x,y pairs we obtain the point graph of Figure 10-1. Figure 10-1 is called a *scatter diagram* for the paired data values of Table 10-1.

FIGURE 10-1 Scatter Diagram for the Botany Experiment

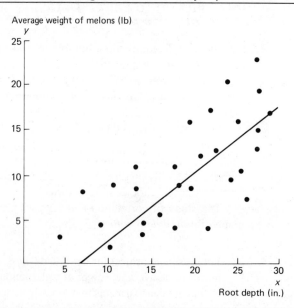

b) By inspecting Figure 10-1 we see that to some extent larger values of x tend to be associated with larger y values, and smaller x values tend to be associated with smaller y values. Roughly speaking, the general trend seems to be reasonably well represented by the line segment shown in Figure 10-1.

Of course, it is possible to draw many curves in Figure 10-1, but the straight line is the simplest and most widely used curve for elementary studies of paired data. We can draw many lines in Figure 10-1, but in some sense the "best" line should be the one that comes closest to each of the points of the scatter diagram. To single out one line as the "best fitting line" we must find a mathematical criterion for this line and a formula representing the line. This will be done in Section 10.2 by the *method of least squares*.

In some sense there is another problem which precedes that of finding the "best fitting line." That is the problem of determining how well the points of the scatter diagram are suited for fitting *any* line. Certainly if the points are a very poor fit to *any* line, there is little use in trying to find the "best" line. This problem will be dealt with in Section 10.3 by use of the Pearson product-moment coefficient of correlation.

If the points of a scatter diagram are located so that *no* line is realistically a "good" fit, we say that the points possess *no linear correlation*. In Figure 10-2 we see some scatter diagrams for which there is no linear correlation.

FIGURE 10-2   Scatter Diagrams with No Linear Correlation

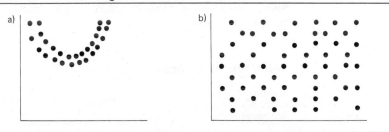

On the other hand, if all the points do in fact lie on a line, then we have perfect *linear correlation*. In Figure 10-3 we see some diagrams with perfect linear correlation. In statistical applications perfect linear correlation almost never occurs.

FIGURE 10-3   Scatter Diagrams with Perfect Linear Correlation

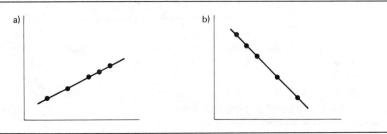

_____ Exercise 1 _____

FIGURE 10-4 Scatter Diagrams

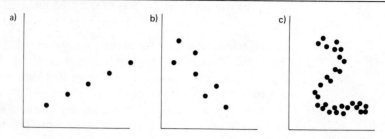

Which of the scatter diagrams in Figure 10-4 do you think
  i) has no linear correlation?
 ii) has perfect linear correlation?
iii) can be reasonably fitted by a straight line?

Figure 10-4a has perfect linear correlation and can be fitted exactly by a straight line.

Figure 10-4b can be reasonably fitted by a straight line.

Figure 10-4c has no linear correlation. No straight-line fit should be attempted.

_____ Exercise 2 _____

A large industrial plant has seven divisions that do the same type of work. A safety inspector visits each division of 20 workers quarterly. The number of work-hours devoted to safety training and the number of work-hours lost due to industry-related accidents are recorded for each separate division in Table 10-2.

TABLE 10-2 Safety Report

Division	x (No. of Work-Hours in Safety Training)	y (No. of Work-Hours Lost Due to Accidents)
1	10.0	80
2	19.5	65
3	30.0	68
4	45.0	55
5	50.0	35
6	65.0	10
7	80.0	12

a) Make a scatter diagram for these pairs. Use the x values on the horizontal axis and the y values on the vertical one.

b) As the number of hours spent on safety training increases, what happens in general to the number of hours lost due to industry-related accidents?

c) Does a line fit the data reasonably well?

a) Figure 10-5 shows the scatter diagram.

b) In general, as the number of hours in safety training goes up the number of hours lost due to accidents goes down.

c) A line fits reasonably well.

FIGURE 10-5   Scatter Diagram for
Safety Report

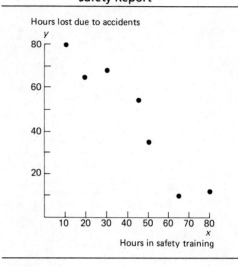

d) Draw a line which you think "fits best."

d) Any line which seems the best fit to you is correct. Later you will see the equation of a line that is a "best fit."

We say a scatter diagram has *high* linear correlation if the points lie close to a straight line. If the points are not close to a straight line, we say the correlation is *moderate* or *low*. If the points fit no straight line, we say there is *no* linear correlation.

Section Problems 10.1

In problems 1–6 look at the scatter diagrams and state which of the following conditions you think is true for each diagram:

i) High linear correlation
ii) Moderate or low linear correlation
iii) No linear correlation

1.   2.

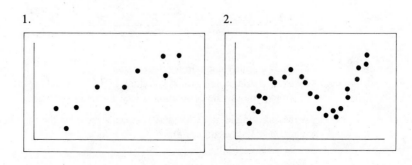

3.

4.

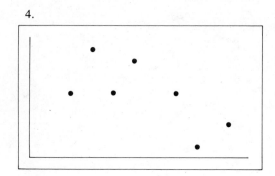

5.

6.

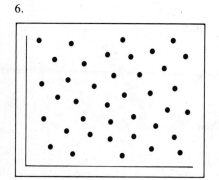

7. Urban travel times and distances are important factors in the analysis of traffic flow patterns. A traffic engineer in Los Angeles obtained the following data from area freeways where x = miles traveled and y = time in minutes for a passenger car.

x (miles)	5	9	3	11	20	15	12	25
y (min)	9	13	6	16	28	21	16	31

a) Draw a scatter diagram for the data.
b) Draw a straight line which you think best fits the data.
c) Would you say the correlation is low, moderate, or high?

8. A random sample of 15 married couples had their heights measured with the following results (x = man's height, y = woman's height):

x (in.)	53	61	57	59	55	63	60	65	56	59	68	67	71	74	58
y (in.)	53	58	56	57	56	61	57	58	59	56	62	60	65	62	56

a) Draw a scatter diagram for these data.
b) Draw a straight line which you think best fits the data points.
c) Would you say the correlation is low, moderate, or high?

9. A random sample of 11 Math-101 students provided the following data (x = midterm score, y = final exam score):

x	66	71	80	76	63	73	81	68	84	83	79
y	130	132	135	134	129	133	136	132	138	136	135

a) Draw a scatter diagram for these data.
b) Draw a straight line which you think best fits the data points.
c) Would you say the correlation is low, moderate, or high?

10. An ecology class did a project to determine the distance a car will travel on 1 gal of gas at different rates of speed. A local dealer loaned the class a new car. The car was given 1 gal of gasoline and driven around a level track at a constant speed until it ran out of gas. Let x be the speed (mph) and y be the distance the car ran. The following data were obtained:

x	30	35	40	45	50	55	60	65	70	75	80	85
y	36	33	34	32.5	31	32	30	29.5	26.5	27	24.5	24

a) Draw a scatter diagram for the data.
b) Draw a straight line which you think best fits the data.
c) Would you say the correlation is low, moderate, or high?

SECTION 10.2   Linear Regression and Confidence Bounds for Prediction

Anyone who has been outdoors on a summer evening has probably heard crickets. Did you know that it is possible to use the cricket as a thermometer? Crickets tend to chirp more frequently as temperatures increase. A Harvard physics professor made a detailed study of this phenomenon; using sophisticated equipment, Professor George W. Pierce studied the striped ground cricket and compiled the data in Table 10-3.

Do the data indicate a linear relation between chirping frequency and temperature? Is there a way we can use the data to predict the temperature that corresponds to a chirp frequency that is not listed in the table? For instance, how can we use the data to predict the temperature for $x = 19$ chirps per second? Let us first make a scatter diagram (Figure 10-6) for the data of Table 10-3.

Looking at the scatter diagram of Figure 10-6, we ask two questions: can we find a relationship between x and y; and, if so, how strong is the relationship?

The first step in answering these questions is to try to express the relationship as a mathematical equation. There are many possible equations, but the simplest and most widely used is the linear equation, or the equation of a straight line.

Our job is to find a linear equation that is the "best" linear equation representing the points of the scatter diagram. For our criterion of best fitting line we use the *least squares criterion,* which says the line we fit to the data points must be such that *the sum of the squares of the vertical distances from the points to the line be made as small as possible.* The least squares criterion is illustrated in Figure 10-7.

TABLE 10-3 Chirping Frequency and Temperature for the Striped Ground Cricket

x (chirps/sec)	y (temp., °F)
20.0	88.6
16.0	71.6
19.8	93.3
18.4	84.3
17.1	80.6
15.5	75.2
14.7	69.7
17.1	82.0
15.4	69.4
16.2	83.3
15.0	79.6
17.2	82.6
16.0	80.6
17.0	83.5
14.4	76.3

Reprinted by permission of the publisher from *The Songs of Insects* by George W. Pierce (Cambridge: Harvard University Press, 1948).

FIGURE 10-6 Scatter Diagram for Table 10-3

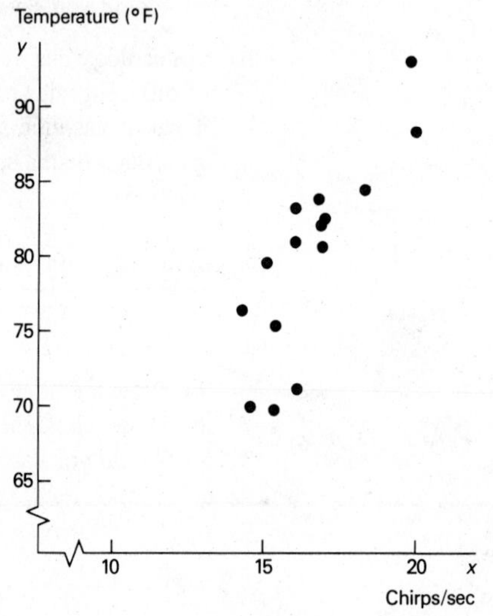

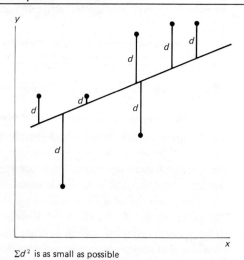

FIGURE 10-7 Least Squares Criterion

Σd^2 is as small as possible

It is important to observe that we are going to minimize the sum of the vertical distances of Figure 10-7 over all points in the scatter diagram. The unsquared quantity d will be positive for points above the line and negative for points below the line. As a result, the sum of the unsquared quantities can be small even if the points are widely spread in the scatter diagram. However, the squares d^2 cannot be negative. By minimizing the sum of the squares, we are in effect not allowing positive and negative distances to "cancel out" one another.

Techniques of calculus can be applied to show that the line which meets the least squares criterion is as follows:

- **Least Squares Line**

$$y = a + bx \qquad (1)$$

where

$$b = \frac{SS_{xy}}{SS_x} \qquad (2)$$

$$a = \bar{y} - b\bar{x} \qquad (3)$$

and $\bar{y}$ = mean of y values in scatter diagram
$\bar{x}$ = mean of x values in scatter diagram

$$SS_{xy} = \Sigma xy - \frac{(\Sigma x)\,(\Sigma y)}{n} \qquad (4)$$

$$SS_x = \Sigma x^2 - \frac{(\Sigma x)^2}{n} \qquad (5)$$

n = number of points in scatter diagram
In formulas (4) and (5) the sums are taken over all x or y values in the scatter diagram.

The simplest way to find a and b is to organize your work into a table. In Table 10-4 we use the data relating rate of cricket chirps to temperature to obtain the sums needed for formulas (4) and (5). We will then use these sums to derive the formula for the least squares line.

The formulas used to compute the slope and y-intercept of the least squares line are sensitive to rounding. Answers to the exercises at the end of each section and at the end of the chapter are computer-generated, so you may expect slight differences in your answers depending on how you round.

- **Comment:** The notation SS_x is the same expression used in the computation for the standard deviation of x values (see Section 3.3). Recall that the formula for SS_x is simply a faster way of computing $\Sigma(x - \bar{x})^2$. Likewise, the formula for SS_{xy} is a more efficient way to compute $\Sigma(x - \bar{x})(y - \bar{y})$.

TABLE 10-4 Sums for Computing $\bar{x}$, $\bar{y}$, SS_x and SS_{xy}

x (chirps/sec)	y (°F)	x^2	xy
20.0	88.6	400.0	1,772.0
16.0	71.6	256.0	1,145.6
19.8	93.3	392.0	1,847.3
18.4	84.3	338.6	1,551.1
17.1	80.6	292.4	1,378.3
15.5	75.2	240.3	1,165.6
14.7	69.7	216.1	1,024.6
17.1	82.0	292.4	1,402.2
15.4	69.4	237.2	1,068.8
16.2	83.3	262.4	1,349.5
15.0	79.6	225.0	1,194.0
17.2	82.6	295.8	1,420.7
16.0	80.6	256.0	1,289.6
17.0	83.5	289.0	1,419.5
14.4	76.3	207.4	1,098.7
$\Sigma x = 249.8$	$\Sigma y = 1,200.6$	$\Sigma x^2 = 4,200.6$	$\Sigma xy = 20,127.5$

From Table 10-4 we have

$$SS_x = \Sigma x^2 - \frac{(\Sigma x)^2}{n} = 4,200.6 - \frac{(249.8)^2}{15} = 40.6$$

and

$$SS_{xy} = \Sigma xy - \frac{(\Sigma x)(\Sigma y)}{n} = 20,127.5 - \frac{(249.8)(1,200.6)}{15} = 133.5$$

We also find

$$\bar{x} = \frac{\Sigma x}{n} = \frac{249.8}{15} = 16.7$$

$$\bar{y} = \frac{\Sigma y}{n} = \frac{1,200.6}{15} = 80.0$$

Therefore, using formulas (2) and (3), we find a and b in the equation of the least squares line.

$$b = \frac{SS_{xy}}{SS_x} = \frac{133.5}{40.6} = 3.3$$

$$a = \bar{y} - b\bar{x} = (80.0) - (3.3)(16.7) = 24.9$$

We conclude that the least squares line for the data of Table 10-3 is

$$y = a + bx \tag{6}$$

$$y = 24.9 + 3.3x$$

To graph the least squares line (6) we have several options available. The slope-intercept method of College Algebra is probably the quickest. The slope is $b = 3.3$ and the y-intercept is $a = 24.9$. However, if you don't remember this method, it is almost as easy to plot two points and connect them with a straight line. For x values we usually use any two values in the range of x data values. Corresponding y values are computed from the equation of the least squares line.

The value of $\bar{x}$ will always be in the range of x values. When we use $x = \bar{x}$ in equations (1) and (3), we see that the corresponding y value is $\bar{y}$. Therefore, $(\bar{x}, \bar{y})$ will always be on the least squares line. Since we have already computed these values, the point $(\bar{x}, \bar{y})$ is a convenient choice for one of the two points we use to graph the least squares line. From Table 10-3 we see that $x = 20$ is also in the range of x values. We compute the corresponding y value by using the equation of the least squares line.

x	$y = 24.9 + 3.3x$
When we choose $x = \bar{x} = 16.7$	$y = 24.9 + 3.3(16.7) = 80.0 = \bar{y}$
When we choose $x = 20.0$	$y = 24.9 + 3.3(20.0) = 90.9$

The line going through the points (16.7, 80.0) and (20.0, 90.9) is the least squares line for the scatter diagram of Figure 10-6. This line is shown in Figure 10-8.

FIGURE 10-8 Least Squares Line $y = 24.9 + 3.3x$

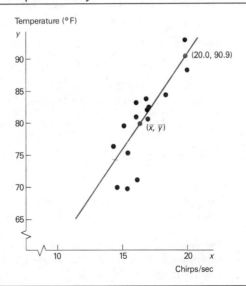

Now suppose we find a striped ground cricket and, with a listening device, discover it chirps at the rate of 19.0 chirps per second. What should we predict for the temperature? We could read the y value above $x = 19.0$ from the least squares line graphed in Figure 10-8. But a more accurate estimate can be obtained by using the value $x = 19.0$ in the equation of the least squares line and computing the corresponding y.

$y = 24.9 + 3.3x$	equation of least squares line
$y = 24.9 + 3.3(19.0)$	use 19.0 in place of x
$y = 87.6\,°\text{F}$	evaluate y

Rounded to the nearest whole number, we should predict the temperature to be 88 °F. Of course this is just a prediction, and we would be quite happy if the temperature turned out to be relatively close to our prediction. This brings up the natural question: How *good* are predictions based on the least squares line? This is a fairly difficult question, and much of the answer requires advanced mathematics; however, a partial answer will be given later in this section.

_____ Exercise 3 _____

The Quick Sell car dealership has been using one-minute spot ads on a local TV station. The ads always occur during the evening hours and advertise the different models and price ranges of cars on the lot that week. During a ten-week period the Quick Sell dealer kept a weekly record of the number of TV ads versus the number of cars sold. The results are given in Table 10-5.

TABLE 10-5

x (No. of Ads in a Week)	y (No. of Cars Sold That Week)
6	15
20	31
0	10
14	16
25	28
16	20
28	40
18	25
10	12
8	15

The manager decided that Quick Sell can only afford 12 ads per week. At that level of advertisement, how many cars can Quick Sell expect to sell each week? We'll answer this question in several steps.

a) Draw a scatter diagram for the data.

b) Look at equations 1–5 pertaining to the least squares line. Two of the quantities we need to find b are (Σx) and (Σxy). List the others.

a) The scatter diagram is shown in Figure 10-9.

b) We also need n, (Σy), (Σx^2), and $(\Sigma x)^2$.

FIGURE 10-9   Scatter Diagram and Least Squares Line for Table 10-5

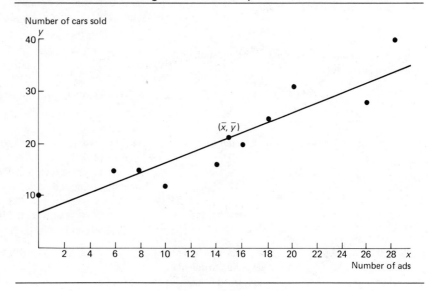

c) Complete Table 10-6.

TABLE 10-6

x	y	x^2	xy
6	15	36	90
20	31	400	620
0	10	0	0
14	16	196	224
25	28	625	700
16	20	256	320
28	40	_____	_____
18	25	_____	_____
10	12	_____	_____
8	15	64	120
$\Sigma x = 145$	$\Sigma y = 212$	$\Sigma x^2 =$ ____	$\Sigma xy =$ ____

c) The missing table entries are:

x^2	xy
$(28)^2 = 784$	$28(40) = 1{,}120$
$(18)^2 = 324$	$18(25) = 450$
$(10)^2 = 100$	$10(12) = 120$
$\Sigma x^2 = 2{,}785$	$\Sigma xy = 3{,}764$

d) Use Table 10-6 to compute SS_x, SS_{xy}, $\bar{x}$, and $\bar{y}$.

d)

$$SS_x = \Sigma x^2 - \frac{(\Sigma x)^2}{n} = 2{,}785 - \frac{(145)^2}{10} = 682.5$$

$$SS_{xy} = \Sigma xy - \frac{(\Sigma x)(\Sigma y)}{n} = 3{,}764 - \frac{(145)(212)}{10} = 690.0$$

$$\bar{x} = \frac{\Sigma x}{n} = \frac{145}{10} = 14.5$$

$$\bar{y} = \frac{\Sigma y}{n} = \frac{212}{10} = 21.2$$

e) Compute a and b in the formula

$$y = a + bx$$

for the least squares line. What is the equation of the least squares line?

e) $b = \dfrac{SS_{xy}}{SS_x} = \dfrac{690.0}{682.5} \approx 1.0$

$a = \bar{y} - b\bar{x}$

$\quad = 21.2 - 1.0(14.5)$

$\quad = 6.7$

The equation for the least squares line is

$$y = 6.7 + 1.0x$$

f) Plot the least squares line on your scatter diagram.

f) The least squares line goes through the point $(\bar{x}, \bar{y})$ $= (14.5, 21.2)$. To get another point on the line, select a value for x and compute the corresponding y value using the equation $y = 6.7 + 1.0x$. For $x = 20$, we get $y = 6.7 + 1.0(20) = 26.7$ so the point $(20, 26.7)$ is also on the line. The least squares line is shown in Figure 10-9.

g) Read the y value for $x = 12$ from your graph. Then use the equation of the least squares line to calculate y when $x = 12$. How many cars can the manager expect to sell if 12 ads per week are aired on TV?

g) The graph gives $y \approx 19$. From the equation we get

$y = 6.7 + 1.0x$

$\quad = 6.7 + 1.0(12)$ use 12 in place of x

$\quad = 18.7$

To the nearest whole number the manager can expect to sell 19 cars when 12 ads are aired on TV each week.

We have used the least squares line to predict y values for x values that were *between* x values observed in the experiment. Predicting y values for x values that are between x values of points in the scatter diagram is called *interpolation*. The least squares lines can be used for interpolation. Predicting y values for an x value beyond the range of observed x values is a complex problem that is not treated in this book. Prediction beyond the range of observations is called *extrapolation*.

The formula

$$y = a + bx$$

can only be used to predict y values from specified x values. If you wish to begin with y values and predict corresponding x values, you must use a different formula. For our purposes we'll always arrange to predict y values from given x values.

Sometimes a scatter diagram clearly indicates the existence of a linear relationship between x and y, but it can happen that the points are widely scattered around the least squares line. We need a method (other than just looking) for measuring the spread of a set of points about the least squares line. There are two commonly used methods of measuring the spread. One involves the *standard error of estimate*. The other, which will be studied in the next section, involves the *coefficient of correlation*.

For the standard error of estimate we use a measure of spread that is in some ways like the standard deviation of measurements of a single variable. Let

$$y_p = a + bx$$

be the predicted value of y from the least squares line. Then $y - y_p$ is the vertical distance from the *point* (x, y) of the scatter diagram to the *least squares line* at (x, y_p), as shown in Figure 10-10. To avoid the difficulty of having some positive and some negative values, we square the quantity $(y - y_p)$. Then we sum the squares and, for technical reasons, divide this sum by $n - 2$. Finally, we take the square root to obtain the *standard error of estimate,* which we denote by S_e.

$$standard\ error\ of\ estimate = S_e = \sqrt{\frac{\Sigma(y - y_p)^2}{n - 2}}$$

$$where\ n \geqslant 3$$

(7)

Note: To compute the standard error of estimate we require that there be at least three points on the scatter diagram. If we had only two points, the line would be a perfect fit since two points determine a line. In such a case there would be no need to compute S_e.

The nearer the scatter points lie to the least squares line, the smaller S_e will be. In fact, if $S_e = 0$, it follows that each $y - y_p$ is also zero. This means that all the scatter points lie *on* the least squares line if $S_e = 0$. The larger S_e becomes, the more scattered the points become.

FIGURE 10-10   The Distance Between Points (x, y) and (x, y_p)

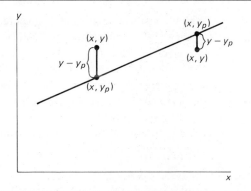

The formula for the standard error of estimate is reminiscent of the formula for the standard deviation. It too is a measure of dispersion. However, the standard deviation involves differences of data values from a mean, while the standard error of estimate involves the differences between experimental and predicted y values for a given x.

The actual computation of S_e using formula (7) is quite long since the formula requires us to use the least squares line equation to compute a predicted value y_p for *each x* value in the data pairs. There is a computational formula which we strongly recommend that you use. However, as with all the computation formulas we have used, be careful about rounding. The formula is sensitive to rounding, and you should carry as many digits as seems reasonable for your problem. Again, answers will vary depending on rounding used.

• **Formula to Calculate S$_e$**

$$S_e = \sqrt{\frac{SS_y - bSS_{xy}}{n-2}}$$ (8)

where

$$SS_y = \Sigma y^2 - \frac{(\Sigma y)^2}{n}$$

$$SS_{xy} = \Sigma xy - \frac{(\Sigma x)(\Sigma y)}{n}$$

$$SS_x = \Sigma x^2 - \frac{(\Sigma x)^2}{n}$$

$$b = \frac{SS_{xy}}{SS_x}$$

n = number of points in scatter diagram

Use caution in rounding.

With a considerable amount of algebra, formulas (7) and (8) can be shown to be mathematically equivalent. Formula (7) shows the strong similarity between the standard error of estimate and standard deviation. Formula (8) is a short cut calculation formula because it involves few subtractions and uses quantities SS_x, b, and SS_{xy} which are needed to determine the least squares line.

In the next example we show you how to compute the standard error of estimate using the computation formula. Then, in the following example and exercise we will show you how to use S_e to create confidence intervals for the y value corresponding to a given x value.

EXAMPLE 2

June and Jim are partners in the chemistry lab. Their assignment is to determine how much copper sulfate ($CuSO_4$) will dissolve in water at 10, 20, 30, 40, 50, 60, and 70 °C. Their lab results are shown in Table 10-7 where y is the weight in grams of copper sulfate which will dissolve in 100 g of water at x °C.

Sketch a scatter diagram, find the equation of the least squares line, and compute S_e.

TABLE 10-7 Lab Results (x = °C,
y = amount of $CuSO_4$)

x	y
10	17
20	21
30	25
40	28
50	33
60	40
70	49

Figure 10-11 on page 400 includes a scatter diagram for the data of Table 10-7. To find the equation of the least squares line and the value of S_e, we set up the following computational table.

TABLE 10-8 Computational Table

x	y	x^2	y^2	xy
10	17	100	289	170
20	21	400	441	420
30	25	900	625	750
40	28	1,600	784	1,120
50	33	2,500	1,089	1,650
60	40	3,600	1,600	2,400
70	49	4,900	2,401	3,430
$\Sigma x = 280$	$\Sigma y = 213$	$\Sigma x^2 = 14,000$	$\Sigma y^2 = 7,229$	$\Sigma xy = 9,940$

$$SS_x = \Sigma x^2 - \frac{(\Sigma x)^2}{n} = 14,000 - \frac{(280)^2}{7} = 2,800$$

$$SS_{xy} = \Sigma xy - \frac{(\Sigma x)(\Sigma y)}{n} = 9,940 - \frac{(280)(213)}{7} = 1,420$$

$$SS_y = \Sigma y^2 - \frac{(\Sigma y)^2}{n} = 7,229 - \frac{(213)^2}{7} = 747.714$$

$$b = \frac{SS_{xy}}{SS_x} = \frac{1,420}{2,800} = 0.507143 \approx 0.51$$

$$\bar{x} = \frac{280}{7} = 40$$

$$\bar{y} = \frac{213}{7} = 30.43$$

$$a = \bar{y} - b\bar{x} = \frac{213}{7} - (0.507143)\left(\frac{280}{7}\right) \approx 10.14$$

The equation of the least squares line is

$$y = a + bx$$
$$y = 10.14 + 0.51x$$

FIGURE 10-11   Scatter Diagram and Least Squares Line for Chemistry Experiment

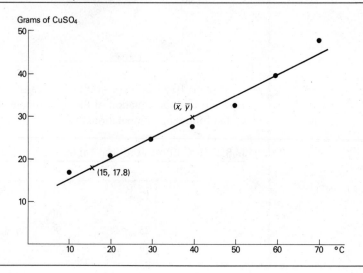

The graph of the least squares line is shown in Figure 10-11. Notice that it passes through the point $(\bar{x}, \bar{y}) = (40, 30.4)$. Another point on the line can be found by using $x = 15$ in the equation of the line $y = 10.14 + 0.51x$. When we use 15 in place of x, we obtain $y = 10.14 + 0.51(15) = 17.8$. The point $(15, 17.8)$ is the other point we used to graph the least squares line in Figure 10-11.

The standard error of estimate is computed using the computational formula

$$S_e = \sqrt{\frac{SS_y - bSS_{xy}}{n-2}}$$

$$= \sqrt{\frac{747.714 - (0.507143)(1,420)}{5}}$$

$$\approx 2.35$$

The least squares line gives us a predicted value y_p for a specified x value. However, we used sample data to get the equation of the line. The line derived from the population of all data pairs is likely to have a slightly different slope, which we designate by the symbol β for population slope and a slightly different y-intercept which we designate by the symbol α, for population intercept. In addition, there is some random error ϵ, so the true y value would be

$$y = \alpha + \beta x + \epsilon$$

Because of the random variable ϵ, for each x value there is a corresponding distribution of y values. The methods of linear regression were developed so that the distribution of y values for a given x is centered on the population regression line. Furthermore, the distributions of y values corresponding to each x value all have the same standard deviation, which we estimate by the standard error of estimate, S_e.

Using all of this background, the theory tells us that for a specific x, a c confidence interval for y is given by the next formula.

- *c* **Confidence Interval for** *y*

$$y_p - E \le y \le y_p + E$$

where

$$E = t_c S_e \sqrt{1 + \frac{1}{n} + \frac{(x - \bar{x})^2}{SS_x}}$$

and

y_p = the predicted value of y from the least squares line for the specified x value

t_c = the value from the Student's t distribution for
   a c confidence level using $n - 2$ degrees of freedom

S_e = the standard error of estimate (see formula (8))

$SS_x = \Sigma x^2 - \dfrac{(\Sigma x)^2}{n}$

n = number of data pairs

The formulas involved in the computation of a c confidence interval look complicated. However, they involve quantities we have already computed or values we can easily look up in tables. The next example illustrates this point.

EXAMPLE 3

Using the data of Table 10-7, find a 95% confidence interval for the amount of copper sulfate that will dissolve in 100 g of water at 45 °C. First we need to find y_p for $x = 45$ °C. We use the equation of the least squares line which we found in Example 2.

$$y = 10.14 + 0.51x \qquad \text{from Example 2}$$

$$y_p = 10.14 + 0.51(45) \qquad \text{use 45 in place of } x$$

$$y_p \simeq 33$$

A 95% confidence interval is then

$$33 - E \leqslant y \leqslant 33 + E$$

where

$$E = t_{0.95} S_e \sqrt{1 + \frac{1}{n} + \frac{(x - \bar{x})^2}{SS_x}}$$

Using $n - 2 = 7 - 2 = 5$ degrees of freedom, we find from Table 6 of Appendix I that $t_{0.95} = 2.571$. We computed S_e, SS_x, and $\bar{x}$ in Example 2. Therefore,

$$E = (2.571)(2.35) \sqrt{1 + \frac{1}{7} + \frac{(45 - 40)^2}{2,800}}$$

$$= (2.571)(2.35) \sqrt{1.15179}$$

$$\approx 6.5$$

A 95% confidence interval for y is

$$33 - 6.5 \leqslant y \leqslant 33 + 6.5$$

$$26.5 \leqslant y \leqslant 39.5$$

This means that we are 95% sure that the actual amount of copper sulfate that will dissolve in 100 g of water at 45 °C is between 26.5 g and 39.5 g. The interval is fairly wide, but would decrease with more sample data.

_____ Exercise 4 _____

Let's use the data of Example 2 to compute a 95% confidence interval for y = amount of copper sulfate that will dissolve at $x = 15$ °C.

a) From Example 2 we have

$$y = 10.14 + 0.51x$$

Evaluate y_p for $x = 15$.

b) The bound E on the error of estimate is

$$E = t_c S_e \sqrt{1 + \frac{1}{n} + \frac{(x - \bar{x})^2}{SS_x}}$$

From Example 2 we know that $S_e = 2.35$, $SS_x = 2,800$ and $\bar{x} = 40$. Recall that there were $n = 7$ data pairs. Find $t_{0.95}$ and compute E.

c) Find the 95% confidence interval

$$y_p - E \leqslant y \leqslant y_p + E$$

a) $y_p = 10.14 + 0.51x$

$$= 10.14 + 0.51(15)$$

$$\approx 17.8$$

b) $t_{0.95} = 2.571$

$$E = (2.571)(2.35) \sqrt{1 + \frac{1}{7} + \frac{(15 - 40)^2}{2,800}}$$

$$= (2.571)(2.35) \sqrt{1.366071}$$

$$\approx 7.1$$

c) The confidence interval is

$$17.8 - 7.1 \leqslant y \leqslant 17.8 + 7.1$$

$$10.7 \leqslant y \leqslant 24.9$$

As we compare the results of Example 3 and Exercise 4, we notice that the 95% confidence interval of y values for $x = 15$ °C is 7.1 units above and below the least squares line, while the 95% confidence interval of y values for $x = 45$ °C is only 6.5 units above and below the least squares line. This comparison reflects the general property that confidence intervals for y are narrower the nearer we are to the mean of the x values. As we move near the extremes of the x distribution, the confidence intervals for y become wider. This is another reason that we should not try to use the least squares line to predict y values for x values beyond the data extremes of the sample x distribution.

If we were to compute a 95% confidence interval for all x values in the range of the sample x values, the confidence interval band would curve away from the least squares line as shown in Figure 10-12.

FIGURE 10-12 95% Confidence Band for Predicted Values y_p

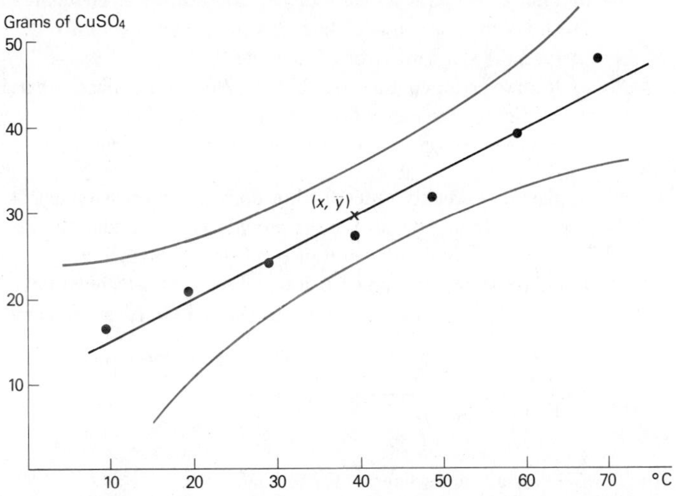

Section Problems 10.2

1. The number of workers on an assembly line varies due to the level of absenteeism on any given day. In a random sample of production output from several days of work the following data were obtained where x = number of workers absent from assembly line and y = number of defects coming off the line.

x	3	5	0	2	1	2	7
y	16	20	9	12	10	13	25

a) Draw a scatter diagram for the data.
b) Find $\bar{x}$, $\bar{y}$, and b. What is the equation of the least squares line?

c) Graph the least squares line on your scatter diagram.
d) Find the standard error of estimate S_e.
e) On a day when four workers are absent, what would the least squares line predict for the number of defects coming off the line?
f) Find a 95% confidence interval for the number of defects when four workers are absent.

2. In a random sample of eight military contracts involving cost overruns, the following information was obtained where x = bid price of contract (in millions of dollars) and y = cost of overrun (expressed as a percent of the bid price).

x	6	10	3	5	9	18	16	21
y	31	25	39	35	29	12	17	8

a) Draw a scatter diagram for the data.
b) Find $\bar{x}$, $\bar{y}$, and b. What is the equation of the least squares line?
c) Graph the least squares line on your scatter diagram.
d) Find the standard error of estimate S_e.
e) If an overrun contract was bid at 12 million dollars, what does the least squares line predict for the percent of overrun?
f) Find a 90% confidence interval for the cost overrun when the bid is 12 million dollars.

3. In placing a weekly order for hot dogs the concessionaire at a large baseball stadium needs to estimate the size of the crowd that will attend the game. Advanced ticket sales often give a good indication of expected attendance at games. Data from six previous weeks of games are shown below, where x = advanced ticket sales (in thousands) and y = number of hot dogs purchased (in thousands) at the game.

x	45	64	37	58	41	29
y	32	46	25	44	32	18

a) Draw a scatter diagram for the data.
b) Find $\bar{x}$, $\bar{y}$, and b. What is the equation of the least squares line?
c) Graph the least squares line on your scatter diagram.
d) Find the standard error of estimate S_e.
e) If advanced ticket sales are 55 thousand this week, how many thousand hot dogs do you recommend the concessionaire be prepared to sell? (Use the predicted value from the least squares line.)
f) Find a 99% confidence interval for the number of hot dogs to be prepared when advanced ticket sales are 55 thousand.

4. On some basketball teams there are certain special players who have a reputation of "coming through" with points when the team especially needs the points. Such situations arise in the fourth quarter of very close games. The following data contain information on seven such players where x = first quarter score and y = fourth quarter score of the player. For each game the final margin of victory was four or fewer points.

x	14	10	18	15	12	16	13
y	20	16	20	22	19	23	21

a) Draw a scatter diagram for the data.
b) Find $\bar{x}$, $\bar{y}$, and b. What is the equation of the least squares line?
c) Graph the least squares line on your scatter diagram.
d) Find the standard error of estimate S_e.
e) If one of the "special" players scores 11 points in the first quarter and the game is very close, what does the least squares line predict the fourth quarter score will be for this player?
f) Find a 90% confidence interval for the score made in the fourth quarter by a "special" player who scores 11 points in the first quarter.

5. The central office of a large manufacturing company manages ten similar plants in different locations. The following data were obtained where x = percent of operating capacity being used at the plant and y = profits in hundreds of thousands of dollars at the same plant.

x	50	61	77	80	82	85	88	81	95	99
y	24	26	35	31	35	34	39	40	38	42

a) Draw a scatter diagram for the data.
b) Find $\bar{x}$, $\bar{y}$, and b. What is the equation of the least squares line?
c) Graph the least squares line on your scatter diagram.
d) Find the standard error of estimate S_e.
e) If a plant is working at 75 percent of its operating capacity what does the least squares line predict for profits coming from that plant?
f) Find a 95% confidence interval for the range of profits of a plant working at 75 percent capacity.

6. A child psychiatrist is studying the mental development of children. A random sample of nine children were given a standard set of questions appropriate to the age of each child. The number of irrelevant responses to the questions was recorded for each child. In the following data x = age of child in years and y = number of irrelevant responses.

x	2	3	4	5	7	9	10	11	12
y	15	15	12	13	11	10	8	6	5

a) Draw a scatter diagram for the data.
b) Find $\bar{x}$, $\bar{y}$, and b. What is the equation of the least squares line?
c) Graph the least squares line on your scatter diagram.
d) Find the standard error of estimate S_e.
e) If a child is 9.5 years old, what does the least squares line predict for the number of irrelevant responses?
f) Find a 99% confidence interval for the number of irrelevant responses for a child who is 9.5 years old.

7. As director of personnel for a prosperous company you have just hired a new public relations person. The final salary arrangements are negotiated depending on the number of years of experience the new public relations person brings to the company. After checking with several other companies in your area, you obtain the following data where x = number of years of experience for a person in public relations and y = annual salary in thousands of dollars.

x	1	2	15	11	9	6
y	25	29	53	49	37	34

a) Draw a scatter diagram for the data.

b) Find $\bar{x}$, $\bar{y}$, and b. What is the equation of the least squares line?

c) Graph the least squares line on your scatter diagram.

d) Find the standard error of estimate S_e.

e) If your candidate for the new public relations position has eight years of experience, what would the least squares line suggest for the annual salary?

f) Estimate a salary range for the candidate so that approximately 90% of the public relations officers with eight years of experience will be in this range.

8. You are the general manager of a car rental service at the airport. In an effort to estimate the maintenance costs you obtain the following data from a random sample of 11 cars. In this data x = number of miles (in thousands) on the car and y = total maintenance cost in dollars.

x	3	5	12	8	3	16	12	9	2	4	10
y	90	150	320	208	98	390	275	240	75	115	265

a) Draw a scatter diagram for the data.

b) Find $\bar{x}$, $\bar{y}$, and b. What is the equation of the least squares line?

c) Graph the least squares line on your scatter diagram.

d) Find the standard error of estimate S_e.

e) If you have a car with 7 thousand miles, what does the least squares line predict for the maintenance cost on this car?

f) Estimate a cost range for maintenance on a car with 7 thousand miles so that you can be 95% sure maintenance costs will fall in this range.

9. For Sherlock Holmes the "science of deduction" was the ultimate key to solving every mystery. Distance between footprints (length of stride) told the master detective the height of a criminal he was pursuing. What kind of relation is there between a person's height and length of stride? A random sample of 12 people provided the following data:

x (Length of Stride, in.)	15	17	19	11	16	22	17	25	12	15	25	23
y (Height, in.)	56	57	64	52	60	66	63	73	54	58	69	71

a) Draw a scatter diagram for the data.

b) Find $\bar{x}$, $\bar{y}$, and b. What is the equation of the least squares line?

c) Graph the least squares line on your scatter diagram.

d) Find the standard error of estimate S_e.

e) If footprints indicated that the suspect had a stride of 20 in., what would the least squares line predict for the suspect's height?

f) Find a 95% confidence interval for the height of a suspect with a 20-in. stride.

10. A government economist obtained the following data from a random sample of ten private companies in the natural gas exploration business. In this case, x is the capital investment the company put into exploration and development (in millions of dollars), and y is the

volume of natural gas discovered (adjusted to sea level pressure and in billions of cubic feet).

x	5.8	8.3	1.7	7.7	5.6	2.4	15.8	10.1	12.4	7.9
y	9.1	17.4	4.7	12.3	3.8	4.6	18.1	19.3	16.9	12.6

a) Draw a scatter diagram for the data.
b) Find $\bar{x}$, $\bar{y}$, and b. What is the equation of the least squares line?
c) Graph the least squares line on your scatter diagram.
d) Find the standard error of estimate S_e.
e) If a company had invested 6.5 million dollars, what would the least squares line predict for the volume of natural gas discovered?
f) Find a 90% confidence interval for the volume of natural gas we would expect to be discovered by a company that invested 6.5 million dollars.

11. The United States Navy did a study of its recruiting efforts one year. They wanted to examine the relationship between the number of recruiting offices located in a particular city and the number of people recruited in that city. Recruiting data (over a six-month period) from 11 cities is given in Table 10-9.
a) Draw a scatter diagram for the data.
b) Find $\bar{x}$, $\bar{y}$, and b. What is the equation of the least squares line?
c) Graph the least squares line on your scatter diagram.
d) Find the standard error of estimate S_e.
e) If a city has six recruiting offices, what would you predict for the number of recruits enlisted?
f) Find a 90% confidence interval for the number of recruits to be enlisted in a city with six recruiting offices.

TABLE 10-9 Recruiting Data for 11 Cities

City	x (No. of Recruiting Offices)	y (No. of Recruits Enlisted)
Salt Lake City	2	60
Denver	3	100
Kansas City	1	20
Atlanta	4	80
Philadelphia	9	120
Honolulu	8	180
Seattle	5	100
Dallas	7	140
Detroit	12	220
Los Angeles	15	280
St. Louis	10	240

12. The Food and Drug Administration is examining the effect of different doses of a new drug on the pulse rate of human subjects. The results of the study on six people are given in Table 10-10.

a) Draw a scatter diagram for the data.
b) Find $\bar{x}$, $\bar{y}$, b, and the equation of the least squares line.
c) Graph the least squares line on your scatter diagram.
d) Find the standard error of estimate S_e.
e) If $x = 2.75$, what is the predicted value of y?
f) Find a 95% confidence interval for the y values when $x = 2.75$.

TABLE 10-10 Drug Effects

x (Drug Dose in mg/kg of Body Wt)	y (Drop in Pulse Rate/Min)
2.50	8
3.00	11
3.50	9
4.50	16
5.50	19
6.00	20

SECTION 10.3

The Linear Correlation Coefficient

Most of us have heard someone say that more intelligent people tend to do better in school. Is this always true? Experienced teachers know it is only partially true. Students with higher IQs (intelligence quotients) often do better schoolwork, but factors other than IQ can affect academic success. For the individual student, academic success is a function of many relevant "human" factors.

Let's look more carefully at a sample of student IQ scores and respective cumulative grade point averages (CGA). The principal of Delta High School chose 12 students from the senior class at random and found the data shown in Table 10-11.

From Figure 10-13 we see that the scatter diagram for the data in Table 10-11 does not indicate a perfect linear relation between $x = $ IQ and $y = $ CGA. However, for a moment suppose there is a perfect linear correlation (i.e., all scatter points are on the least squares line). Then it can be shown mathematically (although the proof is somewhat complicated) that if the relation between x and y is perfectly linear, then as we go *over* one standard deviation s_x (of the x values), we should go *up* one standard deviation s_y (of the y values). Therefore, the slope of a perfect linear relation would be (*see* Figure 10-14, page 410)

$$\text{slope} = \frac{s_y}{s_x}$$

Most scatter diagrams do not indicate a perfect linear relation between x and y values; in fact, many scatter diagrams look something like that in Figure 10-13. If

TABLE 10-11 Intelligence Quotient (IQ) Versus Cumulative Grade Average (CGA) for 12 High School Seniors

IQ	CGA (4-Point Scale)
117	3.7
92	2.6
102	3.3
115	2.2
87	2.4
76	1.8
107	2.8
108	3.2
121	3.8
91	3.0
113	4.0
98	3.5

FIGURE 10-13   Scatter Diagram for Table 10-11

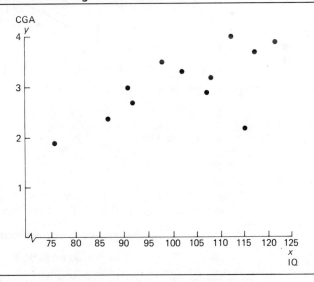

the relation between x and y is *not* a perfect linear relation, then as we go *over* one standard deviation s_x on the x-axis, we will go *up* only a fractional part of the standard deviation s_y before we meet the least squares line. This is shown in Figure 10-15.

FIGURE 10-14   Slope of a Perfect Linear Relationship

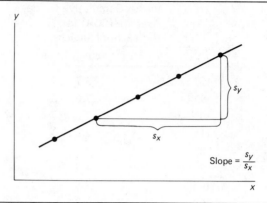

$$\text{Slope} = \frac{s_y}{s_x}$$

FIGURE 10-15   Usually Only a Fractional Part of s_y Can Be Attributed to the Relationship Between x and y

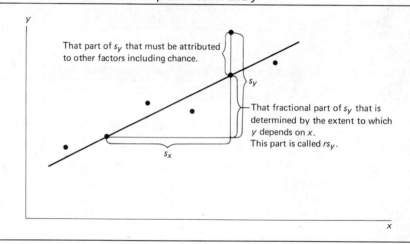

That part of s_y that must be attributed to other factors including chance.

That fractional part of s_y that is determined by the extent to which y depends on x. This part is called rs_y.

In Figure 10-15 we see that only a fractional part of s_y can be attributed to the relation between x and y. If we call that fractional part r, then we can write

$$rs_y = \text{that part of } s_y \text{ that is determined by the extent}$$

$$\text{to which } y \text{ depends on } x$$

From Figure 10-15 it can be seen that the slope of the least squares line is then

$$\text{slope} = \frac{rs_y}{s_x}$$

To mathematically determine r we must use *all* the data points in the scatter diagram. Karl Pearson (1857–1936) was an English statistician who developed a formula for r. Equation (9) is an equivalent formula for r.

- **Formula to Calculate r**

$$r = \frac{SS_{xy}}{\sqrt{SS_x \, SS_y}} \qquad (9)$$

where

$$SS_{xy} = \Sigma xy - \frac{(\Sigma x)(\Sigma y)}{n}$$

$$SS_x = \Sigma x^2 - \frac{(\Sigma x)^2}{n}$$

$$SS_y = \Sigma y^2 - \frac{(\Sigma y)^2}{n}$$

$$n = \text{number of data pairs in scatter diagram}$$

Note: This formula for r is sensitive to rounding. As in other formulas involving SS_x, SS_{xy}, and SS_y, you want to carry as many digits as is reasonable for your problem until the last step. Again, depending on the rounding process used, answers will vary slightly. The answers for the end-of-section and end-of-chapter exercises were computer-generated, so your answers might differ from them slightly and still be essentially correct.

The quantity r is called the *Pearson product-moment correlation coefficient*, or simply the *correlation coefficient*.

Let us delay an example showing how to compute r until we know a little more about the meaning of the correlation coefficient. It can be shown mathematically that r is always a number which is between $+1$ and -1 ($-1 \leq r \leq +1$). Table 10-12 gives a quick summary of some basic facts about r.

TABLE 10-12 Some Facts About the Correlation Coefficient

If r Is	Then	The Scatter Diagram Might Look Something Like
0	There is no linear relation for the points of the scatter diagram.	

TABLE 10-12 *continued*

1 or −1	There is a perfect linear relation between x and y values; all points lie on the least squares line.

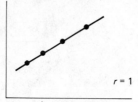

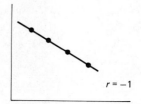

Between 0 and 1 $(0 < r < 1)$	The x and y values have a *positive correlation*. By this we mean that *large x* values are associated with *large y* values and *small x* values are associated with *small y* values.

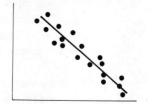

As we go from left to right the least squares line goes *up*.

Between −1 and 0 $(-1 < r < 0)$	The x and y values have a *negative correlation*. By this we mean *large x* values are associated with *small y* values and *small x* values are associated with *large y* values.

As we go from left to right the least squares line goes *down*.

_____ Exercise 5 _____

Match the appropriate statement about r to each scatter diagram in Figure 10-16.

1) $r = 0$
2) $r = 1$
3) $r = -1$
4) r is between 0 and 1
5) r is between −1 and 0

a) $r = 1$ since all the points are on the line and the line goes up from left to right.

b) $r = 0$ since there is no apparent linear relation among the points.

c) r is between −1 and 0 since the points are fairly close to the line, and as we read from left to right the least squares line goes down.

FIGURE 10-16   Scatter Diagrams

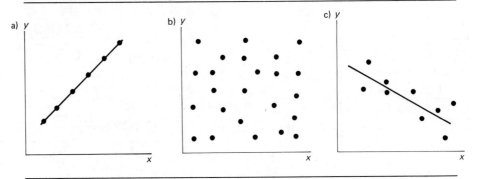

EXAMPLE 4

Now return to the 12 seniors from Delta High School (Table 10-11). Let's find the correlation coefficient when x = IQ and y = CGA. To find r we must compute Σx, Σy, Σx^2, Σy^2, and Σxy. The values for $(\Sigma x)^2$ and $(\Sigma y)^2$ can be obtained from Σx and Σy. It is easiest to organize our work into a table of five columns (Table 10-13). The first two columns are just a repetition of Table 10-11.

TABLE 10-13 Information Necessary to Compute r

x(IQ)	y(CGA)	x^2	y^2	xy
117	3.7	13,689	13.7	432.9
92	2.6	8,464	6.8	239.2
102	3.3	10,404	10.9	336.6
115	2.2	13,225	4.8	253.0
87	2.4	7,569	5.8	208.8
76	1.8	5,776	3.2	136.8
107	2.8	11,449	7.8	299.6
108	3.2	11,664	10.2	345.6
121	3.8	14,641	14.4	459.8
91	3.0	8,281	9.0	273.0
113	4.0	12,769	16.0	452.0
98	3.5	9,604	12.3	343.0

$\Sigma x = 1{,}227$ $\Sigma y = 36.3$ $\Sigma x^2 = 127{,}535$ $\Sigma y^2 = 114.9$ $\Sigma xy = 3{,}780.3$
$(\Sigma x)^2 = 1{,}505{,}529$ $(\Sigma y)^2 = 1{,}317.7$

To use formula (9) to calculate r we will first compute SS_{xy}, SS_x, and SS_y.

$$SS_{xy} = \Sigma xy - \frac{(\Sigma x)(\Sigma y)}{n}$$

$$= 3{,}780.3 - \frac{(1{,}227)(36.3)}{12}$$

$$= 68.63$$

$$SS_x = \Sigma x^2 - \frac{(\Sigma x)^2}{n}$$

$$= 127{,}535 - \frac{(1{,}227)^2}{12}$$

$$= 2{,}074.25$$

$$SS_y = \Sigma y^2 - \frac{(\Sigma y)^2}{n}$$

$$= 114.9 - \frac{(36.3)^2}{12}$$

$$= 5.09$$

Therefore, the correlation coefficient is

$$r = \frac{SS_{xy}}{\sqrt{SS_x\,SS_y}}$$

$$= \frac{68.63}{\sqrt{(2{,}074.25)(5.09)}}$$

$$= 0.6679$$

$$\approx 0.67$$

Our correlation coefficient is $r \approx 0.67$. Let's make sure this answer agrees with what we expect from a quick glance at the scatter diagram (Figure 10-13). In Figure 10-13 the general trend is upward as we read from left to right, so we would expect a positive value for r. The value 0.67 is in the expected range—that is, it is between 0 and 1.

It is quite a task to compute r for even 12 data pairs. The use of columns as in Example 4 is extremely helpful. Your value for r should always be between -1 and 1. Use a scatter diagram to get a rough idea of the value of r. If your computed value of r is outside the allowable range or if it disagrees quite a bit with the scatter diagram, recheck your calculations. Be sure you distinguish between expressions such as (Σx^2) and $(\Sigma x)^2$. Negligible rounding errors may occur, depending on how you (or your calculator) round.

_____ Exercise 6 _____

In one of the Boston city parks there has been a problem with muggings in the summer months. A police cadet took a random sample of 10 days (out of the 90-day summer) and compiled the following data. For each day, x represents the number of patrolmen on duty in the park and y represents the number of reported muggings on that day.

x	10	15	16	1	4	6	18	12	14	7
y	5	2	1	9	7	8	1	5	3	6

a) Construct a scatter diagram of x and y values.

a) Figure 10-17 shows the scatter diagram.

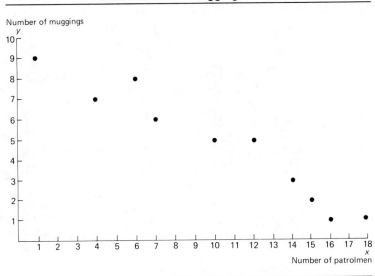

FIGURE 10-17   Scatter Diagram for Number of Patrolmen Versus Number of Muggings

b) From the scatter diagram, do you think the computed value of r will be positive, negative, or zero? Explain.

b) r will be negative. The general trend is that large x values are associated with small y values and vice versa. From left to right the least squares line goes down.

c) Complete Table 10-14.

c)

TABLE 10-14

x	y	x^2	y^2	xy
10	5	100	25	50
15	2	225	4	30
16	1	256	1	16
1	9	1	81	9
4	7	16	49	28
6	8	——	——	——
18	1	——	——	——
12	5	——	——	——
14	3	——	——	——
7	6	49	36	42

$\Sigma x = 103$ $\Sigma y = 47$ $\Sigma x^2 = \_$ $\Sigma y^2 = \_$ $\Sigma xy = \_$
$(\Sigma x)^2 = \_$ $(\Sigma y)^2 = \_$

TABLE 10-15 Completion of Table 10-14

x	y	x^2	y^2	xy
6	8	36	64	48
18	1	324	1	18
12	5	144	25	60
14	3	196	9	42

$\Sigma x^2 = 1,347$ $\Sigma y^2 = 295$ $\Sigma xy = 343$
$(\Sigma x)^2 = 10,609$ $(\Sigma y)^2 = 2,209$

d) Compute SS_{xy}, SS_x, SS_y, and then r.

d) $SS_{xy} = \Sigma xy - \dfrac{(\Sigma x)(\Sigma y)}{n}$

$\quad = 343 - \dfrac{(103)(47)}{10}$

$\quad = -141.1$

$SS_x = \Sigma x^2 - \dfrac{(\Sigma x)^2}{n}$

$\quad = 1,347 - \dfrac{(103)^2}{10}$

$\quad = 286.1$

$SS_y = \Sigma y^2 - \dfrac{(\Sigma y)^2}{n}$

$\quad = 295 - \dfrac{(47)^2}{10}$

$\quad = 74.1$

$r = \dfrac{SS_{xy}}{\sqrt{SS_x\, SS_y}}$

$\quad = \dfrac{-141.1}{\sqrt{(286.1)(74.1)}}$

$\quad = -0.9691$

$\quad \approx -0.97$

The correlation coefficient can be thought of as another measure of how "good" the least squares line fits the data points of the scatter diagram. (Recall that the standard error of estimate S_e in Section 10.2 is another such measure.) The closer r is to $+1$ or -1, the better the least squares line "fits" the data. Values of r close to 0 indicate a poor "fit."

Usually our scatter diagram does not contain *all* possible data points that could be gathered. Most scatter diagrams represent only a *random sample* of data pairs taken from a very large population of possible pairs. Since r is computed by formula (9) on the basis of a random sample of (x, y) pairs, we expect the values of r to vary from one sample to the next (much as sample means $\bar{x}$ varied from sample to sample). This brings up the question of the *significance* of r. Or put another way, what are the chances that our random sample of data pairs indicates a high correlation when in fact the population x and y values are not so strongly correlated? Right now let us just say the significance of r is a separate issue that is left to the next section.

--- Exercise 7 ---

a) In Example 4 we found the correlation coefficient r of the relationship between IQ and cumulative grade averages. In that case r was 0.67. How would you describe the strength of this relationship?

a) The correlation coefficient $r = 0.67$ is moderate but not extremely high. It seems that other factors besides IQ are significant in determining a cumulative grade point average.

b) In Exercise 6, dealing with the relation between the number of patrolmen in the park and the number of muggings, we found r to be -0.97. How would you describe the strength of this relationship? Do you think the city is justified in asking for more patrolmen to be assigned to park duty?

b) $r = -0.97$ is a high correlation. The relation between the number of patrolmen in the park and the number of muggings in the park is a strong and dependable negative correlation. The authors feel the city would be wise to hire more patrolmen to patrol the park. But many other aspects of the situation must be considered. Perhaps more crimes would be prevented by putting those patrolmen elsewhere.

The correlation coefficient is a mathematical tool for measuring the strength of the linear relationship between two variables. As such it makes no implication about cause or effect. Just because two variables tend to increase or decrease together does not mean a change in one is *causing* a change in the other. A strong correlation between x and y is sometimes due to other (either known or unknown) variables.

EXAMPLE 5

Over a period of years a certain town observed that the correlation between x, the number of people attending churches, and y, the number of people in the city jail, was $r = 0.90$.

Does going to church cause people to go to jail? We hope not. During this period there was a steady increase in population. Therefore, it is not too surprising that both the number of people attending churches and the number of people in jail increased together. The high correlation between x and y is due to the common effect of the increase in the general population.

Section Problems 10.3

1. Over the past ten years there has been a high positive correlation between the number of South Dakota safety inspection stickers issued and the number of South Dakota traffic accidents.
   a) Do safety inspection stickers cause traffic accidents?
   b) What third factor might cause traffic accidents and the number of safety stickers to increase together?

2. There is a high positive correlation in the United States between teachers' salaries and annual consumption of liquor.
   a) Do you think increasing teachers' salaries has caused increased liquor consumption?
   b) As teachers' salaries have been going up, most other salaries have been going up too. To some extent this means an upward trend in buying power for everyone. How might this explain the high correlation between teachers' salaries and liquor consumption?

3. Over the past 30 years in the United States there has been a strong negative correlation between number of infant deaths at birth and number of people over age 65.
   a) Is the fact that people are living longer causing a decrease in infant mortalities at birth?
   b) What third factor might be decreasing infant mortalities and at the same time increasing life span?

4. Over the past few years there has been a strong positive correlation between the annual consumption of diet soda pop and the number of unwed teenage mothers.
   a) Do you think that an increasing consumption of diet pop has led more teenage girls to be unwed mothers?
   b) What third factor or factors might be causing both the annual consumption of diet pop and the number of unwed teenage mothers to increase together?

5. A certain psychologist counsels people who are getting divorced. A random sample of six of her patients provided the following data where x = number of years of courtship before marriage and y = number of years of marriage before divorce.

x	3	0.5	2	1.5	5	1
y	9	6	14	10	20	6.5

   a) Draw a scatter diagram for the data.
   b) From the scatter diagram would you say r is closest to 1, 0, or −1?
   c) Find r.

6. Seven children in the third grade were given a verbal test to study the relationship between the number of words used in a response and the silence interval before the response. A psychologist asked each child a number of questions, and the total number of words used in answering and the time before the child began answering were recorded. The following data were obtained where x = total silence time in seconds and y = total number of words in response.

x	25	16	19	27	38	42	31
y	60	55	50	64	73	70	65

   a) Draw a scatter diagram for the data.
   b) From the scatter diagram would you say that r is closest to 1, 0, or −1?
   c) Find r.

7. A new stock called ATA has appeared on the New York Stock Exchange. To determine if ATA price changes tend to follow general market trends a random sample of 10-day trend studies were made. For each 10-day trading period x = number of days the Dow Jones 30 Industrials went up and y = number of days ATA went up.

x	0	5	8	3	7	4	6	8	9	1
y	9	6	2	8	1	5	3	1	5	6

   a) Draw a scatter diagram for the data.
   b) From the scatter diagram would you say r is closest to 1, 0, or −1?
   c) Find r.

8. The systems manager for a large computer center is studying terminal response time. This is the length of time it takes the computer to respond to a command sent to the computer from a terminal by pressing one of the terminal's program function keys. The following data were obtained where x = number of simultaneous users on the computer and y = terminal response time in seconds.

x	53	37	45	62	81	116	90	75
y	0.4	0.2	0.3	0.6	0.9	1.4	1.1	0.8

a) Draw a scatter diagram for the data.
b) From the scatter diagram would you say *r* is closest to 1, 0, or −1?
c) Find *r*.

9. The state criminology laboratory must sometimes estimate a person's height from partial skeleton remains. How strong a correlation is there between body weight and bone size? A random sample of eight adult male X-rays gave the following information where *x* = length of femur (thigh bone) and *y* = body height.

x (in.)	17.5	20	21	19	15.5	18.5	16	18
y (in.)	50	80	78	73	63	71	64	71

a) Draw a scatter diagram for the data.
b) From the scatter diagram, would you say that *r* is closest to 1, 0, or −1?
c) Find *r*.

10. An efficiency expert developed a test measuring job satisfaction of civil service clerks. The following information was obtained from a random sample of ten clerks.

x (Job Satisfaction Index)	92	32	56	20	72	16	56	76	80	48
y (No. of Days Absent from Work in 1 yr)	8	14	10	14	6	17	8	12	7	15

a) Draw a scatter diagram for the data.
b) From the scatter diagram, would you say that *r* is closest to 1, 0, or −1?
c) Find *r*.

11. An English skills placement test is given to all entering freshmen at Walla Woo College. Placement scores and final grades for 12 randomly selected students who took English-101 are shown below.

x (Placement Test Score)	18	28	47	36	40	16	42	24	12	44	38	30
y (Course Grade)	45	50	93	45	90	30	60	35	15	70	75	65

a) Draw a scatter diagram for the data.
b) From the scatter diagram, would you say that *r* is closest to 1, 0, or −1?
c) Find *r*.

12. Some children experience "sugar highs," a form of hypertension caused by a large influx of sugar into the blood stream. In an effort to study this phenomenon, Dr. Adams gave different amounts of sugar to a random sample of nine fourth-grade children. The results follow, where *x* is the blood sugar level in milligrams per milliliter of blood, and *y* is brain wave activity in microvolts.

x	5.2	6.5	9.5	11.0	5.0	5.8	10.7	8.2	7.7
y	9.1	15.2	32.3	38.7	7.8	11.7	36.5	24.4	22.0

a) Draw a scatter diagram for the data.
b) From the scatter diagram, would you say that r is closest to 1, 0, or -1?
c) Find r.

13. A medical doctor specializing in gerontology (study of aging) suspects that older people have more trouble getting necessary vitamins from the food they eat and may require vitamin supplements. A random sample of eight people over 50 gave the following information, where x is the age and y is the measure of vitamin B12 in the blood stream expressed as a percent of B12 that normally should be present. No patient in the sample took vitamin pills.

x	68	57	92	63	77	86	62	81
y	60	80	55	92	64	55	84	60

a) Draw a scatter diagram for the data.
b) From the scatter diagram, would you say that r is closest to 1, 0, or -1?
c) Find r.

14. Pilots who eject from a high-velocity fighter jet receive a tremendous jerk as their parachute opens. A new type of parachute has been developed that automatically increases deceleration time, according to the air speed of the opening chute. A random sample of ten pilots tested the new parachute by ejecting at randomly selected speeds over 300 miles per hour. The results follow, where x is ejection speed and y is deceleration time in seconds.

x	450	510	545	345	395	600	315	410	360	480
y	9	15	17	8	10	19	7	9	5	16

a) Draw a scatter diagram for the data.
b) From the scatter diagram, would you say that r is closest to 1, 0, or -1?
c) Find r.

SECTION 10.4   Testing the Correlation Coefficient

A basic assumption in the study of economics is that people will spend more if they earn more. Economists claim that there is a high positive linear correlation between x = amount earned and y = amount spent. How could you test this claim? One way would be to obtain all possible (x,y) pairs for all people in the United States with an income. If you did this impossible task you would have the *population* of all possible (x,y) pairs. You could then compute the *population correlation coefficient,* which we call ρ (a lowercase Greek letter spelled *rho* and pronounced like *row*). If $\rho = 1$, then you have a perfect positive linear correlation. If $\rho = 0$, then you have no linear correlation.

Most people would not even attempt to take the entire population of all incomes and corresponding amounts spent. Usually we take a random sample and compute the correlation coefficient of the sample. We call the *sample correlation coefficient r*. If *r* is near 1, then we have evidence that ρ, the population coefficient, is near 1, or at least greater than 0.

Different random samples will give different values of *r*. We need a test to decide when a sample value of *r* is far enough from zero to indicate correlation in the population.

For simplicity, we will *assume that both the x and y variables are normally distributed*. To test if the (*x,y*) values are correlated *in the population*, we will set up the null hypothesis that they are not correlated. The choice of the alternate hypothesis depends on the belief that the correlation is positive, negative, or simply not zero. (*See* Table 10-16.)

TABLE 10-16 Alternate Hypotheses

If You Think	Then Use
$\rho > 0$	$H_0 : \rho = 0$ $H_1 : \rho > 0$
$\rho < 0$	$H_0 : \rho = 0$ $H_1 : \rho < 0$
$\rho \neq 0$	$H_0 : \rho = 0$ $H_1 : \rho \neq 0$

When $\rho = 0$ the distribution of sample correlation coefficients (*r* values) will be symmetric about $r = 0$. Figure 10-18 shows the distribution for some values of *n* where *n* is the number of data pairs used to compute *r*.

The type of test (left-tail, right-tail, or two-tail) depends on the choice of H_1, as shown in Table 10-17.

To find the critical values for a test we use Table 7 of Appendix I. The entries in Table 7 are critical values of *r* corresponding to given *n*, the number of data points, and α, the level of significance. Each critical value is listed without a sign. The choice of sign, + or −, depends on the type of test used. A right-tail test uses positive critical values, a left-tail test uses negative critical values, and a two-tail test uses both. Example 6 shows how to use Table 7 when we have a right-tail test.

FIGURE 10-18   Distribution of *r* when $\rho = 0$

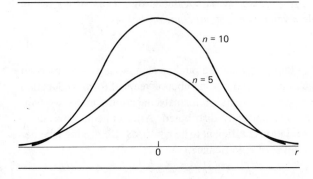

TABLE 10-17 Type of Test to be Used

If H_1 is	Then Use
$H_1 : \rho < 0$	A left-tail test
$H_1 : \rho > 0$	A right-tail test
$H_1 : \rho \neq 0$	A two-tail test

EXAMPLE 6

In Section 10.2 we examined the relation between $x =$ number of cricket chirps per second and $y =$ temperature (°F). Using the data of Table 10-3, we can find the sample correlation coefficient to be $r = 0.84$. In this case it seems that if the population data of x and y values are in fact correlated, then $\rho > 0$ (the least squares line of Figure 10-6 slopes upward). Therefore, we will test

$$H_0 : \rho = 0$$
$$H_1 : \rho > 0$$

and so use a right-tail test.

Let us choose $\alpha = 0.01$ as our level of significance. Table 10-3 has $n = 15$ data entries which are used to find our sample correlation coefficient r. Therefore, Table 7 gives 0.59 as the critical value to be used in a right-tail test. (We use a positive critical value because we have a right-tail test.) Figure 10-19 illustrates the critical region.

FIGURE 10-19 Critical Region for $n = 15$

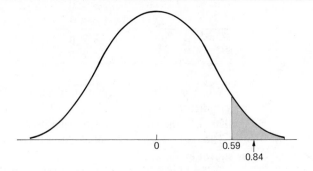

The region to the right of 0.59 is the critical, or rejection, region. Since $r = 0.84$ is in the critical region, we reject $H_0 : \rho = 0$ and conclude that $\rho > 0$. (We conclude that the correlation coefficient of the population is positive.)

Whenever we reject $H_0 : \rho = 0$, we say our r value is *significant*. If we cannot reject H_0, we say our sample r value is *not significant*.

EXAMPLE 7

For her sociology class Zelma interviewed a random sample of $n = 30$ married couples. For each couple Zelma found $x =$ number of years of formal education for the man and $y =$ number of years of formal education for the woman. Assume that both x and y are normally distributed. After collecting the data, Zelma worked out the correlation coefficient to be $r = 0.28$. Determine whether r is significant at the 5% level of significance.

a) Our null hypothesis is $H_0 : \rho = 0$. Since Zelma had no reason to believe ρ is either positive or negative, the alternate hypothesis is $H_1 : \rho \neq 0$. So we are testing

$$H_0 : \rho = 0$$
$$H_1 : \rho \neq 0$$

b) The alternate hypothesis $H_1 : \rho \neq 0$ indicates a two-tail test. Since $\alpha = 0.05$ and $n = 30$, Table 7 gives the critical value of 0.36. The critical regions are shown in Figure 10-20.

FIGURE 10-20   Critical Regions for $n = 30$

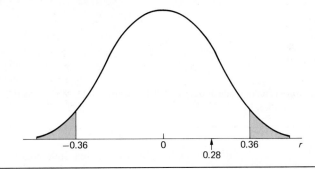

c) The value of the correlation coefficient is $r = 0.28$. This value does not fall in the critical region. Therefore, we cannot reject $H_0 : \rho = 0$, and we conclude that $r = 0.28$ is *not* significant at the 0.05 level.

_____ Exercise 8 _____

Big Rock Life Insurance Company has reason to believe that among white-collar workers, people who earn more money tend to die younger. The Big Rock Company took a random sample of $n = 30$ white-collar workers who died in the past year. For each person $x =$ income in dollars per year and $y =$ age of that person at death. The correlation coefficient was found to be $r = -0.33$. Determine whether r is significant at the 0.05 level of significance.

a) What is the null hypothesis?

$$H_0 : \rho > 0 \quad H_0 : \rho = 0$$
$$H_0 : \rho < 0 \quad H_0 : \rho \neq 0$$

a) $H_0 : \rho = 0$

b) Which alternate hypothesis reflects the statement that people who earn more tend to die younger?

$$H_1 : \rho = 0 \quad H_1 : \rho > 0 \quad H_1 : \rho < 0$$

b) The statement indicates a negative correlation between salary and age at death. So we want the alternate hypothesis to claim there is a negative correlation. The appropriate alternate hypothesis is $H_1 : \rho < 0$.

c) Using the alternate hypothesis selected in part (b), what kind of test should we use: right-tail, left-tail, two-tail?

c) Use a left-tail test.

d) What critical value should be used? Sketch the critical region.

d) The critical value is -0.31.

FIGURE 10-21 Critical Region

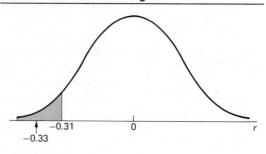

e) Is $r = -0.33$ significant or not?

e) Since $r = -0.33$ is in the critical region, we must reject H_0 and conclude that $r = -0.33$ is significant. We accept the alternate hypothesis.

f) What assumptions were made about the x distribution and the y distribution?

f) We assumed that both x and y had normal distributions.

Even though a significance test indicates the existence of a correlation between x and y in the population, it does not signify a cause-and-effect relationship. Even after a significant correlation between variables has been established, the cause of the correlation must be identified. In addition, we must also decide if the correlation is high enough to be of practical value for the particular application in which it is used.

Problems that require calculating and testing the correlation coefficient must use a *random sample* of data points. This statement means that both x and y are statistical variables whose numerical values are obtained only after the random sample is drawn from the population of all possible data pairs. This procedure is somewhat different from our preceding regression (e.g., least squares line) problems where the x values can be chosen in advance. In this case the y values may depend heavily upon the choice of x values. The least squares methods can be applied to random sample data or to data in which the x values are specified in advance. But the interpretation of r as a measure of the linear correlation between x and y *requires* us to use a *random sample* of data points in which both x and y are statistical variables.

Section Problems 10.4

For each problem do the following:

a) State the null and alternate hypotheses.
b) Find the critical value or values.
c) Sketch the critical region and the acceptance region.

d) Locate the sample correlation coefficient r on the sketch of part (c).

e) Decide whether you should accept or reject the null hypothesis.

In the following problems assume both x and y are normally distributed.

1. Dr. Young is an economist. She claims that the prime interest rate and the unemployment rate are positively correlated. To support her claim, she took a random sample of six x and y value pairs from the past ten years, where $x =$ prime interest rate and $y =$ unemployment rate. The sample correlation coefficient turned out to be $r = 0.73$. Determine if r is significant at the 1% level of significance.

2. A personal finance advisor claims that a person's income and level of life insurance are positively correlated. A random sample of 24 of his clients gave $x =$ personal income and $y =$ total dollar value of life insurance for that individual. After going through the calculations, he found the sample correlation coefficient to be $r = 0.38$. Is r significant at the 1% level of significance?

3. A college counselor claims that there is a negative correlation between $x =$ number of days absent from class and $y =$ score on midterm exam. A random sample of 30 French students provided x and y values. The sample correlation coefficient was computed to be $r = -0.67$. Is the counselor's claim justified at the 1% level of significance?

4. Max is a sales manager for a company that sells its products over the phone. For a random sample of 16 phone calls he recorded $x =$ length of time spent on the phone (in minutes) and $y =$ dollar value of sale. When the correlation coefficient was calculated, it was $r = 0.35$. Does this result indicate that there is a correlation, either positive or negative, between x and y values? Use a 5% level of significance.

5. Big Mike claims there is a negative correlation between the number of speeding tickets and the number of CB radios being sold. Mike randomly selected 20 towns of about the same size in his general locality. For each town he found $x =$ number of CB radios sold that year and $y =$ number of speeding tickets given that year. Then Mike used his calculator and found the correlation coefficient to be $r = -0.40$. Determine if r is significant at the 1% level of significance.

6. Nancy and Pete own the Walred Discount Store. Nancy claims there is a positive correlation between the length of time a customer shops in her store and the amount of money spent by the customer. Behind her one-way mirrors, Nancy observed a random sample of 12 customers. For each customer she determined $x =$ length of time in the store and $y =$ amount of money spent in the store. The correlation coefficient turned out to be $r = 0.75$. Is this correlation coefficient significant at the 5% level of significance?

7. Pete (of problem 6) claims there is a positive correlation between the amount of display space allotted to a product and the weekly sales (in dollars) of that product. Each week for a period of 15 weeks Pete randomly allotted space ($x =$ square feet) to the chocolate marshmallow display and recorded sales ($y =$ dollars) for chocolate marshmallows. The correlation coefficient turned out to be $r = 0.62$. Is this significant at the 1% level?

8. Betty is an industrial psychologist who developed a test to measure aggressiveness in salesmen. A random sample of 14 salesmen took the test. For each salesman the test scores (x) and last year's annual sales (y) were recorded. For the type of sales work involved (funeral insurance), Betty does not know whether to expect positive or negative correlation. However, she found the correlation coefficient to be $r = -0.60$. Is r significant at the 5% level?

9. Professor Stanislaw is directing an oceanography study in which ocean water is tested for oxygen content at different depths. For 30 randomly selected depths (x) the professor determined the percentage of oxygen content (y) at that depth. The correlation coefficient turned out to be $r = -0.63$. Use a 1% level of significance to test the professor's claim that there is a negative correlation between depth and oxygen content.

10. The state medical school has developed a new drug for breaking up cholesterol deposits in blood vessels. A random sample of nine patients with severe cholesterol conditions volunteered to use the new drug. The dosage (x = mg/kg of body weight) and results (y = cholesterol index on a scale of 0 to 100) were determined for each patient. (Each patient started with a cholesterol index above 90.) The correlation coefficient for the x, y values was $r = -0.60$. Use a 1% level of significance to test the claims that there is a negative correlation between drug dosage and cholesterol index.

11. Mr. Tonguetwist owns the Ship Shape Shirt Shop which makes tailored blouses and shirts. The supervisor of the shop is afraid that the percentage of defective items passing her assembly line during a 20-min period increases during later hours of the day. Therefore, the supervisor will not give the employees overtime work. However, the employees claim that the more they work the better they get at it, and they want to work overtime.

    To settle the dispute Mr. Tonguetwist randomly selected sixteen 20-min periods throughout the day. For each of the 16 periods the percent (y) of defective items was determined as well as the number of hours (x) the employees were on the job at the start of the time period. The correlation coefficient for the x, y values was $r = 0.48$.

    Mr. Tonguetwist says he will take the supervisor's advice if a test indicates $\rho > 0$ at the 1% level of significance. Otherwise, he will let the employees work overtime. What should Mr. Tonguetwist do?

12. A marine biologist is studying the echo effect of whales "talking" to each other. It seems that a whale will send a message on one frequency, and another whale will respond by use of an echo message on another frequency. Let x be the frequency of the first whale and y be the frequency of the second whale. A random sample of 23 pairs of whales were studied, and for this sample the correlation coefficient was found to be $r = 0.71$. Using a 1% level of significance, test the claim that there is a positive correlation between communication frequencies used by whales.

13. Lois is studying a possible relationship between the market price of gold and industrial diamonds. Let x be the market price of one ounce of gold, and let y be the market price of industrial diamonds (price per carat). A random sample of six end-of-day closing prices on the New York Exchange gave a sample correlation of $r = 0.70$. Using a 5% level of significance, test the claim that there exists a positive correlation between the market price of gold and industrial diamonds.

14. A government economist is studying a possible relation between x = tax increase in hundreds of dollars per average taxpayer, and y = Dow Jones stock market index for the immediate period after the tax increase. A random sample of eight observations over the past decade gave a correlation of $r = -0.58$. Using a 5% level of significance, test the claim that there is a negative correlation between tax increases and stock market index.

Summary

We can often get information about a population of y values by attempting to predict them from known values of another population x. If data pairs (x, y) are obtained from an overall population of such data pairs, we may use the equation of the least squares line

$$y = a + bx$$

to predict y values from given x values. In this equation a and b are computed from equations (2)–(5) of Section 10.2.

If two sets of data are obtained from the same sample, the extent to which they are related can be measured by the Pearson product-moment correlation coefficient given in equation (9) of Section 10.3. The significance of the correlation coefficient is tested using the null hypothesis

$$H_0 : \rho = 0 \quad \text{(i.e., there is no correlation)}$$

against an appropriate alternate hypothesis (either $\rho \neq 0$, $\rho > 0$, or $\rho < 0$).

Scatter diagrams of (x, y) values are useful to help determine visually if there is any linear relation between the x and y values and, if so, how strong the relation might be.

Important Words and Symbols

	Section		*Section*
Paired data values	10.1	Confidence interval for y	10.2
Scatter diagram	10.1	Correlation coefficient r (later	
Perfect linear correlation	10.1	called sample correlation	
No linear correlation	10.1	coefficient)	10.2
Regression	10.2	Linear correlation	10.3
Method of least squares	10.2	Positive, negative, and zero	
Least squares criterion	10.2	correlation	10.3
Least squares line		Population correlation	
$y = a + bx$	10.2	coefficient ρ	10.4
Interpolation and extrapolation	10.2	A significant sample correlation	
Standard error of estimate S_e	10.2	coefficient r	10.4

Chapter Review Problems

When solving chapter problems involving the standard error of estimate or testing the correlation coefficient, make the assumption that x and y are normally distributed random variables.

1. Some students claim they can tell the cost of a textbook just by looking at the thickness of the book. To test this claim they picked six hardbound books of the same height and width at random. The cost and thickness relation is

x (Thickness, cm)	1	2	0.5	1.3	2.7	1.7
y (Cost, $)	19	25	17	16	27	20

a) Draw a scatter diagram.
b) Find the equation of the least squares line.
c) Find r.
d) Test the claim that the population correlation coefficient is not zero (use a 5% level of significance).

2. A sociologist is interested in the relation between x = number of job changes and y = annual salary (in thousands of dollars) for people living in the Nashville area. A random sample of ten people employed in Nashville provided the following information:

x (No. of Job Changes)	4	7	5	6	1	5	9	10	10	3
y (Salary in $1,000)	33	37	34	32	32	38	43	37	40	33

a) Draw a scatter diagram for the data.
b) Find $\bar{x}$, $\bar{y}$, and b, and then the equation of the least squares line.
c) Graph the least squares line on your scatter diagram.
d) If someone had x = 2 job changes, what does the least squares line predict for y, the annual salary?
e) Find the standard error of estimate S_e.
f) Find a 90% confidence interval for the annual salary of an individual with two job changes.
g) Looking at the scatter diagram and least squares line, do you think the correlation coefficient will be positive, negative, or zero?
h) Find r.
i) Test the claim that the population correlation coefficient is positive (use a 5% level of significance).

3. Modern medical practice tells us not to encourage babies to become too fat. Medical research indicates that there is a positive correlation between the weight x of a 1-year-old baby and the weight y of a mature adult (30 years old). A random sample of medical files produced the following information for 14 females:

x (lb)	21	25	23	24	20	15	25	21	17	24	26	22	18	19
y (lb)	125	125	120	125	130	120	145	130	130	130	130	140	110	115

a) Draw a scatter diagram for the data.
b) Find $\bar{x}$, $\bar{y}$, b, and the equation of the least squares line.
c) Graph the least squares line on your scatter diagram.
d) If a female baby weighs 20 lb at one year, what would you predict she would weigh at 30 years of age?
e) Find the standard error of estimate S_e.
f) Find a 95% confidence interval for the weight at age 30 of a female who weighed 20 lb at one year of age.
g) Looking at the scatter diagram, do you think the correlation coefficient is positive, negative, or zero?

h) Find r.

i) Test the claim that the population correlation coefficient is positive (use a 1% level of significance).

4. Dorothy Kelly sells life insurance for the Prudence Insurance Company. She sells insurance by making visits to her clients' homes. Dorothy believes that the number of sales should depend to some degree on the number of visits made. For the past several years she kept careful records of the number of visits (x) she made each week and the number of people (y) who bought insurance that week. For a random sample of 15 such weeks the x and y values follow:

x	11	19	16	13	28	5	20	14	22	7	15	29	8	25	16
y	3	11	8	5	8	2	5	6	8	3	5	10	6	10	7

a) Draw a scatter diagram for the data.

b) Find $\bar{x}$, $\bar{y}$, b, and the equation of the least squares line.

c) Graph the least squares line on your scatter diagram.

d) Find the standard error of estimate.

e) On a week in which Dorothy made 18 visits, how many people would you predict would buy insurance from her?

f) Find a 90% confidence interval for the number of sales Dorothy would make in a week in which she made 18 visits.

g) Find the correlation coefficient.

h) Test the claim that the population correlation coefficient is positive (use a 1% level of significance).

5. Each box of Healthy Crunch breakfast cereal contains a coupon entitling you to a free package of garden seeds. At the Healthy Crunch home office they use the weight of incoming mail to determine how many of their employees are to be assigned to collecting coupons and mailing out seed packages on a given day. (Healthy Crunch has a policy of answering all its mail the day it is received.)

Let x = weight of incoming mail and y = number of employees required to process the mail in one working day. A random sample of eight days gave the following data:

x (lb)	11	20	16	6	12	18	23	25
y (No. of Employees)	6	10	9	5	8	14	13	16

a) Draw a scatter diagram for the data.

b) Find $\bar{x}$, $\bar{y}$, and b, and the equation of the least squares line.

c) Graph the least squares line on your scatter diagram.

d) If Healthy Crunch receives 15 lb of mail, how many employees should be assigned mail duty?

e) Find the standard error of estimate S_e.

f) Find a 95% confidence interval for the number of employees required to process mail for 15 lb of mail.

g) Find r.

h) Test the claim that the population correlation coefficient is positive (use a 1% level of significance).

USING COMPUTERS

Professional statisticians in industry and research use computers to help them analyze and process statistical data. There are many computer programs available for statistics. Some commonly used statistical packages include Minitab, SAS, and SPSS. There are many others as well. Any of these statistical packages may be used with this text.

  The authors of this text have also written a computer statistics package called ComputerStat. The programs in ComputerStat are available on disks with an accompanying manual that explains how to use the programs and gives additional computer applications. The disks are intended for the Apple IIe, IIc, or the IBM PC. Programs are written in the computer language BASIC and are designed for the beginning statistics student with no previous computer experience. ComputerStat is available as a supplement from the publisher of this text.

  Problems in this section may be done using ComputerStat or other statistical computer programs.

The data in this section are taken from the following reference:

  King, Cuchlaine A. M. *Physical Geography.* Oxford: Basil Blackwell, 1980. 77–86, 196–206. Reprinted with permission of Basil Blackwell Limited, Oxford, England.

Throughout the world natural ocean beaches are beautiful sights to see. If you have visited natural beaches, you may have noticed that when the gradient or drop-off is steep, the grains of sand tend to be larger. In fact, a manmade beach with the "wrong" size granules of sand tends to be washed away and eventually replaced when the proper size grain is selected by the action of the ocean and the gradient of the bottom. Since manmade beaches are expensive, this is an important consideration.

  In the data below x = median diameter (in millimeters) of granules of sand and y = gradient of beach slope in degrees on natural ocean beaches.

  In the ComputerStat menu of topics, select Linear Regression and Correlation. Then use program S19, Linear Regression and Correlation, to solve the following problems.

1. We have nine data pairs (x, y). Enter these data as directed by the computer.
2. Scan the information summary. What is the value of $\bar{x}$? of $\bar{y}$? What are the values of the slope and intercept of the least squares line? What is the value of the standard error of estimate? What is the value of the sample correlation coefficient?
3. Select the graphing option. First graph the data points. Just looking at the scatter diagram, would you expect moderately high correlation and a good fit for the least squares line? Press return and graph the least squares line.
4. Next select the option to test the correlation coefficient. Use a 1% level of significance to test the claim that the population correlation coefficient is positive. The computer program utilizes the P value to conduct the test. For a discussion of P values see Section 9.7, The P Value in Hypothesis Testing. What is the smallest level of significance that will result in our accepting the claim that the population correlation coefficient is positive?

x	0.17	0.19	0.22	0.235	0.235	0.30	0.35	0.42	0.85
y	0.63	0.70	0.82	0.88	1.15	1.50	4.40	7.30	11.30

5. Suppose you have a truckload of sifted sand in which the median size of the granules is 0.38 mm. If you want to put this sand on a beach and you don't want the sand to wash away, then what does the least squares line predict for the angle of the beach? *Note:* Heavy storms that produce abnormal waves may also wash out the sand. However, in the long run the size of sand granules that remain on the beach or are brought back to the beach by long-term wave action are determined to a large extent by the angle at which the beach drops off.

   To solve this problem select the option to predict y values from x values. Enter 0.38 for your x value. Next find a 90% confidence interval for your predicted y value. What range of angles should the beach have if we want to be 90% confident that we are matching the size of our sand granules (0.38 mm) to the proper angle of the beach?

6. Repeat problem 5 for confidence levels $c = 80\%, 70\%, 60\%, 50\%$. What happens to the length of the confidence interval as the level of confidence decreases?

7. Suppose we now have a truckload of sifted sand where the median size of the granules is 0.45 mm. Repeat problems 5 and 6 for this new load of sand. What range of angles should the beach have if we want to be 70% confident that we are matching the size of our sand granules (0.45 mm) to the proper angle of the beach?

Problems 8–15 utilize the following data. Plate tectonics and the spreading of the ocean floor are very important in modern studies of earthquakes and earth science in general. The following data give x = age of volcanic islands in the Atlantic and Indian Oceans and y = distance of the island from the center of the mid-oceanic ridge. As you can see, the oldest islands are the farthest from the ridge crest. This fact is explained by the spreading of the ocean floor on which the islands stand. (The following data are adapted by permission of W. H. Freeman and Company.)

As in Problems 1–7, we will use program S19, Linear Regression and Correlation, to solve the following problems.

8. We have 19 data pairs (x, y). Enter these data as directed by the computer.

9. Scan the information summary. What is the value of $\bar{x}$? of $\bar{y}$? What are the values of the slope and intercept of the least squares line? What is the value of the standard error of estimate? What is the value of the sample correlation coefficient?

10. Select the graphing option. First graph the scatter diagram. Do you think we will have a good fit for the least squares line? Press return and graph the least squares line.

11. Select the option to test the correlation coefficient. Use an 8% level of significance to test the claim that the population correlation coefficient is positive. Again, the computer uses the P value to conclude the test. For an explanation of P values used in hypothesis testing, see Section 9.7. What is the smallest level of significance that will result in our accepting the claim that the population correlation coefficient is positive?

12. Select the option to predict y values from x values. If an island is 70 million years old, how far does the least squares line predict it will be from the mid-oceanic ridge crest? Find a 95% confidence interval for this predicted distance.

13. Repeat problem 12 for confidence levels $c = 75\%, 60\%, 30\%$. What happens to the length of the confidence interval as the level of confidence decreases?

14. Repeat problems 12 and 13 for an island that is estimated to be 100 million years old.

15. Consider the number b, which is the slope of the least squares line. How can this number be used to estimate the rate at which the ocean floor is moving? In one year how many centimeters will the ocean floor be expected to move? (*Hint:* Express the units of b in centimeters per year.)

x (age $\times 10^6$ years)

120	120	120	83	60	50	50	50	35	35	30	20	20	20	17	10	2	1	0
3.0	2.2	2.0	1.6	1.55	1.45	0.6	0.2	2.2	1.6	1.8	1.2	0.7	0.3	0.0	0.0	0.0	0.2	0.0

y (distance $\times 10^3$ km)

11 Chi-Square and *F* Distributions

Statistical thinking will one day be as necessary for efficient citizenship as the ability to read and write.

H. G. Wells

H. G. Wells was an outstanding author of science fiction. At the time his stories were published, they seemed to be completely fictional. Landings on the moon, and interplanetary travel were strictly for the imagination. However, humans have walked on the moon and space exploration is now in progress. There is no doubt that life in the future will be much more technical than it is today. Effective citizenship means that we must use technology to improve *everyone's* life. Anyone who claims citizenship in modern society will be required to make important decisions based on technical information. These authors feel that effective citizenship requires statistical thinking as much as the ability to read and write.

Chi Square: Tests of Independence

Innovative Machines Incorporated has developed two new letter arrangements for typewriter keyboards. They wish to see if there is any relationship between the arrangement of letters on the keyboard and the number of hours it takes a new typing student to learn to type at 20 words per minute. Or, from another point of view, is the time it takes a student to learn to type *independent* of the arrangement of the letters on a typewriter keyboard?

To answer questions of this type, we test the hypotheses

H_0: Keyboard arrangement and learning times *are independent*

H_1: Keyboard arrangement and learning times are *not independent*

In problems of this sort we are testing the *independence* of two factors rather than means or proportions. The probability distribution we use to make the decision is the *chi-square distribution* (*chi* is pronounced like the first two letters of the word *kite*). Chi is a Greek letter denoted by the symbol χ, so chi square is denoted by χ^2.

Because the distribution is of chi-*square* values, the χ^2 values begin at 0 and then are all positive. The graph of the χ^2 distribution is not symmetrical, and like the Student's *t* distribution, it depends on the number of degrees of freedom. Figure 11-1 shows the χ^2 distribution for several degrees of freedom (*d.f.*).

FIGURE 11-1   The χ^2 Distribution

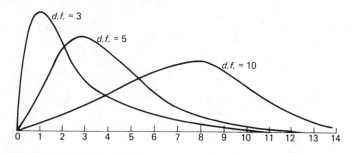

As the degrees of freedom increase, the graph of the chi-square distribution becomes more bell-like and begins to look more and more like the normal distribution. Notice that the mode, or high point, of the graph with *n* degrees of freedom occurs over $n - 2$ (for $n \geq 3$).

When we use the chi-square distribution to test independence, we always use a *right-tail test*. Critical values χ_α^2 for a level of significance α, can be found in Table 8 of Appendix I. The critical value depends not only on the level of significance α, but also on the degrees of freedom *d.f.*

Now let's return to the problem of typewriter keyboard arrangement and the number of hours it takes a student to learn to type at 20 words per minute (wpm). To test the independence of these two variables we must compute a statistic χ^2 for the sample of typing students used by Innovative Machines Incorporated.

Innovative Machines took a random sample of 300 beginning typing students and randomly assigned them to learn to type on one of three typewriter keyboards. The learning times for this sample are shown in Table 11-1.

TABLE 11-1   Keyboard Versus Time to Learn to Type at 20 WPM

Keyboard	21–40 hr	41–60 hr	61–80 hr	Row Total
A	#1 25	#2 30	#3 25	80
B	#4 30	#5 71	#6 19	120
Standard	#7 35	#8 49	#9 16	100
Column Total	90	150	60	300 Sample Size

Table 11-1 is called a *contingency table*. The *shaded boxes* which contain observed frequencies are called *cells*. The row and column totals are not considered to be cells. This contingency table is of size 3 × 3 (read, three-by-three) since there are three rows of cells and three columns. When giving the size of a contingency table we always list the number of *rows first*.

_____ Exercise 1 _____

Give the size of the contingency tables in Figure 11-2 and Figure 11-3. Also count the number of cells in each table. (Remember, each cell is a shaded box.)

a) **FIGURE 11-2** Contingency Table

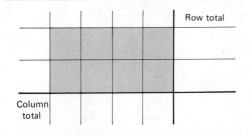

a) There are two rows and four columns so this is a 2 × 4 table. There are eight cells.

b) **FIGURE 11-3** Contingency Table

Row total

Column total

b) Here we have three rows and two columns so this is a 3 × 2 table with six cells.

We are testing the null hypothesis that the keyboard arrangement and the time it takes a student to learn to type are *independent*. We use this hypothesis to determine the *expected frequency* of each cell.

For instance, to compute the expected frequency of cell 1, we observe that cell 1 consists of all the students in the sample who learned to type on keyboard A and who mastered the skill at the 20 word per minute level in 21 to 40 hours. By the assumption (null hypothesis) that the two events are independent, we use the multiplication law to obtain the probability that a student is in cell 1.

$$P(\text{cell 1}) = P(\text{keyboard A } and \text{ skill in } 21\text{-}40\,\text{hr})$$
$$= P(\text{keyboard A}) \cdot P(\text{skill in } 21\text{-}40\,\text{hr})$$

Since there are 300 students in the sample and 80 used keyboard A,

$$P(\text{keyboard A}) = \frac{80}{300}$$

Also, 90 of the 300 students learned to type in 21–40 hours, so

$$P(\text{skill in } 21\text{-}40\,\text{hr}) = \frac{90}{300}$$

Using these two probabilities and the assumption of independence,

$$P(\text{keyboard } A \text{ } and \text{ skill in } 21\text{-}40\,\text{hr}) = \frac{80}{300} \cdot \frac{90}{300}$$

Finally, since there are 300 students in the sample, we have the *expected frequency E* for cell 1.

$$E = P(\text{student in cell 1}) \cdot (\text{no. of students in sample})$$
$$= \frac{80}{300} \cdot \frac{90}{300} \cdot 300$$
$$= \frac{80 \cdot 90}{300}$$
$$= 24$$

We can repeat this process for each cell. However, the next to the last step yields an easier formula for the expected frequency E.

- **Formula for expected frequency E:**

$$E = \frac{(\text{row total})(\text{column total})}{\text{sample size}}$$

Note: If the expected value is not a whole number, do *not* round it to the nearest whole number.

Let's use this formula in Example 1 to find the expected frequency for cell 2.

EXAMPLE 1

Find the expected frequency for cell 2 of contingency Table 11-1.

Solution: Cell 2 is in row 1 and column 2. The row total is 80 and the column total is 150. The size of the sample is still 300.

$$E = \frac{(\text{row total})(\text{column total})}{\text{sample size}} = \frac{(80)(150)}{300} = 40$$

_____ Exercise 2 _____

Table 11-2 contains the *observed frequences, O*, and expected frequencies, *E*, for the contingency table giving keyboard arrangement and number of hours it takes a student to learn to type at 20 words per minute. Fill in the missing expected frequencies.

TABLE 11-2 Complete Contingency Table of Keyboard Arrangement and Time to Learn to Type

Keyboard	21–40 hr	40–60 hr	61–80 hr	Row Total
A	#1 $O = 25$ $E = 24$	#2 $O = 30$ $E = 40$	#3 $O = 25$ $E = \underline{\hphantom{xx}}$	80
B	#4 $O = 30$ $E = 36$	#5 $O = 71$ $E = \underline{\hphantom{xx}}$	#6 $O = 19$ $E = \underline{\hphantom{xx}}$	120
Standard	#7 $O = 35$ $E = \underline{\hphantom{xx}}$	#8 $O = 49$ $E = 50$	#9 $O = 16$ $E = 20$	100
Column Total	90	150	60	300 Sample Size

For cell 3 we have
$$E = \frac{(80)(60)}{300} = 16$$

For cell 5 we have
$$E = \frac{(120)(150)}{300} = 60$$

For cell 6 we have
$$E = \frac{(120)(60)}{300} = 24$$

For cell 7 we have
$$E = \frac{(100)(90)}{300} = 30$$

Now we are in a position to compute the statistic χ^2 for the sample of typing students. The χ^2 value is a measure of the sum of differences between observed frequency, O, and expected frequency, E, in each cell. These differences are listed in Table 11-3 for the nine cells.

TABLE 11-3 Difference Between the Observed and Expected Frequencies

Cell	Observed O	Expected E	Difference $O - E$
1	25	24	1
2	30	40	-10
3	25	16	9
4	30	36	-6
5	71	60	11
6	19	24	-5
7	35	30	5
8	49	50	-1
9	16	20	-4
			$\Sigma (O - E) = 0$

As we see, if we sum the differences between the observed frequencies and the expected frequencies of the cells, we get the value zero. This total certainly does not reflect the fact that there were differences between the observed and expected frequencies. To obtain a measure whose sum does reflect the magnitude of the differences, we square the differences and work with the quantities $(O - E)^2$. But instead of using the terms $(O - E)^2$, we use the values $(O - E)^2/E$. The reason we use this expression is that a small difference between the observed and expected frequency is not nearly as important if the expected frequency is large as it is if the expected frequency is small. For instance, for both cells 1 and 8 the squared difference $(O - E)^2$ is 1. However, this difference is more meaningful in cell 1 where the expected frequency is 24 than it is in cell 8 where the expected frequency is 50. When we divide the quantity $(O - E)^2$ by E, we take the size of the difference with respect to the size of the expected value. We use the sum of these values to form the sample statistic χ^2:

$$\chi^2 = \Sigma \frac{(O - E)^2}{E}$$

where the sum is over all cells in the contingency table.

- **Comment:** If you look up the word *irony* in a dictionary you will find one of its meanings is described as "the difference between actual (or observed) results and expected results." Since irony is so prevalent in much of our human experience, it is not surprising that statisticians have incorporated a related chi square distribution into their work. As we will soon see, the chi square distribution has many applications in social science, business administration, and natural science.

─────────── Exercise 3 ───────────

a) Complete Table 11-4 from the data of Table 11-3.

a) The last two rows of Table 11-4 are

TABLE 11-4

Cell	O	E	$O - E$	$(O - E)^2$	$(O - E)^2/E$
1	25	24	1	1	0.04
2	30	40	-10	100	2.50
3	25	16	9	81	5.06
4	30	36	-6	36	1.00
5	71	60	11	121	2.02
6	19	24	-5	25	1.04
7	35	30	5	25	0.83
8	49	50	____	____	____
9	16	20	____	____	____

$$\Sigma \frac{(O - E)^2}{E} = \_\_$$

Cell	O	E	$(O - E)$	$(O - E)^2$	$(O - E)^2/E$
8	49	50	-1	1	0.02
9	16	20	-4	16	0.80

$$\Sigma \frac{(O - E)^2}{E} = \text{total of last column} = 13.31$$

b) Compute the statistic χ^2 for this sample.

b) Since $\chi^2 = \Sigma \dfrac{(O - E)^2}{E}$, $\chi^2 = 13.31$.

Notice that when the observed frequency and the expected frequency are very close, the quantity $(O - E)^2$ is close to zero, and so the statistic χ^2 is near zero. As the difference increases, the statistic χ^2 also increases. To determine how large the statistic can be before we must reject the null hypothesis of independence, we find a critical value χ^2_α in Table 8 of Appendix I for the specified level of significance α and the number of degrees of freedom in the sample.

As we saw earlier, the chi-square distribution changes as the degrees of freedom change. To find a critical value χ^2_α, we need to know the degrees of freedom in the sample as well as the level of significance designated. To test independence, the degrees of freedom $d.f.$ of a sample are determined by the following formula.

- **Degrees of freedom for independence:**

  degrees of freedom = (number of rows − 1) · (number of columns − 1)

  or

  $$d.f. = (R - 1)(C - 1)$$

  where R = number of cell rows

  C = number of cell columns

_____ Exercise 4 _____

Determine the number of degrees of freedom in the example of typewriter keyboard arrangements (*see* Table 11-1). Recall that the contingency table had three rows and three columns.

$d.f. = (R - 1)(C - 1)$

$\quad = (3 - 1)(3 - 1) = (2)(2) = 4$

To test the hypothesis that the letter arrangement on a typewriter keyboard and the time it takes to learn to type at 20 words per minute are independent at the $\alpha = 0.05$ level of significance, we look up the critical value $\chi^2_{0.05}$ in Table 8 of Appendix I. For $d.f. = 4$ and $\alpha = 0.05$, we see $\chi^2_{0.05} = 9.49$. When we compare the sample statistic $\chi^2 = 13.31$ with the critical value $\chi^2_{0.05}$, we see that the sample statistic is larger. Since it is larger, we *reject* the null hypothesis of independence and conclude that keyboard arrangement and learning time are *not* independent. (*See* Figure 11-4.)

Let's summarize how we use the chi-square distribution to test the independence of two variables.

Step 1: Set up the hypotheses:

H_0: The variables *are* independent

H_1: The variables *are not* independent

FIGURE 11-4   Comparison of Critical Value and Sample Statistic (*d.f.* = 4)

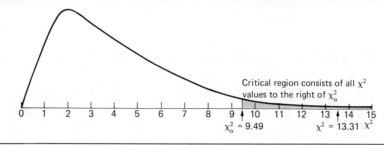

Step 2: Compute the expected frequency for each cell in the contingency table by use of the formula

$$E = \text{expected frequency} = \frac{(\text{row total})(\text{column total})}{\text{sample size}}$$

Step 3: Compute the statistic χ^2 for the sample:

$$\chi^2 = \Sigma \frac{(O - E)^2}{E}$$

where *O* is the observed frequency,
  E is the expected frequency, and
  the sum Σ is over all cells.

Step 4: Find the critical value χ_α^2 in Table 8 of Appendix I. Use the level of significance α and the number of degrees of freedom *d.f.* to find the critical value.

$$d.f. = (R - 1)(C - 1)$$

where *R* is the number of rows, and
  C is the number of columns of cells in the contingency table.

The critical region is all values of χ^2 to the *right* of the critcal value χ_α^2.

Step 5: Compare the sample statistic χ^2 of Step 3 with the critical value χ_α^2 of Step 4. If the sample statistic is *larger,* reject the null hypothesis of independence. Otherwise, do not reject the null hypothesis.

- **Note:** We compare the sample statistic χ^2 with the critical value χ_α^2. But the distribution of sample statistics is only approximately the same as the theoretical distribution whose critical values χ_α^2 are found in Table 8 of Appendix I. In order to safely use critical values χ_α^2, we must be sure that all of the cells have an *expected frequency* larger than or equal to 5. If this condition is not met, the sample size should be increased.

──────────── Exercise 5 ────────────────────────────────

Super Vending Machines Company is to install soda pop machines in elementary schools and high schools. The market analysts wish to know if flavor preference and school level are independent. A random sample of 200 students was taken. Their school level and soda pop preference are given in Table 11-5. Is independence indicated at the $\alpha = 0.01$ level of significance?

Step 1: State the null and alternate hypotheses.

Step 1: H_0 : School level and soda pop preference are independent

H_1 : School level and soda pop preference are not independent

Step 2: Complete the contingency Table 11-5 by filling in the required expected frequencies.

Step 2: The expected frequency

for cell 5 is $\dfrac{(40)(80)}{200} = 16$

for cell 6 is $\dfrac{(40)(120)}{200} = 24$

for cell 7 is $\dfrac{(20)(80)}{200} = 8$

for cell 8 is $\dfrac{(20)(120)}{200} = 12$

Note: In this example the expected frequencies are all whole numbers. If the expected frequency has a decimal part such as 8.45, do *not* round the value to the nearest whole number; rather, give the expected frequency as the decimal number.

TABLE 11-5 School Level and Soda Pop Preference

Soda Pop	High School	Elementary School	Row Total
Kula Kola	$O = 33$ #1 $E = 36$	$O = 57$ #2 $E = 54$	90
Mountain Mist	$O = 30$ #3 $E = 20$	$O = 20$ #4 $E = 30$	50
Jungle Grape	$O = 5$ #5 $E = \underline{}$	$O = 35$ #6 $E = \underline{}$	40
Diet Pop	$O = 12$ #7 $E = \underline{}$	$O = 8$ #8 $E = \underline{}$	20
Column Total	80	120	200 Sample Size

Step 3: Fill in Table 11-6 and use the table to find the sample statistic χ^2.

Step 3: The last three rows of Table 11-6 read as follows:

TABLE 11-6 Computational Table for χ^2

Cell	O	E	$O - E$	$(O - E)^2$	$(O - E)^2/E$
1	33	36	-3	9	0.25
2	57	54	3	9	0.17
3	30	20	10	100	5.00
4	20	30	-10	100	3.33
5	5	16	-11	121	7.56
6	35	24	11	___	___
7	12	8	___	___	___
8	8	12	___	___	___

Cell	O	E	$O - E$	$(O - E)^2$	$(O - E)^2/E$
6	35	24	11	121	5.04
7	12	8	4	16	2.00
8	8	12	-4	16	1.33

$$\chi^2 = \text{total of last column}$$
$$= \Sigma \frac{(O - E)^2}{E} = 24.68$$

Step 4: What is the size of the contingency table? Use the number of rows and columns to determine the number of degrees of freedom. For $\alpha = 0.01$ use Table 8 of Appendix I to find the critical value $\chi^2_{0.01}$.

Step 4: The contingency table is of size 4×2. Since there are four rows and two columns,

$$d.f. = (4 - 1)(2 - 1) = 3$$

For $\alpha = 0.01$ the critical value χ^2_α is 11.34.

Step 5: Do we accept or reject the null hypothesis that school level and soda pop flavor preference are independent?

Step 5: Since the statistic χ^2 is larger than the critical value χ^2_α we reject the null hypothesis of independence and conclude that school level and soda pop preference are dependent.

Section Problems 11.1

For each of the problems, do the following:

a) State the null and alternate hypotheses.
b) Find the value of the chi-square statistic from the sample.
c) Find the degrees of freedom and the appropriate critical chi-square value.
d) Sketch the critical region and the acceptance region and locate your sample chi-square value and critical chi-square value on the sketch.
e) Decide whether you should accept or reject the null hypothesis.

Use the expected values, E, to the hundredths place.

1. The personnel department of Jupiter Scientific Labs is doing a study about job satisfaction. A random sample of 310 employees were given a test designed to diagnose the level of job satisfaction. Each employee's salary was also recorded. (*See* Table 11-7 below.) Use a chi-square test to determine if salary and job satisfaction are independent at the 0.05 level of significance.

TABLE 11-7 Salary Versus Job Satisfaction

Satisfaction	Under $25,000	$25,000–$35,000	Over $35,000	Row Total
High	20	20	10	50
Medium	100	65	35	200
Low	40	15	5	60
Column Total	160	100	50	310

2. The counseling unit of Woodrock College is interested in the relationship between anxiety level and the need to succeed. A random sample of 200 college freshmen was taken. The freshmen were given tests to measure their anxiety level and their need to succeed. The results are given in Table 11-8. Test the hypothesis that anxiety level and need to succeed are independent at the 0.01 level of significance.

TABLE 11-8 Anxiety Level Versus Need to Succeed

Need	High Anxiety	Medium Anxiety	Low Anxiety	Row Total
High	30	40	5	75
Medium	17	50	33	100
Low	3	10	12	25
Column Total	50	100	50	200

3. Mr. Acosta, a sociologist, is doing a study to see if there is a relationship between the age of a young adult (18 to 35 years old) and the type of movie preferred. A random sample of 93 adults revealed the data in Table 11-9. Test if age and type of movie preferred are independent at the 0.05 level.

TABLE 11-9 Person's Age Versus Movie Preference

Movie	18–23 yr	24–29 yr	30–35 yr	Row Total
Musical	8	15	11	34
Science Fiction	12	10	8	30
Comedy	9	8	12	29
Column Total	29	33	31	93

4. Ms. Angel is doing a study to see if the type of driving a person prefers to do and the sex of that person are related. A random sample of 100 drivers gave the data in Table 11-10. Test the hypothesis that sex and type of driving preferred are independent at the 0.01 level.

TABLE 11-10 Person's Sex Versus Driving Preference

Sex	Freeway	Multi-Lane with Traffic Lights	Two-Lane with Traffic Lights	Row Total
Female	18	16	11	45
Male	22	19	14	55
Column Total	40	35	25	100

5. Reading Nook Bookstore has 750 retail outlets across the country. The sales director wanted to see if Christmas music affects book sales in December. She randomly assigned some of the outlets to pipe in music and others not to. Then sales records for the month of December were kept. The results are shown in Table 11-11. Test the hypothesis that sales and Christmas music are independent. Use a 0.05 level of significance.

6. After a large fund drive to help the Boston City Library, the following information was obtained from a random sample of contributors to the library fund (Table 11-12). Using a 1% level of significance, test the claim that the amount contributed to the library fund is independent of ethnic group.

TABLE 11-11 Number of Books Sold Versus Use of Christmas Music

Outlet	Less than 10,000	10,000–20,000	More than 20,000	Row Total
With music	5	18	7	30
Without music	10	7	3	20
Column Total	15	25	10	50

TABLE 11-12 Contributors to the Library Fund

Ethnic Group	Number of People Making Contribution					Row Total
	$1–5	$6–10	$11–15	$16–20	Over $20	
A	310	715	201	105	42	1,373
B	619	511	312	97	22	1,561
C	402	624	217	88	35	1,366
D	544	571	309	79	29	1,532
Column Total	1,875	2,421	1,039	369	128	5,832

7. Blue Bird Consolidated Theaters has more than 600 theaters located across the country. Each theater has four separate screens, and a customer can choose from one of four different movies. The president of Blue Bird Consolidated wants to know if a variety of shows (spy, comedy, horror, children's) or a coordinated bill (all spy, all comedy, all horror, all children's) has any effect on total ticket sales at a theater. The president randomly assigned 47 theaters to use a variety of shows and 53 other theaters to use a coordinated bill of shows. For all theaters, total ticket sales for one week were recorded. Using the data of Table 11-13 and a 5% level of significance, test the claim that total ticket sales are independent of the four shows being varied or coordinated.

8. "Buddies" is a volunteer social service group that works with disadvantaged children. Adults 18 years and older volunteer from one to six hours each week to take a disadvantaged child to the zoo, to a museum, to a movie, fishing, ice skating, or some other activity. The Buddies program recruits adults from three main groups: college students, nonstudents

TABLE 11-13 Total Ticket Sales Versus Type of Billing

Type of Billing	Ticket Sales for One Week				Row Total
	Less than 1,000	1,000 to 2,000	2,001 to 3,000	More than 3,000	
Variety	10	12	18	7	47
Coordinated	6	16	22	9	53
Column Total	16	28	40	16	100

living in the inner city, nonstudents living in the suburbs. A random sample of adult volunteers gave the information in Table 11-14. Using a 5% level of significance, test the claim that the number of hours volunteered is independent of the type of volunteer.

TABLE 11-14 Number of Hours per Week Volunteered to the Buddies Program

Type of Volunteer	Hours Volunteered			Row Total
	1–2	3–4	5–6	
College Student	115	93	47	255
Inner City Resident	88	150	56	294
Suburban Resident	95	133	60	288
Column Total	298	376	163	837

9. A random sample of congressmen in Washington, D.C., gave the following information about party affiliation and number of dollars spent on federal projects in their home districts (Table 11-15). Using a 1% level of significance, test the claim that federal spending level in home districts is independent of party affiliation.

TABLE 11-15 Party Affiliation and Dollars Spent in Home District

Party	Dollars Spent on Federal Projects in Home Districts			Row Total
	Less than 5 Million	5 to 10 Million	More than 10 Million	
Democrat	8	15	22	45
Republican	12	19	16	47
Column Total	20	34	38	92

SECTION 11.2   Chi Square: Goodness of Fit

Last year the labor union bargaining agents listed five items and asked each employee to mark the *one* most important to her or him. The items and corresponding percentage of favorable responses are shown in Table 11-16. The bargaining agents need to determine if the distribution of responses *now* "fits" last year's distribution or if it is different.

In questions of this type, we are asking if a population follows a specified distribution. In other words, we are testing the hypotheses

> H_0 : The population fits the given distribution
> H_1 : The population has a different distribution

We use the chi-square distribution to test "goodness-of-fit" hypotheses.

TABLE 11-16 Bargaining Items

Item	Percentage of Favorable Responses
Vacation time	4%
Salary	65%
Safety regulations	13%
Health and retirement benefits	12%
Overtime policy and pay	6%

Just as with tests of independence, we compute the sample statistic:

$$\chi^2 = \Sigma \frac{(O - E)^2}{E}$$

where E = expected frequency

O = observed frequency

$\dfrac{(O - E)^2}{E}$ is summed for each item in the distribution

and compare it with an appropriate critical value χ^2_α from Table 8, Appendix I. In the case of a *goodness-of-fit test,* we use the null hypothesis to compute the expected values. Let's look at the bargaining item problem to see how this is done.

In the bargaining item problem the two hypotheses are

H_0 : The present distribution of responses is the same as last year's

H_1 : The present distribution of responses is different

The null hypothesis tells us that the expected frequencies of the present response distribution should follow the percentages indicated in last year's survey. To test this hypothesis a random sample of 500 employees was taken. If the null hypothesis is true, then there should be 4%, or 20 responses, out of the 500 rating vacation time as the most important bargaining issue. Table 11-17 gives the other expected values and all the information necessary to compute the sample statistic χ^2.

TABLE 11-17 Observed and Expected Frequencies for Bargaining Items

Item	O	E	$(O - E)^2$	$(O - E)^2/E$
Vacation time	30	4% of 500 = 20	100	5.00
Salary	290	65% of 500 = 325	1,225	3.77
Safety	70	13% of 500 = 65	25	0.38
Health and retirement	70	12% of 500 = 60	100	1.67
Overtime	40	6% of 500 = 30	100	3.33
	$\Sigma O = 500$	$\Sigma E = 500$		$\Sigma \dfrac{(O - E)^2}{E} = 14.15$

We see the sample statistic is

$$\chi^2 = \Sigma \frac{(O - E)^2}{E} = 14.15$$

Again, larger values of the sample statistic χ^2 indicate greater differences between the proposed probability distribution and the one followed by the sample. The critical value χ^2_α tells us how large the sample statistic can be before we reject the null hypothesis that the population does follow the distribution proposed in that hypothesis.

To find the critical value χ^2_α we need to know the level of significance α and the number of degrees of freedom $d.f.$ In the case of a goodness-of-fit test the degrees of freedom are found by the following formula:

> • **Degrees of freedom for goodness-of-fit test:**
>
> $$d.f. = (\text{number of } E \text{ entries}) - 1$$

Notice that when we compute the expected values E, we must use the null hypothesis to compute all but the last one. To compute the last one, we can subtract the previous expected values from the sample size. For instance, for the bargaining issues we could have found the number of responses for overtime policy by adding up the other expected values and subtracting that sum from the sample size 500. We would again get an expected value of 30 responses. The degrees of freedom, then, is the number of E values that *must* be computed by using the null hypothesis.

For the bargaining issues we have

$$d.f. = 5 - 1 = 4$$

To test the hypothesis at the 0.05 level of significance, we find the critical value $\chi^2_{0.05}$ for four degrees of freedom in Table 8 of Appendix I. The critical value, 9.49, and the sample statistic, 14.15, are shown in Figure 11-5. As shown in the figure, the sample statistic χ^2 is in the critical region. Note that the critical region is always to the right of the critical value χ^2_α. Since the sample statistic is in the critical region, we reject the null hypothesis and conclude that the distribution of responses to the bargaining issues now is different from the distribution of last year.

One important application of goodness-of-fit tests is to genetic theories. Such an application is shown in Exercise 6.

FIGURE 11-5 Critical Value and Test Statistic ($d.f. = 4$)

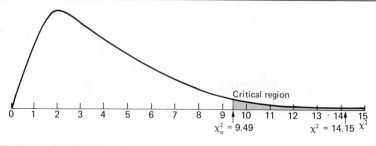

_____ Exercise 6 _____

According to genetics theory, red-green colorblindness in humans is a recessive sex-linked characteristic. In this case the gene is carried on the X chromosome only. We will denote an X chromosome with the gene by X_c and one without the gene by X_n. Women have two X chromosomes, and they will be red-green colorblind only if both chromosomes have the gene, designated $X_c X_c$. A woman can have normal vision but still carry the colorblind gene if only one of the chromosomes has the gene, designated $X_c X_n$. A man carries an X and Y chromosome; if the X chromosome carries the colorblind gene $(X_c Y)$, the man is colorblind.

According to genetics theory, if a man with normal vision $(X_n Y)$ and a woman carrier $(X_c X_n)$ have a child, the probabilities that the child will have red-green colorblindness, have normal vision and not carry the gene, or have normal vision and carry the gene are given by the equally likely events in Table 11-18.

TABLE 11-18 Red-Green Colorblindness

Mother \ Father	X_n	Y
X_c	$X_c X_n$	$X_c Y$
X_n	$X_n X_n$	$X_n Y$

$$P(\text{child has normal vision and is not a carrier}) = P(X_n Y) + P(X_n X_n) = \frac{1}{2}$$

$$P(\text{child has normal vision and is a carrier}) = P(X_c X_n) = \frac{1}{4}$$

$$P(\text{child is red-green colorblind}) = P(X_c Y) = \frac{1}{4}$$

To test this genetics theory, Genetics Labs took a random sample of 200 children whose mothers were carriers of the colorblind gene and whose fathers had normal vision. The results are in Table 11-19. We wish to test the hypothesis that the population follows the distribution predicted by the genetics theory (*see* Table 11-18).

a) State the null and alternate hypotheses.

b) Fill in the rest of Table 11-19 and use the table to compute the sample statistic χ^2.

TABLE 11-19 Colorblindness Sample

Event	O	E	$(O - E)^2$	$(O - E)^2/E$
Red-green colorblind	35	50	225	4.50
Normal vision, noncarrier	105	___	___	___
Normal vision, carrier	60	___	___	___

a) H_0 : The population fits the distribution predicted by genetics theory

H_1 : The population does not fit the distribution predicted by genetics theory.

b)

TABLE 11-20 Completion of Table 11-19

Event	O	E	$(O - E)^2$	$(O - E)^2/E$
Red-green colorblind	35	50	225	4.50
Normal vision, noncarrier	105	100	25	0.25
Normal vision, carrier	60	50	100	2.00

The sample statistic is $\chi^2 = \Sigma \dfrac{(O - E)^2}{E} = 6.75$

c) There are three expected frequencies listed in Table 11-19. Use this information to compute the degrees of freedom.

c) $d.f. = $ (no. of E values) $- 1 = 3 - 1 = 2$

d) Find the critical value $\chi^2_{0.01}$ for a 0.01 level of significance. Can we accept the hypothesis that the population follows the distribution predicted by genetics theory?

d) From Table 8 of Appendix I we see that for $d.f. = 2$ and level of significance 0.01, the critical value is $\chi^2_{0.01} = 9.21$. Since the sample statistic $\chi^2 = 6.75$ is less than the critical value, we do not reject the null hypothesis that the population follows the distribution predicted by genetics theory.

Section Problems 11.2

For each of the problems, do the following:
  a) State the null and alternate hypotheses.
  b) Find the value of the chi-square statistic from the sample.
  c) Find the degrees of freedom and the appropriate critical chi-square value.
  d) Sketch the critical region and the acceptance region and locate your sample chi-square value and critical chi-square value on the sketch.
  e) Decide whether you should accept or reject the null hypothesis.

1. In a study done ten years ago *Market Trends* magazine discovered that 20% of the new-car buyers planned to keep their new cars more than five years, 30% planned to keep them between two and five years, and 50% intended to sell the cars in less than two years. This year the magazine did a similar study. A random sample of 200 new-car buyers was taken. In this sample 48 people planned to keep their cars more than five years, 75 said they would keep them between two and five years, and 77 indicated that they planned to sell the cars in less than two years. Test the hypothesis that the present buyers plan to keep their cars the same length of time as buyers ten years ago. Use an 0.01 level of significance.

2. Jimmy Nuts Company advertises that their nut mix contains 40% cashews, 15% brazil nuts, 20% almonds, and only 25% peanuts. The truth-in-advertising investigators took a random sample (of size 20 lb) of the nut mix and found the distribution to be as follows:

Cashews	Brazil Nuts	Almonds	Peanuts
6 lb	3 lb	5 lb	6 lb

At the 0.01 level of significance, is the claim made by Jimmy Nuts true?

3. Twenty years ago a poll was taken which showed that 70% of the citizens of Alkan City opposed the unionization of city workers, 20% favored it, and 10% had no opinion. Recently a random sample of 197 Alkan City citizens showed that 145 opposed the unionization, 35 favored it, and 17 had no opinion. At the 0.05 level, has there been a change in opinion about the unionization of city employees?

4. The Fish and Game Department stocked Lake Lulu with fish in the following proportions: 30% catfish, 15% bass, 40% bluegill, and 15% pike. Five years later they sampled the lake to see if the distribution of fish had changed. They found the 500 fish in the sample were distributed as follows:

Catfish	Bass	Bluegill	Pike
120	85	220	75

In the five-year interval did the distribution of fish change at the 0.05 level?

5. Dr. Gordon is doing a study of the effect of noise on the litter size of coyotes. When the coyotes were subjected to the noise level found in their natural environment, the litter sizes were as follows:

One	Two	Three	Four	Five	More than Five
7%	13%	15%	25%	28%	12%

Under a similar environment, but with the noise level increased by 30 decibels, a random sample of 100 litters showed the sizes to be as follows:

One	Two	Three	Four	Five	More than Five
15	19	22	20	12	12

Is the distribution of litter sizes different under the noise increase? Use a 0.01 level of significance.

6. The director of library services at Walla Woo College did a survey of types of books (by subject) in the circulation library. Then she used library records to take a random sample of 4,217 books checked out last term and classified the books in the sample by subject. The results are shown below.

Subject Area	% of Books in Circulation Library on This Subject	Number of Books in Sample on This Subject
Business	32%	1,210
Humanities	25%	956
Natural science	20%	940
Social science	15%	814
All other subjects	8%	297

Using a 5% level of significance, test the claim that the subject distribution of books in the library fits the distribution of books checked out by students.

7. Public Service Company planning office is doing a study of electricity consumption in the north Denver suburbs. In the table below the second column gives the percent of total consumption of north Denver suburbs based on data five years old. The last column gives the most recent consumption estimates (in millions of kilowatt hours) based on a random sample of readings taken last month.

Area	Old Rate of Total Consumption	Present Consumption Estimates
North Glenn	20%	201
Westminster	10%	97
Thornton	14%	115
Broomfield	22%	245
Commerce City	15%	140
Arvada	12%	105
Henderson	7%	70
		973 total

Using a 1% level of significance, test the claim that the old distribution of electricity consumption is the same as the present distribution of electricity consumption. (*Note:* We are only testing to see if the *distribution* of consumption is the same; the *level* of consumption has of course increased because of population growth in the entire region.)

8. The community hospital is studying its distribution of patients. A random sample of 317 patients presently in the hospital gave the following information:

Type of Patient	Old Rate of Occurrence of These Types of Patients	Present Number of This Type Patient in Sample
Maternity ward	20%	65
Cardiac ward	32%	100
Burn ward	10%	29
Children's ward	15%	48
All other wards	23%	75

Using a 5% level of significance, test the claim that the distribution of patients in these wards has not changed.

9. The accuracy of a census report on a city in southern California was questioned by some government officials. A random sample of 1,215 people living in the city was used to check the report and the results are shown here:

Ethnic Origin	Census %	Sample Result
Black	10%	127
Asian	3%	40
Anglo	38%	480
Spanish American	41%	502
American Indian	6%	56
All others	2%	10

Using a 1% level of significance, test the claim that census distribution and sample distribution agree.

10. Snoop Incorporated is a firm that does market surveys. The Rollum Record Company hired Snoop to study the age distribution of people who buy records. To check the Snoop report, Rollum used a random sample of 519 customers and obtained the following data:

Customer Age in Years	% of Customers from Snoop Report	Number of Customers in Sample
Less than 14	12%	88
14–18	29%	135
19–23	11%	52
24–28	10%	40
29–33	14%	76
More than 33	24%	128

Using a 1% level of significance, test the claim that the distribution of customer ages in the Snoop report agrees with the sample report.

SECTION 11.3   Testing and Estimating Variances and Standard Deviations

Many problems arise that require us to make decisions about variability. In this section we will study two kinds of problems: (1) we will test hypotheses about the variance (or standard deviation) of a population, and (2) we will find confidence intervals for the variance (or standard deviation) of a population. It is customary to talk about variance instead of standard deviation because our techniques employ the sample variance rather than the standard deviation. Of course, the standard deviation is just the square root of the variance, so any discussion about variance is easily converted to a similar discussion about standard deviation.

Let us consider a specific example in which we might wish to test a hypothesis about the variance. Almost everyone has had to wait in line. In a grocery store, bank, post office, or registration center, there are usually several check-out or service areas. Frequently, each service area has its own independent line. However, many businesses and government offices are adopting a "single-line" procedure.

In a single-line procedure there is only one waiting line for everyone. As any service area becomes available, the next person in line gets served. The old independent-lines procedure has a line at each service center. An incoming customer simply picks the shortest line and hopes it will move quickly. In either procedure, the number of clerks and the rate at which they work is the same, so the average waiting time is the *same*. What is the advantage of the single-line procedure? The difference is in the *attitudes* of people who wait in the lines. A lengthy waiting line will be more acceptable if the variability of waiting times is smaller, even though the average waiting time is the same. When the variability is small, the inconvenience of waiting (although it might not be reduced) does become more predictable. This means impatience is reduced and people are happier.

To test the hypothesis that variability is less in a single-line process, we use the chi-square distribution. The next theorem tells us how to use the sample and population variance to compute values of χ^2.

- **Theorem 11.1:** If we have a normal population with variance σ^2, and a random sample of n measurements is taken from this population with sample variance s^2, then

$$\chi^2 = \frac{(n-1)s^2}{\sigma^2}$$

has a chi-square distribution with degrees of freedom $d.f. = n - 1$.

Recall that the chi-square distribution is *not* symmetric, and that there are different chi-square distributions for different degrees of freedom. Table 8 in Appendix I gives chi-square values for which the area α is to the *right* of the given chi-square value.

EXAMPLE 2

a) Find the χ^2 value so that the area to the right of χ^2 is 0.05 when $d.f. = 10$.

Since the area to the *right* of χ^2 is to be 0.05, we look in the $\alpha = 0.05$ column and the row with $d.f. = 10$. $\chi^2 = 18.31$ (*see* Figure 11-6a).

FIGURE 11-6 χ^2 Distribution with *d.f.* = 10

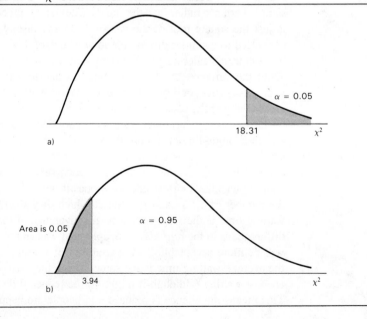

b) Find the χ^2 value so that the area to the *left* of χ^2 is 0.05 when $d.f. = 10$.

When the area left of χ^2 is 0.05, the corresponding area to the *right* is $1 - 0.05 = 0.95$, so we look in the $\alpha = 0.95$ column and the row with $d.f. = 10$. We find $\chi^2 = 3.94$ (*see* Figure 11-6b).

_____ Exercise 7 _____

a) Find the χ^2 value so that 1% of the area under the χ^2 curve is to the right of χ^2 when $d.f. = 20$.

a) The χ^2 value is in the column under $\alpha = 0.01$ and the row with $d.f. = 20$. $\chi^2 = 37.57$ (*see* Figure 11-7).

FIGURE 11-7 χ^2 Distribution with $d.f. = 20$

b) Find the χ^2 value so that 1% of the area under the curve is to the *left* of χ^2 when $d.f. = 20$.

1) First find the corresponding area α to the *right* of the desired χ^2 value.

2) Then look up the χ^2 value with $d.f. = 20$ and $\alpha = 0.99$.

b) The area to the right of the desired χ^2 value is
$$1 - 0.01 = 0.99$$
The χ^2 value is in the column under $\alpha = 0.99$ and the row with $d.f. = 20$. $\chi^2 = 8.26$ (*see* Figure 11-8).

FIGURE 11-8 χ^2 Distribution with $d.f. = 20$

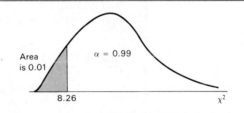

Now let's use Theorem 11.1 and our knowledge of the chi-square distribution to determine if a single-line procedure has less variance of waiting times than independent lines.

EXAMPLE 3

A large discount hardware store in San Antonio has been using the independent-lines procedure to check out customers. After long observation, the manager knows that the standard deviation of waiting times is 7 minutes. The manager decided to introduce the single-line procedure on a trial basis to see if a reduction in waiting time variability would occur. A random sample of 25 customers was monitored and their waiting times for check-out were determined. The sample standard deviation was $s = 4$ minutes. We will use a 5% level of significance to test the claim that the variance of waiting times has been reduced.

As a null hypothesis we assume the variance in waiting times is the same as that of the former independent lines procedure. The alternate hypothesis is that the variance for the single-line procedure is less than that for the independent lines. If we let σ be the standard deviation of waiting times for the single-line procedure, then σ^2 is the variance and we have

$$H_0 : \sigma^2 = 49 \qquad (\text{use } 7^2 = 49)$$

$$H_1 : \sigma^2 < 49$$

We use the chi-square distribution to test the hypotheses. Assuming the waiting times are normally distributed, we compute our observed value of χ^2 by using Theorem 11.1. Since

$$n = 25$$

$$s = 4 \quad \text{so} \quad s^2 = 16 \qquad (\text{observed})$$

$$\sigma = 7 \quad \text{so} \quad \sigma^2 = 49 \qquad (\text{from } H_0 : \sigma^2 = 49)$$

$$\chi^2 = \frac{(n-1)s^2}{\sigma^2} = \frac{(25-1)16}{49} = 7.8 \qquad (\text{by Theorem 11.1})$$

The critical value is obtained from Table 8, using the degrees of freedom and level of significance. By the alternate hypothesis $H_1 : \sigma^2 < 49$, we want a *left* tail with area 0.05. Therefore, the area in the corresponding right tail is $\alpha = 1 - 0.05 = 0.95$. Since $d.f. = n - 1 = 25 - 1 = 24$, the desired critical value is $\chi^2_{0.95} = 13.85$ (*see* Figure 11-9).

Since the observed value of $\chi^2 = 7.8$ is in the critical or rejection region, we reject $H_0 : \sigma^2 = 49$ and accept $H_1 : \sigma^2 < 49$. The variance of the single-line procedure is less than 49.

FIGURE 11-9 Critical and Acceptance Regions for H_0

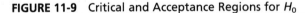

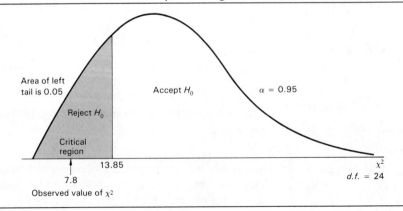

Area of left tail is 0.05

Accept H_0

$\alpha = 0.95$

Reject H_0

Critical region

13.85

χ^2

$d.f. = 24$

7.8

Observed value of χ^2

_____ Exercise 8 _____

Certain industrial machines require overhaul when wear on their parts introduces too much variability to pass inspection. A DOE (Department of Energy) official is visiting a dentist's office to inspect the operation of an X-ray machine. If the machine emits too little radiation, clear photographs cannot be obtained. However, too much radiation can be harmful to the patient. Government regulations specify an average emission of 60 milliRads with standard deviation σ of 12 milliRads, and the machine has been set for these readings. After examining the machine, the inspector is satisfied that the average emission is still 60 milliRads. However, there is wear on certain mechanical parts. To test variability, the inspector takes a random sample of 30 X-ray emissions and finds the sample standard deviation to be $s = 15$ milliRads. Does this support the claim that the variance is too high (i.e., the machine should be overhauled?) Use a 1% level of significance.

Let σ be the (population) standard deviation of emissions (in milliRads) of the machine in its present condition.

a) Which of the following shall we use for the null hypothesis? Explain.

$$H_0 : \sigma^2 = 12 \quad H_0 : \sigma^2 = 144 \quad H_0 : \sigma^2 > 144$$

a) $H_0 : \sigma^2 = 144$. We use $\sigma = 12$ and the initial claim that the variance is still what it should be according to specifications.

b) Which of the following shall we use for the alternate hypothesis? Explain.

$$H_1 : \sigma > 12 \quad H_1 : \sigma^2 \neq 144 \quad H_1 : \sigma^2 > 144$$

b) $H_1 : \sigma^2 > 144$. We want to test the claim that the variance is too large.

c) What is the observed value of χ^2?

c) $\chi^2 = \dfrac{(n-1)s^2}{\sigma^2} = \dfrac{(30-1)15^2}{144} = 45.3$

(since $n = 30$, $s = 15$ and by H_0, $\sigma^2 = 144$)

d) What are the degrees of freedom? Are we using a left-, right-, or two-tail test? Use Table 8 to find the chi-square critical value.

d) $d.f. = n - 1 = 30 - 1 = 29$. Since H_1 is $\sigma^2 > 144$, we use a right-tail test. The problem calls for $\alpha = 0.01$ and an area of 0.01 is to the right of $\chi^2 = 49.59$ when $d.f. = 29$. The critical value is $\chi^2_{0.01} = 49.59$.

e) Sketch the acceptance and critical region on a chi-square curve. Locate the observed chi-square value on the sketch.

e) Figure 11-10 shows the critical region.

FIGURE 11-10 Critical and Acceptance Regions with $d.f. = 29$

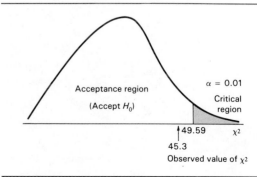

Acceptance region
(Accept H_0)

$\alpha = 0.01$

Critical region

49.59

45.3

χ^2

Observed value of χ^2

f) Do we accept or reject H_0? Should the inspector recommend that the machine be overhauled?

f) Since the observed chi-square value 45.3 is in the acceptance region, we accept H_0 and conclude that the machine does not need an overhaul at this time.

Sometimes it is important to have a confidence interval for the variance or standard deviation. Let us look at another example.

Mr. Wilson is a truck farmer in California who makes his living on a large single-vegetable crop of green beans. Because of modern machinery being used, the entire crop must be harvested at once. Therefore, it is important to plant a variety of green beans that mature all at once. This means he wants a small standard deviation between maturing times of individual plants. A seed company is trying to develop a new variety of green beans with a small standard deviation of maturing times. To test their new variety, Mr. Wilson planted 30 of the new seeds and carefully observed the number of days required for each plant to arrive at its peak of maturity. The maturing times for these plants had a sample standard deviation of $s = 3.4$ days. How can we find a 95% confidence interval for the population standard deviation of maturing times of this variety of green bean? The answer to this question is based on the following theorem.

> • **Theorem 11.2:** Let a random sample of size n be taken from a normal population with population standard deviation σ, and let c be a chosen confidence level $(0 < c < 1)$. Then
>
> $$P\left(\frac{(n-1)s^2}{\chi_U^2} < \sigma^2 < \frac{(n-1)s^2}{\chi_L^2} \right) = c$$
>
> and
>
> $$P\left(\sqrt{\frac{(n-1)s^2}{\chi_U^2}} < \sigma < \sqrt{\frac{(n-1)s^2}{\chi_L^2}} \right) = c$$
>
> where n = sample size
>
> $$s = \sqrt{\frac{\Sigma(x - \bar{x})^2}{n-1}} \qquad \text{is the sample standard deviation}$$
>
> χ_U^2 = chi-square value from Table 8 using $d.f. = n - 1$
> and $\alpha = (1 - c)/2$
>
> χ_L^2 = chi-square value from Table 8 using $d.f. = n - 1$
> and $\alpha = (1 + c)/2$

From Figure 11-11 we see that a c confidence level on a chi-square distribution with equal probability in each tail does not center the middle of the corresponding

FIGURE 11-11   Area Representing a c Confidence Level on a
Chi-Square Distribution with $d.f. = n - 1$

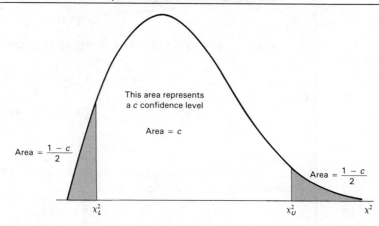

interval over the peak of the curve. This is to be expected since a chi-square curve is skewed to the right.

Let us summarize Theorem 11.2 in the following way.

A c confidence interval for σ^2 is

$$\frac{(n-1)s^2}{\chi_U^2} < \sigma^2 < \frac{(n-1)s^2}{\chi_L^2} \qquad (*)$$

and a c confidence interval for σ is

$$\sqrt{\frac{(n-1)s^2}{\chi_U^2}} < \sigma < \sqrt{\frac{(n-1)s^2}{\chi_L^2}} \qquad (**)$$

Now let us finish our example regarding the variance of maturing times for green beans.

EXAMPLE 4

A random sample of $n = 30$ plants has a sample standard deviation of $s = 3.4$ days for maturity. To find a 95% confidence interval for the population variance σ^2 we use the following.

$$c = 0.95 \qquad \text{confidence level}$$
$$n = 30 \qquad \text{sample size}$$
$$d.f. = n - 1 = 30 - 1 = 29 \qquad \text{degrees of freedom}$$
$$s = 3.4 \qquad \text{sample standard deviation}$$

To find the value of χ_U^2 we use Table 8 in Appendix I with $d.f. = 29$ and $\alpha = (1 - c)/2 = (1 - 0.95)/2 = 0.025$. From Table 8 we get

$$\chi_U^2 = 45.72$$

To find χ_L^2 we use Table 8 with $d.f. = 29$ and $\alpha = (1 + c)/2 = (1 + 0.95)/2 = 0.975$. From Table 8 we get

$$\chi_L^2 = 16.05$$

Formula (*) tells us our desired 95% confidence interval for σ^2 is

$$\frac{(n-1)s^2}{\chi_U^2} < \sigma^2 < \frac{(n-1)s^2}{\chi_L^2}$$

$$\frac{(30-1)(3.4)^2}{45.72} < \sigma^2 < \frac{(30-1)(3.4)^2}{16.05}$$

$$7.33 < \sigma^2 < 20.89$$

To find a 95% confidence interval for σ we simply take square roots; therefore, a 95% confidence interval for σ is

$$\sqrt{7.33} < \sigma < \sqrt{20.89}$$

$$2.71 < \sigma < 4.57$$

_____ Exercise 9 _____

A few miles off the Kona coast of the island of Hawaii a research vessel lies anchored. This ship makes electrical energy from the solar temperature differential of (warm) surface water versus (cool) deep water. The basic idea is that the warm water is flushed over coils to vaporize a special fluid. The vapor is under pressure and drives electrical turbines. Then some electricity is used to pump up cold water to cool the vapor back to a liquid, and the process is repeated. Even though some electricity is used to pump up the cold water, there is plenty left to supply a moderate size Hawaiian town. The subtropic sun always warms up surface water to a reliable temperature, but ocean currents can change the temperature of the deep, cooler water. If the deep-water temperature is too variable, the power plant cannot operate efficiently, or possibly not operate at all. To estimate the variability of deep ocean water temperatures, a random sample of 25 near-bottom readings gave a sample standard deviation of 7.3 °C. Find a 99% confidence interval for the variance (σ^2) and standard deviation (σ) of deep-water temperatures.

a) Determine the following values: $c = $ _____; $n = $ _____; $d.f. = $ _____; $s = $ _____.

a) $c = 0.99$; $n = 25$; $d.f. = 24$; $s = 7.3$

b) What is the value of χ_U^2? _____ of χ_L^2? _____

b) We use Table 8 of Appendix I with $d.f. = 24$
For χ_U^2, $\alpha = (1 - 0.99)/2 = 0.005$

$$\chi_U^2 = 45.56$$

For χ_L^2, $\alpha = (1 + 0.99)/2 = 0.995$

$$\chi_L^2 = 9.89$$

c) Find a 99% confidence interval for σ^2.

c) $\dfrac{(n-1)s^2}{\chi_U^2} < \sigma^2 < \dfrac{(n-1)s^2}{\chi_L^2}$

$$\frac{(24)(7.3)^2}{45.56} < \sigma^2 < \frac{24(7.3)^2}{9.89}$$

$$28.07 < \sigma^2 < 129.32$$

d) Find a 99% confidence interval for σ.

d) $\sqrt{28.07} < \sqrt{\sigma^2} < \sqrt{129.32}$

$$5.30 < \sigma < 11.37$$

Section Problems 11.3

Whenever a test is employed in any of the problems, do the following:

a) State the null and alternate hypotheses
b) Find the degrees of freedom and appropriate critical value or critical values.
c) Find the appropriate chi-square value using the sample standard deviation.
d) Sketch the critical region and acceptance region showing the critical chi-square value and the value of part (c).
e) Decide whether to accept or reject the null hypothesis.

In each of the following problems assume a normal population distribution.

1. A new kind of typhoid shot is being developed by a medical research team. The old typhoid shot was known to protect the population for a mean of 36 months with a standard deviation of 3 months. To test the variability of the new shot, a random sample of 24 people were given the new shot. Regular blood tests showed that the sample standard deviation of protection times was 1.9 months.
   a) Using a 0.05 level of significance, test the claim that the new typhoid shot has a smaller variance of protection times.
   b) Find a 99% confidence interval for the population standard deviation of protection times for the new shot.

2. In a Connecticut ship-building yard, a new and faster method of riveting is being used in the production of ship hulls. Although the new method is faster, it is feared there is greater variability in the shear strength of the new rivets. It is known that for the old method, the rivets had a mean shear strength of 17.2 tons with standard deviation of 4.8 tons. A random sample of 51 new rivets gave a sample standard deviation of 6.1 tons shearing strength.
   a) Using a 0.01 level of significance, test the claim that the new riveting yields rivets with a higher variance of shear strength.
   b) Find a 90% confidence interval for the population variance of shear strengths for the new method.

3. A new car is being advertised as having a highway average of 45 mpg (miles per gallon). However, the ad also says that "your average mileage may vary." A consumer agency decided to estimate the variability a customer might expect. The agency took a random sample of 101 people in the United States who had purchased one of these new cars. Each person reported his or her average highway mileage, and the sample standard deviation was found to be 9.7 miles per gallon.
   a) Find a 95% confidence interval for the population variance of the highway average mpg for the new car.
   b) Use part (a) to find a 95% confidence interval for the population standard deviation.

4. Kula Cola bottling plant has a machine that fills 12-oz pop cans. The machine is set to put a mean of 12 oz in each can. A variance of 0.007 oz is considered acceptable. However, if the variance becomes too large we say the machine is out of control. This means that some cans will run over while other cans will not get enough Kula Cola. During the bottling process random samples are taken from the production line. Recently, a sample of size 28 gave a sample variance of 0.016.

   Use a 0.05 level of significance to test the claim that the variance is now larger than 0.007 (i.e., the machine has slipped out of control).

5. The fan blades on commercial jet engines must be replaced when wear on these parts indicates too much variability to pass inspection. If a single fan blade broke during operation, it could severely endanger a flight. A large engine contains thousands of fan blades, and safety regulations require that variability measurements on the population of all blades not exceed $\sigma^2 = 0.15$ mm. An engine inspector took a random sample of 61 fan blades from an engine. She measured each blade and found a sample variance of 0.27 mm. Using a 0.01 level of significance, is the inspector justified in claiming that all the engine fan blades must be replaced?

6. Jim Hartman is a veterinarian who visits a Vermont farm to examine prize bulls. In order to examine a bull, Jim first gives the animal a tranquilizer shot. The effect of the shot is supposed to last an average of 65 min, and it usually does. However, Jim sometimes gets chased out of the pasture by a bull that recovers too soon, and other times he becomes worried about prize bulls that take too long to recover. By reading his journals, Jim found that the tranquilizer should have a mean duration of 65 min with standard deviation of 15 min. A random sample of ten of Jim's bulls had a mean of close to 65 min but a standard deviation of 24 min.

   a) At the 0.01 level of significance, is Jim justified in the claim that the variance is larger than that stated in his journal?

   b) Using the data Jim collected, find a 95% confidence interval for the population variance.

   c) Find a 95% confidence interval for the population standard deviation.

7. Happy Acres Apple Farm is considering spraying their apple trees with a plant hormone intended to stabilize the size of apples a tree produces. Very small and very large apples cannot be sold as fresh fruit because the distributor will cull them out as seconds for apple cider. Happy Acres knows that their species of tree will produce apples of mean diameter 8.3 cm with standard deviation 1.7 cm. After spraying a typical test tree with the hormone, a random sample of 71 apples gave a sample standard deviation of 1.5 cm for apple diameter.

   a) Using a 0.05 level of significance, test the claim that the hormone had no effect on variance against the claim that it made a difference (either way).

   b) Find a 90% confidence interval for the population variance of apple diameters when the hormone is used.

   c) Find a 90% confidence interval for the standard deviation of part (b).

8. A factor in determining the usefulness of an exam as a measure of demonstrated ability is the amount of spread that occurs in the grades. If the spread or variation of exam scores is very small, it usually means the exam was either too hard or too easy. However, if the variance of scores is moderately large, then there is a definite difference in scores between "better," "average," and "poorer" students. A group of attorneys in a midwest state has been given the task of making up this year's bar exam for the state. The exam has 500 total possible points, and from the history of past exams, it is known that a standard deviation of around 75 points is desirable. Of course, too large or too small a standard

deviation is not good. The attorneys want to test their exam to see how good it is. A preliminary version of the exam (with slight modification to protect the integrity of the real exam) is given to a random sample of 24 newly graduated law students. Their scores give a sample standard deviation of 72 points.

a) Using a 0.01 level of significance, test the claim that the population standard deviation for the new exam is 75 against the claim that the population standard deviation is different from 75.

b) Find a 99% confidence interval for the population variance.

c) Find a 99% confidence interval for the population standard deviation.

9. A set of solar batteries is used in a research satellite. The satellite can run on only one battery, but it runs best if more than one battery is used. The variance (σ^2) of lifetimes of these batteries affects the useful lifetime of the satellite before it goes dead. If the variance is too small, all the batteries will tend to die at once. Why? If the variance is too large, the batteries are simply not dependable. Why? Engineers have determined a variance of $\sigma^2 = 15$ months is most desirable for these batteries. A random sample of 22 batteries gave a sample variance of 14.3 months.

a) Using a 0.05 level of significance, test the claim that $\sigma^2 = 15$ against the claim that σ^2 is different from 15.

b) Find a 90% confidence interval for the population variance σ^2.

c) Find a 90% confidence interval for the population standard deviation σ.

SECTION 11.4   ANOVA: Comparing Several Sample Means

Introduction

In our past work to determine the existence (or nonexistence) of a significant difference between population means, we restricted our attention to only two data groups representing the means in question. Many statistical applications in psychology, social science, business administration, and natural science involve many means and many data groups. Questions commonly asked are: Which of *several* alternative methods yields the best results in a particular setting? Which of *several* treatments leads to highest incidence of patient recovery? Which of *several* teaching methods leads to greatest student retention? Which of *several* investment schemes leads to greatest economic gain?

Using our previous methods (Sections 9.3, 9.4, and 9.5) of comparing only *two* means would require many tests of significance to answer the above questions. For example, if we had only five variables, we would be required to perform ten tests of significance in order to compare each variable to each of the other variables. If we had the time and patience we could perform all ten tests; but what about the risk of accepting a difference where there really is no difference (a type I error)? If the risk of a type I error on each test is $\alpha = 0.05$, then on ten tests we expect the number of tests with a type I error to be 10(0.05) or 0.5 (see expected value, Section 5.3). This situation may not seem too serious to you, but remember that in a "real world" problem and with the aid of a high-speed computer, a researcher may want to study the effect of 50 variables on the outcome of an experiment. Using a little mathematics we could show that the study would require 1,225 separate tests to check *each pair* of variables for a significant difference of means. At the $\alpha = 0.05$ level of significance on each test, we could expect (1,225)(0.05) or 61.25 of the tests to have a type I error. In other words, these 61.25 tests would say there are differences between means when there really are no differences.

To avoid such problems, statisticians have developed a method called *analysis of variance* (abbreviated ANOVA). We will study single factor analysis of variance (also called one-way ANOVA). With appropriate modification, methods of single-factor ANOVA generalize to *n*-dimensional ANOVA, but we leave that topic to more advanced studies.

Example with Explanation of Notation and Formulas

EXAMPLE 5

A psychologist is studying the effect of dream deprivation on a person's anxiety level during waking hours. Brain waves, heart rate, and eye movements can be used to determine if a sleeping person is about to enter into a dream period. Three groups of subjects were randomly chosen from a large group of college students who volunteered to participate in the study. Group I subjects had their sleep interrupted four times each night but never during or immediately before a dream. Group II subjects had their sleep interrupted four times also, but on two occasions they were at the onset of a dream. Group III subjects were wakened four times, each time at the onset of a dream. This procedure was repeated ten nights, and each day all subjects were given a test to determine level of anxiety. The data in Table 11-21 record the total of the test scores for each person over the entire project. Higher totals mean higher anxiety levels.

TABLE 11-21 Dream Deprivation Study

Group I $n_1 = 6$ Subjects		Group II $n_2 = 7$ Subjects		Group III $n_3 = 5$ Subjects	
x_1	x_1^2	x_2	x_2^2	x_3	x_3^2
9	81	10	100	15	225
7	49	9	81	11	121
3	9	11	121	12	144
6	36	10	100	9	81
5	25	7	49	10	100
8	64	6	36		
		8	64		
$\Sigma x_1 = 38$	$\Sigma x_1^2 = 264$	$\Sigma x_2 = 61$	$\Sigma x_2^2 = 551$	$\Sigma x_3 = 57$	$\Sigma x_3^2 = 671$

$N = n_1 + n_2 + n_3 = 18$
$\Sigma x_{TOT} = \Sigma x_1 + \Sigma x_2 + \Sigma x_3 = 156$
$\Sigma x_{TOT}^2 = \Sigma x_1^2 + \Sigma x_2^2 + \Sigma x_3^2 = 1,486$

From Table 11-21 we see that Group I had $n_1 = 6$ subjects, Group II had $n_2 = 7$ subjects, and Group III had $n_3 = 5$ subjects. For each subject, the anxiety score (x value) and the square of the test score (x^2 value) are also shown. In addition, special sums are shown.

We will outline the procedure for single-factor ANOVA in six steps. Each step will contain general methods and rationale appropriate to all single-factor ANOVA tests. As we proceed, we will use the data of Table 11-21 for a specific reference example.

Basic Assumptions of ANOVA

Our application of ANOVA require three basic assumptions. In a general problem with k groups:

1. We assume each of our k groups of measurements is obtained from a population with a *normal* distribution.
2. Each group is randomly selected and is *independent* of all other groups. In particular, this means we will not use the same subjects in more than one group and that the scores of one subject will not have an effect on the scores of another subject.
3. We assume the variables from each group come from distributions with approximately the *same standard deviation*.

Step I: Determine the Null and Alternate Hypotheses   The purpose of an ANOVA test is to determine the existence (or nonexistence) of a statistically significant difference *among* the group means. In a general problem with k groups, we call the (population) mean of the first group μ_1, the population mean of the second group μ_2, and so forth. The null hypothesis is simply that *all* the group population means are the same. Since our basic assumptions say each of the k groups of measurements come from normal, independent distributions with common standard deviation, the null hypothesis says that all the sample groups come from *one and the same* population. The alternate hypothesis is that *not all* the group population means are equal. Therefore, in a problem with k groups we have

$$H_0 : \mu_1 = \mu_2 = \cdots = \mu_k$$

H_1 : At least two of the means $\mu_1, \mu_2, \ldots, \mu_k$ are not equal

Notice that the alternate hypothesis claims *at least* two of the means are not equal. If more than two of the means are unequal, the alternate hypothesis is of course satisfied.

In our dream problem, we have $k = 3$; μ_1 is the population mean of Group 1, μ_2 is the population mean of Group II, and μ_3 is the population mean of Group III. Therefore,

$$H_0 : \mu_1 = \mu_2 = \mu_3$$

$$H_1 : \text{At least two of the means } \mu_1, \mu_2, \mu_3 \text{ are not equal}$$

We will test the null and alternate hypotheses using an $\alpha = 0.05$ or $\alpha = 0.01$ level of significance. Notice that only one test is being performed even though we have $k = 3$ groups and three corresponding means. Using ANOVA avoids the problem of using multiple tests mentioned earlier.

Step II: Find SS_{TOT}   The concept of a *sum of squares* is very important in all of statistics. We used a sum of squares in Chapter 3 to compute the sample standard deviation and sample variance.

$$s = \sqrt{\frac{\Sigma(x - \bar{x})^2}{n - 1}} \qquad \text{sample standard deviation}$$

$$s^2 = \frac{\Sigma(x - \bar{x})^2}{n - 1} \qquad \text{sample variance}$$

The numerator of the sample variance is a special sum of squares that plays a central role in ANOVA. Since this numerator is so important, we give it a special name SS (for sum of squares).

$$SS = \Sigma (x - \bar{x})^2 \tag{1}$$

Using some college algebra, it can be shown that the following simpler formula is equivalent to (1) and involves fewer calculations:

$$SS = \Sigma x^2 - \frac{(\Sigma x)^2}{n} \tag{2}$$

where n is the sample size.

In future reference to SS we will use formula (2), since it is easier to use than formula (1).

The *total sum of squares* SS_{TOT} can be found by using the entire collection of all data values in all groups.

$$SS_{TOT} = \Sigma x_{TOT}^2 - \frac{(\Sigma x_{TOT})^2}{N} \tag{3}$$

where $N = n_1 + n_2 + \cdots + n_k$ is the total sample size from all groups.

$$\Sigma x_{TOT} = \text{sum of all data} = \Sigma x_1 + \Sigma x_2 + \cdots + \Sigma x_k$$
$$\Sigma x_{TOT}^2 = \text{sum of all data squares} = \Sigma x_1^2 + \cdots + \Sigma x_k^2$$

Using the specific data of Table 11-21, for the dream example we have

$k = 3$ Total number of groups

$N = n_1 + n_2 + n_3 = 6 + 7 + 5 = 18$ Total number of subjects

Σx_{TOT} = total sum of x values $= \Sigma x_1 + \Sigma x_2 + \Sigma x_3 = 38 + 61 + 57 = 156$

Σx_{TOT}^2 = total sum of x^2 values $= \Sigma x_1^2 + \Sigma x_2^2 + \Sigma x_3^2 = 264 + 551 + 671 = 1,486$

Therefore, using equation (3) we have

$$SS_{\text{TOT}} = \Sigma x_{\text{TOT}}^2 - \frac{(\Sigma x_{\text{TOT}})^2}{N}$$

$$= 1,486 - \frac{(156)^2}{18}$$

$$= 134$$

The numerator for the total variation for all groups in our dream example is $SS_{\text{TOT}} = 134$. What interpretation can we give to SS_{TOT}? If we let $\bar{x}_{\text{TOT}}$ be the mean of all x values for all groups, then

$$\text{mean of all } x \text{ values} = \bar{x}_{\text{TOT}} = \frac{\Sigma x_{\text{TOT}}}{N}$$

Under the null hypothesis (all groups come from the same normal distribution) $SS_{\text{TOT}} = \Sigma(x_{\text{TOT}} - \bar{x}_{\text{TOT}})^2$ represents the numerator of the sample variance for all groups. Therefore, SS_{TOT} represents total variability of the data. Total variability can occur in two ways:

1. Scores may differ from one another because they belong to *different groups* with different means (recall that the alternate hypothesis says the means are not all equal). This difference is called *between group variability* and is denoted SS_{Bet}.

2. Inherent differences unique to each subject and differences due to chance may cause a particular score to be different from the mean of its *own group*. This difference is called *within group variability* and is denoted SS_{W}.

Since total variability SS_{TOT} is a sum of between group variability SS_{Bet} and within group variability SS_{W}, we may write

$$SS_{\text{TOT}} = SS_{\text{Bet}} + SS_{\text{W}}$$

As we will see, SS_{Bet} and SS_{W} are going to help us decide to accept or reject the null hypothesis. Therefore, our next two steps are to compute these two quantities.

Step III: Find SS_{Bet}   Recall that $\bar{x}_{\text{TOT}}$ is the mean of all x values from all groups. Between group variability (SS_{Bet}) measures the variability of group means. Since different groups may have different numbers of subjects, we must "weight" the variability contribution from each group by the group size n_i.

$$SS_{Bet} = \sum_{\text{all groups}} n_i(\bar{x}_i - \bar{x}_{TOT})^2$$

where n_i is the sample size of group i

$\bar{x}_i$ is the same mean of group i

$\bar{x}_{TOT}$ is the mean for values from all groups

If we use algebraic manipulations, we can write the formula for SS_{Bet} in the following computationally easier form:

$$SS_{Bet} = \sum_{\text{all groups}} \left(\frac{(\Sigma x_i)^2}{n_i} \right) - \frac{(\Sigma x_{TOT})^2}{N} \qquad (4)$$

where, as before, $N = n_1 + n_2 + \cdots + n_k$

Σx_i = sum of data in group i

Σx_i^2 = sum of data squared in group i

Σx_{TOT} = sum of data from all groups

Using data from Table 11-21 for the dream example we have

$$SS_{Bet} = \sum_{\text{all groups}} \left(\frac{(\Sigma x_i)^2}{n_i} \right) - \frac{(\Sigma x_{TOT})^2}{N}$$

$$= \frac{(\Sigma x_1)^2}{n_1} + \frac{(\Sigma x_2)^2}{n_2} + \frac{(\Sigma x_3)^2}{n_3} - \frac{(\Sigma x_{TOT})^2}{N}$$

$$= \frac{(38)^2}{6} + \frac{(61)^2}{7} + \frac{(57)^2}{5} - \frac{(156)^2}{18}$$

$$= 70.038$$

Therefore, the numerator of between group variation is $SS_{Bet} = 70.038$.

Step IV: Find SS_W   We could find the value of SS_W by using the formula relating SS_{TOT} with SS_{Bet} and SS_W and solving for SS_W

$$SS_W = SS_{TOT} - SS_{Bet}$$

However, we prefer to compute SS_W a different way and use the above formula as a check on our calculations.

SS_W is the numerator of the variation within groups. Inherent differences unique to each subject and differences due to chance create the variability assigned to SS_W. In a general problem with k groups, the variability within the ith group could be represented by

$$SS_i = \Sigma (x_i - \bar{x}_i)^2$$

or by the mathematically equivalent formula

$$SS_i = \Sigma x_i^2 - \frac{(\Sigma x_i)^2}{n_i} \qquad (5)$$

Since SS_i represents the variation within the ith group and we are seeking SS_W, the variability within *all* groups, we simply add SS_i for all groups

$$SS_W = SS_1 + SS_2 + \cdots + SS_k \qquad (6)$$

Using formulas (5) and (6) and the data of Table 11-21 with $k = 3$ we have

$$SS_1 = \Sigma x_1^2 - \frac{(\Sigma x_1)^2}{n_1}$$

$$= 264 - \frac{(38)^2}{6} = 23.333$$

$$SS_2 = \Sigma x_2^2 - \frac{(\Sigma x_2)^2}{n_2}$$

$$= 551 - \frac{(61)^2}{7} = 19.429$$

$$SS_3 = \Sigma x_3^2 - \frac{(\Sigma x_3)^2}{n_3}$$

$$= 671 - \frac{(57)^2}{5} = 21.200$$

$$SS_W = SS_1 + SS_2 + SS_3$$

$$= 23.333 + 19.429 + 21.200 = 63.962$$

Let us check our calculation by using SS_{TOT} and SS_{Bet}.

$$SS_{TOT} = SS_{Bet} + SS_W$$
$$134 = 70.038 + 63.962 \qquad \text{(from Steps II and III)}$$

We see that our calculation checks.

Step V: Find Variance Estimates (Mean Squares) In Steps III and IV we found SS_{Bet} and SS_W. Although these quantities represent variability between groups and within groups, they are not yet the variance estimates we need for our ANOVA test. You may recall our study of Student's t distribution in which we introduced the concept of degrees of freedom. Degrees of freedom represent the number of values that are free to vary once we have placed certain restrictions on our data. In ANOVA there are two types of degrees of freedom: $d.f._{Bet}$, representing the degrees of freedom between groups, and $d.f._{W}$, representing degrees of freedom within groups. A theoretical discussion beyond the scope of this text would show:

$$d.f._{Bet} = k - 1 \quad \text{where } k \text{ is the number of groups}$$
$$d.f._{W} = N - k \quad \text{where } N \text{ is the total sample size}$$

(*Note:* $d.f._{Bet} + d.f._{W} = N - 1$.)

The variance estimates we are looking for are designated as follows:

MS_{Bet}, the variance between groups (read *mean square between*)

MS_{W}, the variance within groups (read *mean square within*)

In the literature of ANOVA, the variances between and within groups are usually referred to as *mean squares* between and within groups, respectively. We will use the mean square notation because it is so commonly used. However, remember that the notation MS_{Bet} and MS_{W} both refer to *variances*, and you might occasionally see the variance notation S^2_{Bet} and S^2_{W} used for these quantities. The formulas for the variances between and within samples follow the pattern of the basic formula for sample variance.

$$\text{sample variance} = s^2 = \frac{\Sigma(x - \bar{x})^2}{n - 1} = \frac{SS}{n - 1}$$

Instead of using $n - 1$ in the denominator for MS_{Bet} and MS_{W} variances, we use their respective degrees of freedom.

$$MS_{Bet} = \frac{SS_{Bet}}{d.f._{Bet}} = \frac{SS_{Bet}}{k - 1}$$
$$MS_{W} = \frac{SS_{W}}{d.f._{W}} = \frac{SS_{W}}{N - k}$$

Using these two formulas and the data of Table 11-21, we find the mean squares within and between variances for the dream deprivation example.

$$MS_{Bet} = \frac{SS_{Bet}}{k - 1} = \frac{70.038}{3 - 1} = 35.019$$
$$MS_{W} = \frac{SS_{W}}{N - k} = \frac{63.962}{18 - 3} = 4.264$$

Step VI: Find the F Ratio and Complete the ANOVA Test   The logic of our ANOVA test rests on the fact that one of the variances, MS_{Bet}, *can* be influenced by population differences among means of the several groups while the other variance, MS_{W}, *cannot* be so influenced. For instance, in the dream deprivation and anxiety study, the variance between groups MS_{Bet} will be affected if any of the treatment groups has a population mean anxiety score *different* from any other group. On the other hand, the variance within groups MS_{W} compares anxiety scores of each treatment group to its own group anxiety mean, and the fact that group means might differ *does not* affect the MS_{W} value.

Recall that the null hypothesis claims that all the groups are samples from populations having the *same* (normal) distributions. The alternate hypothesis says that at least two of the sample groups come from populations with *different* (normal) distributions.

If the *null* hypothesis is *true*, MS_{Bet} and MS_{W} should both estimate the *same* quantity. Therefore, if H_0 is true, the F ratio

$$F = \frac{MS_{\text{Bet}}}{MS_{\text{W}}}$$

should be approximately 1, and variations away from 1 should occur only because of sampling errors. The variance within groups MS_{W} is a good estimate of the overall population variance, but the variance between groups MS_{Bet} consists of the population variance *plus* an additional variance stemming from the differences between samples. Therefore, if the *null* hypothesis is *false*, MS_{Bet} will be larger than MS_{W}, and the F ratio will tend to be *larger* than 1.

The decision of whether or not to reject the null hypothesis is determined by the relative size of the F ratio. The F ratio and its corresponding probability distribution were invented by the English statistician Sir Ronald Fisher (1890–1962). Table 9 of Appendix I gives critical F values for levels of significance $\alpha = 0.05$ and $\alpha = 0.01$.

For our example about dreams, the computed F ratio is

$$F = \frac{MS_{\text{Bet}}}{MS_{\text{W}}} = \frac{35.019}{4.264} = 8.213$$

Since larger F values tend to discredit the null hypothesis, our rejection (critical) region will be a *right tail* of the F distribution. To find a critical value we use Table 9 in Appendix I. The table requires us to know *degrees of freedom for the numerator* and *degrees of freedom for the denominator* as well as the level of significance. The degrees of freedom are simply

$$d.f._{\text{Bet}} = \text{degrees of freedom for the numerator}$$

$$d.f._{\text{W}} = \text{degrees of freedom for the denominator}$$

From Step V, we know

$$d.f._{\cdot Bet} = k - 1 = 3 - 1 = 2$$

$$d.f._{\cdot W} = N - k = 18 - 3 = 15$$

Using $\alpha = 0.05$ as our level of significance, Table 9 gives $F_{0.05} = 3.68$ as our critical value (*see* Table 11-22). To find the critical value for $\alpha = 0.05$ and $d.f.$ numerator $= 2$ and $d.f.$ denominator $= 15$, we look in the column headed by 2 and the row headed by 15. Since $\alpha = 0.05$, we use the top entry in roman type. (*See* Figure 11-12).

As we can see, our observed F value of 8.213 lies in the critical or rejection region. Therefore, at the $\alpha = 0.05$ level of significance, we reject H_0 and conclude that not all the means are equal. The amount of dream deprivation *does* make a difference in mean anxiety level.

TABLE 11-22 *F* Distribution (Excerpt from Table 9 of Appendix I)
For $\alpha = 0.05$, entries for F are in Roman Type
For $\alpha = 0.01$, entries for F are in Boldface Type

Degrees of Freedom for Denominator	Degrees of Freedom for Numerator					
	1	2	3	$\cdots$	12	$\cdots$
1	161	200	216	$\cdots$	244	
	4052	**4999**	**5403**	$\cdots$	**6106**	
2	18.51	19.00	19.16	$\cdots$	19.41	
	98.49	**99.01**	**99.17**	$\cdots$	**99.42**	
$\vdots$	$\vdots$	$\vdots$	$\vdots$		$\vdots$	
15	4.54	3.68	3.29	$\cdots$	2.48	
	8.68	**6.36**	**5.42**	$\cdots$	**3.67**	

FIGURE 11-12 The *F* distribution $d.f._{\cdot Bet} = 2$ and $d.f._{\cdot W} = 15$, $\alpha = 0.05$

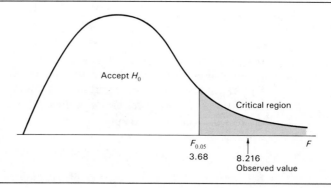

Accept H_0

Critical region

$F_{0.05}$
3.68

8.216
Observed value

F

This completes our single factor ANOVA test. Before we consider another example, let's summarize the main points.

Summary: If we have k independent data groups, each group belonging to a normal distribution and all groups having (approximately) the same standard deviation, then

1. $H_0 : \mu_1 = \mu_2 = \cdots = \mu_k$

   H_1 : not all of $\mu_1, \mu_2, \cdots, \mu_k$ are equal

   where μ_i is the population mean of group i

2. $SS_{\text{TOT}} = \Sigma x_{\text{TOT}}^2 - \dfrac{(\Sigma x_{\text{TOT}})^2}{N}$

   where Σx_{TOT} is the sum of all data elements from all groups

   $\qquad \Sigma x_{\text{TOT}}^2$ is the sum of all data elements squared from all groups

   $\qquad N$ is the total sample size

3. $SS_{\text{TOT}} = SS_{\text{Bet}} + SS_{\text{W}}$

   where $SS_{\text{Bet}} = \displaystyle\sum_{\text{all groups}} \left(\dfrac{(\Sigma x_i)^2}{n_i} \right) - \dfrac{(\Sigma x_{\text{TOT}})^2}{N}$

   $\qquad n_i$ is the number of data elements in group i

   $\qquad \Sigma x_i$ is the sum of the data elements in group i

   $$SS_{\text{W}} = \sum_{\text{all groups}} \left(\Sigma x_i^2 - \dfrac{(\Sigma x_i)^2}{n_i} \right)$$

4. $MS_{\text{Bet}} = \dfrac{SS_{\text{Bet}}}{d.f._{\text{Bet}}}$, where $d.f._{\text{Bet}} = k - 1$

   $MS_{\text{W}} = \dfrac{SS_{\text{W}}}{d.f._{\text{W}}}$, where $d.f._{\text{W}} = N - k$

5. $F = \dfrac{MS_{\text{Bet}}}{MS_{\text{W}}}$

6. We use a right-tail test with critical region and critical values from Table 9 of Appendix I, the F distribution. To find the critical value we use

   a) the given level of significance
   b) degrees of freedom for numerator $= d.f._{\text{Bet}} = k - 1$
   c) degrees of freedom for denominator $= d.f._{\text{W}} = N - k$

Since an ANOVA test requires a number of calculations, we recommend that you summarize your results in a table such as Table 11-23.

TABLE 11-23 Summary of ANOVA Results

Source of Variation	Sum of Squares	Degrees of Freedom	Mean Square (Variance)	F Ratio	F Critical Value	Test Decision
a) Basic Model						
Between groups	SS_{Bet}	$d.f._{Bet}$	MS_{Bet}	$\dfrac{MS_{Bet}}{MS_W}$	From table	Reject
Within groups	SS_W	$d.f._W$	MS_W			H_0 or accept
Total	SS_T	$N - 1$				H_o
b) Summary of ANOVA Results from Dream Experiment						
Between groups	70.038	2	35.019	8.213	3.68	Reject H_0
Within groups	63.962	15	4.264			
Total	134	17				

_____ Exercise 10 _____

A psychologist is studying pattern recognition skills under four laboratory settings. In each setting, a fourth-grade child is given a pattern recognition test with ten patterns to identify. In setting I, the child is given *praise* for each correct answer and no comment about wrong answers. In setting II the child is given *criticism* for each wrong answer and no comment about correct answers. In setting III the child is given no praise or criticism but the observer expresses *interest* in what the child is doing. In setting IV the observer remains *silent* in an adjacent room watching the child through a one-way mirror. A random sample of fourth-grade children was used, and each child participated in the test only once. The test scores (number correct) for each group follow. (*See* Table 11-24.)

TABLE 11-24 Pattern Recognition Experiment

Group I (Praise) $n_1 = 5$		Group II (Criticism) $n_2 = 4$		Group III (Interest) $n_3 = 6$		Group IV (Silence) $n_4 = 5$	
x_1	x_1^2	x_2	x_2^2	x_3	x_3^2	x_4	x_4^2
9	81	2	4	9	81	5	25
8	64	5	25	3	9	7	49
8	64	4	16	7	49	3	9
9	81	3	9	8	64	6	36
7	49			5	25	7	49
				6	36		
$\Sigma x_1 = 41$		$\Sigma x_2 = 14$		$\Sigma x_3 = 38$		$\Sigma x_4 = 28$	
$\Sigma x_1^2 = 339$		$\Sigma x_2^2 = 54$		$\Sigma x_3^2 = 264$		$\Sigma x_4^2 = 168$	
$\Sigma x_{TOT} = \underline{\quad}$		$\Sigma x_{TOT}^2 = \underline{\quad}$		$N = \underline{\quad}$		$k = \underline{\quad}$	

a) Fill in the missing entries of Table 11-24.

$\Sigma x_{TOT} = $ _____ $\Sigma x^2_{TOT} = $ _____

$N = $ _____ $k = $ _____

b) What assumptions are we making about the data to apply a single-factor ANOVA test?

c) What are the null and alternate hypotheses?

d) Find the value of SS_{TOT}.

e) Find SS_{Bet}.

f) Find SS_W and check your calculations using the formula

$$SS_{TOT} = SS_{Bet} + SS_W$$

a) $\Sigma x_{TOT} = \Sigma x_1 + \Sigma x_2 + \Sigma x_3 + \Sigma x_4$

$= 41 + 14 + 38 + 28 = 121$

$\Sigma x^2_{TOT} = \Sigma x_1^2 + \Sigma x_2^2 + \Sigma x_3^2 + \Sigma x_4^2$

$= 339 + 54 + 264 + 168 = 825$

$N = n_1 + n_2 + n_3 + n_4 = 5 + 4 + 6 + 5 = 20$

$k = 4$ groups

b) Since each of the groups comes from independent random samples (no child was tested twice), we need assume only that each group of data came from a normal distribution, and all the groups came from distributions with about the same standard deviation.

c) $H_0 : \mu_1 = \mu_2 = \mu_3 = \mu_4$

In words, all the groups have the same population means, and this hypothesis together with the basic assumptions part (b) says that all the groups come from the same population.

H_1 : not all the means μ_1, μ_2, μ_3, μ_4 are equal

In words, not all the groups have the same population means, so at least one group did not come from the same population as the others.

d) $SS_{TOT} = \Sigma x^2_{TOT} - \dfrac{(\Sigma x_{TOT})^2}{N} = 825 - \dfrac{(121)^2}{20} = 92.950$

e) $SS_{Bet} = \displaystyle\sum_{\text{all groups}} \left(\dfrac{(\Sigma x_i)^2}{n_i} \right) - \dfrac{(\Sigma x_{TOT})^2}{N}$

$SS_{Bet} = \dfrac{(41)^2}{5} + \dfrac{(14)^2}{4} + \dfrac{(38)^2}{6} + \dfrac{(28)^2}{5} - \dfrac{(121)^2}{20}$

$= 50.617$

f) $SS_W = \displaystyle\sum_{\text{all groups}} \left(\Sigma x_i^2 - \dfrac{(\Sigma x_i)^2}{n_i} \right)$

$SS_W = SS_1 + SS_2 + SS_3 + SS_4$

$SS_1 = \Sigma x_1^2 - \dfrac{(\Sigma x_1)^2}{n_1} = 339 - \dfrac{(41)^2}{5} = 2.800$

$SS_2 = \Sigma x_2^2 - \dfrac{(\Sigma x_2)^2}{n_2} = 54 - \dfrac{(14)^2}{4} = 5.000$

$SS_3 = \Sigma x_3^2 - \dfrac{(\Sigma x_3)^2}{n_3} = 264 - \dfrac{(38)^2}{6} = 23.333$

$SS_4 = \Sigma x_4^2 - \dfrac{(\Sigma x_4)^2}{n_4} = 168 - \dfrac{(28)^2}{5} = 11.200$

$SS_W = 42.333$

Check: $SS_{TOT} = SS_{Bet} + SS_W$

$92.950 = 50.617 + 42.333$ checks

g) Find $d.f._{\text{Bet}}$ and $d.f._{\text{w}}$.

h) Find the mean squares MS_{Bet} and MS_{w}.

i) Find the F ratio.

j) Find $F_{0.01}$, the critical value for an $\alpha = 0.01$ level of significance. Sketch the critical region and your observed F ratio. Does the test indicate we should accept or reject the null hypothesis? Explain.

g) $d.f._{\text{Bet}} = k - 1 = 4 - 1 = 3$

  $d.f._{\text{w}} = N - k = 20 - 4 = 16$

  Check: $N - 1 = d.f._{\text{Bet}} + d.f._{\text{w}}$

  $20 - 1 = 3 + 16$ checks

h) $MS_{\text{Bet}} = \dfrac{SS_{\text{Bet}}}{d.f._{\text{Bet}}} = \dfrac{50.617}{3} = 16.872$

  $MS_{\text{w}} = \dfrac{SS_{\text{w}}}{d.f._{\text{w}}} = \dfrac{42.333}{16} = 2.646$

i) $F = \dfrac{MS_{\text{Bet}}}{MS_{\text{w}}} = \dfrac{16.872}{2.646} = 6.376$

j) degrees of freedom for numerator $= d.f._{\text{Bet}} = 3$

  degrees of freedom for denominator $= d.f._{\text{w}} = 16$

  By Table 9 of the Appendix,

  $$F_{0.01} = 5.29$$

  The critical region is shown in Figure 11-13.

FIGURE 11-13   Critical Region

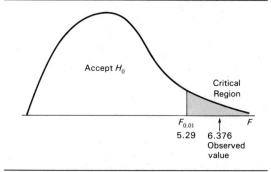

Since the observed F ratio is in the critical region, we reject H_0 and conclude that there is a significant difference in population means of the four groups. The laboratory setting *does* affect the mean scores.

k) Make a summary table of this ANOVA test.

k) (*See* Table 11-25.)

TABLE 11-25 Summary of ANOVA Results for Pattern Recognition Experiment

Source of Variation	Sum of Squares	Degrees of Freedom	Mean Square (Variation)	F Ratio	F Critical Value	Test Decision
Between groups	50.617	3	16.872	6.376	5.29	Reject H_0
Within groups	42.333	16	2.646			
Total	92.950	19				

Section Problems 11.4

In each problem assume that the distributions are normal and have approximately the same population standard deviations.

In each problem do the following:
  a) State the null and alternate hypotheses.
  b) Find SS_{TOT}, SS_{Bet}, SS_W and check that $SS_{TOT} = SS_{Bet} + SS_W$.
  c) Find $d.f._{Bet}$ and $d.f._W$.
  d) Find MS_{Bet} and MS_W.
  e) Find the F ratio.
  f) Find the critical value $F_{0.01}$ or $F_{0.05}$ as the problem requires.
  g) Decide if the null hypothesis is to be accepted or rejected.
  h) Make a summary table for your ANOVA test.

1. Anthropologists working in Central America have found three burial mounds that are somewhat removed from each other, but in the same region. Prevailing burial customs did not permit outsiders to be buried in the same mound. Anthropologists know that several tribes lived in the region and have classified them according to length of skull. There were four skulls found in each mound, and their measurements are given in Table 11-26. The question is: Were all three mounds made by the same or different tribes?

TABLE 11-26 Lengths of Skulls in 3 Burial Mounds (cm)

Mound I	Mound II	Mound III
22.3	20.5	25.6
19.1	22.1	25.9
22.5	24.7	26.8
20.7	24.9	22.5

Use an $\alpha = 0.05$ level of significance. Shall we accept or reject the claim that all three mounds were made by the same tribe?

2. The quantity of dissolved oxygen is a measure of water pollution in lakes, rivers, and streams. Water samples were taken at four different locations in a river in an effort to determine if water pollution varied from location to location. Location I was 500 m above an industrial plant water discharge and near the shore. Location II was 200 m above the discharge point and in midstream. Location III was 50 m downstream from the discharge point and near the shore. Location IV was 200 m downstream from the discharge point and in midstream. Table 11-27 shows the results. Lower dissolved oxygen readings mean

TABLE 11-27 Ecology Experiment in River Pollution

Location I	Location II	Location III	Location IV
7.3	6.6	4.2	4.4
6.9	7.1	5.9	5.1
7.5	7.7	4.9	6.2
6.8	8.0	5.1	
6.2		4.5	

more pollution. Because of the difficulty in getting midstream samples, ecology students collecting the data had fewer of these samples. Use an $\alpha = 0.05$ level of significance. Do we accept or reject the claim that the quantity of dissolved oxygen does not vary from one location to another?

3. An executive at the home office of Big Rock Life Insurance is considering three branch managers as candidates for promotion to vice president. The branch reports include records showing sales volume for each salesperson in the branch. A random sample of these records was selected for salespersons in each branch. All three branches are located in cities where per capita income is the same. The executive wishes to compare these samples to see if there is a significant difference in performance of salespersons in the three different branches. If so, the information will be used to determine which of the managers to promote. (*See* Table 11-28.)

TABLE 11-28 Monthly Volume of Insurance Sales by Individual Salespersons (in hundred thousands)

Branch Managed by Adams	Branch Managed by McDale	Branch Managed by Vasquez
7.2	8.8	6.9
6.4	10.7	8.7
10.1	11.1	10.5
11.0	9.8	11.4
9.9		
10.6		

Use an $\alpha = 0.01$ level of significance. Shall we accept or reject the claim that there is no difference among the salespersons in the different branches?

4. Four fraternities each took a random sample of brothers and recorded their grade averages for the past term. The results are shown in Table 11-29.

TABLE 11-29 Grade Average of Four Fraternities

Fraternity 1	Fraternity 2	Fraternity 3	Fraternity 4
2.11	2.61	2.61	3.77
1.80	1.75	3.76	3.41
2.81	3.20	3.95	3.01
1.66	1.80	2.50	2.22
3.31	2.20	2.40	3.15

The Dean of Students wishes to know if there is a statistical difference in academic performance of brothers among the four fraternities. Using an $\alpha = 0.01$ level of significance, shall we accept or reject the claim that there is no difference in grade averages among the fraternities?

5. A sociologist studying New York City ethnic groups wants to determine if there is a difference in income for immigrants from four different countries during their first year in the city. She obtained the data in Table 11-30 from a random sample of immigrants from these countries (incomes in thousands of dollars). Use a 0.05 level of significance to test the claim that there is no difference in the earnings of immigrants from the four different countries.

TABLE 11-30 Income of Immigrants (in thousands of dollars)

Country I	Country II	Country III	Country IV
12.7	8.3	20.3	17.2
9.2	17.2	16.6	8.8
10.9	19.1	22.7	14.7
8.9	10.3	25.2	21.3
16.4		19.9	19.8

6. Paul has three different fishing reels. To determine if the mean casting distance is the same for these reels, each reel is tested. The same pole, casting weight, and type of line are used for each reel. The casting distances (in meters) of four casts from each reel follow:

Reel 1	Reel 2	Reel 3
20	22	18
24	25	23
19	31	35
26	21	27

Use a 0.05 level of significance to test the claim that the mean casting distance is the same for all three reels.

7. A standardized mathematics test was given to fourth-grade children in five schools in each of three different Wisconsin school districts. The mean score for each school and district follows:

District 1	District 2	District 3
59	58	55
68	63	79
77	72	52
70	55	77
79	80	66

Use the 0.05 level of significance to test the claim that there are no differences in the mean scores from each school district.

8. To compare the strengths of rolled aluminum produced from five experimental alloys, three specimens of rolled aluminum alloy plate were prepared for each alloy. Each plate was alike in every respect except for the alloy composition. The following data show the compression load in tons per square inch at breakdown.

Alloy 1	Alloy 2	Alloy 3	Alloy 4	Alloy 5
1.1	1.3	1.1	1.5	1.6
1.7	1.4	1.2	1.4	1.4
0.8	0.9	0.7	1.3	1.3

Use a 0.01 level of significance to test the claim that there is no difference in the mean breakdown load of the five alloys.

9. The following data show the number of arrests made last year for drug abuse in rural, suburban, and urban areas in four midwestern states. Entries in the table are number of arrests per 10,000 inhabitants.

Rural	Suburban	Urban
15	19	40
14	27	36
21	20	22
17	33	27

At the 0.05 level of significance, can we say the evidence is sufficient to reject the hypothesis that the mean rates of arrest in these states are the same in rural, suburban, and urban areas?

Summary

In this chapter we introduced applications of two new probability distributions: the chi-square and F distributions. We used the chi-square distribution to test for independence, goodness of fit, to estimate the variance σ^2, and to test hypotheses involving the variance σ^2.

We used a method called analysis of variance to test differences among means of several groups of data. There are two ways of estimating population variance, one called variance between samples and the other called variance within samples. The quotient of these two variances gives the F ratio. Critical F ratios from the F distribution (with appropriate degrees of freedom) are compared with the computed F ratio. If the computed F ratio is greater than the critical F ratio, we reject the null hypothesis that the means are all equal.

Important Words and Symbols

	Section		*Section*
Independence test	11.1	Hypotheses tests about σ^2	11.3
Chi-square distribution, χ^2	11.1	Confidence interval c for σ^2	11.3
Degrees of freedom, $d.f. =$		χ_U^2, χ_L^2	11.3
$(R-1)(C-1)$ for χ^2		ANOVA	11.4
distribution and tests		Sum of squares, SS	11.4
of independence	11.1	SS_{Bet}, SS_{W}, SS_{TOT}	11.4
Contingency table with cells	11.1	MS_{Bet}, MS_{W}	11.4
Expected frequency of a cell, E	11.1	F ratio	11.4
Observed frequency of a cell, O	11.1	F distribution	11.4
Critical chi-square values χ_α^2	11.1	Summary Table for ANOVA	11.4
Row total	11.1	Degrees of freedom $d.f.$ for	
Column total	11.1	numerator distribution for F	11.4
Goodness-of-fit test	11.2	Degrees of freedom $d.f.$ for	
Degrees of freedom $d.f.$ for χ^2		denominator for F distribution	11.4
distribution and goodness-of-fit			
tests	11.2		

Chapter Review Problems

Before you solve a problem, first classify the problem as one of the following:
 a) Chi-square test of independence
 b) Chi-square goodness of fit
 c) Chi-square for testing or estimating σ^2 or σ
 d) ANOVA

Then, in each of the problems when a test is to be performed, do the following:
 i) State the null and alternate hypotheses.
 ii) Find all critical values.
 iii) Sketch the rejection region, acceptance region, and location of the critical value and sample statistic.
 iv) Decide whether you should accept or reject the null hypothesis.
 v) If the problem is an ANOVA problem, make a summary table.

1. The makers of Country Boy Corn Flakes are thinking about changing the packaging of the cereal with the hope of improving sales. In an experiment five stores of similar size in the same region sold Country Boy Corn Flakes in different-shaped containers for two weeks. Total packages sold are given in Table 11-31. Using a 0.05 significance, shall we accept or reject the hypothesis that the mean sales are the same, no matter which shape box is used?

TABLE 11-31 Country Boy Corn Flakes Package Shapes

	Package Shape		
Cube	Cylinder	Pyramid	Rectangular
120	110	74	165
88	115	62	98
65	180	110	125
95	96	66	87
71	85	83	118

2. Professor Fair believes extra time does not improve grades on exams. He randomly divided a group of 300 students into two groups and gave them all the same test. One group had exactly one hour in which to finish the test, and the other group could stay as long as desired. The results are shown in Table 11-32. Test at the 0.01 level of significance that time to do a test and test results are independent.

TABLE 11-32 Exam Grades Versus Exam Time

Time	A	B	C	F	Row Total
1 hr	23	42	65	12	142
Unlimited	17	48	85	8	158
Column Total	40	90	150	20	300

3. A consumer agency is investigating the blow-out pressures of Soap Stone tires. A Soap Stone tire is said to blow out when it separates from the wheel rim due to impact forces

usually caused by hitting a rock or a hole in the road. A random sample of 30 Soap Stone tires were inflated to the recommended pressure, and then forces measured in foot-pounds were applied to each tire (one foot-pound is the force of one pound dropped from a height of one foot). The customer complaint is that some Soap Stone tires blow out under small impact forces, while other tires seem to be well made and don't have this fault. For the 30 test tires, the sample standard deviation of blow-out forces was 1,353 foot-pounds.

a) Soap Stone claims their tires will blow out at an average pressure of 20,000 foot-pounds with a standard deviation of 900 foot-pounds. The average blow-out force is not in question, but the variability of blow-out forces is in question. Using a 0.01 level of significance, test the claim that the variance of blow-out pressures is more than Soap Stone claims it is.

b) Find a 95% confidence interval for the variance of blow-out pressures, using the information from the random sample.

4. Anela is a computer scientist who is formulating a large and complicated program for a type of data processing. She has three ways of storing and retrieving data: cards, tape, and disks. As an experiment she sets up her program three different ways: one using cards, one using tape, and the other using disks. Then four test runs of this type of data processing are made on each program. The time required to execute each program is shown in Table 11-33 (in min). Use a 0.01 level of significance to test the hypothesis that the mean processing time is the same for each method.

TABLE 11-33 Results of Four Test Runs of Three Programs

Cards	Tape	Disks
8.7	7.2	7.0
9.3	9.1	6.4
7.9	7.5	9.8
8.0	7.7	8.2

5. Professor Stone complains that student teacher-ratings depend on the grade the student received. In other words, according to Professor Stone, a teacher who gives good grades gets good ratings, and a teacher who gives bad grades gets bad ratings. To test this claim the Student Assembly took a random sample of 300 teacher ratings on which the student's grade for the course was also indicated. The results are in Table 11-34. Test the hypothesis that teacher ratings and student grades are independent at the 0.01 level of significance.

TABLE 11-34 Teacher Ratings and Student Grades

Rating	A	B	C	F (or withdrawal)	Row Total
Excellent	14	18	15	3	50
Average	25	35	75	15	150
Poor	21	27	40	12	100
Column Total	60	80	130	30	300

6. A machine that puts corn flakes in boxes is adjusted to put an average of 15 oz in each box with standard deviation of 0.25 oz. If a random sample of 12 boxes gave a sample

standard deviation of 0.38 oz, does this data support the claim that the variance has increased and the machine needs to be brought back into adjustment? (Use a 0.05 level of significance.)

7. A sociologist is studying the age of the population in Blue Valley. Ten years ago the population was such that 20% were under 20 years old, 15% were in the 20–35-year-old bracket, 30% were between 36 and 50, 25% were between 51 and 65, and 10% were over 65. A study done this year used a random sample of 210 residents. This sample showed:

Under 20	20–35	36–50	51–65	Over 65
15	25	70	80	20

At the 0.01 level of significance, has the age distribution of the population of Blue Valley changed?

USING COMPUTERS

Professional statisticians in industry and research use computers to help them analyze and process statistical data. There are many computer programs available for statistics. Some commonly used statistical packages include Minitab, SAS, and SPSS. There are many others as well. Any of these statistical packages may be used with this text.

The authors of this text have also written a computer statistics package called ComputerStat. The programs in ComputerStat are available on disks with an accompanying manual that explains how to use the programs and gives additional computer applications. The disks are intended for the Apple IIe, IIc, or the IBM PC. Programs are written in the computer language BASIC and are designed for the beginning statistics student with no previous computer experience. ComputerStat is available as a supplement from the publisher of this text.

Problems in this section may be done using ComputerStat or other statistical computer programs.

1. A study involving people who are food processors gave the following information about work shift and number of sick days.

TABLE 11-35   Shift Versus Number of Sick
Days for Food Processors

Shift \ Number Sick days	0	1	2	3	4 or more
Day	134	44	24	10	61
Aft/Ev	90	39	23	18	99
Night	107	37	21	20	82
Rotating	56	20	14	17	92

Source: United States Department of Health, Education, and Welfare, NIOSH Technical Report. Tasto, Colligan, *et al. Health Consequences of Shift Work.* Washington: GPO, 1978. 29.

(*Note:* This table was adapted from Table 7 on page 25 of the source.)

In the ComputerStat menu of topics, select Hypothesis Testing. Then use program S20, Chi-Square Test for Independence, to solve the following problems.

a) We view the preceding table as a contingency table with four rows and five columns. Starting in the upper left part of the table, cell #1 has an observed frequency of 134. Cell #2 has an observed frequency of 44 and so on. Cell #20 is in row four and column five. The observed frequency of cell #20 is 92. Enter these data into the computer as directed by program S20. Use a 1% level of significance to test the null hypothesis that work shift and number of sick days are independent against the alternate hypothesis that they are not independent.

b) When the data have been entered press return to list the cell number, observed frequency, and corresponding expected frequency. In your own words explain how the computer calculates the expected frequencies.

c) Press return and scan the Information Summary. What is the sample chi-square value? What is the *P* value? Press return again. Shall we accept or reject the statement that work shift and number of sick days are independent? What is the smallest level of significance at which we would reject the statement that work shift and number of sick days are independent? (*Note:* For more information about the *P* value, see Section 9.7.)

2. In the following table the 1984 Denver crime rate (expressed as a percent of total reported crimes) is compared with the 1985 counts (expressed in tens of crimes). All data are taken from October of one year to October of the next year.

Crime	1984	1985 (in tens)
Murder	0.23%	6
Forcible rape	1.44%	42
Aggravated robbery	3.25%	127
Simple robbery	1.67%	56
Aggravated assault	5.28%	168
Burglary	39.77%	1,383
Grand larceny	32.42%	984
Auto theft	14.40%	468
Arson	1.54%	53
Total	100.00%	3,287

Source: Adapted from the table "Denver Crime Statistics," *Rocky Mountain News* [Denver], 5 November 1985. 25.

In the ComputerStat menu of topics, select Hypothesis Testing. Then use program S21, Chi-Square Test for Goodness-of-Fit, to solve the following problems.

a) We have nine different types of crimes listed. Consequently the sample distribution consists of nine items. Let's use a 5% level of significance to test the null hypothesis that the distribution of Denver crimes for 1984 is the same as for 1985 against the alternate hypothesis that the distributions are different. Use information from the preceding table to enter data into the computer as directed by program S21.

b) Scan the Information Summary. What is the sample chi-square value? What is the *P* value? Press return. Shall we accept or reject the statement that the distribution of crimes was the same for 1984 and 1985? Explain.

c) What is the smallest level of significance at which we can say the Denver crime distributions for 1984 and 1985 are different? Notice that we are only studying the data to see if the crime distributions are the same or different. If we conclude that they are different, then a follow-up study would be needed to determine which areas are different. An interested student could follow this up by using Hypothesis Testing program S17, which tests proportions.

3. The following data is a winter mildness/severity index for three European locations near 50° north latitude. For each decade the number of unmis-

Decade	Britain	Germany	Russia
1800	−2	−1	+1
1810	−2	−3	−1
1820	0	0	0
1830	−3	−2	−1
1840	−3	−2	+1
1850	−1	−2	+3
1860	+8	+6	+1
1870	0	0	−3
1880	−2	0	+1
1890	−3	−1	+1
1900	+2	0	+2
1910	+5	+6	+1
1920	+8	+6	+2
1930	+4	+4	+5
1940	+1	−1	−1
1950	0	+1	+2

Table is based on data from *Exchanging Climate* by H. H. Lamb, copyright © 1966. Reprinted by permission of Methuen & Co., London.

takably mild months minus the number of unmistakable severe months for December, January, and February is recorded.

In the ComputerStat menu of topics, select Hypothesis Testing. Then use program S22, ANOVA (analysis of variance), to solve the following problems.

a) We wish to test the null hypothesis that the mean winter index for Britain, Germany, and Russia are all equal against the alternate hypothesis that they are not all equal. We are using three groups: Britain, Germany, Russia. For each group we have 16 decades for which we have data. Enter the information from the above table into the computer as directed by program S22. Use a 5% level of significance.

b) Scan the Information Summary. What is the sum of squares between groups? Within groups? What is the sample F ratio? What is the P value? Press return. Shall we accept or reject the statement that the mean winter indexes for these locations in Britain, Germany, and Russia are the same?

c) What is the smallest level of significance at which we could conclude that the mean winter indexes for these locations are not all equal?

12 Nonparametric Statistics

Make everything as simple as possible, but no simpler.

Albert Einstein

The Sign Test

There are many situations where very little is known about the population from which samples are drawn. Therefore, we cannot make assumptions about the population distribution, such as the distribution is normal or binomial. In this chapter we will study methods which come under the heading of *nonparametric statistics*. These methods are called nonparametric because they require no assumptions about the population distributions from which samples are drawn. The obvious advantages of these tests are that they are quite general and (as we shall see) not difficult to apply. The disadvantages are that they tend to waste information and tend to result in acceptance of the null hypothesis more often than they should; nonparametric tests are sometimes less sensitive than other tests.

The easiest of all the nonparametric tests is probably the *sign test*. The sign test is used when we compare sample means from two populations which are *not independent*. This occurs when we measure the same sample twice, as done in "before-and-after" studies. The following example shows how the sign test is constructed and used.

As part of their training police cadets took a special course on identification awareness. To determine how the course affects a cadet's ability to identify a suspect, the 15 cadets were first given an identification awareness exam, then after the course they were tested again. The police school would like to use the results of the two tests to see if the identification awareness course *improves* a cadet's score. Table 12-1 gives the scores for each identification awareness exam.

485

TABLE 12-1 Scores for 15 Police Cadets

Cadet	Postcourse Score	Precourse Score	Sign of Difference
1	93	76	+
2	70	72	−
3	81	75	+
4	65	68	−
5	79	65	+
6	54	54	No difference
7	94	88	+
8	91	81	+
9	77	65	+
10	65	57	+
11	95	86	+
12	89	87	+
13	78	78	No difference
14	80	77	+
15	76	76	No difference

The sign of the difference is obtained by subtracting the precourse score from the postcourse score. If the difference is positive, we say the sign of the difference is +, and if the difference is negative, we indicate it with −. No sign is indicated if the scores are identical; in essence such scores are ignored when using the sign test.

To use the sign test, we need to compute the *proportion, r, of plus signs* to all signs. This is done in Exercise 1.

——————————— Exercise 1 ———————————

Look at Table 12-1 under the *sign of difference* column.

a) How many plus signs do you see?

b) How many plus and minus signs do you see?

c) The *proportion of plus signs* is

$$r = \frac{\text{number of plus signs}}{\text{total number of plus and minus signs}}$$

Use parts (a) and (b) to find *r*.

a) 10

b) 12

c) $r = \dfrac{10}{12} = \dfrac{5}{6} = 0.833$

We let μ_2 be the population mean of all precourse scores and μ_1 be the population mean of all postcourse scores (for all police cadets). The null hypothesis

$$H_0 : \mu_1 = \mu_2$$

means the identification awareness course does *not* affect the mean scores. Under the null hypothesis we expect that the number of plus signs and minus signs should be about equal. So *r*, the proportion of plus signs, should be approximately 0.5 under the null hypothesis.

The police department wants to see if the course *improves* a cadet's score. Therefore, the alternate hypothesis will be

$$H_1 : \mu_1 > \mu_2$$

To test the null hypothesis $H_0 : \mu_1 = \mu_2$ against the alternate hypothesis $H_1 : \mu_1 > \mu_2$, we let *p* be the population proportion of plus signs for all police cadets. Then we use methods of Section 9.6 for tests involving proportions. As in Section 9.6, we will assume all our samples are sufficiently large to permit a good normal approximation to the binomial distribution. For most practical work this will be the case if the total number of plus and minus signs is 12 or more ($n \geq 12$).

To find the critical values we simply use those of Table 9-19 (binomial distribution) with $p = 0.5$ and $q = 1 - 0.5 = 0.5$. For convenience the critical values for a sign test are specified in Table 12-2 on page 488. If *r* is in the critical region, we reject the null hypothesis, and if *r* is not in the critical region, we accept it.

In the police cadet example there are $n = 12$ total plus and minus signs. (Note that of the 15 cadets in the sample, three had no difference in precourse and postcourse test scores, so there are no signs for those three.) The null hypothesis and alternate hypothesis are

$$H_0 : \mu_1 = \mu_2 \qquad H_1 : \mu_1 > \mu_2$$

If we want to use an $\alpha = 0.05$ level of significance, Table 12-2 tells us the critical value will be

$$c = 0.5 + 1.645 \sqrt{\frac{0.25}{n}} = 0.5 + 1.645 \sqrt{\frac{0.25}{12}} = 0.737$$

The critical region is indicated in Figure 12-1.

Since our computed *r* value, $r = 0.833$ (from Exercise 1c), is in the critical region, we reject H_0 and conclude that the identification awareness course improves the cadets' mean scores.

FIGURE 12-1 Critical Region for $\alpha = 0.05$

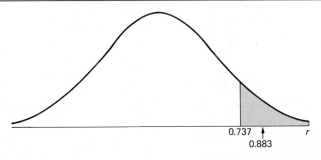

TABLE 12-2 Critical Values and Critical Regions for the Sign Test

Alternate Hypothesis	Level of Significance		Critical Region
	$\alpha = 0.05$	$\alpha = 0.01$	
$H_1 : \mu_1 > \mu_2$	$c = 0.5 + 1.645\sqrt{\dfrac{0.25}{n}}$	$c = 0.5 + 2.33\sqrt{\dfrac{0.25}{n}}$	r to right of c
$H_1 : \mu_1 < \mu_2$	$c = 0.5 - 1.645\sqrt{\dfrac{0.25}{n}}$	$c = 0.5 - 2.33\sqrt{\dfrac{0.25}{n}}$	r to left of c
$H_1 : \mu_1 \neq \mu_2$	$c_1 = 0.5 - 1.96\sqrt{\dfrac{0.25}{n}}$ and $c_2 = 0.5 + 1.96\sqrt{\dfrac{0.25}{n}}$	$c_1 = 0.5 - 2.58\sqrt{\dfrac{0.25}{n}}$ and $c_2 = 0.5 + 2.58\sqrt{\dfrac{0.25}{n}}$	r to left of c_1 or to right of c_2

n = total number of plus and minus signs ($n \geqslant 12$)
c = critical value

_____ Exercise 2 _____

Dr. Kick-a-poo's Traveling Circus made a stop at Middlebury, Vermont, where the doctor opened a booth and sold bottles of Dr. Kick-a-poo's Magic Gasoline Additive. The additive is supposed to increase gas mileage when used according to instructions. Twenty local people purchased bottles of the additive and used it according to instructions. These people carefully recorded their mileage with and without the additive. The results are shown in Table 12-3.

a) In Table 12-4 complete the column headed *sign of difference*. How many plus signs are there? How many total plus and minus signs are there? What is r, the proportion of plus signs?

a) There are seven plus signs and 17 total plus and minus signs. The proportion of plus signs is

$$r = \frac{7}{17} = 0.41$$

TABLE 12-3 Mileage Before and After Kick-a-poo's Additive

Car	With Additive	Without Additive	Sign of Difference
1	17.1	16.8	+
2	21.2	20.1	+
3	12.3	12.3	No difference (N.D.)
4	19.6	21.0	−
5	22.5	20.9	+
6	17.0	17.9	_____
7	24.2	25.4	_____
8	22.2	20.1	_____
9	18.3	19.1	_____
10	11.0	12.3	_____
11	17.6	14.2	_____
12	22.1	23.7	_____
13	29.9	30.2	_____
14	27.6	27.6	_____
15	28.4	27.7	_____
16	16.1	16.1	_____
17	19.0	19.5	_____
18	38.7	37.9	_____
19	17.6	19.7	_____
20	21.6	22.2	_____

TABLE 12-4 Completion of Table 12-3

Car	Sign of Difference
6	−
7	−
8	+
9	−
10	−
11	+
12	−
13	−
14	N.D.
15	+
16	N.D.
17	−
18	+
19	−
20	−

b) Most people claim the additive had no significant effect. Let's use a 0.05 level of significance to test this claim against the alternative hypothesis that the additive did have an effect (one way or the other). Use Table 12-2 to find critical values, and shade the critical region. Does your r value from part (a) fall in the critical region? Shall we accept or reject the claim that the additive had no effect?

b) $n = 17$ since there are 17 total plus and minus signs. From Table 12-2

$$c_1 = 0.5 - 1.96 \sqrt{\frac{0.25}{n}} = 0.5 - 1.96 \sqrt{\frac{0.25}{17}}$$

$$= 0.5 - 0.24 = 0.26$$

$$c_2 = 0.5 + 1.96 \sqrt{\frac{0.25}{n}} = 0.5 + 1.96 \sqrt{\frac{0.25}{17}}$$

$$= 0.5 + 0.24 = 0.74$$

Since $r = 0.41$ is not in the critical region, we conclude the additive had no effect (at the 0.05 level of significance). (*See* Figure 12-2.)

FIGURE 12-2   Critical Regions for $\alpha = 0.05$

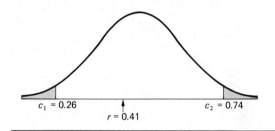

$c_1 = 0.26$ $c_2 = 0.74$

$r = 0.41$

Section Problems 12.1

For each of the problems do the following:

a) State the null and alternate hypotheses.
b) Find all critical values.
c) Sketch the critical region, the acceptance region, the location of the critical values, and the sample value of r (proportion of plus signs to all signs).
d) Decide whether to accept or reject the null hypothesis.

1. The Daisy Pen Company has developed a new tip for their felt-tip pens. Twelve pens are filled with ink and fitted with the old tip. Each pen is attached to one of 12 motor-driven paper-covered drums and the writing life of each pen is determined (in hours). Then each pen is refilled with ink and fitted with a new type tip. Again the writing life of each pen is determined. The results follow (in hours):

Pen	New Tip	Old Tip	Pen	New Tip	Old Tip
1	52	50	7	47	46
2	47	55	8	57	53
3	56	51	9	56	52
4	48	45	10	46	40
5	51	57	11	56	49
6	59	54	12	47	51

Use a 0.05 level of significance to test the hypothesis that the mean writing life of the new tip is longer than that of the old tip.

2. A psychologist claims that students' pulse rates tend to increase just before an exam. To test this claim she used a random sample of 14 psychology students and took their pulses before an ordinary class meeting and then again before a class meeting that consisted of an examination. The results follow:

Student	Pulse Rate Before Exam	Pulse Rate Before Ordinary Class
1	88	81
2	77	77
3	72	75
4	74	79
5	81	79
6	70	68
7	75	77
8	80	73
9	68	71
10	75	73
11	82	76
12	61	66
13	77	68
14	64	60

Use a 0.05 level of significance to test the psychologist's claim.

3. A high-school social science teacher decided to give a series of lectures on current events. To determine if the lectures had any effect on student awareness of current events, an exam was given to the class before the lectures, and another similar exam was given after the lectures. The scores follow on the next page. Use a 0.05 level of significance to test the claim that the lectures made no difference against the claim that the lectures did make some difference (either up or down).

Student	After Lectures	Before Lectures
1	107	111
2	115	110
3	120	93
4	78	75
5	83	88
6	56	56
7	71	75
8	89	73
9	77	83
10	44	40
11	119	115
12	130	101
13	91	110
14	99	90
15	96	98
16	83	76
17	100	100
18	118	109

4. A manufacturer of flashbulbs is using two production lines to make flashbulbs for cameras. The same production process is used on each line and the only difference is in the employees working the lines. Employees on line A are experienced at their work, whereas those on line B are new on the job. The number of defective bulbs produced by each line for a period of 15 days follows:

Day	Line A	Line B	Day	Line A	Line B
1	389	517	9	300	222
2	412	610	10	444	357
3	509	430	11	392	412
4	420	420	12	306	580
5	471	415	13	319	289
6	171	310	14	510	505
7	460	370	15	240	350
8	650	618			

Use a 0.01 level of significance to test the claim that there is no difference in defective production against the claim that there is a difference (one way or the other).

5. To compare two elementary schools in teaching of reading skills, 12 sets of identical twins were used. In each case one child was selected at random and sent to school A and his or her twin was sent to school B. Near the end of fifth grade, an achievement test was given to each child. The results follow:

Twin Pair	School A	School B	Twin Pair	School A	School B
1	177	86	7	86	93
2	150	135	8	111	77
3	112	115	9	110	96
4	95	110	10	142	130
5	120	116	11	125	147
6	117	84	12	89	101

Use a 0.05 level of significance to test the hypothesis that the two schools have the same effectiveness in teaching reading skills against the alternate hypothesis that the schools are not equally effective.

6. A chemical company is testing two types of food preservatives to be used in bread. Twenty bakeries each bake two similar batches of bread, one with preservative A and the other with preservative B. The shelf life for each batch was determined as follows (shelf life in days):

Bakery	Preservative A	Preservative B
1	5	6
2	7	3
3	3	3
4	5	7
5	5	6
6	3	5
7	5	4
8	6	8
9	5	7
10	6	4
11	7	6
12	5	8
13	6	4
14	9	7
15	3	4
16	5	7
17	8	6
18	4	6
19	5	8
20	7	3

Use a 0.05 level of significance to test the claim that bread with preservative B has a longer shelf life.

7. One program to help people stop smoking cigarettes uses the method of posthypnotic suggestion to remind subjects to avoid smoking. A random sample of 18 subjects agreed to test the program. All subjects counted the number of cigarettes they usually smoke a day; then they counted the number of cigarettes smoked the day after hypnosis. (*Note:* It usually takes several weeks for the subject to stop smoking completely, and the method does not work for everyone.) The results follow:

Subject	Number of Cigarettes Smoked Per Day	
	After Hypnosis	Before Hypnosis
1	28	28
2	15	35
3	2	14
4	20	20
5	31	25
6	19	40
7	6	18
8	17	15
9	1	21
10	5	19
11	12	32
12	20	42
13	30	26
14	19	37
15	0	19
16	16	38
17	4	23
18	19	24

Using a 1% level of significance, test the claim that the mean number of cigarettes smoked per day was less after hypnosis.

8. Some fishermen claim that eating a clove of garlic will help keep mosquitoes away from your skin. To test this claim, Gary collected a large, clear plastic tube of hungry mosquitoes. He took a random sample of 16 students, each of whom agreed to put a bare arm in the tube and count the number of mosquitoes that landed on the skin during a 2-min time interval. Then each student ate a clove of garlic. After 3 hr (when the effect of the garlic was in the blood stream), each student again put his or her arm in the tube and counted mosquitoes landing on the skin during a 2-min time interval. The results follow at the top of page 495. Using a 5% level of significance, test the claim that garlic tends to reduce the mean number of mosquitoes on the skin.

9. A pediatrician is studying the pulse rate of babies before and after birth. A random sample of 17 babies gave the information at the bottom of page 495 (pulse rate = heart beats per minute). Using a 1% level of significance, test the claim that the mean pulse rates are different (either up or down).

Student	Number of Mosquitoes Landing on Skin	
	After Garlic	Before Garlic
1	15	75
2	8	63
3	30	42
4	16	91
5	56	56
6	44	39
7	12	88
8	42	72
9	0	94
10	40	37
11	19	48
12	16	66
13	0	58
14	12	77
15	42	110
16	51	49

Baby	Pulse Rate	
	24 Hours After Birth	24 Hours Before Labor Starts
1	70	61
2	73	72
3	82	80
4	80	83
5	58	58
6	88	77
7	80	80
8	60	65
9	71	73
10	60	52
11	63	59
12	58	64
13	79	67
14	71	71
15	73	60
16	72	75
17	85	81

10. The Know-It-All Encyclopedia Company sent a team of 15 sales people to Garden City, Kansas, for door-to-door sales. After one week of sales efforts, the company decided to buy local spot TV ads for their encyclopedia and then continue the sales effort another week. The results follow:

Salesperson	Number of Sales	
	After TV Ads	Before TV Ads
1	4	3
2	1	0
3	0	1
4	3	4
5	3	2
6	0	0
7	6	5
8	4	3
9	3	2
10	0	1
11	1	0
12	4	3
13	4	4
14	5	6
15	3	2

Using a 5% level of significance, test the claim that mean sales before and after the TV ads are different (either up or down).

SECTION 12.2   The Rank Sum Test

The sign test is used when we have paired data values coming from dependent samples as in "before-and-after" studies. However, if the data values are *not* paired, the sign test should *not* be used.

For the situation where we draw independent random samples from two populations there is another nonparametric method for testing the difference between sample means; it is called the *rank sum test* (also called the *Mann-Whitney* test). The rank sum test can be used when assumptions about *normal* populations are not satisfied or when assumptions about *equal population variances* are not satisfied. To fix our thoughts on a definite problem, let's consider the following example.

When a scuba diver makes a deep dive, nitrogen builds up in the diver's blood. After returning to the surface the diver must wait in a decompression chamber until the nitrogen level of the blood returns to normal. A physiologist working with the Navy has invented a pill that a diver takes one hour before diving. The pill is supposed to have the effect of reducing the waiting time spent in the decompression chamber. Nineteen Navy divers volunteered to help the physiologist determine if the pill has any effect. The divers were randomly divided into two groups: group A had nine divers who each took the pill, and group B had ten divers who did not take the pill. All the divers worked the same length of time on a deep salvage operation and returned to the decompression chamber. A monitoring device in the decompression chamber measured the waiting time for each diver's nitrogen level to return to normal. These times are recorded in Table 12-5.

TABLE 12-5 Decompression Times for 19 Navy Divers (in min)

Group A (had pill)	41	56	64	42	50	70	44	57	63	
				mean time = 54.1 min						
Group B (no pill)	66	43	72	62	55	80	74	75	77	78
				mean time = 68.2 min						

The means of our two samples are 54.1 and 68.2 minutes. We will use the rank sum test to decide whether the difference between the means is significant. First we arrange the two samples jointly in order of increasing time. This means we use the data of groups A and B as if they were one sample. The times, groups, and ranks are shown in Table 12-6.

TABLE 12-6 Ranks for Decompression Time

Time	Group	Rank	Time	Group	Rank
41	A	1	64	A	11
42	A	2	66	B	12
43	B	3	70	A	13
44	A	4	72	B	14
50	A	5	74	B	15
55	B	6	75	B	16
56	A	7	77	B	17
57	A	8	78	B	18
62	B	9	80	B	19
63	A	10			

Group A occupies the ranks 1, 2, 4, 5, 7, 8, 10, 11, 13, and group B occupies the ranks 3, 6, 9, 12, 14, 15, 16, 17, 18, 19. We add up the ranks of the group with the *smaller* sample size, in this case group A.

The sum of the ranks is denoted by R.

$$R = 1 + 2 + 4 + 5 + 7 + 8 + 10 + 11 + 13 = 61$$

Let n_1 be the size of the *smaller sample* and n_2 be the size of the *larger sample*. In the case of the divers $n_1 = 9$ and $n_2 = 10$. So R is the sum of the ranks from the smaller sample. If both samples are of the same size, then $n_1 = n_2$ and R is the sum of the ranks of either group (but not both groups).

When both n_1 and n_2 are sufficiently large (each of size eight or more), advanced mathematical statistics can be used to show that R is approximately normally distributed with mean

$$\mu_R = \frac{n_1(n_1 + n_2 + 1)}{2}$$

and standard deviation

$$\sigma_R = \sqrt{\frac{n_1 n_2(n_1 + n_2 + 1)}{12}}$$

_____ Exercise 3 _____

For the Navy divers compute μ_R and σ_R. (Recall that $n_1 = 9$ and $n_2 = 10$.)

$$\mu_R = \frac{n_1(n_1 + n_2 + 1)}{2} = \frac{9(9 + 10 + 1)}{2} = 90$$

$$\sigma_R = \sqrt{\frac{n_1 n_2(n_1 + n_2 + 1)}{12}} = \sqrt{\frac{9 \cdot 10(9 + 10 + 1)}{12}}$$

$$= \sqrt{150}$$

$$= 12.25$$

To determine if the difference between sample means is significant we use a (two-sided) rank sum test. Table 12-7 indicates the critical values to be used.

Let's use the rank sum test with a 0.05 level of significance for the problem of the Navy divers. From Exercise 3 we have $\mu_R = 90$ and $\sigma_R = 12.25$. From Table 12-7 the critical values are

$$c_1 = \mu_R - 1.96\sigma_R = 90 - 1.96(12.25) = 65.99$$
$$c_2 = \mu_R + 1.96\sigma_R = 90 + 1.96(12.25) = 114.01$$

The critical region is shown in Figure 12-3.

TABLE 12-7 Critical Values for the Rank Sum Test

When α is	We Use the Critical Values	And the Critical Region is
0.05	$c_1 = \mu_R - 1.96\sigma_R$ and $c_2 = \mu_R + 1.96\sigma_R$	
0.01	$c_1 = \mu_R - 2.58\sigma_R$ and $c_2 = \mu_R + 2.58\sigma_R$	

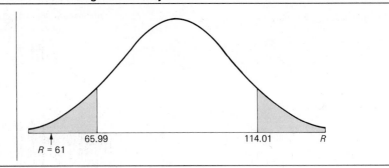

FIGURE 12-3 Critical Region for Navy Divers

65.99   114.01   R

R = 61

Using the ranks from Table 12-7 we found $R = 61$, which is in the critical region. Therefore, we reject the null hypothesis (the pill makes no difference) and conclude at the 0.05 level of significance that the pill does make a difference in decompression time.

_____ Exercise 4 _____

A biologist is doing research on elk in their natural Colorado habitat. Two regions are under study, both with about the same amount of forage and natural cover. However, region A seems to have more predators than region B. To determine if elk tend to live longer in either region a sample of ten elk from each region are tranquilized and have a tooth removed. A laboratory examination of the teeth reveals the ages of the elk. Results for each sample are given in Table 12-8.

TABLE 12-8 Ages of Elk

Group A	11	6	21	23	16	1	3	10	13	8
Group B	12	7	5	24	4	14	19	22	18	2

a) Make a table showing the ages, groups, and ranks for the combined data.

a) **TABLE 12-9** Ranks of Elk

Age	Group	Rank	Age	Group	Rank
1	A	1	12	B	11
2	B	2	13	A	12
3	A	3	14	B	13
4	B	4	16	A	14
5	B	5	18	B	15
6	A	6	19	B	16
7	B	7	21	A	17
8	A	8	22	B	18
10	A	9	23	A	19
11	A	10	24	B	20

b) Find μ_R, σ_R, and R.

b) Since $n_1 = 10$ and $n_2 = 10$,

$$\mu_R = \frac{(10)(10 + 10 + 1)}{2} = 105$$

and

$$\sigma_R = \sqrt{\frac{10 \cdot 10(10 + 10 + 1)}{12}} = 13.23$$

Since $n_1 = n_2 = 10$ we can use either the sum of the ranks of the A group or the B group. Let's use the sum of the ranks of the A group. The A group ranks are 1, 3, 6, 8, 9, 10, 12, 14, 17, and 19. Therefore,

$$R = 1 + 3 + 6 + 8 + 9 + 10 + 12$$
$$+ 14 + 17 + 19$$
$$= 99$$

c) Using an $\alpha = 0.05$ level of significance, what critical values should we use for a (two-sided) rank sum test? Sketch the critical region.

c) For the 0.05 level of significance, we have

$$c_1 = \mu_R - 1.96\sigma_R = 105 - 1.96(13.23) = 79.07$$
$$c_2 = \mu_R + 1.96\sigma_R = 105 + 1.96(13.23) = 130.93$$

(*See* Figure 12-4).

FIGURE 12-4   Critical Region for $\alpha = 0.05$

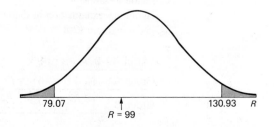

d) Do the data support the null hypothesis (mean life-times are different for the two regions)?

d) Since $R = 99$ is not in the critical region we cannot reject the null hypothesis. The data do not indicate that elk live longer in one region.

• **Note:** In Exercise 4 there were no ties for any rank. If a tie does occur, then each of the tied observations is given the *mean* of the ranks that they occupy. For example, if we rank the numbers

$$41 \qquad 42 \qquad 44 \qquad 44 \qquad 44 \qquad 44$$

we see that 44 occupies ranks three, four, five, and six. Therefore, we give each of the 44s a rank which is the mean of 3, 4, 5, 6:

$$\text{mean of ranks} = \frac{3 + 4 + 5 + 6}{4} = 4.5$$

The final ranking would then be that shown in Table 12-10.

TABLE 12-10

Observation	Rank
41	1
42	2
44	4.5
44	4.5
44	4.5
44	4.5

For samples where n_1 or n_2 is less than eight, there are statistical tables which give appropriate critical values for the rank sum test. Most libraries contain such tables, and the interested reader can find such information by looking under the *Mann-Whitney U Test*.

Section Problems 12.2

In problems 1–5 use a 0.05 level of significance to test the null hypothesis that there is no difference between average performance of the groups against the alternate hypothesis that there is a difference either way.

1. Two groups of ninth-grade students are given a reading comprehension exam. Group A students are from Windy Heights Public School and group B students are from Califf, a neighboring private school. The following table shows the results:

Group A	71	65	70	44	81	73	50	60	88	
Group B	69	45	66	85	75	90	63	84	77	55

2. A psychologist tested two adult groups for boredom tolerance. Group A consisted of 12 females, and group B consisted of 10 males. Their scores are in the following table:

Group A	73	68	41	103	92	88	50	111	120	66	75	115
Group B	150	99	85	77	35	69	100	135	54	72		

3. A horse trainer teaches horses to jump by using two methods of instruction. Horses being taught by method A have a lead horse that accompanies each jump. Horses being taught by method B have no lead horse. The table shows the number of training sessions required before each horse would do the jumps properly.

Method A	28	35	19	41	37	31	38	40	25
Method B	42	33	26	24	44	46	34	20	48	39

4. A French teacher teaches verbs using two different methods. Two groups of ten students were taught a list of verbs using the two different methods, one method for each group. The time required to learn the list using each method is shown in the following table (in min):

Method A	18	25	41	22	56	20	15	30	33	44
Method B	15	28	19	46	55	30	63	58	40	29

5. A cognitive aptitude test consists of putting together a puzzle. Nine people in group A took the test in a competitive setting (first and second to finish received a prize). Twelve people in group B took the test in a noncompetitive setting. The results follow (in minutes required to complete the puzzle):

Group A	7	12	10	15	22	17	18	13	8			
Group B	9	16	30	11	33	28	19	14	24	27	31	29

In problems 6–10, use a 0.01 level of significance to test the null hypothesis that there is no difference between average performance of the groups against the alternate hypothesis that there is a difference either way.

6. A psychologist has developed a mental alertness test. She wishes to study the effects (if any) of type of food consumed on mental alertness. Twenty-one volunteers were randomly divided into two groups. Both groups were told to eat the amount they usually eat for lunch at noon. At 2:00 P.M. all subjects were given the alertness test. Group A had a low-fat lunch with no red meat, lots of vegetables, carbohydrates, and fiber. Group B had a high-fat lunch with red meat, vegetable oils, and low fiber. The only drink for both groups was water. The test scores are shown below:

Group A	76	93	52	81	68	79	88	90	67	85	60
Group B	44	57	60	91	62	86	82	65	96	42	

7. Dr. Winchester is studying the effect of Vitamin B52 on the common cold. A group of 19 Army personnel with common colds was randomly divided into two groups. Group A subjects were given 500-milligram doses of B52 three times a day. Group B subjects were given the same doses of placebos (sugar pills). For each group the duration of the subject's cold is given (in days).

Group A	14	19	12	21	25	16	20	28	10
Group B	9	15	24	26	18	17	31	8	22	11

8. A group of 18 cross-country ski racers was randomly divided into two groups. Group A used modern no-wax "fish-scale" type Teflon ski bottoms. Group B used the traditional wood-bottom skis with tar and wax. The times for these skiers to complete a 10-km run (with hills) are shown below in minutes:

Group A	45	41	52	47	58	55	40	38	33
Group B	44	39	30	61	37	42	36	50	31

9. Sixteen fourth-grade children were randomly divided into two groups. Group A was taught spelling by a phonetic method. Group B was taught spelling by a memorization method. At the end of the fourth grade, all children were given a standard spelling exam. The scores are shown as follows:

Group A	77	95	83	69	85	92	61	79
Group B	62	90	70	81	63	75	80	72

10. Dr. Hansen, an industrial chemist, has discovered a new catalyst that may affect the setting time of wet cement. Two groups of test slabs of cement were studied. Group A had no catalyst, and Group B used the catalyst. The setting time for each slab was measured by Dr. Hansen. The results follow (in hours):

Group A	2.7	2.4	1.9	2.9	3.4	1.6	3.6	4.1
Group B	2.5	1.8	1.6	2.2	4.0	3.8	1.4	2.8

SECTION 12.3   Spearman Rank Correlation

Data given in ranked form are different from data given in measurement form. For instance, if we compare the test performance of three students and, say, Elizabeth did the best, Joel did next best, and Sally did the worst, we are giving the information in ranked form. We cannot say how much better Elizabeth did than Sally or Joel, but we do know how the three scores compare. If the actual test scores for the three tests were given, we would have data in measurement form and could tell exactly how much better Elizabeth did than Joel or Sally. In Chapter 10 we studied linear correlation of data in measurement form. In this section we will study correlation of data in ranked form.

As a specific example of a situation in which we might want to compare ranked data from two sources, consider the following. Hendricks College has a new faculty position in its political science department. A national search to fill this position has resulted in a large number of qualified candidates. The political science faculty reserves the right to make the final hiring decision. However, the faculty is interested in comparing its opinion with student opinion about the teaching ability of the candidates. A random sample of nine equally qualified candidates was asked to give a classroom presentation to a large class of students. Both faculty and students attended the lectures. At the end of each lecture, both faculty and students filled out a questionnaire about the teaching performance of the candidate. Based on these questionnaires, each candidate was given an overall rank from the faculty and an overall rank from the students. The results are shown in Table 12-11. Higher ranks mean better teaching performance.

TABLE 12-11 Faculty and Student Ranks of Candidates

Candidate	Faculty Rank	Student Rank
1	3	5
2	7	7
3	5	6
4	9	8
5	2	3
6	8	9
7	1	1
8	6	4
9	4	2

Using data in ranked form, how can we answer the following questions:

1. Do candidates getting higher ranks from the faculty tend to get higher ranks from students?
2. Is there any relation between faculty rankings and student rankings?
3. Do candidates getting higher ranks from faculty tend to get lower ranks from students?

We will use the Spearman rank correlation to answer such questions. In the early 1900s Charles Spearman of the University of London developed the techniques that now bear his name. The Spearman test of rank correlation requires us to use *ranked variables*. Because we are only using ranks, we cannot use the Spearman test to check on the existence of a linear relationship between the variables as we

FIGURE 12-5   Examples of Monotone Relations

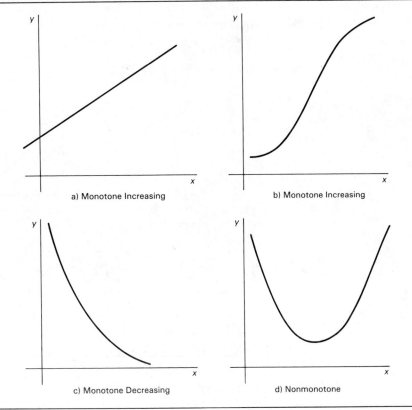

did with the Pearson correlation coefficient (Section 10.3). The Spearman test only checks on the existence of a *monotone* relationship between the variables. (*See* Figure 12-5.) By a *monotone relationship** between variables x and y we mean a relationship in which

1. as x increases, y also increases or
2. as x increases, y decreases.

The relationship shown in Figure 12-5d is a nonmonotone relationship because as x increases, y at first decreases, but later starts to increase. Remember, for a relation to be monotone, as x increases, y must *always* increase or *always* decrease. In a nonmonotone relation, as x increases, y sometimes increases and sometimes decreases or stays unchanged.

---

*Some advanced texts call the monotone relationship we describe *strictly monotone*.

————————————— Exercise 5 —————————————

Identify each of the relations in Figure 12-6 as monotone increasing, monotone decreasing, or nonmonotone.

FIGURE 12-6

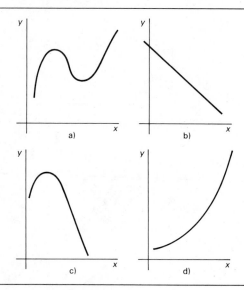

Answers: a) nonmonotone b) monotone decreasing c) nonmonotone
d) monotone increasing

Before we can complete the solution of our problem about the political science department at Hendricks College, we need the following information.

If we have a sample of size n of randomly obtained ordered pairs (x, y) where both x and y values are from *ranked variables,* and if there are no ties in the ranks, then the Pearson product moment correlation coefficient (Section 10.3) can be reduced to the simpler equation.

$$r_S = 1 - \frac{6\Sigma d^2}{n(n^2 - 1)} \quad \text{where} \quad d = x - y$$

We call r_S the *Spearman rank correlation coefficient.*

The Spearman rank correlation coefficient r_S has the following important properties:

1. $-1 \le r_S \le 1$ If $r_S = -1$, the relation between x and y is perfectly monotone decreasing. If $r_S = 0$, there is no monotone relation between x

and y. If $r_S = 1$, the relation between x and y is perfectly monotone increasing. Values of r_S close to 1 or -1 indicate a strong tendency for x and y to have a monotone relationship (increasing or decreasing as the case may be), and values of r_S close to 0 indicate a very weak (or perhaps nonexistent) monotone relationship.

2. The probability distribution of r_S depends on the sample size n. Table 10 in Appendix I gives critical values for certain left- and right-tail tests of r_S. It is important to note that we make no assumptions that x and y are normally distributed variables and we make no assumption about the x and y relationship being linear.

3. The Spearman rank correlation coefficient r_S is our *sample* estimate for ρ_S, the *population* Spearman rank correlation coefficient. We will construct a test of significance for the Spearman rank correlation coefficient in much the same way that we tested the Pearson correlation coefficient (Section 10.4). The null hypothesis is

$$H_0 : \rho_S = 0$$

In effect, the null hypothesis says there is no monotone relation (either increasing or decreasing). The alternate hypothesis depends on the type of test we want to use.

$H_1 : \rho_S < 0$ (Left-tail test)   The alternate hypothesis claims that there is a monotone-decreasing relation between x and y. The critical region is shown in Figure 12-7.

$H_1 : \rho_S > 0$ (Right-tail test)   The alternate hypothesis claims that there is a monotone-increasing relation between x and y. The critical region is shown in Figure 12-8.

FIGURE 12-7   Left-Tail Test

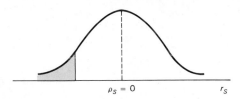

FIGURE 12-8   Right-Tail Test

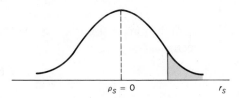

$H_1 : \rho_S \neq 0$ (Two-tail test)   The alternate hypothesis claims that there is a monotone relation (either increasing or decreasing) between x and y. Figure 12-9 shows the corresponding critical regions.

FIGURE 12-9   Two-Tail Test

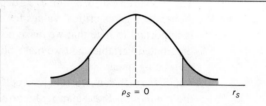

To find critical values and critical regions of the r_S distribution, we use Table 10 of Appendix I with the level of significance, sample size, and type of test (left-, right-, or two-tail) employed.

EXAMPLE   1

Using the information about the Spearman rank correlation coefficient, let's finish our problem about the search for a new member of the political science department at Hendricks College. Our work is organized in Table 12-12, where the rankings given by students and faculty are listed for each of the nine candidates.

TABLE 12-12   Student and Faculty Ranks of Candidates and Calculations for the Spearman Rank Correlation Test

Candidate	Faculty Rank x	Student Rank y	$d = x - y$	d^2
1	3	5	−2	4
2	7	7	0	0
3	5	6	−1	1
4	9	8	1	1
5	2	3	−1	1
6	8	9	−1	1
7	1	1	0	0
8	6	4	2	4
9	4	2	2	4
				$\Sigma d^2 = 16$

Since the sample size is $n = 9$, and $\Sigma d^2 = 16$, the Spearman rank correlation coefficient is

$$r_S = 1 - \frac{6\Sigma d^2}{n(n^2 - 1)}$$

$$r_S = 1 - \frac{6(16)}{9(81 - 1)} = 0.867$$

Let's test the claim that faculty and students tend to agree about the candidate's teaching ability. This means the x and y variables should be monotone-increasing (as x increases, y increases). Since ρ_S is the population Spearman rank correlation coefficient, we have

$$H_0 : \rho_S = 0 \qquad \text{(There is no monotone relation.)}$$
$$H_1 : \rho_S > 0 \qquad \text{(There is a monotone-increasing relation.)}$$

If we use $\alpha = 0.005$ as our level of significance for a right-tail test, then Table 10 with $n = 9$ gives a critical value of 0.834. Figure 12-10 shows the critical and acceptance regions.

FIGURE 12-10   Acceptance and Critical Regions for r_S.

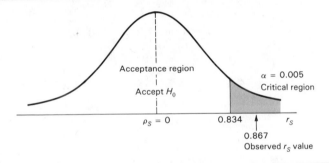

Since the observed r_S value is in the rejection region, we reject $H_0 : \rho_S = 0$ and conclude that the relation between faculty and student ranks is monotonic increasing $\rho_S > 0$. This means faculty and students tend to rank the teaching performance of candidates in the same way.

_____ Exercise 6 _____

Fishermen in the Adirondack Mountains are complaining that acid rain caused by air pollution is killing fish in their region. To study this claim, a biology research team studied a random sample of 12 lakes in the region. For each lake they measured the level of acidity of rain in the drainage leading into the lake and the density of fish in the lake (number of fish per acre foot of water). Then they did a ranking of x = acidity, and y = density of fish. The results are shown in Table 12-13. Higher x ranks mean more acidity, and higher y ranks mean higher density of fish.

TABLE 12-13 Acid Rain and Density of Fish

Lake	Acidity x	Fish Density y	$d = x - y$	d^2
1	5	8	-3	9
2	8	6	2	4
3	3	9	-6	36
4	2	12	-10	100
5	6	7	-1	1
6	1	10	-9	81
7	10	2	8	64
8	12	1	___	___
9	7	5	___	___
10	4	11	___	___
11	9	4	___	___
12	11	3	___	___
			$\Sigma d^2 =$ ___	

a) Complete the entries in the d and d^2 columns of Table 12-13, and find Σd^2.

a)

Lake	x	y	d	d^2
8	12	1	11	121
9	7	5	2	4
10	4	11	-7	49
11	9	4	5	25
12	11	3	8	64
			$\Sigma d^2 =$	558

b) Compute r_S.

b) $r_S = 1 - \dfrac{6 \Sigma d^2}{n(n^2 - 1)}$

$= 1 - \dfrac{6(558)}{12(144 - 1)}$

$= -0.951$

c) The fishermen are claiming that more acidity means lower density of fish. Would this claim say that x and y have a monotone-increasing or monotone-decreasing relation or no monotone relation?

c) The claim says that as x increases y decreases, so the claim is that the relation of x and y is monotone-decreasing.

d) To test the fishermen's claim, what should we use for the null hypothesis? What should we use for the alternate hypothesis?

d) $H_0 : \rho_S = 0$ (no monotone relation)

   $H_1 : \rho_S < 0$ (monotone-decreasing relation)

e) Using a 0.001 level of significance, find a critical value and sketch the critical region for the test of part (d).

e) $n = 12; \alpha = 0.001$

By Table 10 the critical value for a left-tail test is $r_S = -0.826$. Figure 12-11 shows the critical region.

FIGURE 12-11   Critical Region

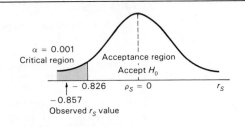

f) Do the data indicate that we should accept or reject the claim that more acidity means fewer fish? Explain.

f) Since the observed value $r_S = -0.951$ is in the critical region, we reject H_0 and conclude that the level of acidity has a very significant ($\alpha = 0.001$) monotone-decreasing relation with fish density.

If ties occur in the assignment of ranks, we follow the usual method of averaging tied ranks. This method was discussed in Section 12.2 (the Rank Sum Test). The next example illustrates the method.

• **Comment:** Technically, the use of the given formula for r_S requires that there be no ties in rank. However, if the number of ties in rank is small relative to the number of ranks, the formula can be used with quite a bit of reliability.

EXAMPLE 2

Do people who smoke more tend to drink more cups of coffee? The following data were obtained from a random sample of $n = 10$ cigarette smokers who also drink coffee.

Person	Cigarettes Smoked Per Day	Cups of Coffee Per Day
1	8	4
2	15	7
3	20	10
4	5	3
5	22	9
6	15	5
7	15	8
8	25	11
9	30	18
10	35	18

To use the Spearman rank correlation test, we need to rank the data. It does not matter if we rank from smallest to largest or largest to smallest. The only requirement is that we be consistent in our rankings. Let us rank from smallest to largest.

First we rank data as though there were no ties; then we average the ties.

TABLE 12-14 Rankings of Cigarettes Smoked Per Day

Person	Cigarettes Smoked Per Day	Rank		Average Rank x	
4	5	1		1	
1	8	2		2	
2	15 ⎫	3 ⎫		4 ⎫	
6	15 ⎬ ties	4 ⎬ Average		4 ⎬ Use the average	
7	15 ⎭	5 ⎭ rank is 4.		4 ⎭ rank for tied data.	
3	20	6		6	
5	22	7		7	
8	25	8		8	
9	30	9		9	
10	35	10		10	

TABLE 12-15 Rankings of Cups of Coffee Per Day

Person	Cups of Coffee Per Day	Rank		Average Rank y	
4	3	1		1	
1	4	2		2	
6	5	3		3	
2	7	4		4	
7	8	5		5	
5	9	6		6	
3	10	7		7	
8	11	8		8	
9	18 ⎫ ties	9 ⎫ Average		9.5 ⎫ Use the average	
10	18 ⎭	10 ⎭ rank is 9.5.		9.5 ⎭ rank for tied data.	

Next we compute the observed value of r_S.

$$r_S = 1 - \frac{6\Sigma d^2}{n(n^2 - 1)}$$

$$r_S = 1 - \frac{6(4.5)}{10(100 - 1)} = 0.973$$

TABLE 12-16 Ranks to Be Used for a Spearman Rank Correlation Test

Person	Cigarette Rank x	Coffee Rank y	$d = x - y$	d^2
1	2	2	0	0
2	4	4	0	0
3	6	7	-1	1
4	1	1	0	0
5	7	6	1	1
6	4	3	1	1
7	4	5	-1	1
8	8	8	0	0
9	9	9.5	-0.5	0.25
10	10	9.5	0.5	0.25
				$\Sigma d^2 = 4.5$

Using 0.001 as a level of significance, test the claim that x and y have a monotone-increasing relationship. In other words, test the claim that people who tend to smoke more tend to drink more cups of coffee.

$$H_0 : \rho_S = 0 \quad \text{(there is no monotone relation)}$$
$$H_1 : \rho_S > 0 \quad \text{(right-tail test)}$$

From Table 12-10 we find that when $n = 10$ and $\alpha = 0.001$, the critical value is $r_S = 0.879$. The resulting critical and acceptance regions are shown in Figure 12-12.

FIGURE 12-12 Critical and Acceptance Regions

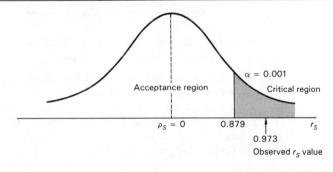

Since the observed value $r_S = 0.973$ is in the rejection region, we reject H_0 and conclude that x and y have a monotone-increasing relation. People who smoke more cigarettes tend to drink more coffee.

Section Problems 12.3

1. A data processing company has a training program for new sales people. After completing the training program, each trainee is ranked by his or her instructor. After a year of sales, the same class of trainees is again ranked by a company supervisor according to net value of the contracts they have acquired for the company. The results for a random sample of 11 sales people trained in the last year follow, where x is rank in training class and y is rank in sales after one year. Lower ranks mean higher standing in class and higher net sales.

Person	1	2	3	4	5	6	7	8	9	10	11
x Rank	6	8	11	2	5	7	3	9	1	10	4
y Rank	4	9	10	1	6	7	8	11	3	5	2

Using a 0.05 level of significance, test the claim that the relation between x and y is monotone (either increasing or decreasing).

2. As an economics class project, Debbie studied a random sample of 14 stocks. For each of these stocks she found the cost per share (in dollars) and ranked each of the stocks according to cost. After three months she found the earnings per share on each stock (in dollars). Again Debbie ranked each of the stocks according to earnings. The way Debbie ranked, higher ranks mean higher cost and higher earnings. The results follow where x is the rank in cost, and y is the rank in earnings.

Stock	1	2	3	4	5	6	7	8	9	10	11	12	13	14
x	5	2	4	7	11	8	12	3	13	14	10	1	9	6
y	5	13	1	10	7	3	14	6	4	12	8	2	11	9

Using a 0.01 level of significance, test the claim that there is a monotone relation, either way, between ranks of cost and earnings.

3. A psychology professor is studying the relation between overcrowding and violent behavior in a rat colony. Eight colonies with different degrees of overcrowding are being used. By using a television monitor, lab assistants record incidents of violence. Each colony has been ranked for crowdedness and violence. A rank of 1 means most crowded or most violent. The results for the eight colonies are in the following table with x being the population density rank and y the violence rank.

Colony	1	2	3	4	5	6	7	8
x Rank	3	5	6	1	8	7	4	2
y Rank	1	3	5	2	8	6	4	7

Using a 0.05 level of significance, test the claims that lower crowding ranks mean lower violence ranks (i.e., the variables have a monotone-increasing relation).

4. A history professor claims that students who finish exams quicker tend to get higher scores. The following data show the order of finish and score for ten students selected at random for a recent exam.

Student	1	2	3	4	5	6	7	8	9	10
Order of Finish	5	7	3	1	6	2	8	4	10	9
Score	73	90	82	95	65	82	78	75	80	55

   a) Ranking order of finish with 1 as first to finish, and ranking score with 1 as highest score, construct a table of ranks to be used for a Spearman rank correlation test.
   b) Using a 0.05 level of significance, test the claim that there is a monotone-increasing relation between rank of finish and rank of score.

5. An army psychologist gave a random sample of seven soldiers a test to measure sense of humor and another test to measure aggressiveness. High scores mean greater sense of humor or more aggressiveness.

Soldier	1	2	3	4	5	6	7
Score on Humor Test	60	85	78	90	93	45	51
Score on Aggressiveness Test	78	42	68	53	62	50	76

   a) Ranking the data with highest rank for highest score on a test, make a table of ranks to be used in a Spearman rank correlation test.
   b) Using a 0.05 level of significance, test the claim that rank in humor has a monotone-decreasing relation to rank in aggressiveness.

6. A consumer research group examined a random sample of eight stereo speaker systems selected from among major brands. They ranked each of the systems for overall quality of equipment and sound produced. They obtained the manufacturer's recommended price for each brand. A rank of 1 means highest quality.

System	1	2	3	4	5	6	7	8
Quality Rank	4	8	5	2	7	6	1	3
Price, $	690	175	1,200	970	225	785	470	850

   a) Using a rank of 1 for the highest price, make a table of ranks to be used in a Spearman rank correlation test.
   b) Using a 0.05 level of significance, test the claim that there is a monotone relation (either way) between rank of quality and rank of price.

7. A group of 11 cadets selected at random were given a flying aptitude test before they went to flight training school. After graduation from training school, their commanding officer ranked each cadet according to his or her flying ability (higher ranks mean greater ability). The results were:

Cadet	1	2	3	4	5	6	7	8	9	10	11
Aptitude Score	720	390	710	480	970	480	517	830	690	850	480
Performance Rank	7	1	8	4	10	2	5	11	6	9	3

a) Using a rank of 1 for the lowest aptitude score, make a table of ranks to be used in a Spearman rank correlation test.
b) Using a 0.005 level of significance, test the claim that there is a monotone-increasing relation between aptitude rank and performance rank.

8. At the Big Rock Insurance Company branch office, a pool of six secretaries work under two different managers. At the end of the year, both managers are asked to rank the secretaries. A rank of 1 means best secretary.

Secretary	1	2	3	4	5	6
Manager A Rank	3	5	2	1	6	4
Manager B Rank	1	3	6	2	5	4

Using a 0.05 level of significance, test the claim that x and y have a monotone-increasing relationship (i.e., the managers agree).

9. Big Rock Insurance Company did a study of per capita income and volume of insurance sales in eight midwest cities. The volume of sales in the cities was ranked with 1 being the largest volume. The per capita income was rounded to the nearest thousand dollars.

City	1	2	3	4	5	6	7	8
Volume of Insurance Sales Rank	6	7	1	8	3	2	5	4
Per Capita Income in $1,000	17	18	19	11	16	20	15	19

a) Using a rank of 1 for the highest per capita income, make a table of ranks to be used for a Spearman rank correlation test.
b) Using a 0.01 level of significance, test the claim that there is a monotone relation (either way) between rank in volume of sales and rank of per capita income.

Summary

When we cannot assume that data come from a normal, binomial, or Student's t distribution, we can employ tests that make no assumptions about data distribution. Such tests are called nonparametric tests. We studied three widely used tests: the sign test, the rank sum test, and the Spearman rank correlation coefficient test. Nonparametric tests have the advantage of being easy to use; however, they do tend to waste information and to be less sensitive than standard tests. It is usually good advice to use standard tests when possible, keeping nonparametric tests for situations where assumptions about the data distributions cannot be made.

Important Words and Symbols

	Section
Nonparametric statistics	12.1
Sign test	12.1
Rank sum test	12.2
Spearman rank correlation coefficient r_S	12.3

Chapter Review Problems

For each problem do the following:

a) Decide whether you should use a sign test, rank sum test, or Spearman test.
b) State the null and alternate hypotheses.
c) Find all critical values.
d) Sketch the critical region, the acceptance region, the critical values, and the sample statistic value.
e) Decide whether you should accept or reject the null hypothesis.

1. In the production of synthetic motor lubricant from coal, a new catalyst has been discovered that seems to affect the viscosity index of the lubricant. In an experiment consisting of 21 production runs, 10 used the new catalyst and 11 did not. After each production run the viscosity index of the lubricant was determined to be as follows.

With Catalyst	1.6	3.2	2.9	4.4	3.7	2.5	1.1	1.8	3.8	4.2	
Without Catalyst	3.9	4.6	1.5	2.2	2.8	3.6	2.4	3.3	1.9	4.0	3.5

Use a 0.05 level of significance to test the null hypothesis that the mean viscosity index is unchanged by the catalyst against the alternate hypothesis that the mean viscosity index has changed.

2. Professor Adams wrote a book called *Improving Your Memory*. The professor claims that if you follow the program outlined in the book, your memory will definitely improve. Fifteen people took the professor's course in which the book and its program were used. On the first day of class everyone took a memory exam and on the last day everyone took a similar exam. Their scores were as follows.

Last Exam	225	120	115	275	85	76	114	200	99	135	170	110	216	280	78
First Exam	175	110	115	200	60	85	160	190	70	110	140	10	190	200	92

Use a 0.05 level of significance to test the null hypothesis that the mean scores are the same whether or not people have taken the course against the alternate hypothesis that the mean scores of people who have taken the course are higher.

3. A chain of hardware stores is trying to sell more paint by mailing pamphlets describing the paint. In 15 communities containing one of these hardware stores the paint sales (in dollars) were recorded for the month before and the month after the ads were sent out. The results are given in Table 12-17.

TABLE 12-17 Sales Before and After Advertising Campaign

Sales After	Sales Before	Sales After	Sales Before
610	460	500	370
150	216	118	118
790	640	265	117
288	250	365	360
715	685	93	93
465	430	217	291
280	220	280	430
640	470		

Use a 0.01 level of significance to test the null hypothesis that the advertising had no effect on average sales against the alternate hypothesis that it improved sales.

4. An obedience school for dogs experimented with two methods of training. One method involved rewards (food, praise); the other involved no rewards. The number of sessions required for training each of 19 dogs follows.

With Rewards	12	17	15	10	16	20	9	23	8	14
No Rewards	19	22	11	18	13	25	24	28	21	

Use a 0.05 level of significance to test the hypothesis that the mean number of sessions was the same for the two groups against the alternate hypothesis that they were not the same.

5. At McDouglas Hamburger stands, each employee must undergo a training program before he or she is hired. A group of nine people went through the training program and were hired to work in the Teton Park McDouglas Hamburger stand. Rankings in performance for the training program and after one month on the job are shown (a rank of 1 is for best performance).

Employee	1	2	3	4	5	6	7	8	9
Rank, Training Program	8	9	7	3	6	4	1	2	5
Rank on Job	9	8	6	7	5	1	3	4	2

Using a 0.05 level of significance, test the claim that there is a monotone-increasing relation between rank in the training program and rank in performance on the job.

6. Two expert French chefs judged chocolate mousse made by students in a Paris cooking school. Each chef ranked the best chocolate mousse as 1.

Student	1	2	3	4	5
Rank by Chef Pierre	4	2	3	1	5
Rank by Chef André	4	1	2	3	5

Use a 0.10 level of significance to test the claim that there is a monotone relation (either way) for ranks given by Chef Pierre and Chef André.

Appendix I Tables

1 Random Numbers
2 Factorials
3 Binomial Coefficients
4 Binomial Probabilities
5 Areas of a Standard Normal Distribution

6 Student's t Distribution
7 Critical Values of Pearson Product-Moment Correlation, r
8 The χ^2 Distribution
9 The F Distribution
10 Critical Values for Spearman Rank Correlation, r_s

TABLE 1 Random Numbers

85796	43265	33460	59731	66850	92726	21852	32338
45464	50110	37363	74112	42640	25483	97133	93006
05511	82121	26251	10481	58899	57286	08966	25274
58263	43355	06948	38729	25024	92509	24649	44444
02320	48686	94087	99407	83329	25478	93307	20090
40811	58304	83589	49585	76322	82229	59875	08054
31167	07249	99440	31576	99171	38121	55072	26364
86185	41059	55868	69154	50615	83098	61125	39313
79609	68355	91700	22183	43337	92621	78620	85061
05684	72181	70029	99171	38398	69022	07595	84157
35734	41765	19595	12730	25425	58129	15106	01737
57164	93149	03372	79769	60556	76177	07324	39253
32411	98336	25394	37581	40354	96347	73305	28845
78346	59193	43148	38210	23950	85859	23728	41002
10466	50586	57359	78513	58425	00082	42059	56015
67455	32880	37826	28232	58464	37726	51581	43138
30411	72660	04269	38533	73194	03771	83313	05359
02728	53844	92952	31352	38923	75755	24575	95796
93746	53709	68778	45229	54363	69987	53441	24550
79210	89959	80848	66369	87957	30184	72572	05926
83343	42354	94017	02803	91736	71996	62184	95675
21931	28394	63374	96480	61986	96656	12906	25603
76540	35400	92026	65259	30628	25433	63642	68723
86262	39727	03266	25967	19545	81967	37027	17771
93141	92033	75331	36118	30271	73522	73843	36886
92514	30926	36194	87681	77170	72539	69505	29663
47313	21503	96052	54492	89173	36351	06078	78975
06874	66167	76252	00493	42807	61228	41870	04035
18157	28373	52801	42056	53496	73967	77561	31768
36375	23733	44274	61699	09986	84603	25247	39827
30141	82436	15897	68474	08421	29395	60291	15716
61332	44956	27053	51218	18706	64198	37321	41987

TABLE 1 *continued*

97945	60182	08761	13513	74675	83910	51913	40366
06251	70303	97748	76487	47229	75560	99542	81579
60478	99570	21612	79525	67147	06392	44256	47018
39026	55609	26907	52180	15538	56277	54190	10910
97564	11278	03772	83834	57300	21769	78972	05007
19561	91610	00432	08299	63480	04119	61624	35625
60506	26926	65005	67372	44503	50946	77305	30805
50828	72347	52216	34855	86094	24552	80704	92097
82264	30250	03883	01064	23756	59468	10522	81602
88478	20494	73577	36184	79937	62994	10954	22482
35327	37909	92085	46495	17380	56864	08622	37989
54031	64589	39172	29795	39811	69193	92461	59174
24089	76885	38316	05492	71773	16622	00875	03573

TABLE 2   Factorials

n	$n!$
0	1
1	1
2	2
3	6
4	24
5	120
6	720
7	5040
8	40320
9	362880
10	3628800
11	39916800
12	479001600
13	6227020800
14	87178291200
15	1307674368000
16	20922789888000
17	355687428096000
18	6402373705728000
19	121645100408832000
20	2432902008176640000

TABLE 3 Binomial Coefficients $C_{n,r}$

n \ r	0	1	2	3	4	5	6	7	8	9	10
1	1	1									
2	1	2	1								
3	1	3	3	1							
4	1	4	6	4	1						
5	1	5	10	10	5	1					
6	1	6	15	20	15	6	1				
7	1	7	21	35	35	21	7	1			
8	1	8	28	56	70	56	28	8	1		
9	1	9	36	84	126	126	84	36	9	1	
10	1	10	45	120	210	252	210	120	45	10	1
11	1	11	55	165	330	462	462	330	165	55	11
12	1	12	66	220	495	792	924	792	495	220	66
13	1	13	78	286	715	1,287	1,716	1,716	1,287	715	286
14	1	14	91	364	1,001	2,002	3,003	3,432	3,003	2,002	1,001
15	1	15	105	455	1,365	3,003	5,005	6,435	6,435	5,005	3,003
16	1	16	120	560	1,820	4,368	8,008	11,440	12,870	11,440	8,008
17	1	17	136	680	2,380	6,188	12,376	19,448	24,310	24,310	19448
18	1	18	153	816	3,060	8,568	18,564	31,824	43,758	48,620	43,758
19	1	19	171	969	3,876	11,628	27,132	50,388	75,582	92,378	92,378
20	1	20	190	1,140	4,845	15,504	38,760	77,520	125,970	167,960	184,756

TABLE 4 Binomial Probabilities $C_{n,r}p^r q^{n-r}$

n	r	.01	.05	.10	.15	.20	.25	.30	.35	.40	.45	.50	.55	.60	.65	.70	.75	.80	.85	.90	.95
2	0	.980	.902	.810	.723	.640	.563	.490	.423	.360	.303	.250	.203	.160	.123	.090	.063	.040	.023	.010	.002
	1	.020	.095	.180	.255	.320	.375	.420	.455	.480	.495	.500	.495	.480	.455	.420	.375	.320	.255	.180	.095
	2		.002	.010	.023	.040	.063	.090	.123	.160	.203	.250	.303	.360	.423	.490	.563	.640	.723	.810	.902
3	0	.970	.857	.729	.614	.512	.422	.343	.275	.216	.166	.125	.091	.064	.043	.027	.016	.008	.003	.001	
	1	.029	.135	.243	.325	.384	.422	.441	.444	.432	.408	.375	.334	.288	.239	.189	.141	.096	.057	.027	.007
	2		.007	.027	.057	.096	.141	.189	.239	.288	.334	.375	.408	.432	.444	.441	.422	.384	.325	.243	.135
	3			.001	.003	.008	.016	.027	.043	.064	.091	.125	.166	.216	.275	.343	.422	.512	.614	.729	.857
4	0	.961	.815	.656	.522	.410	.316	.240	.179	.130	.092	.062	.041	.026	.015	.008	.004	.002	.001		
	1	.039	.171	.292	.368	.410	.422	.412	.384	.346	.300	.250	.200	.154	.112	.076	.047	.026	.011	.004	
	2	.001	.014	.049	.098	.154	.211	.265	.311	.346	.368	.375	.368	.346	.311	.265	.211	.154	.098	.049	.014
	3		.001	.004	.011	.026	.047	.076	.112	.154	.200	.250	.300	.346	.384	.412	.422	.410	.368	.292	.171
	4				.001	.002	.004	.008	.015	.026	.041	.062	.092	.130	.179	.240	.316	.410	.522	.656	.815
5	0	.951	.774	.590	.444	.328	.237	.168	.116	.078	.050	.031	.019	.010	.005	.002	.001				
	1	.048	.204	.328	.392	.410	.396	.360	.312	.259	.206	.156	.113	.077	.049	.028	.015	.006	.002		
	2	.001	.021	.073	.138	.205	.264	.309	.336	.346	.337	.312	.276	.230	.181	.132	.088	.051	.024	.008	.001
	3		.001	.008	.024	.051	.088	.132	.181	.230	.276	.312	.337	.346	.336	.309	.264	.205	.138	.073	.021
	4				.002	.006	.015	.028	.049	.077	.113	.156	.206	.259	.312	.360	.396	.410	.392	.328	.204
	5						.001	.002	.005	.010	.019	.031	.050	.078	.116	.168	.237	.328	.444	.590	.774
6	0	.941	.735	.531	.377	.262	.178	.118	.075	.047	.028	.016	.008	.004	.002	.001					
	1	.057	.232	.354	.399	.393	.356	.303	.244	.187	.136	.094	.061	.037	.020	.010	.004	.002			
	2	.001	.031	.098	.176	.246	.297	.324	.328	.311	.278	.234	.186	.138	.095	.060	.033	.015	.006	.001	
	3		.002	.015	.042	.082	.132	.185	.236	.276	.303	.312	.303	.276	.236	.185	.132	.082	.042	.015	.002
	4			.001	.006	.015	.033	.060	.095	.138	.186	.234	.278	.311	.328	.324	.297	.246	.176	.098	.031
	5					.002	.004	.010	.020	.037	.061	.094	.136	.187	.244	.303	.356	.393	.399	.354	.232
	6						.001	.001	.002	.004	.008	.016	.028	.047	.075	.118	.178	.262	.377	.531	.735

TABLE 4 *continued*

n	r	.01	.05	.10	.15	.20	.25	.30	.35	.40	.45	.50	.55	.60	.65	.70	.75	.80	.85	.90	.95
7	0	.932	.698	.478	.321	.210	.133	.082	.049	.028	.015	.008	.004	.002	.001						
	1	.066	.257	.372	.396	.367	.311	.247	.185	.131	.087	.055	.032	.017	.008	.004	.001				
	2	.002	.041	.124	.210	.275	.311	.318	.299	.261	.214	.164	.117	.077	.047	.025	.012	.004	.001		
	3		.004	.023	.062	.115	.173	.227	.268	.290	.292	.273	.239	.194	.144	.097	.058	.029	.011	.003	
	4			.003	.011	.029	.058	.097	.144	.194	.239	.273	.292	.290	.268	.227	.173	.115	.062	.023	.004
	5				.001	.004	.012	.025	.047	.077	.117	.164	.214	.261	.299	.318	.311	.275	.210	.124	.041
	6						.001	.004	.008	.017	.032	.055	.087	.131	.185	.247	.311	.367	.396	.372	.257
	7								.001	.002	.004	.008	.015	.028	.049	.082	.133	.210	.321	.478	.698
8	0	.923	.663	.430	.272	.168	.100	.058	.032	.017	.008	.004	.002	.001							
	1	.075	.279	.383	.385	.336	.267	.198	.137	.090	.055	.031	.016	.008	.003	.001					
	2	.003	.051	.149	.238	.294	.311	.296	.259	.209	.157	.109	.070	.041	.022	.010	.004	.001			
	3		.005	.033	.084	.147	.208	.254	.279	.279	.257	.219	.172	.124	.081	.047	.023	.009	.003		
	4			.005	.018	.046	.087	.136	.188	.232	.263	.273	.263	.232	.188	.136	.087	.046	.018	.005	
	5				.003	.009	.023	.047	.081	.124	.172	.219	.257	.279	.279	.254	.208	.147	.084	.033	.005
	6					.001	.004	.010	.022	.041	.070	.109	.157	.209	.259	.296	.311	.294	.238	.149	.051
	7							.001	.003	.008	.016	.031	.055	.090	.137	.198	.267	.336	.385	.383	.279
	8									.001	.002	.004	.008	.017	.032	.058	.100	.168	.272	.430	.663
9	0	.914	.630	.387	.232	.134	.075	.040	.021	.010	.005	.002	.001								
	1	.083	.299	.387	.368	.302	.225	.156	.100	.060	.034	.018	.008	.004	.001						
	2	.003	.063	.172	.260	.302	.300	.267	.216	.161	.111	.070	.041	.021	.010	.004	.001				
	3		.008	.045	.107	.176	.234	.267	.272	.251	.212	.164	.116	.074	.042	.021	.009	.003	.001		
	4		.001	.007	.028	.066	.117	.172	.219	.251	.260	.246	.213	.167	.118	.074	.039	.017	.005	.001	
	5			.001	.005	.017	.039	.074	.118	.167	.213	.246	.260	.251	.219	.172	.117	.066	.028	.007	.001
	6				.001	.003	.009	.021	.042	.074	.116	.164	.212	.251	.272	.267	.234	.176	.107	.045	.008
	7						.001	.004	.010	.021	.041	.070	.111	.161	.216	.267	.300	.302	.260	.172	.063
	8								.001	.004	.008	.018	.034	.060	.100	.156	.225	.302	.368	.387	.299
	9										.001	.002	.005	.010	.021	.040	.075	.134	.232	.387	.630

TABLE 4 *continued*

p

n	r	.01	.05	.10	.15	.20	.25	.30	.35	.40	.45	.50	.55	.60	.65	.70	.75	.80	.85	.90	.95
10	0	.904	.599	.349	.197	.107	.056	.028	.014	.006	.003	.001									
	1	.091	.315	.387	.347	.268	.188	.121	.072	.040	.021	.010	.004	.002							
	2	.004	.075	.194	.276	.302	.282	.233	.176	.121	.076	.044	.023	.011	.004	.001					
	3		.010	.057	.130	.201	.250	.267	.252	.215	.166	.117	.075	.042	.021	.009	.003	.001			
	4		.001	.011	.040	.088	.146	.200	.238	.251	.238	.205	.160	.111	.069	.037	.016	.006	.001		
	5			.001	.008	.026	.058	.103	.154	.201	.234	.246	.234	.201	.154	.103	.058	.026	.008	.001	
	6				.001	.006	.016	.037	.069	.111	.160	.205	.238	.251	.238	.200	.146	.088	.040	.011	.001
	7					.001	.003	.009	.021	.042	.075	.117	.166	.215	.252	.267	.250	.201	.130	.057	.010
	8							.001	.004	.011	.023	.044	.076	.121	.176	.233	.282	.302	.276	.194	.075
	9									.002	.004	.010	.021	.040	.072	.121	.188	.268	.347	.387	.315
	10											.001	.003	.006	.014	.028	.056	.107	.197	.349	.599
11	0	.895	.569	.314	.167	.086	.042	.020	.009	.004	.001										
	1	.099	.329	.384	.325	.236	.155	.093	.052	.027	.013	.005	.002	.001							
	2	.005	.087	.213	.287	.295	.258	.200	.140	.089	.051	.027	.013	.005	.002	.001					
	3		.014	.071	.152	.221	.258	.257	.225	.177	.126	.081	.046	.023	.010	.004	.001				
	4		.001	.016	.054	.111	.172	.220	.243	.236	.206	.161	.113	.070	.038	.017	.006	.002			
	5			.002	.013	.039	.080	.132	.183	.221	.236	.226	.193	.147	.099	.057	.027	.010	.002		
	6				.002	.010	.027	.057	.099	.147	.193	.226	.236	.221	.183	.132	.080	.039	.013	.002	
	7					.002	.006	.017	.038	.070	.113	.161	.206	.236	.243	.220	.172	.111	.054	.016	.001
	8						.001	.004	.010	.023	.046	.081	.126	.177	.225	.257	.258	.221	.152	.071	.014
	9							.001	.002	.005	.013	.027	.051	.089	.140	.200	.258	.295	.287	.213	.087
	10									.001	.002	.005	.013	.027	.052	.093	.155	.236	.325	.384	.329
	11												.001	.004	.009	.020	.042	.086	.167	.314	.569
12	0	.886	.540	.282	.142	.069	.032	.014	.006	.002	.001										
	1	.107	.341	.377	.301	.206	.127	.071	.037	.017	.008	.003	.001								
	2	.006	.099	.230	.292	.283	.232	.168	.109	.064	.034	.016	.007	.002	.001						
	3		.017	.085	.172	.236	.258	.240	.195	.142	.092	.054	.028	.012	.005	.001					
	4		.002	.021	.068	.133	.194	.231	.237	.213	.170	.121	.076	.042	.020	.008	.002	.001			
	5			.004	.019	.053	.103	.158	.204	.227	.223	.193	.149	.101	.059	.029	.011	.003	.001		
	6				.004	.016	.040	.079	.128	.177	.212	.226	.212	.177	.128	.079	.040	.016	.004		

TABLE 4 *continued*

												p									
n	*r*	.01	.05	.10	.15	.20	.25	.30	.35	.40	.45	.50	.55	.60	.65	.70	.75	.80	.85	.90	.95
12	7				.001	.003	.011	.029	.059	.101	.149	.193	.223	.227	.204	.158	.103	.053	.019	.004	
	8					.001	.002	.008	.020	.042	.076	.121	.170	.213	.237	.231	.194	.133	.068	.021	.002
	9							.001	.005	.012	.028	.054	.092	.142	.195	.240	.258	.236	.172	.085	.017
	10								.001	.002	.007	.016	.034	.064	.109	.168	.232	.283	.292	.230	.099
	11										.001	.003	.008	.017	.037	.071	.127	.206	.301	.377	.341
	12												.001	.002	.006	.014	.032	.069	.142	.282	.540
15	0	.860	.463	.206	.087	.035	.013	.005	.002												
	1	.130	.366	.343	.231	.132	.067	.031	.013	.005	.002										
	2	.009	.135	.267	.286	.231	.156	.092	.048	.022	.009	.003	.001								
	3		.031	.129	.218	.250	.225	.170	.111	.063	.032	.014	.005	.002							
	4		.005	.043	.116	.188	.225	.219	.179	.127	.078	.042	.019	.007	.002						
	5		.001	.010	.045	.103	.165	.206	.212	.186	.140	.092	.051	.024	.010	.003	.001				
	6			.002	.013	.043	.092	.147	.191	.207	.191	.153	.105	.061	.030	.012	.003	.001			
	7				.003	.014	.039	.081	.132	.177	.201	.196	.165	.118	.071	.035	.013	.003	.001		
	8				.001	.003	.013	.035	.071	.118	.165	.196	.201	.177	.132	.081	.039	.014	.003		
	9					.001	.003	.012	.030	.061	.105	.153	.191	.207	.191	.147	.092	.043	.013	.002	
	10						.001	.003	.010	.024	.051	.092	.140	.186	.212	.206	.165	.103	.045	.010	.001
	11							.001	.002	.007	.019	.042	.078	.127	.179	.219	.225	.188	.116	.043	.005
	12									.002	.005	.014	.032	.063	.111	.170	.225	.250	.218	.129	.031
	13										.001	.003	.009	.022	.048	.092	.156	.231	.286	.267	.135
	14												.002	.005	.013	.031	.067	.132	.231	.343	.366
	15														.002	.005	.013	.035	.087	.206	.463
16	0	.851	.440	.185	.074	.028	.010	.003	.001												
	1	.138	.371	.329	.210	.113	.053	.023	.009	.003	.001										
	2	.010	.146	.275	.277	.211	.134	.073	.035	.015	.006	.002	.001								
	3		.036	.142	.229	.246	.208	.146	.089	.047	.022	.009	.003	.001							
	4		.006	.051	.131	.200	.225	.204	.155	.101	.057	.028	.011	.004	.001						
	5		.001	.014	.056	.120	.180	.210	.201	.162	.112	.067	.034	.014	.005	.001					
	6			.003	.018	.055	.110	.165	.198	.198	.168	.122	.075	.039	.017	.006	.001				
	7				.005	.020	.052	.101	.152	.189	.197	.175	.132	.084	.044	.019	.006	.001			

TABLE 4 continued

p

n	r	.01	.05	.10	.15	.20	.25	.30	.35	.40	.45	.50	.55	.60	.65	.70	.75	.80	.85	.90	.95
16	8				.001	.006	.020	.049	.092	.142	.181	.196	.181	.142	.092	.049	.020	.006	.001		
	9					.001	.006	.019	.044	.084	.132	.175	.197	.189	.152	.101	.052	.020	.005		
	10						.001	.006	.017	.039	.075	.122	.168	.198	.198	.165	.110	.055	.018	.003	
	11							.001	.005	.014	.034	.067	.112	.162	.201	.210	.180	.120	.056	.014	.001
	12								.001	.004	.011	.028	.057	.101	.155	.204	.225	.200	.131	.051	.006
	13									.001	.003	.009	.022	.047	.089	.146	.208	.246	.229	.142	.036
	14										.001	.002	.006	.015	.035	.073	.134	.211	.277	.275	.146
	15												.001	.003	.009	.023	.053	.113	.210	.329	.371
	16														.001	.003	.010	.028	.074	.185	.440
20	0	.818	.358	.122	.039	.012	.003	.001													
	1	.165	.377	.270	.137	.058	.021	.007	.002												
	2	.016	.189	.285	.229	.137	.067	.028	.010	.003											
	3	.001	.060	.190	.243	.205	.134	.072	.032	.012	.001	.001	.001								
	4		.013	.090	.182	.218	.190	.130	.074	.035	.004	.005	.005								
	5		.002	.032	.103	.175	.202	.179	.127	.075	.014	.015	.015	.001							
	6			.009	.045	.109	.169	.192	.171	.124	.036	.036	.037	.005	.001						
	7			.002	.016	.055	.112	.164	.184	.166	.075	.074	.073	.015	.005	.001					
	8				.005	.022	.061	.114	.161	.180	.122	.120	.119	.035	.014	.004	.001				
	9				.001	.007	.027	.065	.116	.160	.162	.160	.159	.071	.034	.012	.003				
	10					.002	.010	.031	.069	.117	.177	.176	.177	.117	.069	.031	.010	.002			
	11						.003	.012	.034	.071	.159	.160	.162	.160	.116	.065	.027	.007	.001		
	12						.001	.004	.014	.035	.119	.120	.122	.180	.161	.114	.061	.022	.005		
	13							.001	.005	.015	.073	.074	.075	.166	.184	.164	.112	.055	.016	.002	
	14								.001	.005	.037	.037	.036	.124	.171	.192	.169	.109	.045	.009	
	15									.001	.015	.015	.014	.075	.127	.179	.202	.175	.103	.032	.002
	16										.005	.005	.004	.035	.074	.130	.190	.218	.182	.090	.013
	17										.001	.001	.001	.012	.032	.072	.134	.205	.243	.190	.060
	18													.003	.010	.028	.067	.137	.229	.285	.189
	19														.002	.007	.021	.058	.137	.270	.377
	20															.001	.003	.012	.039	.122	.358

TABLE 5 Areas of a Standard Normal Distribution

The table entries represent the area under the standard
normal curve from 0 to the specified value of z.

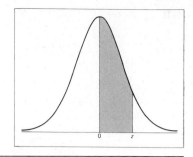

z	.00	.01	.02	.03	.04	.05	.06	.07	.08	.09
0.0	.0000	.0040	.0080	.0120	.0160	.0199	.0239	.0279	.0319	.0359
0.1	.0398	.0438	.0478	.0517	.0557	.0596	.0636	.0675	.0714	.0753
0.2	.0793	.0832	.0871	.0910	.0948	.0987	.1026	.1064	.1103	.1141
0.3	.1179	.1217	.1255	.1293	.1331	.1368	.1406	.1443	.1480	.1517
0.4	.1554	.1591	.1628	.1664	.1700	.1736	.1772	.1808	.1844	.1879
0.5	.1915	.1950	.1985	.2019	.2054	.2088	.2123	.2157	.2190	.2224
0.6	.2257	.2291	.2324	.2357	.2389	.2422	.2454	.2486	.2517	.2549
0.7	.2580	.2611	.2642	.2673	.2704	.2734	.2764	.2794	.2823	.2852
0.8	.2881	.2910	.2939	.2967	.2995	.3023	.3051	.3078	.3106	.3133
0.9	.3159	.3186	.3212	.3238	.3264	.3289	.3315	.3340	.3365	.3389
1.0	.3413	.3438	.3461	.3485	.3508	.3531	.3554	.3577	.3599	.3621
1.1	.3643	.3665	.3686	.3708	.3729	.3749	.3770	.3790	.3810	.3830
1.2	.3849	.3869	.3888	.3907	.3925	.3944	.3962	.3980	.3997	.4015
1.3	.4032	.4049	.4066	.4082	.4099	.4115	.4131	.4147	.4162	.4177
1.4	.4192	.4207	.4222	.4236	.4251	.4265	.4279	.4292	.4306	.4319
1.5	.4332	.4345	.4357	.4370	.4382	.4394	.4406	.4418	.4429	.4441
1.6	.4452	.4463	.4474	.4484	.4495	.4505	.4515	.4525	.4535	.4545
1.7	.4554	.4564	.4573	.4582	.4591	.4599	.4608	.4616	.4625	.4633
1.8	.4641	.4649	.4656	.4664	.4671	.4678	.4686	.4693	.4699	.4706
1.9	.4713	.4719	.4726	.4732	.4738	.4744	.4750	.4756	.4761	.4767
2.0	.4772	.4778	.4783	.4788	.4793	.4798	.4803	.4808	.4812	.4817
2.1	.4821	.4826	.4830	.4834	.4838	.4842	.4846	.4850	.4854	.4857
2.2	.4861	.4864	.4868	.4871	.4875	.4878	.4881	.4884	.4887	.4890
2.3	.4893	.4896	.4898	.4901	.4904	.4906	.4909	.4911	.4913	.4916
2.4	.4918	.4920	.4922	.4925	.4927	.4929	.4931	.4932	.4934	.4936
2.5	.4938	.4940	.4941	.4943	.4945	.4946	.4948	.4949	.4951	.4952
2.6	.4953	.4955	.4956	.4957	.4959	.4960	.4961	.4962	.4963	.4964
2.7	.4965	.4966	.4967	.4968	.4969	.4970	.4971	.4972	.4973	.4974
2.8	.4974	.4975	.4976	.4977	.4977	.4978	.4979	.4979	.4980	.4981
2.9	.4981	.4982	.4982	.4983	.4984	.4984	.4985	.4985	.4986	.4986
3.0	.4987	.4987	.4987	.4988	.4988	.4989	.4989	.4989	.4990	.4990
3.1	.4990	.4991	.4991	.4991	.4992	.4992	.4992	.4992	.4993	.4993
3.2	.4993	.4993	.4994	.4994	.4994	.4994	.4994	.4995	.4995	.4995
3.3	.4995	.4995	.4995	.4996	.4996	.4996	.4996	.4996	.4996	.4997
3.4	.4997	.4997	.4997	.4997	.4997	.4997	.4997	.4997	.4997	.4998
3.5	.4998									
4.0	.49997									
4.5	.499997									
5.0	.4999997									

TABLE 6   Student's *t* Distribution

c is a confidence level:

α' is the level of significance for a one-tail test:

α'' is the level of significance for a two-tail test:

c	0.90	0.95	0.98	0.99
α'	0.05	0.025	0.01	0.005
d.f. α''	0.1	0.05	0.02	0.01
1	6.314	12.706	31.821	63.657
2	2.920	4.303	6.965	9.925
3	2.353	3.182	4.541	5.841
4	2.132	2.776	3.747	4.604
5	2.015	2.571	3.365	4.032
6	1.943	2.447	3.143	3.707
7	1.895	2.365	2.998	3.499
8	1.860	2.306	2.896	3.355
9	1.833	2.262	2.821	3.250
10	1.812	2.228	2.764	3.169
11	1.796	2.201	2.718	3.106
12	1.782	2.179	2.681	3.055
13	1.771	2.160	2.650	3.012
14	1.761	2.145	2.624	2.977
15	1.753	2.131	2.602	2.947
16	1.746	2.120	2.583	2.921
17	1.740	2.110	2.567	2.898
18	1.734	2.101	2.552	2.878
19	1.729	2.093	2.539	2.861
20	1.725	2.086	2.528	2.845
21	1.721	2.080	2.518	2.831
22	1.717	2.074	2.508	2.819
23	1.714	2.069	2.500	2.807
24	1.711	2.064	2.492	2.797
25	1.708	2.060	2.485	2.787
26	1.706	2.056	2.479	2.779
27	1.703	2.052	2.473	2.771
28	1.701	2.048	2.467	2.763
29	1.699	2.045	2.462	2.756
30	1.697	2.042	2.457	2.750

Table 6 is taken from Table III of Fisher and Yates: *Statistical Tables for Biological, Agricultural and Medical Research,* published by Longman Group Ltd., London (previously published by Oliver & Boyd, Ltd. Edinburgh) and by permission of the authors and publishers.

TABLE 7 Critical Values of Pearson Product-Moment Correlation, *r*

For a right-tail test, use a positive *r* value:

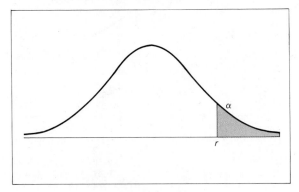

For a left-tail test, use a negative *r* value:

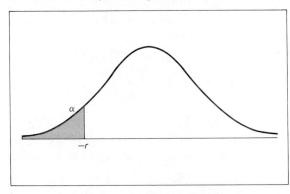

For a two-tail test, use a positive *r* value and negative *r* value:

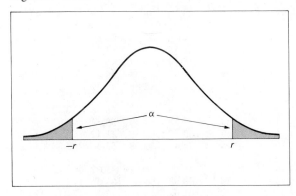

	$\alpha = 0.01$		$\alpha = 0.05$	
n	*one tail*	*two tails*	*one tail*	*two tails*
3	1.00	1.00	.99	1.00
4	.98	.99	.90	.95
5	.93	.96	.81	.88
6	.88	.92	.73	.81
7	.83	.87	.67	.75
8	.79	.83	.62	.71
9	.75	.80	.58	.67
10	.72	.76	.54	.63
11	.69	.73	.52	.60
12	.66	.71	.50	.58
13	.63	.68	.48	.53
14	.61	.66	.46	.53
15	.59	.64	.44	.51
16	.57	.61	.42	.50
17	.56	.61	.41	.48
18	.54	.59	.40	.47
19	.53	.58	.39	.46
20	.52	.56	.38	.44
21	.50	.55	.37	.43
22	.49	.54	.36	.42
23	.48	.53	.35	.41
24	.47	.52	.34	.40
25	.46	.51	.34	.40
26	.45	.50	.33	.39
27	.45	.49	.32	.38
28	.44	.48	.32	.37
29	.43	.47	.31	.37
30	.42	.46	.31	.36

TABLE 8   The χ^2 Distribution

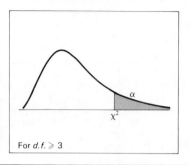

For $d.f. \geqslant 3$

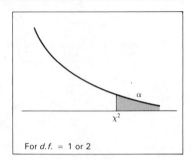

For $d.f. = 1$ or 2

$d.f.\backslash^\alpha$	.995	.990	.975	.950	.900	.100	.050	.025	.010	.005
1	0.0^4393	0.0^3157	0.0^3982	0.0^2393	0.0158	2.71	3.84	5.02	6.63	7.88
2	0.0100	0.0201	0.0506	0.103	0.211	4.61	5.99	7.38	9.21	10.60
3	0.072	0.115	0.216	0.352	0.584	6.25	7.81	9.35	11.34	12.84
4	0.207	0.297	0.484	0.711	1.064	7.78	9.49	11.14	13.28	14.86
5	0.412	0.554	0.831	1.145	1.61	9.24	11.07	12.83	15.09	16.75
6	0.676	0.872	1.24	1.64	2.20	10.64	12.59	14.45	16.81	18.55
7	0.989	1.24	1.69	2.17	2.83	12.02	14.07	16.01	18.48	20.28
8	1.34	1.65	2.18	2.73	3.49	13.36	15.51	17.53	20.09	21.96
9	1.73	2.09	2.70	3.33	4.17	14.68	16.92	19.02	21.67	23.59
10	2.16	2.56	3.25	3.94	4.87	15.99	18.31	20.48	23.21	25.19
11	2.60	3.05	3.82	4.57	5.58	17.28	19.68	21.92	24.72	26.76
12	3.07	3.57	4.40	5.23	6.30	18.55	21.03	23.34	26.22	28.30
13	3.57	4.11	5.01	5.89	7.04	19.81	22.36	24.74	27.69	29.82
14	4.07	4.66	5.63	6.57	7.79	21.06	23.68	26.12	29.14	31.32
15	4.60	5.23	6.26	7.26	8.55	22.31	25.00	27.49	30.58	32.80
16	5.14	5.81	6.91	7.96	9.31	23.54	26.30	28.85	32.00	34.27
17	5.70	6.41	7.56	8.67	10.09	24.77	27.59	30.19	33.41	35.72
18	6.26	7.01	8.23	9.39	10.86	25.99	28.87	31.53	34.81	37.16
19	6.84	7.63	8.91	10.12	11.65	27.20	30.14	32.85	36.19	38.58
20	7.43	8.26	8.59	10.85	12.44	28.41	31.41	34.17	37.57	40.00
21	8.03	8.90	10.28	11.59	13.24	29.62	32.67	35.48	38.93	41.40
22	8.64	9.54	10.98	12.34	14.04	30.81	33.92	36.78	40.29	42.80
23	9.26	10.20	11.69	13.09	14.85	32.01	35.17	38.08	41.64	44.18
24	9.89	10.86	12.40	13.85	15.66	33.20	36.42	39.36	42.98	45.56
25	10.52	11.52	13.12	14.61	16.47	34.38	37.65	40.65	44.31	46.93
26	11.16	12.20	13.84	15.38	17.29	35.56	38.89	41.92	45.64	48.29
27	11.81	12.88	14.57	16.15	18.11	36.74	40.11	43.19	46.96	49.64
28	12.46	13.56	15.31	16.93	18.94	37.92	41.34	44.46	48.28	50.99
29	13.21	14.26	16.05	17.71	19.77	39.09	42.56	45.72	49.59	52.34
30	13.79	14.95	16.79	18.49	20.60	40.26	43.77	46.98	50.89	53.67
40	20.71	22.16	24.43	26.51	29.05	51.80	55.76	59.34	63.69	66.77
50	27.99	29.71	32.36	34.76	37.69	63.17	67.50	71.42	76.15	79.49
60	35.53	37.48	40.48	43.19	46.46	74.40	79.08	83.30	88.38	91.95
70	43.28	45.44	48.76	51.74	55.33	85.53	90.53	95.02	100.4	104.2
80	51.17	53.54	57.15	60.39	64.28	96.58	101.9	106.6	112.3	116.3
90	59.20	61.75	65.65	69.13	73.29	107.6	113.1	118.1	124.1	128.3
100	67.33	70.06	74.22	77.93	82.36	118.5	124.3	129.6	135.8	140.2

TABLE 9 The F Distribution 5% (Roman Type) and 1% (Boldface Type) Points for the Distribution of F

$\alpha = 0.05$ — F (in Roman Type)

$\alpha = 0.01$ — F (in Boldface Type)

Degrees of Freedom for Numerator

Degrees of Freedom for Denominator	1	2	3	4	5	6	7	8	9	10	11	12	14	16	20	24	30	40	50	75	100	200	500	∞
1	161	200	216	225	230	234	237	239	241	242	243	244	245	246	248	249	250	251	252	253	253	254	254	254
	4052	**4999**	**5403**	**5625**	**5764**	**5859**	**5928**	**5981**	**6022**	**6056**	**6082**	**6106**	**6142**	**6169**	**6208**	**6234**	**6258**	**6286**	**6302**	**6323**	**6334**	**6352**	**6361**	**6366**
2	18.51	19.00	19.16	19.25	19.30	19.33	19.36	19.37	19.38	19.39	19.40	19.41	19.42	19.43	19.44	19.45	19.46	19.47	19.47	19.48	19.49	19.49	19.50	19.50
	98.49	**99.01**	**99.17**	**99.25**	**99.30**	**99.33**	**99.34**	**99.36**	**99.38**	**99.40**	**99.41**	**99.42**	**99.43**	**99.44**	**99.45**	**99.46**	**99.47**	**99.48**	**99.48**	**99.49**	**99.49**	**99.49**	**99.50**	**99.50**
3	10.13	9.55	9.28	9.12	9.01	8.94	8.88	8.84	8.81	8.78	8.76	8.74	8.71	8.69	8.66	8.64	8.62	8.60	8.58	8.57	8.56	8.54	8.54	8.53
	34.12	**30.81**	**29.46**	**28.71**	**28.24**	**27.91**	**27.67**	**27.49**	**27.34**	**27.23**	**27.13**	**27.05**	**26.92**	**26.83**	**26.69**	**26.60**	**26.50**	**26.41**	**26.30**	**26.27**	**26.23**	**26.18**	**26.14**	**26.12**
4	7.71	6.94	6.59	6.39	6.26	6.16	6.09	6.04	6.00	5.96	5.93	5.91	5.87	5.84	5.80	5.77	5.74	5.71	5.70	5.68	5.66	5.65	5.64	5.63
	21.20	**18.00**	**16.69**	**15.98**	**15.52**	**15.21**	**14.98**	**14.80**	**14.66**	**14.54**	**14.45**	**14.37**	**14.24**	**14.15**	**14.02**	**13.93**	**13.83**	**13.74**	**13.69**	**13.61**	**13.57**	**13.52**	**13.48**	**13.46**
5	6.61	5.79	5.41	5.19	5.05	4.95	4.88	4.82	4.78	4.74	4.70	4.68	4.64	4.60	4.56	4.53	4.50	4.46	4.44	4.42	4.40	4.38	4.37	4.36
	16.26	**13.27**	**12.06**	**11.39**	**10.97**	**10.67**	**10.45**	**10.27**	**10.15**	**10.05**	**9.96**	**9.89**	**9.77**	**9.68**	**9.55**	**9.47**	**9.38**	**9.29**	**9.24**	**9.17**	**9.13**	**9.07**	**9.04**	**9.02**
6	5.99	5.14	4.76	4.53	4.39	4.28	4.21	4.15	4.10	4.06	4.03	4.00	3.96	3.92	3.87	3.84	3.81	3.77	3.75	3.72	3.71	3.69	3.68	3.67
	13.74	**10.92**	**9.78**	**9.15**	**8.75**	**8.47**	**8.26**	**8.10**	**7.98**	**7.87**	**7.79**	**7.72**	**7.60**	**7.52**	**7.39**	**7.31**	**7.23**	**7.14**	**7.09**	**7.02**	**6.99**	**6.94**	**6.90**	**6.88**
7	5.59	4.74	4.35	4.12	3.97	3.87	3.79	3.73	3.68	3.63	3.60	3.57	3.52	3.49	3.44	3.41	3.38	3.34	3.32	3.29	3.28	3.25	3.24	3.23
	12.25	**9.55**	**8.45**	**7.85**	**7.46**	**7.19**	**7.00**	**6.84**	**6.71**	**6.62**	**6.54**	**6.47**	**6.35**	**6.27**	**6.15**	**6.07**	**5.98**	**5.90**	**5.85**	**5.78**	**5.75**	**5.70**	**5.67**	**5.65**
8	5.32	4.46	4.07	3.84	3.69	3.58	3.50	3.44	3.39	3.34	3.31	3.28	3.23	3.20	3.15	3.12	3.08	3.05	3.03	3.00	2.98	2.96	2.94	2.93
	11.26	**8.65**	**7.59**	**7.01**	**6.63**	**6.37**	**6.19**	**6.03**	**5.91**	**5.82**	**5.74**	**5.67**	**5.56**	**5.48**	**5.36**	**5.28**	**5.20**	**5.11**	**5.06**	**5.00**	**4.96**	**4.91**	**4.88**	**4.86**
9	5.12	4.26	3.86	3.63	3.48	3.37	3.29	3.23	3.18	3.13	3.10	3.07	3.02	2.98	2.93	2.90	2.86	2.82	2.80	2.77	2.76	2.73	2.72	2.71
	10.56	**8.02**	**6.99**	**6.42**	**6.06**	**5.80**	**5.62**	**5.47**	**5.35**	**5.26**	**5.18**	**5.11**	**5.00**	**4.92**	**4.80**	**4.73**	**4.64**	**4.56**	**4.51**	**4.45**	**4.41**	**4.36**	**4.33**	**4.31**
10	4.96	4.10	3.71	3.48	3.33	3.22	3.14	3.07	3.02	2.97	2.94	2.91	2.86	2.82	2.77	2.74	2.70	2.67	2.64	2.61	2.59	2.56	2.55	2.54
	10.04	**7.56**	**6.55**	**5.99**	**5.64**	**5.39**	**5.21**	**5.06**	**4.95**	**4.85**	**4.78**	**4.71**	**4.60**	**4.52**	**4.41**	**4.33**	**4.25**	**4.17**	**4.12**	**4.05**	**4.01**	**3.96**	**3.93**	**3.91**
11	4.84	3.98	3.59	3.36	3.20	3.09	3.01	2.95	2.90	2.86	2.82	2.79	2.74	2.70	2.65	2.61	2.57	2.53	2.50	2.47	2.45	2.42	2.41	2.40
	9.65	**7.20**	**6.22**	**5.67**	**5.32**	**5.07**	**4.88**	**4.74**	**4.63**	**4.54**	**4.46**	**4.40**	**4.29**	**4.21**	**4.10**	**4.02**	**3.94**	**3.86**	**3.80**	**3.74**	**3.70**	**3.66**	**3.62**	**3.60**
12	4.75	3.88	3.49	3.26	3.11	3.00	2.92	2.85	2.80	2.76	2.72	2.69	2.64	2.60	2.54	2.50	2.46	2.42	2.40	2.36	2.35	2.32	2.31	2.30
	9.33	**6.93**	**5.95**	**5.41**	**5.06**	**4.82**	**4.65**	**4.50**	**4.39**	**4.30**	**4.22**	**4.16**	**4.05**	**3.98**	**3.86**	**3.78**	**3.70**	**3.61**	**3.56**	**3.49**	**3.46**	**3.41**	**3.38**	**3.36**
13	4.67	3.80	3.41	3.18	3.02	2.92	2.84	2.77	2.72	2.67	2.63	2.60	2.55	2.51	2.46	2.42	2.38	2.34	2.32	2.28	2.26	2.24	2.22	2.21
	9.07	**6.70**	**5.74**	**5.20**	**4.86**	**4.62**	**4.44**	**4.30**	**4.19**	**4.10**	**4.02**	**3.96**	**3.85**	**3.78**	**3.67**	**3.59**	**3.51**	**3.42**	**3.37**	**3.30**	**3.27**	**3.21**	**3.18**	**3.16**
14	4.60	3.74	3.34	3.11	2.96	2.85	2.77	2.70	2.65	2.60	2.56	2.53	2.48	2.44	2.39	2.35	2.31	2.27	2.24	2.21	2.19	2.16	2.14	2.13
	8.86	**6.51**	**5.56**	**5.03**	**4.69**	**4.46**	**4.28**	**4.14**	**4.03**	**3.94**	**3.86**	**3.80**	**3.70**	**3.62**	**3.51**	**3.43**	**3.34**	**3.26**	**3.21**	**3.14**	**3.11**	**3.06**	**3.02**	**3.00**

*From *Biometrika Tables for Statisticians*, Vol. 1, by permission of the Biometrika Trustees.

TABLE 9 *continued:* α = 0.05 (Roman Type); α = 0.01 (Boldface Type)

Values shown as: α = 0.05 value / α = 0.01 value

Degrees of Freedom for Numerator

Degrees of Freedom for Denominator	1	2	3	4	5	6	7	8	9	10	11	12	14	16	20	24	30	40	50	75	100	200	500	∞
15	4.54/8.68	3.68/6.36	3.29/5.42	3.06/4.89	2.90/4.56	2.79/4.32	2.70/4.14	2.64/4.00	2.59/3.89	2.55/3.80	2.51/3.73	2.48/3.67	2.43/3.56	2.39/3.48	2.33/3.36	2.29/3.29	2.25/3.20	2.21/3.12	2.18/3.07	2.15/3.00	2.12/2.97	2.10/2.92	2.08/2.89	2.07/2.87
16	4.49/8.53	3.63/6.23	3.24/5.29	3.01/4.77	2.85/4.44	2.74/4.20	2.66/4.03	2.59/3.89	2.54/3.78	2.49/3.69	2.45/3.61	2.42/3.55	2.37/3.45	2.33/3.37	2.28/3.25	2.24/3.18	2.20/3.10	2.16/3.01	2.13/2.96	2.09/2.89	2.07/2.86	2.04/2.80	2.02/2.77	2.01/2.75
17	4.45/8.40	3.59/6.11	3.20/5.18	2.96/4.67	2.81/4.34	2.70/4.10	2.62/3.93	2.55/3.79	2.50/3.68	2.45/3.59	2.41/3.52	2.38/3.45	2.33/3.35	2.29/3.27	2.23/3.16	2.19/3.08	2.15/3.00	2.11/2.92	2.08/2.86	2.04/2.79	2.02/2.76	1.99/2.70	1.97/2.67	1.96/2.65
18	4.41/8.28	3.55/6.01	3.16/5.09	2.93/4.58	2.77/4.25	2.66/4.01	2.58/3.85	2.51/3.71	2.46/3.60	2.41/3.51	2.37/3.44	2.34/3.37	2.29/3.27	2.25/3.19	2.19/3.07	2.15/3.00	2.11/2.91	2.07/2.83	2.04/2.78	2.00/2.71	1.98/2.68	1.95/2.62	1.93/2.59	1.92/2.57
19	4.38/8.18	3.52/5.93	3.13/5.01	2.90/4.50	2.74/4.17	2.63/3.94	2.55/3.77	2.48/3.63	2.43/3.52	2.38/3.43	2.34/3.36	2.31/3.30	2.26/3.19	2.21/3.12	2.15/3.00	2.11/2.92	2.07/2.84	2.02/2.76	2.00/2.70	1.96/2.63	1.94/2.60	1.91/2.54	1.90/2.51	1.88/2.49
20	4.35/8.10	3.49/5.85	3.10/4.94	2.87/4.43	2.71/4.10	2.60/3.87	2.52/3.71	2.45/3.56	2.40/3.45	2.35/3.37	2.31/3.30	2.28/3.23	2.23/3.13	2.18/3.05	2.12/2.94	2.08/2.86	2.04/2.77	1.99/2.69	1.96/2.63	1.92/2.56	1.90/2.53	1.87/2.47	1.85/2.44	1.84/2.42
21	4.32/8.02	3.47/5.78	3.07/4.87	2.84/4.37	2.68/4.04	2.57/3.81	2.49/3.65	2.42/3.51	2.37/3.40	2.32/3.31	2.28/3.24	2.25/3.17	2.20/3.07	2.15/2.99	2.09/2.88	2.05/2.80	2.00/2.72	1.96/2.63	1.93/2.58	1.89/2.51	1.87/2.47	1.84/2.42	1.82/2.38	1.81/2.36
22	4.30/7.94	3.44/5.72	3.05/4.82	2.82/4.31	2.66/3.99	2.55/3.76	2.47/3.59	2.40/3.45	2.35/3.35	2.30/3.26	2.26/3.18	2.23/3.12	2.18/3.02	2.13/2.94	2.07/2.83	2.03/2.75	1.98/2.67	1.93/2.58	1.91/2.53	1.87/2.46	1.84/2.42	1.81/2.37	1.80/2.33	1.78/2.31
23	4.28/7.88	3.42/5.66	3.03/4.76	2.80/4.26	2.64/3.94	2.53/3.71	2.45/3.54	2.38/3.41	2.32/3.30	2.28/3.21	2.24/3.14	2.20/3.07	2.14/2.97	2.10/2.89	2.04/2.78	2.00/2.70	1.96/2.62	1.91/2.53	1.88/2.48	1.84/2.41	1.82/2.37	1.79/2.32	1.77/2.28	1.76/2.26
24	4.26/7.82	3.40/5.61	3.01/4.72	2.78/4.22	2.62/3.90	2.51/3.67	2.43/3.50	2.36/3.36	2.30/3.25	2.26/3.17	2.22/3.09	2.18/3.03	2.13/2.93	2.09/2.85	2.02/2.74	1.98/2.66	1.94/2.58	1.89/2.49	1.86/2.44	1.82/2.36	1.80/2.33	1.76/2.27	1.74/2.23	1.73/2.21
25	4.24/7.77	3.38/5.57	2.99/4.68	2.76/4.18	2.60/3.86	2.49/3.63	2.41/3.46	2.34/3.32	2.28/3.21	2.24/3.13	2.20/3.05	2.16/2.99	2.11/2.89	2.06/2.81	2.00/2.70	1.96/2.62	1.92/2.54	1.87/2.45	1.84/2.40	1.80/2.32	1.77/2.29	1.74/2.23	1.72/2.19	1.71/2.17
26	4.22/7.72	3.37/5.53	2.98/4.64	2.74/4.14	2.59/3.82	2.47/3.59	2.39/3.42	2.32/3.29	2.27/3.17	2.22/3.09	2.18/3.02	2.15/2.96	2.10/2.86	2.05/2.77	1.99/2.66	1.95/2.58	1.90/2.50	1.85/2.41	1.82/2.36	1.78/2.28	1.76/2.25	1.72/2.19	1.70/2.15	1.69/2.13
27	4.21/7.68	3.35/5.49	2.96/4.60	2.73/4.11	2.57/3.79	2.46/3.56	2.37/3.39	2.30/3.26	2.25/3.14	2.20/3.06	2.16/2.98	2.13/2.93	2.08/2.83	2.03/2.74	1.97/2.63	1.93/2.55	1.88/2.47	1.84/2.38	1.80/2.33	1.76/2.25	1.74/2.21	1.71/2.16	1.68/2.12	1.67/2.10
28	4.20/7.64	3.34/5.45	2.95/4.57	2.71/4.07	2.56/3.76	2.44/3.53	2.36/3.36	2.29/3.23	2.24/3.11	2.19/3.03	2.15/2.95	2.12/2.90	2.06/2.80	2.02/2.71	1.96/2.60	1.91/2.52	1.87/2.44	1.81/2.35	1.78/2.30	1.75/2.22	1.72/2.18	1.69/2.13	1.67/2.09	1.65/2.06
29	4.18/7.60	3.33/5.42	2.93/4.54	2.70/4.04	2.54/3.73	2.43/3.50	2.35/3.33	2.28/3.20	2.22/3.08	2.18/3.00	2.14/2.92	2.10/2.87	2.05/2.77	2.00/2.68	1.94/2.57	1.90/2.49	1.85/2.41	1.80/2.32	1.77/2.27	1.73/2.19	1.71/2.15	1.68/2.10	1.65/2.06	1.64/2.03
30	4.17/7.56	3.32/5.39	2.92/4.51	2.69/4.02	2.53/3.70	2.42/3.47	2.34/3.30	2.27/3.17	2.21/3.06	2.16/2.98	2.12/2.90	2.09/2.84	2.04/2.74	1.99/2.66	1.93/2.55	1.89/2.47	1.84/2.38	1.79/2.29	1.76/2.24	1.72/2.16	1.69/2.13	1.66/2.07	1.64/2.03	1.62/2.01
32	4.15/7.50	3.30/5.34	2.90/4.46	2.67/3.97	2.51/3.66	2.40/3.42	2.32/3.25	2.25/3.12	2.19/3.01	2.14/2.94	2.10/2.86	2.07/2.80	2.02/2.70	1.97/2.62	1.91/2.51	1.86/2.42	1.82/2.34	1.76/2.25	1.74/2.20	1.69/2.12	1.67/2.08	1.64/2.02	1.61/1.98	1.59/1.96
34	4.13/7.44	3.28/5.29	2.88/4.42	2.65/3.93	2.49/3.61	2.38/3.38	2.30/3.21	2.23/3.08	2.17/2.97	2.12/2.89	2.08/2.82	2.05/2.76	2.00/2.66	1.95/2.58	1.89/2.47	1.84/2.38	1.80/2.30	1.74/2.21	1.71/2.15	1.67/2.08	1.64/2.04	1.61/1.98	1.59/1.94	1.57/1.91

df																								
36	1.55 / 1.87	1.56 / 1.90	1.59 / 1.94	1.62 / 2.00	1.65 / 2.04	1.69 / 2.12	1.72 / 2.17	1.78 / 2.26	1.82 / 2.35	1.87 / 2.43	1.93 / 2.54	1.98 / 2.62	2.03 / 2.72	2.06 / 2.78	2.10 / 2.86	2.15 / 2.94	2.21 / 3.04	2.28 / 3.18	2.36 / 3.35	2.48 / 3.58	2.63 / 3.89	2.86 / 4.38	3.26 / 5.25	4.11 / 7.39
38	1.53 / 1.84	1.54 / 1.86	1.57 / 1.90	1.60 / 1.97	1.63 / 2.00	1.67 / 2.08	1.71 / 2.14	1.76 / 2.22	1.80 / 2.32	1.85 / 2.40	1.92 / 2.51	1.96 / 2.59	2.02 / 2.69	2.05 / 2.75	2.09 / 2.82	2.14 / 2.91	2.19 / 3.02	2.26 / 3.15	2.35 / 3.32	2.46 / 3.54	2.62 / 3.86	2.85 / 4.34	3.25 / 5.21	4.10 / 7.35
40	1.51 / 1.81	1.53 / 1.84	1.55 / 1.88	1.59 / 1.94	1.61 / 1.97	1.66 / 2.05	1.69 / 2.11	1.74 / 2.20	1.79 / 2.29	1.84 / 2.37	1.90 / 2.49	1.95 / 2.56	2.00 / 2.66	2.04 / 2.73	2.07 / 2.80	2.12 / 2.88	2.18 / 2.99	2.25 / 3.12	2.34 / 3.29	2.45 / 3.51	2.61 / 3.83	2.84 / 4.31	3.23 / 5.18	4.08 / 7.31
42	1.49 / 1.78	1.51 / 1.80	1.54 / 1.85	1.57 / 1.91	1.60 / 1.94	1.64 / 2.02	1.68 / 2.08	1.73 / 2.17	1.78 / 2.26	1.82 / 2.35	1.89 / 2.46	1.94 / 2.54	1.99 / 2.64	2.02 / 2.70	2.06 / 2.77	2.11 / 2.86	2.17 / 2.96	2.24 / 3.10	2.32 / 3.26	2.44 / 3.49	2.59 / 3.80	2.83 / 4.29	3.22 / 5.15	4.07 / 7.27
44	1.48 / 1.75	1.50 / 1.78	1.52 / 1.82	1.56 / 1.88	1.58 / 1.92	1.63 / 2.00	1.66 / 2.06	1.72 / 2.15	1.76 / 2.24	1.81 / 2.32	1.88 / 2.44	1.92 / 2.52	1.98 / 2.62	2.01 / 2.68	2.05 / 2.75	2.10 / 2.84	2.16 / 2.94	2.23 / 3.07	2.31 / 3.24	2.43 / 3.46	2.58 / 3.78	2.82 / 4.26	3.21 / 5.12	4.06 / 7.24
46	1.46 / 1.72	1.48 / 1.76	1.51 / 1.80	1.54 / 1.86	1.57 / 1.90	1.62 / 1.98	1.65 / 2.04	1.71 / 2.13	1.75 / 2.22	1.80 / 2.30	1.87 / 2.42	1.91 / 2.50	1.97 / 2.60	2.00 / 2.66	2.04 / 2.73	2.09 / 2.82	2.14 / 2.92	2.22 / 3.05	2.30 / 3.22	2.42 / 3.44	2.57 / 3.76	2.81 / 4.24	3.20 / 5.10	4.05 / 7.21
48	1.45 / 1.70	1.47 / 1.73	1.50 / 1.78	1.53 / 1.84	1.56 / 1.88	1.61 / 1.96	1.64 / 2.02	1.70 / 2.11	1.74 / 2.20	1.79 / 2.28	1.86 / 2.40	1.90 / 2.48	1.96 / 2.58	1.99 / 2.64	2.03 / 2.71	2.08 / 2.80	2.14 / 2.90	2.21 / 3.04	2.30 / 3.20	2.41 / 3.42	2.56 / 3.74	2.80 / 4.22	3.19 / 5.08	4.04 / 7.19
50	1.44 / 1.68	1.46 / 1.71	1.48 / 1.76	1.52 / 1.82	1.55 / 1.86	1.60 / 1.94	1.63 / 2.00	1.69 / 2.10	1.74 / 2.18	1.78 / 2.26	1.85 / 2.39	1.90 / 2.46	1.95 / 2.56	1.98 / 2.62	2.02 / 2.70	2.07 / 2.78	2.13 / 2.88	2.20 / 3.02	2.29 / 3.18	2.40 / 3.41	2.56 / 3.72	2.79 / 4.20	3.18 / 5.06	4.03 / 7.17
55	1.41 / 1.64	1.43 / 1.66	1.46 / 1.71	1.50 / 1.78	1.52 / 1.82	1.58 / 1.90	1.61 / 1.96	1.67 / 2.06	1.72 / 2.15	1.76 / 2.23	1.83 / 2.35	1.88 / 2.43	1.93 / 2.53	1.97 / 2.59	2.00 / 2.66	2.05 / 2.75	2.11 / 2.85	2.18 / 2.98	2.27 / 3.15	2.38 / 3.37	2.54 / 3.68	2.78 / 4.16	3.17 / 5.01	4.02 / 7.12
60	1.39 / 1.60	1.41 / 1.63	1.44 / 1.68	1.48 / 1.74	1.50 / 1.79	1.56 / 1.87	1.59 / 1.93	1.65 / 2.03	1.70 / 2.12	1.75 / 2.20	1.81 / 2.32	1.86 / 2.40	1.92 / 2.50	1.95 / 2.56	1.99 / 2.63	2.04 / 2.72	2.10 / 2.82	2.17 / 2.95	2.25 / 3.12	2.37 / 3.34	2.52 / 3.65	2.76 / 4.13	3.15 / 4.98	4.00 / 7.08
65	1.37 / 1.56	1.39 / 1.60	1.42 / 1.64	1.46 / 1.71	1.49 / 1.76	1.54 / 1.84	1.57 / 1.90	1.63 / 2.00	1.68 / 2.09	1.73 / 2.18	1.80 / 2.30	1.85 / 2.37	1.90 / 2.47	1.94 / 2.54	1.98 / 2.61	2.02 / 2.70	2.08 / 2.79	2.15 / 2.93	2.24 / 3.09	2.36 / 3.31	2.51 / 3.62	2.75 / 4.10	3.14 / 4.95	3.99 / 7.04
70	1.35 / 1.53	1.37 / 1.56	1.40 / 1.63	1.45 / 1.69	1.47 / 1.74	1.53 / 1.82	1.56 / 1.88	1.62 / 1.98	1.67 / 2.07	1.72 / 2.15	1.79 / 2.28	1.84 / 2.35	1.89 / 2.45	1.93 / 2.51	1.97 / 2.59	2.01 / 2.67	2.07 / 2.77	2.14 / 2.91	2.23 / 3.07	2.35 / 3.29	2.50 / 3.60	2.74 / 4.08	3.13 / 4.92	3.98 / 7.01
80	1.32 / 1.49	1.35 / 1.52	1.38 / 1.57	1.42 / 1.65	1.45 / 1.70	1.51 / 1.78	1.54 / 1.84	1.60 / 1.94	1.65 / 2.03	1.70 / 2.11	1.77 / 2.24	1.82 / 2.32	1.88 / 2.41	1.91 / 2.48	1.95 / 2.55	1.99 / 2.64	2.05 / 2.74	2.12 / 2.87	2.21 / 3.04	2.33 / 3.25	2.48 / 3.56	2.72 / 4.04	3.11 / 4.88	3.96 / 6.96
100	1.28 / 1.43	1.30 / 1.46	1.34 / 1.51	1.39 / 1.59	1.42 / 1.64	1.48 / 1.73	1.51 / 1.79	1.57 / 1.89	1.63 / 1.98	1.68 / 2.06	1.75 / 2.19	1.79 / 2.26	1.85 / 2.36	1.88 / 2.43	1.92 / 2.51	1.97 / 2.59	2.03 / 2.69	2.10 / 2.82	2.19 / 2.99	2.30 / 3.20	2.46 / 3.51	2.70 / 3.98	3.09 / 4.82	3.94 / 6.90
125	1.25 / 1.37	1.27 / 1.40	1.31 / 1.46	1.36 / 1.54	1.39 / 1.59	1.45 / 1.68	1.49 / 1.75	1.55 / 1.85	1.60 / 1.94	1.65 / 2.03	1.72 / 2.15	1.77 / 2.23	1.83 / 2.33	1.86 / 2.40	1.90 / 2.47	1.95 / 2.56	2.01 / 2.65	2.08 / 2.79	2.17 / 2.95	2.29 / 3.17	2.44 / 3.47	2.68 / 3.94	3.07 / 4.78	3.92 / 6.84
150	1.22 / 1.33	1.25 / 1.37	1.29 / 1.43	1.34 / 1.51	1.37 / 1.56	1.44 / 1.66	1.47 / 1.72	1.54 / 1.83	1.59 / 1.91	1.64 / 2.00	1.71 / 2.12	1.76 / 2.20	1.82 / 2.30	1.85 / 2.37	1.89 / 2.44	1.94 / 2.53	2.00 / 2.62	2.07 / 2.76	2.16 / 2.92	2.27 / 3.13	2.43 / 3.44	2.67 / 3.91	3.06 / 4.75	3.91 / 6.81
200	1.19 / 1.28	1.22 / 1.33	1.26 / 1.39	1.32 / 1.48	1.35 / 1.53	1.42 / 1.62	1.45 / 1.69	1.52 / 1.79	1.57 / 1.88	1.62 / 1.97	1.69 / 2.09	1.74 / 2.17	1.80 / 2.28	1.83 / 2.34	1.87 / 2.41	1.92 / 2.50	1.98 / 2.60	2.05 / 2.73	2.14 / 2.90	2.26 / 3.11	2.41 / 3.41	2.65 / 3.88	3.04 / 4.71	3.89 / 6.76
400	1.13 / 1.19	1.16 / 1.24	1.22 / 1.32	1.28 / 1.42	1.32 / 1.47	1.38 / 1.57	1.42 / 1.64	1.49 / 1.74	1.54 / 1.84	1.60 / 1.92	1.67 / 2.04	1.72 / 2.12	1.78 / 2.23	1.81 / 2.29	1.85 / 2.37	1.90 / 2.46	1.96 / 2.55	2.03 / 2.69	2.12 / 2.85	2.23 / 3.06	2.39 / 3.36	2.62 / 3.83	3.02 / 4.66	3.86 / 6.70
1000	1.08 / 1.11	1.13 / 1.19	1.19 / 1.28	1.26 / 1.38	1.30 / 1.44	1.36 / 1.54	1.41 / 1.61	1.47 / 1.71	1.53 / 1.81	1.58 / 1.89	1.65 / 2.01	1.70 / 2.09	1.76 / 2.20	1.80 / 2.26	1.84 / 2.34	1.89 / 2.43	1.95 / 2.53	2.02 / 2.66	2.10 / 2.82	2.22 / 3.04	2.38 / 3.34	2.61 / 3.80	3.00 / 4.62	3.85 / 6.66
	1.00 / 1.00	1.11 / 1.15	1.17 / 1.25	1.24 / 1.36	1.28 / 1.41	1.35 / 1.52	1.40 / 1.59	1.46 / 1.69	1.52 / 1.79	1.57 / 1.87	1.64 / 1.99	1.69 / 2.07	1.75 / 2.18	1.79 / 2.24	1.83 / 2.32	1.88 / 2.41	1.94 / 2.51	2.01 / 2.64	2.09 / 2.80	2.21 / 3.02	2.37 / 3.32	2.60 / 3.78	2.99 / 4.60	3.84 / 6.64

TABLE 10 Critical Values for Spearman Rank Correlation, r_s

For a right- (left-) tail test, use the positive (negative) critical value found in the table under significance level for a one-tailed test. For a two-tailed test, use both the positive and negative of the critical value found in the table under significance level for a two-tailed test. n = number of pairs.

	Significance level for a one-tailed test at			
	0.05	0.025	0.005	0.001
	Significance level for a two-tailed test at			
n	0.10	0.05	0.01	0.002
5	0.900	1.000		
6	0.829	0.886	1.000	
7	0.715	0.786	0.929	1.000
8	0.620	0.715	0.881	0.953
9	0.600	0.700	0.834	0.917
10	0.564	0.649	0.794	0.879
11	0.537	0.619	0.764	0.855
12	0.504	0.588	0.735	0.826
13	0.484	0.561	0.704	0.797
14	0.464	0.539	0.680	0.772
15	0.447	0.522	0.658	0.750
16	0.430	0.503	0.636	0.730
17	0.415	0.488	0.618	0.711
18	0.402	0.474	0.600	0.693
19	0.392	0.460	0.585	0.676
20	0.381	0.447	0.570	0.661
21	0.371	0.437	0.556	0.647
22	0.361	0.426	0.544	0.633
23	0.353	0.417	0.532	0.620
24	0.345	0.407	0.521	0.608
25	0.337	0.399	0.511	0.597
26	0.331	0.391	0.501	0.587
27	0.325	0.383	0.493	0.577
28	0.319	0.376	0.484	0.567
29	0.312	0.369	0.475	0.558
30	0.307	0.363	0.467	0.549

Appendix II   Answers to Odd-Numbered Problems

Chapter 1

SECTION 1.1

1. a) Student fees at all colleges and universities in the nation
   b) Student fees at a random sample of 30 colleges and universities in the nation
3. a) Time interval between arrival and clearance of all payment checks coming to the regional office
   b) Time interval between arrival and clearance of 32 payment checks coming to the regional office

SECTION 2.1

1. *See* text.
3. Because the largest number has five digits, use groups of five digits in the random number table. Proceed until you have obtained six numbers between 00001 and 17,431.
5. Because the largest number has two digits, use groups of two digits in the random number table. Proceed until you have seven numbers from 01 to 99.
7. Because the largest number has three digits, use groups of three digits in the random number table. Proceed until you have five numbers between 119 and 964.
9. Use a random number table to select four distinct numbers corresponding to students in your class.
   a) The first student walking into the classroom may make a special effort to get to class on time.
   b) Four students coming in late may have particular majors that require them to take a class that meets in a distant building during the preceding class period.
   c) The students in the back may not wish to be so visible to the professor.
   d) The tallest students might all be male.
11. Use the random number table to select ten distinct numbers between 1 and 150. Number the customers as they come in and select the ones with the numbers you chose from the random number table.
13. Number the stations listed and use a random number table to select seven distinct numbers. Pick the stations with the corresponding numbers.
15. Begin at a random location in the random number table and use single digits. Write down the first ten digits you encounter between 1 and 5. Use corresponding letters for the correct responses.
17. Latitude normally is measured as 0° to 90° north or south of the equator, and longitude is measured as 0° to 180° east or west of 0°. To simplify the problem, measure latitude starting at the South Pole from 0° to 180° and longitude from 0° to 360°. Starting at any point of the random number table, group the digits into 3's. Then find 20 pairs of numbers so that the first number in the pair is between 0 and 180 and the second is between 0 and 360. Let the first number of the pair be the latitude and the second one be the longitude of the random position change.

SECTION 2.2   1. **Cost of Processing Data**

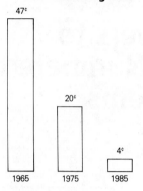

The time to process data is the feature that seems to have dropped most.

Time to Process Data

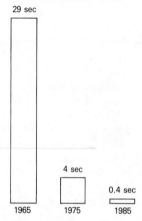

3. **Allergy Level of 1,000 People (At Least One Parent Had Known Allergies)**

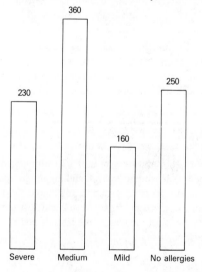

5. **Distribution of Each Dollar Contributed**

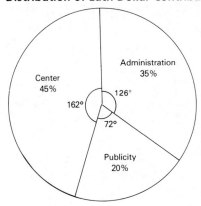

7. **Type of Injury in PE Classes**

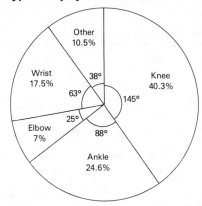

9. **Nations Earning the Most Gold Medals in the 1984 Summer Olympics**

 = 10 gold metals

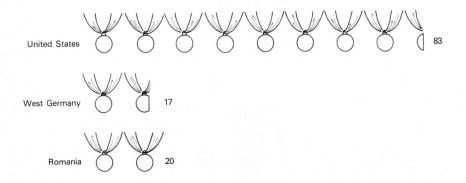

9   **Nations Earning the Most Gold Medals in the 1984 Summer Olympics, continued**

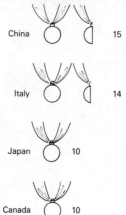

China	15
Italy	14
Japan	10
Canada	10

SECTION 2.3

1. a) Class width is 8

Number of Movies Rented per Day

Class	Frequency	Midpoint
4–11	5	7.5
12–19	4	15.5
20–27	10	23.5
28–35	8	31.5
36–43	5	39.5

b & c) **Number of Movies Rented per Day**

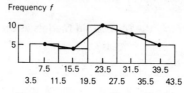

Frequency f

Number of movies

3. a) Class width is 5

Number of Fillings in Adults

Class	Frequency	Midpoint
3–7	12	5
8–12	11	10
13–17	6	15
18–22	4	20
23–27	2	25
28–32	1	30

b & c) **Number of Fillings in Adults**

Frequency *f*

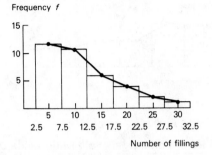

Number of fillings

5. a) Class width is 17

Number of Room Calls

Class	Frequency	Midpoint
18–34	1	26
35–51	2	43
52–68	5	60
69–85	15	77
86–102	12	94

b & c) **Number of Room Calls**

Frequency *f*

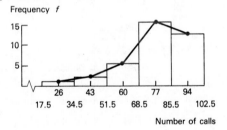

Number of calls

7. a) Class width is 8

Age of Customer

Class	Frequency	Midpoint
16–23	10	19.5
24–31	6	27.5
32–39	6	35.5
40–47	4	43.5
48–55	4	51.5

b & c) **Age of Customer**

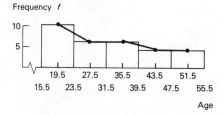

9. a) Class width is 20

Number of Hamburgers Sold Nationwide During August (units of 1,000)

Class	Frequency	Midpoint
15–34	4	24.5
35–54	6	44.5
55–74	12	64.5
75–94	7	84.5
95–114	7	104.5
115–134	3	124.5
135–154	5	144.5
155–174	3	164.5
175–194	1	184.5
195–214	2	204.5

b & c) **Number of Hamburgers Sold Nationwide During August (units of 1,000)**

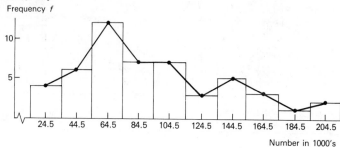

11. a) Class width is 5

Annual Rainfall at Centennial 1887–1955 (to nearest inch)

Class	Frequency	Midpoint
6–10	11	8
11–15	37	13
16–20	15	18
21–25	5	23
26–30	1	28

b & c) **Annual Rainfall at Centennial 1887–1955 (to nearest inch)**

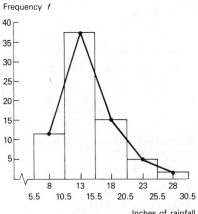

SECTION 2.4

1. **Price of Tickets from Scalpers**

unit = $1
1|6 represents $16

```
1 | 6
2 | 0  5  5  8  4  5  7
3 | 5  1  7  5  5  8  0  5  3  7  6  9  0  5  3  5
4 | 0  0  5  5  0
5 | 0
```

3. **Time to Complete Student Teacher Evaluation Form**

unit = 1 minute
0|8 represents 8 minutes

```
0 | 8
1 | 8  9  2  8  5  7  9  6  7  5  2  8  9  6  2
2 | 0  3  2  1  6  3
3 | 0  5
4 | 4
```

5. **Price of Felicita**

unit = $10
10|83 represents $10,830

```
10 | 83  70  50  23  50  87  20  65  55
11 | 82  24  93  45  03
12 | 09
```

7. **Hours Spent in Meetings**

unit = 1 hour
0*|2 represents 02 hours

```
0*  2  3
0.  9  6  7  5  7  8  9  6
1*  2  3  0  2  0  1  0
1.  5  8  7  9
2*  1  4
2.  6  5
```

9. **Number of Pills per Patient per Day in a Hospital**

unit = 1 pill
0*|4 represents 04 pills

```
0*  4  2  1  1  4  3  4  2  3  2  4  3
0.  8  5  7  9  8  9  7  6  7  5  8  7  6  9  6  8  9  8  9  8  9
1*  3  2  0  2  2  1  2  2  0  2  4  2  0  1
1.  5  5  5
```

Chapter 2 Review Problems

1. **Existing Oil Reserves in Western Hemisphere**

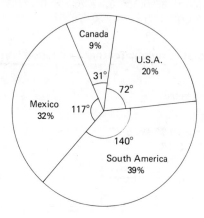

3. a) **Ages of Evening Students Taking Colorado History**

unit = 1 year
1|8 = 18 years

```
1  8  9  9  9
2  8  2  7  9  0  2  9  7  9
3  0  5  1  2  1  9  3
4  1  5  3  6
5  6  5  4  1  3  1  8  3  6
6  3  3  5
```

b) **Frequency Table: Ages of Students Taking Colorado History (5 classes)**

Class	Frequency	Class Midpoint
18–27	9	22.5
28–37	10	32.5
38–47	5	42.5
48–57	8	52.5
58–67	4	62.5

**Ages of Evening Students Taking Colorado History
(Histogram & Frequency Polygon)**

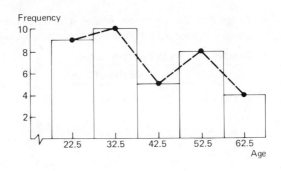

c) **Frequency Table: Ages of Students Taking Colorado History (12 classes)**

Class	Frequency	Class Midpoint
18–21	5	19.5
22–25	2	23.5
26–29	6	27.5
30–33	5	31.5
34–37	1	35.5
38–41	2	39.5
42–45	2	43.5
46–49	1	47.5
50–53	4	51.5
54–57	4	55.5
58–61	1	59.5
62–65	3	63.5

Ages of Evening Students Taking Colorado History (Histogram & Frequency Polygon)

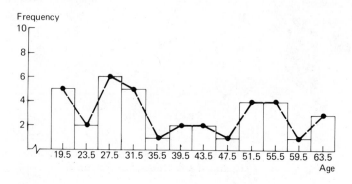

5. a) Assign each car a different number. Use the random number table with the digits grouped appropriately to find the 30 cars to be included in the sample.

b) **Sample of Cars in Parking Lot**

Make	Number of Cars
Chevrolet	8
Ford	7
Pontiac	5
Toyota	3
Chrysler	3
Oldsmobile	3
Cadillac	1

c) **Cars in Parking Lot**

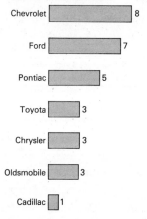

Cars in Parking Lot

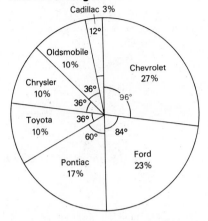

d) **Frequency Table: Model Year**

Class	Frequency	Midpoint
69–72	4	70.5
73–76	10	74.5
77–80	7	78.5
81–84	6	82.5
85–89	3	86.5

Model Year

Frequency *f*

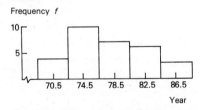

Year

7. a) **Foreign Investments in U.S. Industry**

Year	Amount (billions of dollars)
1955	13.4
1960	18.4
1965	26.3
1970	44.8
1980	160

b) **Foreign Investments in U.S. Industry (in billions of dollars)**

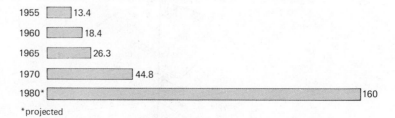

1955 ▭ 13.4
1960 ▭ 18.4
1965 ▭ 26.3
1970 ▭ 44.8
1980* ▭ 160

*projected

c) **Foreign investments in U.S. Industry (in billions of dollars)**

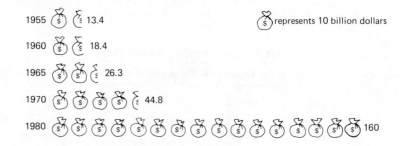

1955 13.4

1960 18.4

1965 26.3

1970 44.8

1980 160

represents 10 billion dollars

9. a) **Frequency Table: Battery Lives (in months)**

Class	Frequency	Class Midpoint
13–17	2	15
18–22	5	20
23–27	10	25
28–32	6	30
33–37	8	35
38–42	5	40
43–47	2	45
48–52	3	50
53–57	1	55

b & c) **Life of Car Batteries**

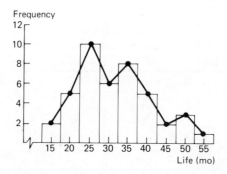

SECTION 3.1

1. Mean = 7.625 min, median = 8 min, mode = 8 min
3. Mean = 4.58, median = 3, mode = 3
5. Mean = 7.13 hr, median = 7 hr, mode = 8 hr
7. a) Mean = 28.83 thousand dollars
   b) Median = 18.5 thousand dollars. The median reflects the salary for more workers.
   c) Mean = 17.3 thousand dollars, median = 17 thousand dollars
   d) Without the salaries for the two executives the mean and median are closer, and both reflect the salary of most of the other workers more accurately. The mean changed quite a bit while the median did not, a difference which indicates that the mean is more sensitive to the absence or presence of extreme values.
9. a) Mean = 14.57, median = 15, mode = 15
   b) Mean = 24.56, median = 15, mode = 15
   c) Extreme values affect the mean far more than they do the median or mode.
11. a) Mean = 16.47, median = 17, mode = 16

SECTION 3.2

1. 82% or less, 18% or less
3. No, the score 82 might have a percentile rank less than 70.

5. a) **Ranking of the Number of Months Nurses Have Been in Current Position**

Months	Rank Up	Rank Down	Months	Rank Up	Rank Down
2	1	20	23	11	10
5	2	19	25	12	9
7	3	18	26	13	8
8	4	17	27	14	7
8	5	16	28	15	6
11	6	15	29	16	5
12	7	14	31	17	4
14	8	13	36	18	3
20	9	12	36	19	2
23	10	11	42	20	1

b) Median rank = (20 + 1)/2 = 10.5, median value = 23
Hinge rank = (10 + 1)/2 = 5.5, lower hinge value = (8 + 11)/2 = 9.5
Upper hinge value = (28 + 29)/2 = 28.5

Months in Current Position for Nurses

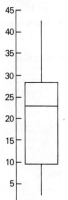

7. a) Median rank = (63 + 1)/2 = 32 and median value = 1
   Hinge rank = (31 + 1)/2 = 16, lower hinge value = 1, the lowest extreme value
   is also 1. The three values are all the same.
   b) Upper hinge value = 3, upper extreme value = 7
   c) **Number of Consecutive Days for Increasing Dow Jones Industrial Average
   (July 1, 1984–July 1, 1985)**

   Consecutive days
   of increase

   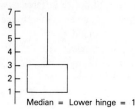

9. a) **Ranking of Number of No-Shows Using Discount Air Fare**

Number	Rank Up	Rank Down	Number	Rank Up	Rank Down	Number	Rank Up	Rank Down
0	1	40	6	15	26	9	29	12
0	2	39	6	16	25	9	30	11
1	3	38	6	17	24	10	31	10
2	4	37	7	18	23	10	32	9
2	5	36	7	19	22	10	33	8
3	6	35	7	20	21	10	34	7
3	7	34	7	21	20	12	35	6
4	8	33	7	22	19	12	36	5
4	9	32	8	23	18	15	37	4
5	10	31	8	24	17	16	38	3
5	11	30	8	25	16	18	39	2
5	12	29	9	26	15	21	40	1
5	13	28	9	27	14			
5	14	27	9	28	13			

   b) Median rank = (40 + 1)/2 = 20.5, median value = (7 + 7)/2 = 7
   Hinge rank = (20 + 1)/2 = 11.5, lower hinge value = (5 + 5)/2 = 5
   Upper hinge value = (9 + 9)/2 = 9

   Number of No-Shows Using Airline Discount Fares

   No shows

   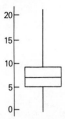

11. a) **Ranking of Age of VCR Purchaser**

Age	Rank Up	Down	Age	Rank Up	Down	Age	Rank Up	Down
17	1	50	32	18	33	41	35	16
19	2	49	32	19	32	42	36	15
21	3	48	32	20	31	42	37	14
22	4	47	32	21	30	43	38	13
23	5	46	33	22	29	43	39	12
24	6	45	33	23	28	45	40	11
25	7	44	34	24	27	45	41	10
26	8	43	35	25	26	46	42	9
27	9	42	35	26	25	47	43	8
27	10	41	36	27	24	47	44	7
27	11	40	37	28	23	47	45	6
27	12	39	37	29	22	51	46	5
27	13	38	37	30	21	51	47	4
28	14	37	38	31	20	52	48	3
31	15	36	38	32	19	55	49	2
31	16	35	39	33	18	55	50	1
31	17	34	40	34	17			

b) Median rank = (50 + 1)/2 = 25.5, median value = (35 + 35)/2 = 35
   Hinge rank = (25 + 1)/2 = 13, lower hinge value = 27
   Upper hinge value = 43

Age of VCR Purchaser

Age

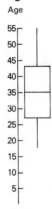

SECTION 3.3

1. a) Range = 5 hr
   b) $\bar{x}$ = 3.8 hr
   c) s = 1.92 hr
3. a) Range = 111 days
   b) $\bar{x}$ = 41.14 days
   c) s = 37.61 days
5. a) Range = 11
   b) $\bar{x}$ = 17.67

     c) $s = 3.24$

7. a) Range $= 43.3$ carats

   b) $\mu = 40.3$ carats

   c) $\sigma = 14.8$ carats

9. a) Range $= 9$, $\bar{x} = 5.36$, $s = 4.29$

   b) Range $= 2$, $\bar{x} = 5.14$, $s = 0.53$

   c) The means are close, but the standard deviations are different. The question about rent control raised a greater diversity of opinion, as reflected by the greater standard deviation and range of answers.

11. a) $\bar{x} = 7.83\ell b$, $s = 2.32\ell b$, range $= 4.8\ell b$

   b) $\bar{x} = 9.95\ell b$, $s = 0.29\ell b$, range $= 0.7\ell b$

   c) The second line had more consistent performance as reflected by the smaller standard deviation.

SECTION 3.4

1. a) $\bar{x} \approx 70.36$ years

   b) $s \approx 1.84$ years

3. a) $\bar{x} \approx 70.46$ years, $s \approx 1.73$ years

   b) $\bar{x} \approx 77.84$ years, $s \approx 1.36$ years

5. a) $\bar{x} \approx 10.38$ years

   b) $s \approx 6.62$ years

7. a) $\bar{x} \approx 27.26$

   b) $s \approx 10.32$

9. a) $\bar{x} = \$280.4$

   b) $s = \$108.5$

11. a) $\bar{x} = 13.67$ microcuries

   b) $s = 6.78$ microcuries

13. a) $\bar{x} = 28.17$ min

   b) $s = 18.73$ min

15. 87.65

17. 8.5

Chapter 3 Review Problems

1. a) $\bar{x} \approx 28$ days

   b) $s \approx 20.2$ days

3. a) **Characters per Second**

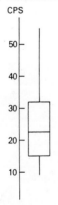

b) $\bar{x} \approx 26.03$ cps, $s \approx 12.42$ cps

c) $\bar{x} = 25.57$ cps, $s = 13.21$ cps

5. $156.25\ell b$

7. a) $\bar{x} = 2.15$ months

   b) $s = 2.40$ months

9. a) **Speed of Cars in Rush-hour Traffic**

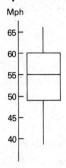

Mph

b) $\bar{x} = 53.3$ mph, median $= 55$ mph, mode $= 60$ mph

c) The mean and median satisfy the criteria, but the mode does not.

11. a) $\bar{x} = 29$ yr, median $= 26.5$ yr, mode $= 27$ yr

   b) Range $= 33$ yr, $s = 11.055$ yr

13. a) It is possible for the range and standard deviation to be the same. For instance, for data values that are all the same, such as 1, 1, 1, 1, 1, the range and standard deviation are both 0.

   b) It is possible for the mean, median, and mode to all be the same. For instance, the data set 1, 2, 3, 3, 3, 4, 5 has mean, median, and mode all equal to 3. The averages can all be different as in the data set 1, 2, 2, 4. In this case the mean is 2.25, the median is 2.5, and the mode is 2.

15. 7.56

SECTION 4.1

1. *See* text.

3. b, d, h

5. a) Intuitive

   b) Not likely; likely

7. a) Superior, average, inferior

   b) $P(\text{superior}) = 0.115$; $P(\text{average}) = 0.519$; $P(\text{inferior}) = 0.366$

   c) Yes, the sum of the probabilities of all events in a sample space should always be one.

9. a) $\frac{3}{5} = 0.6$; $\frac{2}{5} = 0.4$; $\frac{1}{5} = 0.2$

11. a) The sample space consists of the values: 1, 2, 3, 4, 5, 6; yes

   b) The probability of each outcome is $\frac{1}{6}$ and the probabilities of all the outcomes add up to one.

   c) There are 4 numbers less than five, so $P(\text{less than 5}) = \frac{4}{6} = \frac{2}{3}$.

   d) $\frac{1}{3}$

SECTION 4.2

1. a) $P(5 \text{ on green } and \text{ 3 on red}) = P(5) \cdot P(3) = (1/6)(1/6) = 1/36$

   b) $P(3 \text{ on green } and \text{ 5 on red}) = P(3) \cdot P(5) = (1/6)(1/6) = 1/36$

   c) $P((5 \text{ on green } and \text{ 3 on red}) \text{ or } (3 \text{ on green } and \text{ 5 on red})) = (1/36) + (1/36) = 1/18$

3. a) P(sum of 6) $= P(1 \text{ } and \text{ } 5) + P(2 \text{ } and \text{ } 4) + P(3 \text{ } and \text{ } 3) + P(4 \text{ } and \text{ } 2) + P(5 \text{ } and \text{ } 1) = (1/36) + (1/36) + (1/36) + (1/36) + (1/36) = 5/36$

  b) P(sum of 4) $= P(1 \text{ } and \text{ } 3) + P(2 \text{ } and \text{ } 2) + P(3 \text{ } and \text{ } 1) = (1/36) + (1/36) + (1/36) = 3/36 \text{ or } 1/12$

  c) P(sum of 6 *or* sum of 4) $= P$(sum of 6) $+ P$(sum of 4) $= (5/36) + (3/36) = 8/36 \text{ or } 2/9$

5. a) P(ace on 1st *and* king on 2nd) $= P(\text{ace}) \cdot P(\text{king, } given \text{ ace}) = (4/52)(4/51) = 4/663$

  b) P(king on 1st *and* ace on 2nd) $= P(\text{king}) \cdot P(\text{ace, } given \text{ king}) = (4/52)(4/51) = 4/663$

  c) P(ace and king in either order) $= P$(ace on 1st *and* king on 2nd) $+ P$(king on 1st *and* ace on 2nd) $= (4/663) + (4/663) = 8/663$

7. a) P(ace on 1st *and* king on 2nd) $= P(\text{ace}) \cdot P(\text{king}) = (4/52)(4/52) = 1/169$

  b) P(king on 1st *and* ace on 2nd) $= P(\text{king}) \cdot P(\text{ace}) = (4/52)(4/52) = 1/169$

  c) P(ace and king in either order) $= P$(ace on 1st *and* king on 2nd) $+ P$(king on 1st *and* ace on 2nd) $= (1/169) + (1/169) = 2/169$

9. a) There is 1 ace of spades in the deck of cards, so P(ace of spades) $= 1/52$. Also P(ace of spades) $= P$(ace *and* spades) $= P(\text{ace}) \cdot P(\text{spade, } given \text{ ace}) = (4/52)(1/4) = 1/52.$

  b) P(spade) $= 13/52 = 1/4$

  c) P(ace *or* spade) $= P(\text{ace}) + P(\text{spade}) - P(\text{ace } and \text{ spade})$
    $= (4/52) + (13/52) - (1/52) = 16/52 \text{ or } 4/13$

11. P(written *and* physical) $= P(\text{physical})P(\text{written, } given \text{ physical})$
    $= (0.82)(0.58) = 0.48$

13. a) $P(Fa) = 364/653 = 0.557; P(Fa, given F) = 11/34 = 0.324$
      $P(Fa, given S) = 353/619 = 0.570$

  b) $P(F \text{ } and \text{ } Fa) = 11/653 = 0.017$

  c) No

  d) $P(S) = 619/653 = 0.948; P(S, given O) = 191/209 = 0.914$
      $P(S, given Fa) = 353/364 = 0.970; P(S \text{ } given \text{ } N) = 75/80 = 0.938$

  e) $P(S \text{ } and \text{ } Fa) = 353/653 = 0.541; P(S \text{ } and \text{ } O) = 191/653 = 0.292$

  f) No

  g) $P(Fa \text{ } or \text{ } O) = P(Fa) + P(O) = (364/653) + (209/653) = 0.877$
     The events Favor and Oppose are mutually exclusive.

15. a) $P(O) = 267/558 = 0.478; P(O, given L) = 73/283 = 0.258$
      $P(O, given M) = 194/275 = 0.705$

  b) $P(O \text{ } and \text{ } M) = 194/558 = 0.348; P(O \text{ } or \text{ } M) = P(O) + P(M) - P(O \text{ } and \text{ } M)$
       $= (267/558) + (275/558) - (194/558) = 0.624$

  c) $P(R) = 291/558 = 0.522; P(R, given L) = 210/283 = 0.742$
      $P(R, given M) = 81/275 = 0.295$

  d) $P(R \text{ } and \text{ } L) = 210/558 = 0.376; P(R \text{ } or \text{ } L) = P(R) + P(L) - P(R \text{ } and \text{ } L)$
       $= (291/558) + (283/558) - (210/558) = 0.652$

  e) No; no

17. a) $P(A) = 697/1153 = 0.605; P(A, given G) = 406/511 = 0.795$
      $P(A, given D) = 291/642 = 0.453$

  b) $P(B) = 456/1153 = 0.395; P(B, given G) = 105/511 = 0.205$
      $P(B, given D) = 351/642 = 0.547$

c) $P(G) = 511/1153 = 0.443; P(G, \text{ given } A) = 406/697 = 0.582$
   $P(G, \text{ given } B) = 105/456 = 0.230$

d) $P(D) = 642/1153 = 0.557; P(D, \text{ given } A) = 291/697 = 0.418$
   $P(D, \text{ given } B) = 351/456 = 0.770$

e) $P(G \text{ and } A) = 406/1153 = 0.352; P(D \text{ or } B) = P(D) + P(B) - P(D \text{ and } B)$
   $= (642/1153) + (456/1153) - (351/1153) = 0.648$

f) No

g) No

19. a) $P(A) = 0.65.$ b) $P(B) = 0.71.$ c) $P(B, \text{ given } A) = 0.87$

    d) $P(A \text{ and } B) = P(A) \cdot P(B, \text{ given } A) = (0.65)(0.87) \approx 0.57$

    e) $P(A \text{ or } B) = P(A) + P(B) - P(A \text{ and } B) = 0.65 + 0.71 - 0.57 = 0.79$

    f) $P(\text{not close}) = P(\text{profit 1st year } or \text{ profit 2nd year}) = P(A \text{ or } B) = 0.79;$
       $P(\text{close}) = 1 - P(\text{not close}) = 1 - 0.79 = 0.21$

21. a) $P(A) = 0.08.$ b) $P(B) = 0.08.$ c) $P(B, \text{ given } A) = 0.23$

    d) $P(A \text{ and } B) = P(A) \cdot P(B, \text{ given } A) = (0.08)(0.23) = 0.018$

    e) $P(A \text{ or } B) = P(A) + P(B) - P(A \text{ and } B) = 0.08 + 0.08 - 0.018 = 0.142$

    f) $P(\text{no claim}) = 1 - P(A \text{ or } B) = 1 - 0.142 = 0.858$

23. a) $P(\text{TB } and \text{ positive}) = P(\text{TB})(\text{positive, } given \text{ TB})$
       $= (0.04)(0.82) = 0.033$

    b) $P(\text{does not have TB}) = 1 - P(\text{TB}) = 1 - 0.04 = 0.96$

    c) $P(\text{no TB } and \text{ positive}) = P(\text{no TB})P(\text{positive, } given \text{ no TB})$
       $= (0.96)(0.09) = 0.086$

SECTION 4.3

1. a) **Outcomes for Tossing a Coin Three Times**

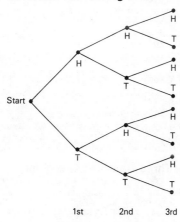

   b) 3
   c) $\frac{3}{8}$

3. a) **Outcomes for Drawing Two Balls (without replacement)**

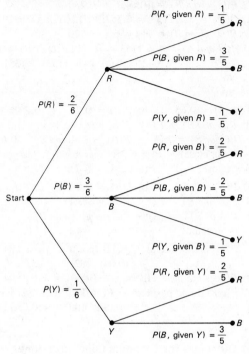

1st Ball   2nd Ball

b) $P(R \text{ and } R) = \frac{2}{6} \cdot \frac{1}{5} = \frac{1}{15}$

$P(R \text{ 1st } and \text{ B 2nd}) = \frac{2}{6} \cdot \frac{3}{5} = \frac{1}{5}$

$P(R \text{ 1st and Y 2nd}) = \frac{2}{6} \cdot \frac{1}{5} = \frac{1}{15}$

$P(B \text{ 1st } and \text{ R 2nd}) = \frac{3}{6} \cdot \frac{2}{5} = \frac{1}{5}$

$P(B \text{ 1st } and \text{ B 2nd}) = \frac{3}{6} \cdot \frac{2}{5} = \frac{1}{5}$

$P(B \text{ 1st } and \text{ Y 2nd}) = \frac{3}{6} \cdot \frac{1}{5} = \frac{1}{10}$

$P(Y \text{ 1st } and \text{ R 2nd}) = \frac{1}{6} \cdot \frac{2}{5} = \frac{1}{15}$

$P(Y \text{ 1st } and \text{ B 2nd}) = \frac{1}{6} \cdot \frac{3}{5} = \frac{1}{10}$

5. a) **Choices for Three True/False Questions**

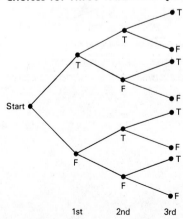

1st   2nd   3rd

b) $\frac{1}{8}$

7. $4 \cdot 3 \cdot 2 \cdot 1 = 24$ ways

9. a) $52 \cdot 52 = 2{,}704$

   b) $4 \cdot 4 = 16$

   c) $16/2{,}704 = 0.006$

11. $4 \cdot 3 \cdot 3 = 36$

13. $P_{5, 2} = (5!/3!) = 5 \cdot 4 = 20$

15. $P_{7, 7} = (7!/0!) = 7! = 5{,}040$

17. $C_{5, 2} = (5!/(2!3!)) = 10$

19. $C_{7, 7} = (7!/(7!0!)) = 1$

21. $15 \cdot 14 \cdot 13 = 2{,}730$

23. a) $8! = 40{,}320$

    b) $8 \cdot 7 \cdot 6 \cdot 5 \cdot 4 = 6{,}720$

25. $5 \cdot 4 \cdot 3 = 60$

27. $C_{15, 5} = (15!/(5!10!)) = (15 \cdot 14 \cdot 13 \cdot 12 \cdot 11/(5 \cdot 4 \cdot 3 \cdot 2 \cdot 1)) = 3{,}003$

29. a) $C_{12, 6} = (12!/(6!6!)) = 924$

    b) $C_{7, 6} = (7!/(6!1!)) = 7$

    c) $7/924 = 0.008$

SECTION 4.4

1. a) Discrete

   b) Continuous

   c) Continuous

   d) Discrete

   e) Continuous

   f) Discrete

3. a) Continuous

   b) Discrete

   c) Discrete

   d) Continuous

   e) Continuous

   f) Discrete

5. a)

x	15	20	25	30	35	40	45
$P(x)$	0.046	0.122	0.184	0.240	0.219	0.130	0.059

   b)

   c) 0.189

   d) 0.352

   e) $30.45 = \mu$

   f) $7.66 = \sigma$

7. a)

x	36	37	38	39	40	41	42
P(x)	0.029	0.048	0.053	0.096	0.125	0.154	0.163

x	43	44	45
P(x)	0.135	0.120	0.077

b)

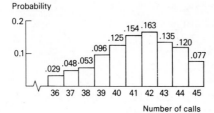

Probability

Number of calls

c) 0.673

d) 0.351

e) $41.288 = \mu$

f) $2.326 = \sigma$

9. a)

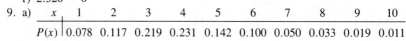

x	1	2	3	4	5	6	7	8	9	10
P(x)	0.078	0.117	0.219	0.231	0.142	0.100	0.050	0.033	0.019	0.011

b) **Histogram**

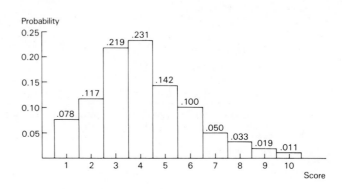

Probability

Score

c) $P(7 \text{ or more}) = P(7) + P(8) + P(9) + P(10) = 0.113$

d) $P(3 \text{ or less}) = P(3) + P(2) + P(1) = 0.414$

e) $\mu = 4.098$ f) $\sigma = 1.942$

11. $\mu = 1,800; \mu + 250 = 2050$

13. a) $P(\text{win}) = 0.006; P(\text{not win}) = 0.994$

b) $\$14.88 = $ expected earnings; $\$25.12 = $ contribution

Chapter 4 Review Problems

1. a) 0.016   b) 0.091   c) 0.21

d) $0.420 + 0.150 + 0.083 + 0.021 + 0.009 = 0.683$

e) $0.150 + 0.420 + 0.210 + 0.091 + 0.016 = 0.887$

3. a) $P(\text{heart } and \text{ heart}) = (13/52)(13/52) = 0.063$
   b) $P(\text{heart } and \text{ heart}) = (13/52)(12/51) = 0.059$
5. a) Drop a fixed number of tacks and count how many land flat side down. Then form the ratio of the number landing flat side down to the total number dropped.
   b) Up, down
   c) $P(\text{up}) = 160/500 = 0.32; P(\text{down}) = 340/500 = 0.68$
7. a) $P(\text{queen}) = 0.067; P(\text{bishop}) = 0.133; P(\text{knight}) = 0.133; P(\text{rook}) = 0.133;$ $P(\text{pawn}) = 0.533$
   b) $\mu = 2.599$
   c) No
9. a)

x	0	1	2	3	4	5	6
$P(x)$	0.05	0.48	0.22	0.11	0.09	0.03	0.02

Histogram

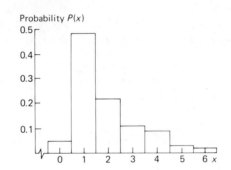

   b) $\mu = 1.88$ c) \$235
11. $C_{8,2} = (8!/(2!6!)) = (8 \cdot 7/2) = 28$
13. $3 \cdot 2 \cdot 1 = 6$
15. $4 \cdot 4 \cdot 4 \cdot 4 \cdot 4 = 1{,}024$ choices; $P(\text{all correct}) = 1/1{,}024 = 0.00098$
17. $10 \cdot 10 \cdot 10 = 1{,}000$

SECTION 5.1

1. Binomial with a) trial $=$ room call; b) $S =$ respond within three minutes, $F =$ respond after three minutes; c) $n = 73, p = 0.80, q = 0.20, r = 62$
3. Binomial with a) trial $=$ reported crime; b) $S =$ crime solved, $F =$ crime not solved; c) $n = 54, p = 0.90, q = 0.10, r = 46$
5. Not binomial because there are more than two outcomes on a trial.
7. Not binomial because the probability of drawing short straws changes.
9. Binomial with a) trial $=$ patient with rocky mountain spotted fever; b) $S =$ recover with drug, $F =$ not recover with drug; c) $n = 50, p = 0.90, q = 0.10, r = 43$
11. Binomial with a) trial $=$ laying hen receiving hormone; b) $S =$ hen lays at least two eggs, $F =$ hen lays fewer than two eggs; c) $n = 500, p = 0.85, q = 0.15, r = 360$

SECTION 5.2

1. $n = 3; p = 0.5$
   a) $P(r = 3) = 0.125$
   b) $P(r = 2) = 0.375$
   c) $P(r = 2 \text{ or } 3) = 0.125 + 0.375 = 0.500$
   d) $P(r = 0) = 0.125$
3. $n = 10; p = 0.2$
   a) $P(r = 10) = 0$

b) $P(r \geqslant 5) = P(r = 5) + P(r = 6) + P(r = 7) + P(r = 8) + P(r = 9) + P(r = 10) = 0.033$

c) $P(r = 0) = 0.107$

d) $P(r \geqslant 3) = P(r = 3) + P(r = 4) + P(r = 5) + P(r = 6) + P(r = 7) + P(r = 8) + P(r = 9) + P(r = 10) = 0.322$

5. $n = 9; p = 0.15$

   a) $P(r = 0) = 0.232$

   b) $P(r \geqslant 1) = 1 - P(r = 0) = 0.768$

   c) $P(r > 2) = P(r = 3) + P(r = 4) + P(r = 5) + P(r = 6) + P(r = 7) + P(r = 8) + P(r = 9) = 0.141$

   d) $P(1 \leqslant r \leqslant 5) = P(r = 1) + P(r = 2) + P(r = 3) + P(r = 4) + P(r = 5) = 0.768$

7. $n = 5; P = 0.45$

   a) $P(r = 5) = 0.019$

   b) $P(r = 0) = 0.050$

   c) $P(r \geqslant 3) = P(r = 3) + P(r = 4) + P(r = 5) = 0.408$

   d) $P(r \leqslant 2) = P(r = 2) + P(r = 1) + P(r = 0) = 0.593$

9. a) $n = 15; p = 0.10; P(r \leqslant 2) = 0.816$

   b) $n = 9; p = 0.10; P(r \geqslant 1) = 0.612$

   c) $n = 12; p = 0.10; P(r > 2) = 0.110$

   d) $n = 6; p = 0.10; P(r > 2) = 0.016$

11. $n = 4; p = 0.20;$

   a) $P(r = 4) = 0.002$

   b) $P(r = 0) = 0.410$

   c) $P(r \geqslant 3) = 0.028$

13. $n = 7; p = 0.75$

   a) $P(r = 7) = 0.133$

   b) $P(r \geqslant 4) = 0.928$

   c) $P(r = 4) = 0.173$

15. $n = 5; p = 0.25$

   a) $P(r = 3) = 0.088$

   b) $P(r \geqslant 3) = 0.104$

   c) $P(r \leqslant 2) = 0.897$

SECTION 5.3   1. $n = 6; p = 0.35$

a) **Histogram**

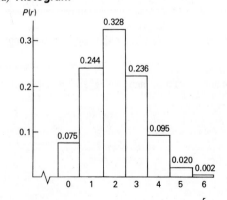

b) $\mu = np = 6(0.35) = 2.10$
c) $\sigma = \sqrt{npq} = \sqrt{6(0.35)(0.65)} = 1.17$

3. $n = 7; p = 0.10$

a) **Histogram**

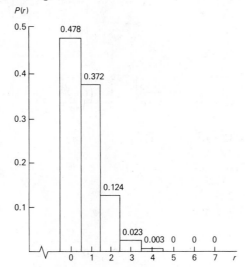

b) $\mu = np = 7(0.10) = 0.70$
c) $\sigma = \sqrt{npq} = \sqrt{7(0.10)(0.90)} = 0.794$

5. $n = 5; p = 0.7$

a)

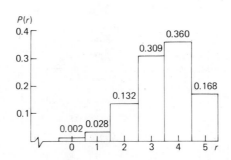

b) $\mu = 3.5$, the expected number of hits is 3.5
c) $\sigma = 1.02$

7. $n = 16; p = 0.80$

a)

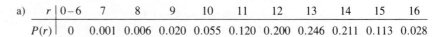

r	0–6	7	8	9	10	11	12	13	14	15	16
$P(r)$	0	0.001	0.006	0.020	0.055	0.120	0.200	0.246	0.211	0.113	0.028

b)

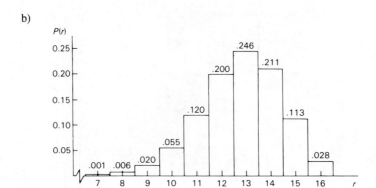

c) $\mu = 12.80$
d) $\sigma = 1.60$

9. $n = 27; p = 0.40$
   a) $\mu = 10.8$
   b) $\sigma = 2.5$

11. $n = 12; p = 0.65$

a)

r	0&1	2	3	4	5	6	7	8	9	10
$P(r)$	0	0.001	0.005	0.020	0.059	0.128	0.204	0.237	0.195	0.109

r	11	12
$P(r)$	0.037	0.006

b)

c) $\mu = 7.80$
d) $\sigma = 1.65$

13. $n = 4$ stations are necessary to be 98% certain. If 4 stations are used, the expected number of stations that will detect an enemy plane is $\mu = 2.6$.

15. $n = 6$ alarms are necessary to be 99% certain at least one alarm will go off. If nine alarms are installed, the expected number that will sound is $\mu = 4.95$ or 5.

Chapter 5 Review Problems

1. a)

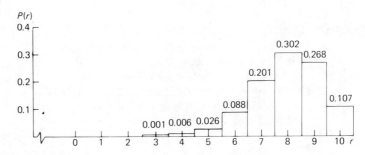

  b) $P(r \geqslant 6) = 0.966$
  c) The expected number is $\mu = 8$.
  d) $\sigma = 1.26$
3. $P(r \leqslant 2) = 0.000$ (to 3 decimal places). The data seem to indicate that the percent favoring the increase in fees is less than 85%.
5. a) $P(r \geqslant 12) = 0.630$
  b) $P(r < 8) = P(r \leqslant 7) = 0.007$
  c) The expected number is $\mu = 12$.

7. a)

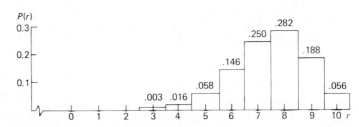

  b) $P(r \geqslant 9) = 0.244$, $P(r \geqslant 1) = 0.999$
  c) The expected number is $\mu = 7.5$.
  d) $\sigma = 1.37$
9. The expected number is $\mu = 102$.

11. a)

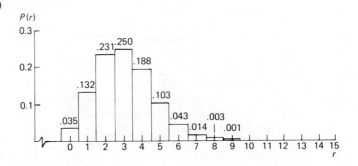

      b) The expected number is $\mu = 3$, $\sigma = 1.55$.

      c) $P(r \geqslant 3) = 0.602$, $P(r = 3) = 0.250$

13. $P(r \leqslant 5) = 0.021$

15. The expected number is $\mu = 325$.

SECTION 6.1

1. a) No, it's skewed.

    b) No, it crosses the horizontal axis.

    c) No, it has three peaks.

    d) No, the curve is not smooth.

3. Figure 6-13 has the larger standard deviation. The mean of Figure 6-13 is $\mu = 10$. The mean of Figure 6-14 is $\mu = 4$.

5. a) 50%

    b) 68.2%

    c) 99.7%

7. a) 50%

    b) 50%

    c) 68.2%

    d) 95.4%

9. a) 95.4%

    b) 68.2%

    c) 50%

SECTION 6.2

1. a) Robert; Jan; Linda

    b) Joel

    c) Susan; John

    d) Robert, 172; Jan, 184; Susan, 110; Joel, 150; John, 134; Linda, 182

3. a) 0.75

    b) 0

    c) -0.83

    d) -1.93

    e) 0.81

5. $-1.90 \leqslant z \leqslant 2.10$

7. a) $z \geqslant 0$

    b) $-0.67 \leqslant z \leqslant 1.00$

    c) $-2.00 \leqslant z \leqslant -0.83$

    d) $z \leqslant 0.83$

    e) $-0.17 \leqslant z \leqslant 0.17$

    f) $z \leqslant -1.00$

9. a) $410.1 \leqslant x \leqslant 529.5$

    b) $x \leqslant 359.7$

    c) $507.6 \leqslant x$ or equivalently, $x \geqslant 507.6$

    d) $381.4 \leqslant x \leqslant 518.6$

    e) $x \leqslant 450$

    f) $494.8 \leqslant x \leqslant 500.4$

11. a) No

    b) For Niko, $z = 2.55$; for Walter, $z = 1.07$

    c) Promote Niko

13. For George, $z = -2.93$ and for Fred, $z = -2.75$. George is likely to be in a little better condition.

SECTION 6.3

1. 0.4993	3. 0.4778	5. 0.8953	7. 0.3471
9. 0.0306	11. 0.5000	13. 0.0643	15. 0.0934
17. 0.8888	19. 0.6736	21. 0.4474	23. 0.2939
25. 0.6704	27. 0.3226	29. 0.8808	31. 0.5000
33. 0.0885	35. 0.4483	37. 0.8849	39. 0.8849

SECTION 6.4

1. $P(3 \leq x \leq 6) = P(-0.5 \leq z \leq 1) = 0.5328$
3. $P(50 \leq x \leq 70) = P(0.67 \leq z \leq 2.0) = 0.2286$
5. $P(8 \leq x \leq 12) = P(-2.19 \leq z \leq -0.94) = 0.1593$
7. $P(x \geq 30) = P(z \geq 2.94) = 0.0016$
9. $P(x \geq 90) = P(z \geq -0.67) = 0.7486$
11. 1.645
13. -1.41
15. -1.555
17. 1.41
19. ± 2.33
21. a) 0.5000
    b) 0.0004
    c) 0.0004
23. a) 0.4599
    b) 0.0043
    c) 0.0043
25. a) 0.8036
    b) 0.0228
    c) 0.1736
27. a) 0.0228
    b) 0.2420
    c) 0.2061
29. 2.6 yr
31. a) 21.19%
    b) 22 months
33. a) 81 months
    b) 0.0526
35. a) 61.8 hr
    b) 58.8 hr
    c) 53.4 hr

Chapter 6 Review Problems

1. a) 0.4599
   b) 0.4015
   c) 0.0384
   d) 0.0104
   e) 0.0250
   f) 0.8413
3. a) 0.9821
   b) 0.3156
   c) 0.2977

5. 1.645

7. $z = \pm 1.96$

9. a) 0.89

   b) 0

   c) 0.2514

11. a) 0.0013

    b) 0.1587

    c) 0.8185

13. a) 0.9772

    b) 17.3 hr

15. a) 0.5812

    b) 0.0668

    c) 0.0122

SECTION 7.1

1. A set of measurements or counts either existing or conceptual. For example, the population of all ages of all people in Colorado; the population of weights of all students in your school; the population count of all antelope in Wyoming.

3. A numerical descriptive measure of a population, such as μ, the population mean; σ, the population standard deviation; σ^2, the population variance.

5. A statistical inference is a conclusion about the value of a population parameter. We will do both estimation and testing.

7. They help us visualize the sampling distribution by using tables and graphs that approximately represent the sampling distribution.

9. Sampling distribution helps us evaluate the reliability of inferences about population parameters.

SECTION 7.2

1. a) $\mu_{\bar{x}} = 15$; $\sigma_{\bar{x}} = 2.0$; $P(15 \leq \bar{x} \leq 17) = P(0 \leq z \leq 1.00) = 0.3413$

   b) $\mu_{\bar{x}} = 15$; $\sigma_{\bar{x}} = 1.75$; $P(15 \leq \bar{x} \leq 17) = P(0 \leq z \leq 1.14) = 0.3729$

   c) The standard deviation is smaller in part b because of the larger sample size. Therefore the distribution about $\mu_{\bar{x}}$ is narrower in part b.

3. a) No; the sample size is only 9 and so is too small.

   b) Yes; the $\bar{x}$ distribution will also be normal with $\mu_{\bar{x}} = 25$; $\sigma_{\bar{x}} = 3.5/3$; $P(23 \leq \bar{x} \leq 26) = P(-1.71 \leq z \leq 0.86) = 0.7615$

5. a) $P(6 \leq \bar{x} \leq 7) = P(-1.69 \leq z \leq 2.53) = 0.9488$

   b) $P(6 \leq \bar{x} \leq 7) = P(-2.39 \leq \bar{x} \leq 3.58) = 0.9914$

   c) Yes

7. a) $P(\bar{x} \leq 19) = P(z \leq -1.20) = 0.1151$

   b) $P(19 \leq \bar{x} \leq 21) = P(-1.20 \leq z \leq 0.65) = 0.6271$

   c) $P(\bar{x} \geq 22) = P(z \geq 1.58) = 0.0571$

9. a) $P(\bar{x} \leq 600) = P(z \leq -3.48) = 0.0003$

   b) $P(\bar{x} \geq 700) = P(z \geq 0.87) = 0.1922$

   c) $P(600 \leq \bar{x} \leq 700) = P(-3.48 \leq z \leq 0.87) = 0.8075$

11. $P(358 \leq \bar{x} \leq 372) = P(-1.48 \leq z \leq 1.48) = 0.8612$

13. a) $P(\bar{x} \geq 5) = P(z \geq 0.97) = 0.1660$

    b) $P(4 \leq \bar{x} \leq 5) = P(-1.46 \leq z \leq 0.97) = 0.7619$

    c) $P(\bar{x} \leq 4) = P(z \leq -1.46) = 0.0721$

15. a) $P(\bar{x} \leq 35.5) = P(z \leq -0.63) = 0.2643$

    b) $P(\bar{x} \leq 35.5) = P(z \leq -2.80) = 0.0026$

    c) Yes

17. a) $P(67 \le x \le 69) = P(-0.33 \le z \le 0.33) = 0.2586$
    b) $P(67 \le x \le 69) = P(-1.00 \le z \le 1.00) = 0.6826$
    c) The standard deviation is smaller for the $\bar{x}$ distribution.
19. a) $P(13.9 \le \bar{x} \le 14.2) = P(-0.25 \le z \le 0.50) = 0.2902$
    b) $P(13.9 \le \bar{x} \le 14.2) = P(-0.87 \le z \le 1.73) = 0.7660$
    c) Yes, the standard deviation of the $\bar{x}$ distribution is smaller.

SECTION 7.3

1. a) $P(r \ge 240) = P(x \ge 239.5) = P(z \ge 0.35) = 0.3632$
   b) $P(r \le 225) = P(x \le 225.5) = P(z \le -1.15) = 0.1251$
   c) $P(220 \le r \le 250) = P(219.5 \le x \le 250.5) = P(-1.79 \le z \le 1.52) = 0.8990$
3. a) $P(r \le 370) = P(x \le 370.5) = P(z \le 1.47) = 0.9292$
   b) $P(r \ge 350) = P(x \ge 349.5) = P(z \ge -1.72) = 0.9573$
   c) $P(345 \le r \le 375) = P(344.5 \le x \le 375.5) = P(-2.48 \le z \le 2.23) = 0.9805$
5. a) $P(r \ge 15) = P(x \ge 14.5) = P(z \ge -2.35) = 0.9906$
   b) $P(r \ge 30) = P(x \ge 29.5) = P(z \ge 0.62) = 0.2676$
   c) $P(25 \le r \le 35) = P(24.5 \le x \le 35.5) = P(-0.37 \le z \le 1.81) = 0.6092$
   d) $P(r \ge 40) = P(x \ge 39.5) = P(z \ge 2.61) = 0.0045$
7. a) $P(r \ge 70) = P(x \ge 69.5) = P(z \ge 1.74) = 0.0409$
   b) $P(50 \le r \le 65) = P(49.5 \le x \le 65.5) = P(-2.36 \le z \le 0.92) = 0.8121$
   c) $P(r \le 55) = P(x \le 55.5) = P(z \le -1.13) = 0.1292$
9. a) $P(r \ge 1400) = P(x \ge 1399.5) = P(z \ge -0.38) = 0.6480$
   b) $P(r \le 1320) = P(x \le 1320.5) = P(z \le -3.27) = 0.0005$
   c) $P(1425 \le r \le 1500) = P(1424.5 \le x \le 1500.5) = P(0.53 \le z \le 3.31) =$
      0.2976
11. a) $P(r \ge 540) = P(x \ge 539.5) = P(z \ge 2.07) = 0.192$
    b) $P(r \le 500) = P(x \le 500.5) = P(z \le -0.67) = 0.2514$
    c) $P(485 \le r \le 525) = P(484.5 \le r \le 525.5) = P(-1.79 \le z \le 1.09) = 0.8254$
13. a) $P(r \ge 85) = P(x \ge 84.5) = P(z \ge 1.13) = 0.1292$
    b) $P(r \le 68) = P(x \le 68.5) = P(z \le -2.88) = 0.0020$
    c) $P(69 \le r \le 84) = P(68.5 \le x \le 84.5) = P(-2.88 \le z \le 1.13) = 0.8688$
15. a) $P(r \ge 100) = P(x \ge 99.5) = P(z \ge -0.67) = 0.7486$
    b) $P(100 \le r \le 120) = P(99.5 \le x \le 120.5) = P(-0.67 \le z \le 1.88) = 0.7185$

Chapter 7 Review Problems

1. a) A normal distribution
   b) The mean μ of the x distribution
   c) $\sigma/\sqrt{n}$, where σ is the standard deviation of the x distribution
   d) They will both be approximately normal with the same mean, but the standard deviations will be $\sigma/\sqrt{50}$ and $\sigma/\sqrt{100}$ respectively.
3. a) $np = 10 > 5$ and $nq = 10 > 5$ so it is appropriate to use the normal approximation to the binomial.
   b) The normal approximation gives 0.846 while the binomial table gives 0.846.
5. a) $P(x \ge 40) = P(z \ge 0.71) = 0.2389$
   b) $P(\bar{x} \ge 40) = P(z \ge 2.14) = 0.0162$
7. a) $P(\bar{x} \ge 15) = P(z \ge -1.56) = 0.9406$
   b) $P(\bar{x} \ge 15) = P(z \ge -1.34) = 0.9099$
   c) The standard deviation of part (a) is smaller.
9. a) $P(r \ge 300) = P(x \ge 299.5) = P(z \ge 2.13) = 0.0166$
   b) $P(260 \le r \le 300) = P(259.5 \le x \le 300.5) = P(-2.24 \le z \le 2.24) = 0.9750$

11. a) $P(x < 7) = P(z < -0.58) = 0.2810$; $P(x > 10) = P(z > 0.79) = 0.2148$;
$P(7 \leqslant x \leqslant 10) = P(-0.58 \leqslant z \leqslant 0.79) = 0.5042$
    b) $P(7 \leqslant \bar{x} \leqslant 10) = P(-1.83 \leqslant z \leqslant 2.49) = 0.9600$
    c) Yes, the standard deviation of part (b) is smaller so the range from $x = 7$ to $x = 10$ has larger probability.
13. $P(98 \leqslant \bar{x} \leqslant 102) = P(-1.33 \leqslant z \leqslant 1.33) = 0.8164$
15. a) $P(\bar{x} \geqslant 50) = P(z \geqslant 0) = 0.5000$
    b) $P(745 \leqslant \bar{x} \leqslant 775) = P(-2.00 \leqslant z \leqslant 2.00) = 0.9544$

SECTION 8.1

1. 7.02 to 7.18
3. 5.28 to 5.66
5. 488.84 to 555.50
7. 165.78 to 180.22
9. 18.09 to 19.11
11. 4.89 to 5.51
13. a)

80%	56.85 to 58.35
90%	56.64 to 58.56
95%	56.45 to 58.75
99%	56.09 to 59.11

   b) 1.50; 1.92; 2.30; 3.02; as c increases, the lengths of the confidence intervals increase
15. a) For $n = 30$, 14.05 to 17.37
    b) For $n = 90$, 14.63 to 16.53
    c) For $n = 300$, 15.07 to 16.11
    d) 3.32; 1.90; 1.04; as the sample size increases, the lengths decrease
17. a) The mean and standard deviation round to the results given.
    b) Using the rounded mean and standard deviation given in part (a), the interval is from 14.54 to 17.36.

SECTION 8.2

1. 2.110
3. 1.721
5. $\bar{x} = 11.98$; $s = 0.69$; Using these values, we obtain a confidence interval from 9.96 to 14.00
7. 13.36 to 15.84
9. 239.5 to 286.5
11. a) The mean and standard deviation round to results given.
    b) Using the rounded mean and standard deviation given in part (a), the interval is from 10.84 to 13.86.
13. a) The mean and standard deviation round to results given.
    b) Using the rounded mean and standard deviation given in part (a), the interval is from 3.20 to 3.56.
15. a) The mean and standard deviation round to results given.
    b) Using the rounded mean and standard deviation given in part (a), the interval is from 1.25 to 1.61.
17. a) The mean and standard deviation round to results given.
    b) Using the rounded mean and standard deviation given in part (a), the interval is from 14.68 to 19.66.

SECTION 8.3

1. a) $r/n = 0.4589$
   b) 0.40 to 0.51
   c) Yes
3. a) $r/n = 0.16$
   b) 0.12 to 0.20
   c) Yes
5. a) $r/n = 0.125$
   b) 0.06 to 0.19
7. a) $r/n = 0.6091$
   b) 0.53 to 0.69
9. a) $r/n = 0.77$
   b) 0.72 to 0.82
11. a) $r/n = 0.6$
    b) 0.55 to 0.65
13. a) $r/n = 0.06$
    b) 0.04 to 0.08
15. a) $r/n = 0.4824$
    b) 0.34 to 0.62
17. a) $r/n = 0.11$
    b) 0.07 to 0.15
19. a) $r/n = 0.1496$
    b) 0.10 to 0.20

SECTION 8.4

1. 291 more
3. 28 more
5. 61 more
7. a) 666
   b) 149 more
9. a) 144
   b) 69 more
11. 409 more
13. a) 385
    b) 163
15. $n = 432$ so we need 392 more
17. a) $\frac{1}{4} - (p - \frac{1}{2})^2 = \frac{1}{4} - (p^2 - p + \frac{1}{4}) = -p^2 + p = p(1 - p)$
    b) Since $(p - \frac{1}{2})^2 \geq 0$, then $\frac{1}{4} - (p - \frac{1}{2})^2 \leq \frac{1}{4}$ because we are subtracting $(p - \frac{1}{2})^2$ from $\frac{1}{4}$.
19. 19 more

Chapter 8 Review Problems

1. *See* text for the definitions.
3. 1.56 to 2.62
5. 11.8 to 13.8
7. 0.04 to 0.12
9. 165.9 to 174.1
11. 346 total or 297 more
13. With preliminary estimate of 0.381 for p, we need a total of 256 or 232 more. With no preliminary estimate, we need a total of 271.

15. 0.03 to 0.19

17. $\bar{x} = 14.67$; $s = 4.18$; the interval is from 11.2 to 18.1

19. 69 total or 39 more

21. Using a preliminary estimate of 0.4 for p, we need a total of 640 or 415 more. With no preliminary study, we need 666.

23. 3.78 to 4.04

25. 68 total

27. a) The mean and standard deviation round to the values given.
    b) 8.3 to 10.8

SECTION 9.1

1. *See* text.

3. No, we simply should not reject it based on the data at hand.

5. Class discussion

7. a) H_0: $\mu = 750$
   b) H_1: $\mu \neq 750$
   c) We also need a critical region and test statistic (i.e., sample $\bar{x}$).

9. a) H_0: $\mu = 885$
   b) H_1: $\mu < 885$
   c) We also need a critical region and test statistic (i.e., sample $\bar{x}$).

SECTION 9.2

1. H_0: $\mu = 3,218$; H_1: $\mu > 3,218$; $\bar{x}_0 = 3,321$. Because the sample statistic $\bar{x} = 3,492$ is in the critical region, we reject H_0. The data indicate that the average number of customers entering the store has increased.

3. H_0: $\mu = 678$; H_1: $\mu \neq 678$; $\bar{x}_0 = 644.7$ or 711.3. Because the observed sample statistic $\bar{x} = 650$ is in the acceptance region, we accept H_0. The average loss is not different.

5. H_0: $\mu = 4.75$; H_1: $\mu > 4.75$; $\bar{x}_0 = 5.12$. Because the observed sample statistic $\bar{x} = 5.25$ falls in the critical region, we reject H_0. Maureen's average tip is higher than reported.

7. H_0: $\mu = 12,000$; H_1: $\mu > 12,000$; $\bar{x}_0 = 12,240$. Because the observed sample statistic $\bar{x} = 12,100$ falls in the acceptance region, we accept H_0. The data do not indicate that the average weight is more than 12,000 lb.

9. H_0: $\mu = 3.3$; H_1: $\mu \neq 3.3$; $\bar{x}_0 = 2.98$ or 3.62. Because the observed sample statistic $\bar{x} = 4.3$ falls in the critical region, we reject H_0. The average size of a family unit in Pleasant View is different from the national average.

11. H_0: $\mu = 38.6$; H_1: $\mu < 38.6$; $\bar{x}_0 = 34.64$. Because the observed sample statistic $\bar{x} = 31.5$ falls in the critical region, we reject H_0. The manufacturer's claim is too high.

13. H_0: $\mu = 0.25$; H_1: $\mu > 0.25$; $\bar{x}_0 = 0.27$. Because the observed sample statistic $\bar{x} = 0.32$ falls in the critical region, we reject H_0. The supplier's claim is too low.

15. H_0: $\mu = 14.2$; H_1: $\mu \neq 14.2$; $\bar{x}_0 = 13.93$ or 14.47. Because the observed sample statistic $\bar{x} = 14.9$ is in the critical region, we reject H_0 and conclude that the mean radio frequency of the signals has changed.

SECTION 9.3

1. H_0: $\mu_1 = \mu_2$; H_1: $\mu_1 < \mu_2$; critical value is -1.45. Because $\bar{x}_1 - \bar{x}_2 = -2.3$ falls in the critical region, we reject H_0. The night workers seem to take more sick leave.

3. H_0: $\mu_1 = \mu_2$; H_1: $\mu_1 < \mu_2$; critical value is -2.79. Because $\bar{x}_1 - \bar{x}_2 = -4.7$ is in the critical region, we reject H_0. The cars seem to burn more fuel at one mile above sea level.

5. H_0: $\mu_1 = \mu_2$; H_1: $\mu_1 \neq \mu_2$; critical values are -0.88 and 0.88. Because $\bar{x}_1 - \bar{x}_2 = 0.3$ does not fall in the critical region, we accept H_0. There does not seem to be a difference.

7. H_0: $\mu_1 = \mu_2$; H_1: $\mu_1 \neq \mu_2$; critical values are -7.16 and 7.16. Because $\bar{x}_1 - \bar{x}_2 = -3$ is not in the critical region, we accept H_0. There is no significant difference in the mean scores of students from the two schools.

9. H_0: $\mu_1 = \mu_2$; H_1: $\mu_1 < \mu_2$; critical value is -2.49. Because $\bar{x}_1 - \bar{x}_2 = -3.4$ is in the critical region, we reject H_0. The mean speed for the ship with modified hull is higher.

11. H_0: $\mu_1 = \mu_2$; H_1: $\mu_1 < \mu_2$; critical value is -0.45. Because $\bar{x}_1 - \bar{x}_2 = -0.3$ is not in the critical region, we accept H_0. The claim is not justified at the 5% level of significance.

13. H_0: $\mu_1 = \mu_2$; H_1: $\mu_1 < \mu_2$; critical value is -0.16. Because $\bar{x}_1 - \bar{x}_2 = -0.34$ is in the critical region, we reject H_0. The claim that mean grocery costs for inner city residents are higher is justified at the 5% level of significance.

15. H_0: $\mu_1 = \mu_2$; H_1: $\mu_1 < \mu_2$; critical value is -0.645. Because $\bar{x}_1 - \bar{x}_2 = -0.61$ does not fall in the critical region, we accept H_0. There is no difference in the speeds of the systems.

SECTION 9.4

1. $t = -1.860$

3. $t = 2.807$

5. $t = 2.201$

7. H_0: $\mu = 2.68$; H_1: $\mu > 2.68$; d.f. $= 14$; $t = 1.761$; critical value is 3.10. Because the observed sample value $\bar{x} = 2.81$ lies in the acceptance region, we accept H_0. The claim is not justified.

9. H_0: $\mu = 35$; H_1: $\mu > 35$; d.f. $= 17$; $t = 1.740$; critical value is 38.32. Because the observed sample mean 39.5 falls in the critical region, we reject H_0. The mean cost per month is more than \$35.00.

11. H_0: $\mu = 10.7$; H_1: $\mu > 10.7$; d.f. $= 11$; $t = 2.718$; critical value is 12.46. Because the observed sample mean 11.81 falls in the acceptance region, we accept H_0. The packages do not weigh more than 10.7 oz.

13. H_0: $\mu = 95$; H_1: $\mu < 95$; d.f. $= 9$; $t = -2.821$; critical value is 75.99. Because the observed value of $\bar{x} = 73.9$ falls in the critical region, we reject H_0. The mean income is less than \$95,000.

15. H_0: $\mu = 9.2$; H_1: $\mu > 9.2$; d.f. $= 3$, $t = 4.541$, the calculated value of $s = 1.18$, the critical value is 11.88. Because the observed sample mean $\bar{x} = 10.1$ is in the acceptance region, we accept H_0. The manufacturer's claim is not justified at the 1% level of significance.

17. H_0: $\mu_1 = \mu_2$; H_1: $\mu_1 \neq \mu_2$; d.f. $= 25$; $t = \pm 2.787$; $s = 1.60$; critical values are -1.73 and 1.73. Because the observed difference $\bar{x}_1 - \bar{x}_2 = -2.6$ is in the critical region, we reject H_0. The mean petal lengths are different.

19. H_0: $\mu_1 = \mu_2$; H_1: $\mu_1 < \mu_2$; d.f. $= 20$; $t = -1.725$; $s = 2.59$; critical value is -1.91. Because the observed difference $\bar{x}_1 - \bar{x}_2 = -3.6$ falls in the critical region, we reject H_0. The mean water temperature has increased.

21. H_0: $\mu_1 = \mu_2$; H_1: $\mu_1 < \mu_2$; d.f. $= 10$; $t = -2.764$; $s = 2.8$; critical value is -4.47. Because the observed difference $\bar{x}_1 - \bar{x}_2 = -2.8$ falls in the acceptance region, we accept H_0. It does not take the second type of extinguisher longer.

SECTION 9.5

1. H_0: $\mu_d = 0$; H_1: $\mu_d > 0$; d.f. $= 9$; $t = 2.821$; $s_d = 1.70$; critical value is 1.52. Because $\bar{d} = 0.08$ is in the acceptance region, we accept H_0. It seems that there is no change in the January deer population.

3. H_0: $\mu_d = 0$; H_1: $\mu_d > 0$; d.f. $= 7$; $t = 2.998$; $s_d = 2.07$; critical value is 2.19. Because $\bar{d} = 0.625$ falls in the acceptance region, we accept H_0. There seems to be no difference in recognition level.

5. H_0: $\mu_d = 0$; H_1: $\mu_d < 0$; d.f. $= 6$; $t = -1.943$; $s_d = 6.73$; critical value is -4.9. Because $\bar{d} = -10$ is in the critical region, we reject H_0. It seems that more books are shelved when there is a coffee break.

7. $H_0: \mu_d = 0$; $H_1: \mu_d < 0$; d.f. = 9; $t = -2.821$; $s_d = 0.96$; critical value is -0.85. Because $\bar{d} = -0.57$ is in the acceptance region, we accept H_0. There seems to be no difference in flavor.

9. $H_0: \mu_d = 0$; $H_1: \mu_d > 0$; d.f. = 6; $t = 1.943$; $s_d = 8.44$; critical value is 6.20. Because $\bar{d} = 9.3$ is in the critical region, we reject H_0. The attitude toward smoking seems less positive after the film.

11. $H_0: \mu_d = 0$; $H_1: \mu_d > 0$; d.f. = 5; $t = 3.365$; $s_d = 7.17$; critical value is 9.85. Because $\bar{d} = 6.33$ falls in the acceptance region, we accept H_0. The number of reported claims has not dropped.

13. $H_0: \mu_d = 0$; $H_1: \mu_d \neq 0$; d.f. = 7; $t = \pm 3.499$; $s_d = 3.72$; critical values are -4.60 and 4.60. Because $\bar{d} = -1.125$ is in the acceptance region, we accept H_0. The pulse rates were the same.

15. $H_0: \mu_d = 0$; $H_1: \mu_d > 0$; d.f. = 5; $t = 3.365$; $s_d = 0.597$; critical value is 0.82. Because $\bar{d} = 0.40$ is in the acceptance region, we accept H_0. The sales remained the same with or without the bonus.

SECTION 9.6

1. $H_0: p = 0.03$; $H_1: p > 0.03$; critical value $p_0 = 0.061$. Because $r/n = 0.067$ is in the critical region, we reject H_0. The proportion of people developing a chemical reaction is higher than 3%.

3. $H_0: p = 0.10$; $H_1: p \neq 0.10$; critical values are 0.073 and 0.127. Because $r/n = 0.105$ does not fall in the critical region, we accept H_0. The percentage of tax returns containing arithmetic errors in excess of \$1,200 is not different from 10%.

5. $H_0: p = 0.55$; $H_1: p < 0.55$; critical value $p_0 = 0.47$. Because $r/n = 0.45$ falls in the critical region, we reject H_0. The percentage of voters favoring the water project seems to be less than 55%.

7. $H_0: p = 0.048$; $H_1: p > 0.048$; critical value $p_0 = 0.074$. Because $r/n = 0.067$ is not in the critical region, we cannot reject H_0. There does not seem to be a greater proportion of left-handed people listed in *Who's Who* than is found in the general population.

9. $H_0: p = 0.68$; $H_1: p > 0.68$; $p_0 = 0.77$. Because $r/n = 0.81$ is in the critical region, we reject H_0 and accept the claim $p > 0.68$.

11. p_1 = proportion of failures for machine A; p_2 = proportion of failures for machine B; $H_0: p_1 = p_2$; $H_1: p_1 < p_2$; $\hat{p} = 0.10$; critical value = -0.05. Because $(r_1/n_1) - (r_2/n_2) = -0.04$ we do not reject H_0. The claim $p_1 < p_2$ is not justified.

13. $H_0: p = 0.15$; $H_1: p \neq 0.15$; critical values are 0.21 and 0.09. Because $r/n = 0.11$ we cannot reject H_0. The claim $p \neq 0.15$ is not justified.

15. p_1 = accident rate during a full moon; p_2 = accident rate during other times; $H_0: p_1 = p_2$; $H_1: p_1 \neq p_2$; $\hat{p} = 0.08$; critical values are ± 0.047. Because $(r_1/n_1) - (r_2/n_2) = 0.009$ is in the acceptance region, we accept H_0. There seems to be no difference in accident rate.

17. $H_0: p_1 = p_2$; $H_1: p_1 < p_2$ where p_1 is the proportion of union brick layers unemployed and p_2 is the proportion of non-union brick layers unemployed. $\hat{p} = 0.0855$; critical value is 0.027. Because $(r_1/n_1) - (r_2/n_2) = 0.018$ is in the acceptance region, we accept H_0. There is no difference in the proportions of unemployed union and non-union brick layers.

19. $H_0: p_1 = p_2$; $H_1: p_1 > p_2$ where p_1 is the proportion of fish killed on foreign tuna boats that are porpoises and p_2 is the proportion on U.S. tuna boats. $\hat{p} = 0.018$; critical value is 0.012. Since $(r_1/n_1) - (r_2/n_2) = 0.028$ is in the critical region, we reject H_0. The proportion of porpoises killed on the foreign tuna boats is higher.

21. $H_0: p_1 = p_2$; $H_1: p_1 \neq p_2$ where p_1 is the proportion of spoiled Bartlett pears and p_2 is the proportion of spoiled LeConte pears. $\hat{p} = 0.053$; critical values are -0.028 and

0.028. Because $(r_1/n_1) - (r_2/n_2) = -0.021$ is not in the critical region, we accept H_0. There is no difference in the proportions of spoiled pears for the two varieties.

SECTION 9.7

1. a) Because $\alpha = 0.01$ is less than P value $= 0.0312$, we accept H_0 at the 1% level of significance.
   b) Because $\alpha = 0.05$ is greater than P value $= 0.0312$, we reject H_0 at the 5% level of significance.
3. a) Because $\alpha = 0.01$ is greater than P value $= 0.0093$, we reject H_0 at the 1% level of significance.
   b) Because $\alpha = 0.05$ is greater than P value $= 0.0093$, we reject H_0 at the 5% level of significance.
5. a) Because $\alpha = 0.01$ is less than P value $= 0.0262$, we accept H_0 at the 1% level of significance.
   b) Because $\alpha = 0.05$ is greater than P value $= 0.0262$, we reject H_0 at the 5% level of significance.
7. The P value is the area to the *right* of the observed sample statistic $\bar{x} = 8$.
9. The P value is the area to the left of the observed sample statistic $\bar{x} = 3$.
11. The P value is the area in the two tails. That is, the area to the right of $\bar{x} = 12$ with that to the left of $\bar{x} = 8$.
13. The P value is the area in the two tails. That is, the area to the left of $\bar{x} = 8$ with that to the right of $\bar{x} = 10$.

Chapter 9 Review Problems

1. a) Single mean   b) Large sample   c) Not applicable
   d) H_0: $\mu = 5.2$; H_1: $\mu \neq 5.2$; critical values are 4.85 and 5.55. Because the observed value $\bar{x} = 5$ is in the acceptance region, we accept H_0. The average time between moves is not different from 5.2 years.
3. a) Difference of means   b) Large samples   c) Independent samples
   d) H_0: $\mu_1 = \mu_2$; H_1: $\mu_1 > \mu_2$; critical value is 0.22. Because $\bar{x}_1 - \bar{x}_2 = 0.10$ is in the acceptance region, we accept H_0. Nippon tubes do not last longer.
5. a) Single mean   b) Small sample   c) Not applicable
   d) H_0: $\mu = 0.8$; H_1: $\mu > 0.8$; critical value is 1.20. Because the observed sample mean $\bar{x} = 1.6$ is in the critical region, we reject H_0. The Toylot claim is too low.
7. a) Difference of means   b) Small samples   c) Independent
   d) H_0: $\mu_1 = \mu_2$; H_1: $\mu_1 > \mu_2$; critical value is 2.1. Because $\bar{x}_1 - \bar{x}_2 = 2.6$ is in the critical region, we reject H_0. The yellow paint is less visible after one year.
9. a) Single mean   b) Large sample   c) Not applicable
   d) H_0: $\mu = 19,800$; H_1: $\mu < 19,800$; critical value is 19,635.48. Because the observed mean $\bar{x} = 19,400$ is in the critical region, we reject H_0. The average salary is less.
11. a) Single proportion   b) Large sample   c) Not applicable
    d) H_0: $p = 0.20$; H_1: $p > 0.20$; critical value is 0.241. Because $r/n = 0.30$ is in the critical region, we reject H_0. The magazine should be continued.
13. a) Single mean   b) Large sample   c) Not applicable
    d) H_0: $\mu = 40$; H_1: $\mu > 40$; critical value is 41.44. Because the observed sample mean $\bar{x} = 43.1$ is in the critical region, we reject H_0. The average number of matches per box is more than 40.
15. a) Difference of means   b) Small samples   c) Independent samples
    d) H_0: $\mu_1 = \mu_2$; H_1: $\mu_1 \neq \mu_2$; critical values are -1.43 and 1.43. Because $\bar{x}_1 - \bar{x}_2 =$

-0.40 is in the acceptance region, we accept H_0. There is no difference in waiting times.

17. a) Difference of means   b) Small samples   c) Independent samples

 d) $H_0: \mu_1 = \mu_2; H_1: \mu_1 < \mu_2$; critical value is -8.15. Because $\bar{x}_1 - \bar{x}_2 = -6.3$ is in the acceptance region, we accept H_0 at the 1% level of significance.

 2nd part: At the 5% level of significance, the critical value is -5.56 and the observed sample statistic -6.3 falls in the critical region, so we reject H_0 at the 5% level of significance.

19. a) Testing single mean   b) Small sample   c) Not applicable

 d) $H_0: \mu = 7; H_1: \mu \neq 7$; critical values are 6.58 and 7.42. Because the observed sample mean $\bar{x} = 7.3$ falls in the acceptance region, we do not reject H_0. The machine does not seem to need adjustment.

21. a) Difference of means   b) Small samples   c) Dependent samples

 d) $H_0: \mu_d = 0; H_1: \mu_d > 0$; critical value is 5.45. Because the observed difference $\bar{d} = 9.83$ falls in the critical region, we reject H_0. The experimental group did promote creative problem solving.

23. a) Single mean   b) Small sample   c) Not applicable

 d) $H_0: \mu = 48; H_1: \mu < 48$; critical value is 41.7. Because the observed sample mean $\bar{x} = 46.2$ falls in the acceptance region, we cannot reject H_0. The manufacturer's claim seems to be valid.

25. a) Because $\alpha = 0.01$ is less than P value $= 0.0213$, we accept H_0 at the 1% level of significance.

 b) Because $\alpha = 0.05$ is greater than P value $= 0.0213$, we reject H_0 at the 5% level of significance.

27. The P value is the area to the left of the observed sample statistic $\bar{x} = 1.3$.

SECTION 10.1

1. Moderate or low linear correlation
3. High linear correlation
5. High linear correlation
7. a)

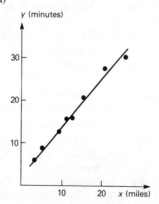

b) Just draw the line you think is best.

c) There seems to be a high linear correlation.

9. a)

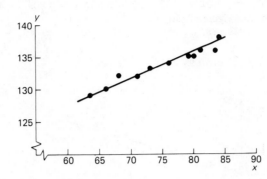

b) Just draw the line you think is best.
c) There seems to be a high correlation.

SECTION 10.2   Note: In this section and the next, answers may vary slightly depending on how many significant digits are used throughout the calculations.

1. a)

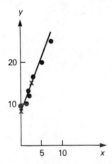

b) $\bar{x} = 2.86$; $\bar{y} = 15$; $b = 2.38$; $y = 2.38x + 8.20$
c) *See* Figure of part (a)
d) $S_e = 0.69$
e) 17.72
f) 15.80 to 19.64

3. a)

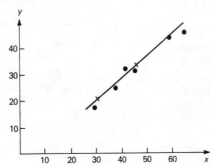

b) $\bar{x} = 45.67$; $\bar{y} = 32.83$; $b = 0.809$; $y = 0.809x - 4.12$
c) *See* Figure of part (a)
d) $S_e = 1.96$
e) 40.39
f) 30.23 to 50.55

5. a)

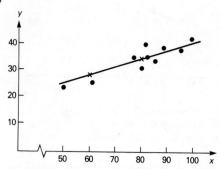

b) $\bar{x} = 79.8$; $\bar{y} = 34.4$; $b = 0.364$; $y = 0.364x + 5.347$
c) *See* Figure of part (a)
d) $S_e = 2.645$
e) 32.65
f) 26.22 to 39.09

7. a)

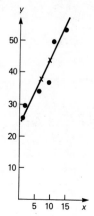

b) $\bar{x} = 7.33$; $\bar{y} = 37.83$; $b = 1.99$; $y = 1.99x + 23.23$
c) *See* Figure of part (a)
d) $S_e = 3.03$
e) 39.16
f) 32.16 to 46.16

9. a)

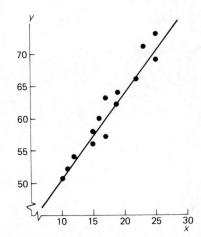

b) $\bar{x} = 18.08$; $\bar{y} = 61.92$; $b = 1.39$; $y = 1.39x + 36.85$

c) *See* Figure of part (a)

d) $S_e = 1.96$

e) 64.57

f) 59.99 to 69.16

11. a)

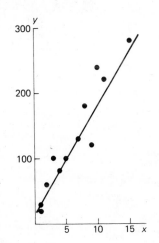

b) $\bar{x} = 6.91$; $\bar{y} = 140$; $b = 17.42$; $y = 17.42x + 19.66$

c) *See* Figure of part (a)

d) $S_e = 28.04$

e) 124.17

f) 70.38 to 177.97

SECTION 10.3

1. a) No
   b) Increase in population
3. a) No
   b) Better medical treatment
5. a)

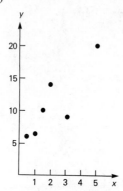

   b) *See* Figure of part (a)
   c) $r = 0.869$

7. a)

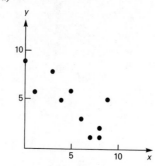

   b) *See* Figure of part (a)
   c) $r = -0.784$

9. a)

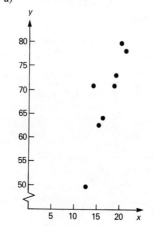

b) *See* Figure of part (a)

c) $r = 0.706$

11. a)

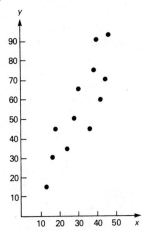

b) *See* Figure of part (a)

c) $r = 0.868$

13. a)

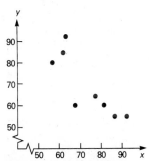

b) *See* Figure of part (a)

c) $r = -0.838$

SECTION 10.4

1. $H_0: \rho = 0$; $H_1: \rho > 0$; the critical value is 0.88. Because the observed r statistic $r = 0.73$ lies in the acceptance region, we accept H_0. At the 1% level of significance, r is not significant.

3. $H_0: \rho = 0$; $H_1: \rho < 0$; the critical value is -0.42. Because the observed r statistic $r = -0.67$ lies in the critical region, we reject H_0. The claim is justified at the 1% level of significance.

5. $H_0: \rho = 0$; $H_1: \rho < 0$; the critical value is -0.52. Because the observed sample statistic $r = -0.40$ lies in the acceptance region, we accept H_0. At the 1% level, r is not significant.

7. $H_0: \rho = 0$; $H_1: \rho > 0$; the critical value is 0.59. Because the observed sample statistic $r = 0.62$ lies in the critical region, we reject H_0. The claim is significant at the 1% level.

9. $H_0: \rho = 0$; $H_1: \rho < 0$; the critical value is -0.42. Because the observed sample statistic $r = -0.63$ falls in the critical region, we reject H_0. The professor's claim is significant at the 1% level.

11. $H_0: \rho = 0; H_1: \rho > 0$; the critical value is 0.57. Because the observed sample statistic $r = 0.48$ falls in the acceptance region, we accept H_0. At the 1% level of significance, there is no correlation; therefore Mr. Tonguetwist should let the employees work overtime.

13. $H_0: \rho = 0; H_1: \rho > 0$; the critical value is 0.73. Because the observed sample statistic $r = 0.70$ lies in the acceptance region, we accept H_0. There does not seem to be any correlation at the 5% level of significance.

Chapter 10 Review Problems

1. a)

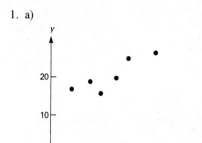

b) $\bar{x} = 1.53; \bar{y} = 20.67; b = 5; y = 5x + 13$

c) $r = 0.88$

d) $H_0: \rho = 0; H_1: \rho \neq 0$; critical values are -0.81 and 0.81. Because the sample statistic $r = 0.88$ falls in the critical region, we reject H_0. There seems to be a positive correlation at the 5% level of significance.

3. a)

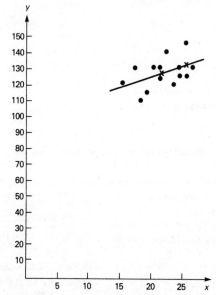

b) $\bar{x} = 21.43; \bar{y} = 126.79; b = 1.28; y = 1.28 + 99.25$

c) *See* Figure of part (a)

d) 124.95

e) $S_e = 8.38$

f) 105.91 to 143.99

g) *See* Figure of part (a)

h) $r = 0.47$

i) H_0: $\rho = 0$; H_1: $\rho > 0$; critical value is 0.61. Because the observed sample statistic $r = 0.47$ falls outside the critical region, we accept H_0. There does not seem to be any significant correlation at the 1% level.

5. a)

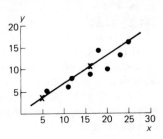

b) $\bar{x} = 16.38$; $\bar{y} = 10.13$; $b = 0.554$; $y = 0.554x + 1.051$

c) *See* line in Figure of part (a)

d) 9.36

e) $S_e = 1.73$

f) 4.87 to 13.86

g) *See* Figure of part (a)

h) $r = 0.91$

i) H_0: $\rho = 0$; H_1: $\rho > 0$; critical value is 0.79. Because the observed sample statistic $r = 0.91$ falls in the critical region, we reject H_0.

SECTION 11.1

1. H_0: job satisfaction and salary are independent; H_1: job satisfaction and salary are not independent; $\chi^2 = 8.91$, $\chi^2_{0.05} = 9.49$. Accept H_0 that they are independent.

3. H_0: age of adult and type of movie preferred are independent; H_1: age of adult and type of movie preferred are not independent; $\chi^2 = 3.62$; $\chi^2_{0.05} = 9.49$. Accept H_0. They are independent.

5. The cell in row 2 column 3 has expected frequency 4, which is too small for accurate results. The sample size should be increased.

7. H_0: total ticket sales and program offered are independent; H_1: total ticket sales and program offered are not independent; $\chi^2 = 1.87$; $\chi^2_{0.05} = 7.81$. Accept H_0 that they are independent.

9. H_0: party affiliation and dollars spent are independent; H_1: party affiliation and dollars spent are not independent; $\chi^2 = 2.17$; $\chi^2_{0.01} = 9.21$. Accept H_0. Party affiliation and dollars spent are independent.

SECTION 11.2

1. H_0: the distributions are the same; H_1: the distributions are different; $\chi^2 = 10.64$; $\chi^2_{0.01} = 9.21$. Reject H_0. The distributions are different.

3. H_0: the distributions are the same; H_1: the distributions are different; $\chi^2 = 1.23$; $\chi^2_{0.05} = 5.99$. Accept H_0. The distributions are the same.

5. H_0: the distributions are the same; H_1: the distributions are different; $\chi^2 = 25.32$; $\chi^2_{0.01} = 15.09$. Reject H_0. The distributions are different.

7. H_0: the distributions are the same; H_1: the distributions are different; $\chi^2 = 9.46$; $\chi^2_{0.01} = 16.81$. Accept H_0. The distributions are the same.

9. H_0: the distributions are the same; H_1: the distributions are different; $\chi^2 = 13.70$; $\chi^2_{0.01} = 15.09$. Accept H_0. The distributions are the same.

SECTION 11.3

1. a) H_0: $\sigma^2 = 9$; H_1: $\sigma^2 < 9$; $\chi^2 = 9.23$; the critical value is $\chi^2_{0.95} = 13.09$. Because the observed value $\chi^2 = 9.23$ falls in the critical region, we reject H_0. The new shot has smaller variance of protection times.
   b) $\chi^2_L = 9.26$; $\chi^2_U = 44.18$; $1.37 < \sigma < 2.99$.

3. a) $\chi^2_U = 129.6$; $\chi^2_L = 74.22$; $72.60 < \sigma^2 < 126.77$
   b) $8.52 < \sigma < 11.26$

5. H_0: $\sigma^2 = 0.15$; H_1: $\sigma^2 > 0.15$; $\chi^2 = 108.0$; the critical value is $\chi^2_{0.01} = 88.38$. Because the observed value $\chi^2 = 108.0$ falls in the critical region, we reject H_0. The variance is too large and the fan blades should be replaced.

7. a) H_0: $\sigma^2 = 2.89$; H_1: $\sigma^2 \neq 2.89$; $\chi^2 = 54.50$; the critical values are 95.02 and 48.76. Because the observed value $\chi^2 = 54.50$ falls in the acceptance region, we accept H_0. The hormones have no effect on variance.
   b) $\chi^2_U = 90.53$; $\chi^2_L = 51.74$; $1.74 < \sigma^2 < 3.04$
   c) $1.32 < \sigma < 1.74$

9. a) H_0: $\sigma^2 = 15$; H_1: $\sigma^2 \neq 15$; $\chi^2 = 20.02$; the critical values are 35.48 and 10.28. Because the observed value $\chi^2 = 20.02$ falls in the acceptance region, we accept H_0. There is no difference in variance.
   b) $\chi^2_U = 32.67$; $x^2_L = 11.59$; $9.19 < \sigma^2 < 25.91$
   c) $3.03 < \sigma < 5.09$

SECTION 11.4

1. a) H_0: $\mu_1 = \mu_2 = \mu_3$; H_1: not all the means are equal
   b–h)

Source of Variation	Sum of Squares	Degrees of Freedom	MS	F ratio	F critical val.	Test Decision
Between groups	32.847	2	16.424	4.678	4.26	Reject H_0
Within groups	31.60	9	3.511			
Total	64.447	11				

3. a) H_0: $\mu_1 = \mu_2 = \mu_3$; H_1: not all the means are equal
   b–h)

Source of Variation	Sum of Squares	Degrees of Freedom	MS	F ratio	F critical val.	Test Decision
Between groups	2.042	2	1.021	0.336	7.20	Accept H_0
Within groups	33.428	11	3.039			
Total	35.470	13				

5. H_0: $\mu_1 = \mu_2 = \mu_3 = \mu_4$
   H_1: not all the means are equal

Source of Variation	Sum of Squares	Degrees of Freedom	MS	F ratio	F critical val.	Test Decision
Between groups	238.225	3	79.408	4.611	3.29	Reject H_0
Within groups	258.340	15	17.223			
Total	496.565	18				

7. H_0: $\mu_1 = \mu_2 = \mu_3$
   H_1: not all the means are equal

Source of Variation	Sum of Squares	Degrees of Freedom	MS	F ratio	F critical val.	Test Decision
Between groups	80.133	2	40.067	0.374	3.88	Accept H_0
Within groups	1285.20	12	107.10			
Total	1365.33	14				

9. H_0: $\mu_1 = \mu_2 = \mu_3$
   H_1: not all the means are equal

Source of Variation	Sum of Squares	Degrees of Freedom	MS	F ratio	F critical val.	Test Decision
Between groups	422.00	2	211.00	5.271	4.26	Reject H_0
Within groups	360.25	9	40.028			
Total	782.25	11				

Chapter 11 Review Problems

1. H_0: $\mu_1 = \mu_2 = \mu_3 = \mu_4$
   H_1: not all the means are equal

Source of Variation	Sum of Squares	Degrees of Freedom	MS	F ratio	F critical val.	Test Decision
Between groups	6149.75	3	2049.917	2.633	3.24	Accept H_0
Within groups	12454.80	16	778.425			
Total	18604.55	19				

3. a) H_0: $\sigma^2 = 810000$; H_1: $\sigma^2 > 810000$; $\chi^2 = 65.54$; critical value is $\chi^2_{0.01} = 49.59$. Because the observed value $\chi^2 = 65.54$ is in the critical region, we reject H_0. The variance is more than claimed.

b) $\chi_U^2 = 45.72$; $\chi_L^2 = 16.05$; $1161147.4 < \sigma^2 < 3307642.4$

5. H_0: student grade and teacher rating are independent; H_1: student grade and teacher rating are not independent; $\chi^2 = 9.80$; $\chi_{0.01}^2 = 16.81$. Accept H_0: they are independent.

7. H_0: the distributions are the same; H_1: the distributions are different; $\chi^2 = 33.93$; $\chi_{0.01}^2 = 13.28$. Reject H_0. The distribution has changed.

SECTION 12.1

1. $r = 0.75$; $n = 12$; $c = 0.74$. Because $0.75 > 0.74$ reject H_0. The new tip has longer mean life. (μ_1 is mean life of new tip; μ_2 is mean life of old tip; H_0: $\mu_1 = \mu_2$; H_1: $\mu_1 > \mu_2$.)

3. $r = 0.63$; $n = 16$; critical values are 0.255 and 0.745. Because $r = 0.63$ is in the acceptance region, we accept the claim that the lectures make no difference. (μ_1 is mean score after lecture; μ_2 is mean score before lecture; H_0: $\mu_1 = \mu_2$; H_1: $\mu_1 \neq \mu_2$.)

5. $r = 0.58$; $n = 12$; critical values are 0.22 and 0.78. Because $r = 0.58$ is in the acceptance region, we conclude that the schools have the same effectiveness. (μ_1 is mean score at school A; μ_2 is mean score at school B; H_0: $\mu_1 = \mu_2$; H_1: $\mu_1 \neq \mu_2$.)

7. $r = 0.188$; $n = 16$; $c = 0.209$. Because $r = 0.188$ is in the critical region, we reject H_0. The mean number of cigarettes smoked after hypnosis is smaller. (μ_1 = mean number of cigarettes smoked after hypnosis; μ_2 = mean number smoked before hypnosis; H_0: $\mu_1 = \mu_2$; H_1: $\mu_1 < \mu_2$.)

9. $r = 0.643$; $n = 14$; critical values are 0.155 and 0.845. Because $r = 0.643$ is in the acceptance region, we accept H_0. The mean pulse rates are the same. (μ_1 = mean pulse rate after birth; μ_2 = mean pulse rate before birth; H_0: $\mu_1 = \mu_2$, H_1: $\mu_1 \neq \mu_2$.)

SECTION 12.2

1. $R = 82$; $\mu_R = 90$; $\sigma_R = 12.25$; critical values are 65.99 and 114.01. Because $R = 82$ is in the acceptance region, we conclude that there is no difference between mean scores for the schools.

3. $R = 80$; $\mu_R = 90$; $\sigma_R = 12.25$; critical values are 65.99 and 114.01. Because $R = 80$ is in the acceptance region, we conclude there is no difference.

5. $R = 66$; $\mu_R = 99$; $\sigma_R = 14.07$; critical values are 71.42 and 126.58. Because $R = 66$ is in the rejection region, we conclude the mean times for the groups are different.

7. $R = 92$; $\mu_R = 90$; $\sigma_R = 12.25$; critical values are 58.40 and 121.61. Because $R = 92$ is in the acceptance region, we conclude that vitamin B_{42} makes no difference.

9. $R = 78$ or $R = 58$; $\mu_R = 68$; $\sigma_R = 9.52$; critical values are 43.44 and 92.56. Because $R = 78$ and $R = 58$ are both in the acceptance region, we conclude that there is no difference in performance.

SECTION 12.3

1. H_0: $\rho_S = 0$; H_1: $\rho_S \neq 0$; $r_S = 0.682$; the critical values are 0.619 and -0.619. Because the observed value $r_S = 0.682$ falls in the critical region, we reject H_0 and conclude that there is a monotone relation (either increasing or decreasing).

3. H_0: $\rho_S = 0$; H_1: $\rho_S > 0$; $r_S = 0.571$; the critical value is 0.620. Because the observed value $r_S = 0.571$ falls in the acceptance region, we accept H_0 and conclude that there is no monotone relation between crowding and violence.

5.

Soldier	1	2	3	4	5	6	7
Humor Rank	5	3	4	2	1	7	6
Aggressiveness Rank	1	7	3	5	4	6	2
d	4	-4	1	-3	-3	1	4

H_0: $\rho_S = 0$; H_1: $\rho_S < 0$; $r_S = -0.214$; the critical value is 0.715. Because the observed value $r_S = -0.214$ falls in the acceptance region, we accept H_0 and conclude that there is no monotone relation.

7.

Cadet	1	2	3	4	5	6	7	8	9	10	11
Aptitude Rank	8	1	7	3	11	3	5	9	6	10	3
Performance Rank	7	1	8	4	10	2	5	11	6	9	3
d	1	0	-1	-1	1	1	0	-2	0	1	0

$H_0: \rho_S = 0$; $H_1: \rho_S > 0$; $r_S = 0.955$; the critical value is 0.764. Because the observed value $r_S = 0.955$ falls in the critical region, we reject H_0 and conclude that there is a monotone increasing relation between aptitude rank and performance rank.

9.

City	1	2	3	4	5	6	7	8
Sales Rank	6	7	1	8	3	2	5	4
Income Rank	5	4	2.5	8	6	1	7	2.5
d	1	3	-1.5	0	-3	1	-2	1.5

$H_0: \rho_S = 0$; $H_1: \rho_S \neq 0$; $r_S = 0.661$; the critical value is 0.881. Because the observed value $r_S = 0.661$ falls in the acceptance region, we accept H_0 and conclude that there is no monotone relation.

Chapter 12 Review Problems

1. $R = 107$; $\mu_k = 110$; $\sigma_k = 14.20$; critical values are 82.17 and 137.83. Because $R = 107$ is not in the critical region we accept H_0. The mean viscosity index is unchanged.
3. $r = 0.77$; $n = 13$; critical value is 0.82. Because 0.77 is in the acceptance region we accept H_0. The advertising had no effect. $H_0: \mu_1 = \mu_2$; $H_1: \mu_1 > \mu_2$ where $\mu_1 = $ mean sales after ads and $\mu_2 = $ mean sales before ads.
5. $H_0: \rho_S = 0$; $H_1: \rho_S > 0$; $\rho_S = 0.616$; critical value is 0.600. Because the observed value $r_S = 0.617$ falls in the critical region, we reject H_0 and conclude that there is a monotone increasing relation in the ranks.

Index

Acceptance region, 303
Additive rules of probability, 107–109
  general rule, 107
  rule for mutually exclusive events, 108
Alpha (α, level of significance), 299
Alpha (α probability of a type I error), 299, 305
Alternate hypothesis H_1, 298
  for difference of several means, 463
  for difference of two means, 323
  for difference of two proportions, 358
  for left-tail test, 311
  for right-tail test, 311
  for tests of correlation coefficient, 421, 507
  for tests of goodness of fit, 455
  for tests of independence, 432, 438
  for two-tail test, 305, 311
  for variance, 453, 454
Analysis of variance (ANOVA), 461
  alternate hypothesis, 463, 464
  degrees of freedom for denominator, 469
  degrees of freedom for numerator, 469
  F distribution, 468–469
  null hypothesis, 464
And (A and B), 101–102; *see also* Probability
Arithmetic mean, x, 53; *see also* Mean
Averages, 50–55
  mean, 53, 79, 137, 167
  median, 52, 61
  mode, 50
  population mean, 74, 225
  sample mean, 54
  weighted, 83

b (slope of least squares line), 391, 408 410
Bar graphs, 17, 18
Bernoulli, 152

Best-fitting line, 389; *see also* Least squares line
Beta (β, probability of a type II error), 299, 305
Binomial, 151, 242, 275, 354
  approximation by normal, 242
  coefficients, 158, 159
  distribution, 164–169
  experiment, 152
  formula for probabilities, 158
  histogram, 164
  mean of binomial distribution, 167
  standard deviation of binomial distribution, 167
  variable (r), 152, 354
Boundaries, class, 27
Box-and-whisker plot, 59, 61

Cause and effect relations, 417, 424
Cells, 434
Central Limit Theorem, 234
  applications of, 234–238
Chi square (χ^2), 433
  calculation of, 437, 445
  degrees of freedom for goodness-of-fit test, 446
  degrees of freedom for independence test, 438
  degrees of freedom for tests of variance, 452
  distribution of, 433
  tests for goodness of fit, 444–448
  tests for independence, 432–441
  tests for variance and standard deviation, 451–456
Circle graph, 22
Class, 26
  boundaries, 27
  frequency, 27

Class *(continued)*
  limits, 27
  midpoint, 27
  width, 26
Coefficient, binomial, 158–159
Coefficient of linear correlation, 410
  formula for, 411
Combination rule, 128
Complement of A, 99
Conclusions (for hypothesis testing),
  304–305
  using P values, 369, 371–372
Conditional probability, 102
Confidence intervals
  for mean, large samples, 260
  for mean, small samples, 270
  for predicted value of y, 401
  for proportions, 278
  for variance and standard deviation, 456
Confidence level c, 255
Confidence prediction band, 401
Contingency tables, 434
Continuity correction factor, 245; *see also*
  Normal approximation to binomial
Continuous random variable, 133
Correlation
  Pearson Product Moment Correlation
    Coefficient, r, 411
    formula for, 411
    interpretation of, 411–412
    testing, 420–424
  Spearman rank correlation r_S
    formula for, 506
    interpretation of, 504–507
    testing, 507
Criterion for least squares line, 389
Critical regions, 303; *see also* Critical values
  for left-tail tests, 309
  for right-tail test, 310
  for two-tail tests, 310
Critical values, 303
  for χ^2, 439, 446, 454
  for comparing several sample means, 469
  for correlation, 421, 508
  for F, 469
  for rank sum test, 499
  for sign test, 488
  for t, 268, 330
  for testing goodness of fit, 446
  for testing independence, 438

  for testing means or differences of means,
    313
  for testing paired differences, 344
  for testing proportions or difference of
    proportions, 354, 359
  for testing variance or standard deviation,
    452, 453
  for z, 257

Data
  continuous, 133
  discrete, 133
  paired, 342, 383
Decision errors, types of, 299–300
Degrees of freedom (*d.f.*)
  for chi-square goodness-of-fit tests, 446
  for chi-square tests of independence, 438
  for estimating variance, 452
  for F distribution, denominator, 469
  for F distribution, numerator, 469
  difference of means, 335, 343
  means, 267, 330
  paired difference test, 343
DeMoivre, 178
Dependent events, 102
Descriptive statistics, 5
Deviation
  population standard, 74, 137, 167, 234,
    235
  sample standard, 68, 74
    computation formula, 72
Difference
  among several means, 461–462
  between two means, 320
  between two proportions, 354
  paired difference test, 342–343
Discrete random variable, 133
Distribution
  bell-shaped, 179, 180
  binomial, 156–169
  chi-square, 333
  F, 469–470
  normal, 178
  probability, 134
  sampling, 225
  skewed, 167
  Student's t, 266–269
  symmetrical, 167
Distribution-free tests, 485; *see also*
  Nonparametric tests

E, maximal error of estimate, 259
  for μ large sample, 259
  for μ small sample, 270
  for p, 276
EDA, 36
Equally likely outcomes, 97
Equation of least squares line, 391
Error of estimate, 255, 259
Errors
  type I, 299, 305
  type II, 299, 305
Estimation, 254
  means, 254–259, 266–268
  proportions, 275–278
Event, probability of, 94
Events
  dependent, 102
  equally likely, 97
  failure F (binomial), 152
  independent, 102
  mutually exclusive, 107
  success S (binomial), 152
Expected frequency for contingency tables E,
  435, 445
Expected value
  for binomial distribution μ, 167
  for general probability distribution μ, 137
Experiment, binomial, 152
Exploratory Data Analysis, 36
Extrapolation, 307

F distribution
  calculations for, 469
  critical regions for, 469–470
  critical values for, 469
  degrees of freedom for, 468, 469
  tests of hypothesis using, 463, 471
F (failure), 152; *see also* Binomial
F ratio, 469
Fisher, R.A., 469
Frequency, 27
  expected, E, 435, 445
  polygon, 30
  relative, 95
Frequency distribution, 78; *see also*
  Histogram
Frequency histogram; *see* Histogram

Gaussian distribution; *see* Normal
  distribution

General probability rule
  for addition, 108
  for multiplication, 102
Goodness-of-fit tests, 444
Gosset, W.S., 266
Graphs, 16–23
  bar, 17–18
  circle, 22
  frequency polygons, 30
  histograms, 25
  pictograms, 21
  scatter diagrams, 384

Hinge, 59
Hinge rank, 50
Histogram, 26
  how to construct, 26–30
Hypothesis tests, in general, 298
  acceptance region, 303
  alternate hypothesis H_1, 298
  conclusion, 304
  conclusion based on P value, 369, 370
  critical region or regions, 303
  critical value, 303
  level of significance α, 299
  null hypothesis H_o, 298
Hypothesis testing (types of tests)
  of correlation coefficient, 420
  of difference of means, 320
  of difference of proportions, 358
  of differences among several means, 461
  of goodness of fit, 444
  of independence, 432
  of means, 309, 329
  of nonparametric, 485
  of paired differences, 342
  of proportions, 354
  rank sum test, 496
  sign test, 485
  of small samples, 329
  of Spearman rank correlation coefficient,
    503
  of variance, 451

Independence tests, 432
Independent events, 102
Independent samples, 321, 358
Independent trials, 152
Inference, statistical, 5

Interpolation, 396
Interval, confidence, 260, 270, 278

Large samples, 235, 243, 259, 260, 312
  for confidence intervals of means, 260
  for confidence intervals of proportions, 275
  for tests involving differences of means, 322
  for tests involving differences of proportions, 358–360
  for tests involving means, 311–312
  for tests involving proportions, 354
Leaf, 37
Least squares criterion, 389
Least squares line, 391
  calculation of, 391
  formula for, 391
  predictions from, 393–394
  slope of, 391, 393
Level of confidence c, 255
Level of significance α, 299, 305
Limits, class, 27
Linear correlation coefficient; *see* Correlation coefficient, r
Linear regression, 389–403
Lower class limit, 27

Mann-Whitney U Test, 496, 501
Maximum error of estimate, 259
Mean $\bar{x}$, 53; *see also* Estimation and Hypothesis testing
  for binomial distribution, 167
  comparison with mode and median, 55
  defined, 53
  formula for grouped data, 79
  formula for ungrouped data, 53
  of probability distribution, 137
Mean square, MS, 468
Median, 52, 61
Median rank, 60
Mode, 40
Monotone relation
  decreasing, 505
  increasing, 505
Mu (μ), 74, 137, 167, 225
Multiplication rule of counting, 124
Multiplication rule of probability
  for general events, 102
  for independent events, 102

Mutually exclusive events, 107

Negative correlation, 412
Nonparametric tests
  rank sum test, 496
  sign test, 485
  Spearman correlation test, 503
Normal approximation to binomial, 242
Normal distribution, 178
  areas under normal curves, 208–215
  normal curves, 179–180
  standard normal, 194
Null hypothesis, 298; *see also* Alternate hypothesis
Number of degrees of freedom; *see* Degrees of freedom ($d.f.$)

Observed frequency (O), 436, 445
One-tail test, 309
Or ($A \ or \ B$), 106

p (probability of success in a binomial trial), 152
P value, 366
Paired data, 342, 383
Paired difference test, 342
Parameter, 137, 254
Pearson, Karl, 410
Pearson product-moment coefficient of correlation r, 410–411
Percentile, 58
Permutation rule, 126
Pie charts, 22
Point estimate, 254
Pooled estimate
  of a proportion, 359
  of a variance, 334
Population
  defined, 2, 224
  mean μ, 74, 137, 167, 225
  standard deviation σ, 74, 137, 169, 225
Population parameter, 225
Positive correlation, 412
Power of a test, 300
Prediction for y given x, 393–394
Probability, 94
  addition rule (general events), 108
  addition rule (mutually exclusive events), 108
  binomial, 158

conditional, 102
defined, 94–95
of an event, 94
multiplication rule (general events), 102
multiplication rule (independent events), 102
Probability distribution, 134
continuous, 134, 178
discrete, 134
mean, 137
standard deviation, 137
Proportion, estimation of, 275
Proportion, pooled estimate, 359
Proportion, tests of, 342, 358

q (probability of failure in a binomial trial), 152

r (Pearson product moment correlation coefficient), 411
r (number of successes in a binomial experiment), 152
r (sign test), 485
r_S (Spearman rank correlation), 506
R (sum of ranks), 497
Random, 10
Random number table, 12
Random sample, 10
Random variable, 133
Range, 68
Rank
hinge, 60
median, 60
Rank sum test, 496–501
Ranked data, 60, 497
Ranks, 59, 496
Raw score, 191
Region, acceptance, 303
Region, rejection or critical, 303
Regression, linear, 383, 389–403
Relative frequency, 95
Rho (ρ), 420

S, success on a binomial trial, 152
Sample, 2, 224
large, 235, 243, 255, 263, 275, 312, 323, 354, 360
mean $\bar{x}$, 53, 79
random, 10

small, 266, 329
standard deviation s, 69, 70, 72
Sample size, determination of, 283–287
Samples
independent, 321, 358
repeated with replacement, 154
repeated without replacement, 154
Sampling distribution, 224–238; *see also* Central Limit Theorem
Scatter diagram, 384
Sigma
σ, 74, 137, 167, 225
Σ, 54, 72
Sign test, 485–490
Significance level α, 299
Simulation, 12
Skewed distribution, 167
Slope of least squares line, 391, 393, 408
Small samples, 266, 329
Spearman rank correlation, 503–513
Standard deviation
for population σ, 74, 137, 167, 225, 451
for samples s, 69, 72, 79
for $\bar{x}$ distribution, 231, 232, 234, 235
testing and estimating, 451–459
Standard error of estimate S_e, 397, 398
Standard normal distribution, 194
Standard unit z, 189
Statistic, 225
Statistics
definition, 1
descriptive, 5
inferential, 5, 254
Stem, 37
Stem-and-leaf display, 36, 39
Student's t distribution, 266
Summation notation Σ, 54, 72

t (critical value notation), 268, 330
t (Student's t distribution), 266–269
Tallies, 27
Test for independence, 432–441
Tests of hypothesis; *see* Hypothesis testing
Trees, 118
Trials, binomial, 152
Two-tail test, 309, 311
Type I error, 299
Type II error, 299

Upper class limit, 27

Variable
  continuous, 133
  discrete, 133
  random, 133
  standard normal, 189; *see also z* value
Variance
  analysis of (ANOVA), 461–474
  between samples, 465
  estimation of, 456
  for ungrouped sample data s^2, 69
  testing of, 451

  within samples, 465
Variation, 68

Weighted average, 83
Whisker, 61

x bar $(\bar{x})$, 53; *see also* Mean

z, critical value notation, 255, 257
z scores, 189
z value, 189

2 3 4 5 6 7 8 9 0

Regression and Correlation

In all these formulas

$$SS_x = \Sigma x^2 - \frac{(\Sigma x)^2}{n}$$

$$SS_y = \Sigma y^2 - \frac{(\Sigma y)^2}{n}$$

$$SS_{xy} = \Sigma xy - \frac{(\Sigma x)(\Sigma y)}{n}$$

Least squares line $\quad y = a + bx$ where $b = \dfrac{SS_{xy}}{SS_x}$ and $a = \bar{y} - b\bar{x}$

Standard error of estimate $\quad S_e = \sqrt{\dfrac{SS_y - bSS_{xy}}{n-2}}$ where $b = \dfrac{SS_{xy}}{SS_x}$

Pearson product-moment correlation coefficient $\quad r = \dfrac{SS_{xy}}{\sqrt{SS_x SS_y}}$

Confidence interval for $y \quad y_p - E < y < y_p + E$ where y_p is the predicted y value for x

$$E = t_c S_e \sqrt{1 + \frac{1}{n} + \frac{(x - \bar{x})^2}{SS_x}}$$

Spearman Rank correlation coefficient $\quad r_S = 1 - \dfrac{6(\Sigma d^2)}{n(n^2 - 1)}$

Other Formulas

Chi square $\quad \chi^2 = \dfrac{\Sigma(O - E)^2}{E}$ where O is the observed value

E is the expected value

ANOVA and Fisher F distribution $\quad$ See Summary, page 471 and Table 11-23, page 472

Critical values for the Sign test $\quad$ See Table 12-2, page 488

Critical values for the Rank Sum test $\quad$ See Table 12-7, page 499